BIOMONITORS AND BIOMARKERS AS INDICATORS OF ENVIRONMENTAL CHANGE 2

A Handbook

ENVIRONMENTAL SCIENCE RESEARCH

Recent Volumes in this Series

Volume 47—GLOBAL ENERGY STRATEGIES: Living with Restricted Greenhouse Gas Emissions
Edited by James C. White

Volume 48—GLOBAL ATMOSPHERIC–BIOSPHERIC CHEMISTRY
Ronald G. Prinn

Volume 49—BIODEGRADATION OF NITROAROMATIC COMPOUNDS
Edited by Jim C. Spain

Volume 50—BIOMONITORS AND BIOMARKERS AS INDICATORS OF ENVIRONMENTAL CHANGE: A Handbook
Edited by Frank M. Butterworth, Lynda D. Corkum, and Judith Guzmán-Rincón

Volume 51—CHEMISTRY FOR THE PROTECTION OF THE ENVIRONMENT 2
Edited by Lucjan Pawłowski, William J. Lacy, Christopher G. Uchrin, and Marzenna R. Dudzińska

Volume 52—PRODUCED WATER 2: Environmental Issues and Mitigation Technologies
Edited by Mark Reed and Ståle Johnsen

Volume 53—EVALUATING CLIMATE CHANGE ACTION PLANS: National Actions for International Commitment
Edited by James C. White

Volume 54—BIOTECHNOLOGY IN THE SUSTAINABLE ENVIRONMENT
Edited by Gary S. Sayler, John Sanseverino, and Kimberly L. Davis

Volume 55—CHEMISTRY FOR THE PROTECTION OF THE ENVIRONMENT 3
Edited by Lucjan Pawłowski, Marjorie A. Gonzales, Marzenna R. Dudzińska, and William J. Lacy

Volume 56—BIOMONITORS AND BIOMARKERS AS INDICATORS OF ENVIRONMENTAL CHANGE 2
Edited by Frank M. Butterworth, Amara Gunatilaka, and Maria Eugenia Gonsebatt

Volume 57—PERCHLORATE IN THE ENVIRONMENT
Edited by Edward Todd Urbansky

Volume 58—THERMAL SOLID WASTE UTILISATION IN REGULAR AND INDUSTRIAL FACILITIES
Edited by Lucjan Pawłowski, Marzenna R. Dudzińska, and Marjorie A. Gonzalez

BIOMONITORS AND BIOMARKERS AS INDICATORS OF ENVIRONMENTAL CHANGE 2

A Handbook

Edited by

Frank M. Butterworth

Institute for River Research International
Rochester, Michigan

Amara Gunatilaka

Donaukraft Engineering
Vienna, Austria

and

Maria Eugenia Gonsebatt

Universidad Nacional Autonoma de Mexico
Cuidad Universitaria, Mexico

Springer Science+Business Media, LLC

Library of Congress Cataloging-in-Publication Data

Proceedings of the 41st Conference of the International Association for Great Lakes Research Symposium on Biomonitors and Biomarkers as Indicators of Environmental Change 2, held May 18–22, 1998 in Hamilton, Ontario, Canada

ISBN 978-1-4613-5488-8 ISBN 978-1-4615-1305-6 (eBook)

DOI 10.1007/978-1-4615-1305-6

Originally published by Kluwer Academic/Plenum Publishers in 2001

Softcover reprint of the hardcover 1st edition 2001

http://www.wkap.nl/

10 9 8 7 6 5 4 3 2 1

A C.I.P. record for this book is available from the Library of Congress

PREFACE

Biomonitors and Biomarkers as Indicators of Environmental Change II

Monitoring the environment is absolutely essential if we are to identify hazards to human health, to assess environmental cleanup efforts, and to prevent further degradation of the ecosystem. Biomonitors and biomarkers combined with chemical monitoring offer, it will be learned, the best approach to making these assessments. The purpose of this book was the same as that of Volume I, to document recent developments and applications in biomonitor and biomarker research. The second volume builds on the first (Butterworth et al., eds., 1995, *Biomonitors and Biomarkers as Indicators of Environmental Change*, Plenum Publishing, New York) with a compilation of methods enriching the list of possible monitoring systems.

The book is intended for researchers who want to incorporate newer and different technologies in their development of specifically-crafted monitors; students who are learning the field of biomonitoring [Note: the National Autonomous University of Mexico (UNAM) is planning to use both volumes in a newly-organized course on biomonitoring]; and regulatory agencies that want to consider newer technologies to replace inadequate and less powerful test regimes.

The book resulted from a symposium of the same title that was held (May 19–22, 1998) as part of the annual conference of the International Association of Great Lakes Research held at the McMaster University in Hamilton, Ontario, Canada. As in the first conference and volume, we searched for systems that would go beyond bioaccumulation of specific pollutant chemicals and their toxic effects on individuals, populations, and communities. We sought candidate biomonitor/biomarker systems that included a range of endpoints in a variety of laboratory and sentinel organisms and systems exhibiting high reliability, short turn-around-time and low cost. There was also a need to detect atmospheric pollution, to use plant test systems, and to apply molecular biotechnology to the construction of biomarkers and biomonitors. There were two major changes over the first volume: one section completely devoted to on-line/automated biomonitoring and another section devoted to recombinogenesis offering new methods and applications. In both cases we sought expertise from Europe.

The objectives for the conference and this book were to: facilitate an exchange of ideas and knowledge on state-of-the-art biomonitoring methods and applications; develop associations between laboratories in signatory countries of the North American Free Trade Agreement (NAFTA) where there will be challenges of increased pollution in already threatened ecosystems; and introduce researchers of the Great Lakes, the greatest natural water resource in North America, to the benefits of recent environmental monitoring technology.

Outcomes of the first volume led to at least two river monitoring projects in Canada and Mexico. In the Canadian project the late Robbin Hough and Carl Freeman, authors in the first volume, instituted with the aid of the First Nation, an automated chemical and biological monitoring system on Walpole Island in the delta of the St. Clair River in Ontario. The Mexican project is a chemical and biomonitoring regime on the heavily industrialized Zahuapan River watershed in Tlaxcala State, Mexico led by Alfredo Delgado with the aid of Rafael Villalobos and Sandra Gomez authors in the first volume. With the success of these projects it is realistic to expect that even more projects will result form the current volume.

Although the systems described in these proceedings were not meant to be exhaustive, they do offer interesting insights, ingenious applications, and opportunities to combine these to form powerful monitoring suites that will be able to detect and also define pollution effects in biological terms. We are grateful to the International Association for Great Lakes Research and McMaster University for their help in establishing the symposium and providing resources and facilities at the conference, and to Patricia Ramirez for her valuable help in the final preparation of the manuscripts for publication.

CONTENTS

SECTION III: NEW APPROACHES AND APPLICATIONS OF ESTABLISHED SYSTEMS: SENTINEL SYSTEMS

LABORATORY-BASED BIOMONITORS

SECTION IV: ABSTRACTS

BIOMONITORS AND BIOMARKERS AS INDICATORS OF ENVIRONMENTAL CHANGE 2

A Handbook

1

INTRODUCTION

Biomonitors and Biomarkers as Indicators of Environmental Change, Volume 2

FRANK M. BUTTERWORTH,[1,4] RAFAEL VILLALOBOS-PIETRINI,[2] AND MARÍA E. GONSEBATT[3]

[1]Institute for River Research International, 920 Ironwood Dr. Suite 344, Rochester, MI 48307; [2]Laboratorio de Citogenética y Mutagénesis Ambientales, Centro de Ciencias de la Atmósfera Circuito Exterior, UNAM, Ciudad Universitaria, C. P. 04510 México, D.F. MEXICO, [3]Biomedical Research Institute, UNAM, Ciudad Universitaria, C. P. 04510 México, D.F. MEXICO; [4]To whom correspondence should be sent

1. NECESSITY OF BIOMONITORING

Rarely has a technology been so desperately needed by society. Until recently, environmental monitoring has been based mainly on chemical observation and analysis, but the approach is problematic. Chemical monitoring being expensive cannot measure all pollutants and, given the complex array of contaminants in the environment, cannot provide good predictors of the effect of pollution on living

---

Email addresses of the conference participants: Daniela Baganz <daniela@igb-berlin.de>, Elke Bluebaum-Gronau <bluebaum-gronau@bfg.de>, Agneta Burton <m.a.burton@herts.ac.uk>, Frank Butterworth <fbirri@ees.eesc.com> <fbirri@prodigy.net.mx>, David Cowell <David.Cowell@uwe.ac.uk>, Manohar Das <das@oakland.edu>, Peter Diehl <peter.diehl@wwv.rpl.de>, Rudolf Fahrig <fahrig@ita.fhg.de>, Almut Gerhardt <limco.int@t-online.de>, Sandra Gomez <slga@ccaunam.atmosfcu.mx>, Maria Eugenia Gonsebatt <margen@servidor.unam.mx>, Amara Gunatilaka <GunatilakaA@verbundplan.at>, Sandra J. Gunselman <sandrag@cvmbs.colostate.edu>, Judith Guzmán <jgr@nuclear.inin.mx>, Michael Hoffmann <michael.hoffmann@bafg.de>, Perwez Hussain <HussainP@intra.nci.nih.gov>, Robert Jaffe <info@envirolab.com>, Klaas Kaag <k.kaag@mep.tno.nl>, Kees J.M. Kramer <mermayde@xs4all.nl>, Lesley Lovett-Doust <b87@uwindsor.ca>, Lydia Marquez Bravo <lmarquez@chac.imta.mx>, Tom Muir <Tom.Muir@ec.gc.ca>, Scott Munro <lis@ebtech.net>, Jim Quinn <quinn@mcmaster.ca>, Patricia Ramos Morales <prm@hp.fciencias.unam.mx>, Roberto Rico Martinez <RRICO@correo.uaa.mx>, Patricia Scully <p.j.scully@livjm.ac.uk>, Hunrich Spieser <hunrich.spieser@gsf.de>, Manuel Uribe Alcocer <muribe@mar.icmyl.unam.mx>, Rafael Villalobos-Pietrini <rvp@ccaunam.atmosfcu.unam.mx>.

Biomonitors and Biomarkers as Indicators of Environmental Change 2, Edited by Butterworth *et al.*
Kluwer Academic/Plenum Publishers, New York, 2000

organisms. Knowing lists of chemical contaminants is important, but not enough. Furthermore the lists imply the contaminants have an additive effect, when in fact we do not know if the effects are additive, subtractive, or synergistic. Thus, even knowing the major pollutant chemicals is not enough. Also, we learn nothing about which species will be most affected, which organ systems most sensitive, etc.

A solution to this problem is to use bioindicators. Thus, interest in biomonitors, biomarkers, and biosensors has steadily grown as the value of such methods is realized in the surveillance of terrestrial and aquatic ecosystems (Waters et al., 1980; Gruber and Diamond, 1988; Jeffrey and Madden, 1991; Huggett et al., 1992; Mathewson and Finley, 1992; Kramer, 1994; Butterworth et al., 1995; Gerhardt, 1999). To this end, a symposium was held where recent advances in biomonitoring, biomarker and biosensor technologies were presented by renowned researchers and practitioners from seven countries sponsored by the Institute for River Research International and the International Association of Great Lakes Research held (May 18–22, 1998) at McMaster University, in Hamilton, Ontario, Canada. The main objective was to introduce scientists and environmental managers of the Great Lakes to recent advances in biomonitoring technology in addition to facilitate information exchange and stimulate collaboration among the participants.

This symposium was a follow-up to one held in 1994 sponsored by the International Association of Great Lakes Research and the National Science Foundation at the University of Windsor, the proceedings of which was also published (Butterworth et al., 1995). A wide variety of organism types including humans, non-human vertebrates, invertebrates including insects, a higher plant, and microorganisms were used mainly in whole organism systems. The systems assay exposure to one of the three media (terrestrial, aquatic, and atmospheric), some with multiple exposures, and some with multiple endpoints including development, biochemistry, genetics, physiology, and behavior.

Generally, a biomonitor is a whole organism system that responds to environmental change with one or more endpoints (e.g. viability, metabolism, behavior, genetic damage etc.) that are easily detectable. Biomarkers are generally subsystems of an organism with a specific endpoint, and a biosensor is an abstracted biological detection/response system with a chemical or physical manifestation.

Many of the systems demonstrate that they do not necessarily have to be sophisticated assays, requiring expensive equipment and training. Therefore, another advantage is that biomonitoring can provide much information at low cost. This feature is particularly relevant to developing countries and less affluent communities.

One of the challenges to the global environment are the increased industrial activities hastened by industrial treaties such as the North American Free Trade Agreement (NAFTA). As business activity and population increase, threats to the dwindling quality of the environment will continue to occur. Thus comprehensive and efficacious environmental monitoring will need to grow, but scientists from the three NAFTA countries need to be mobilized to design and apply newer monitoring technologies.

2. THE PARTICIPANTS

The chapters show what technologies are possible now and fall roughly into three categories: automated monitoring, recombinogenicity testing, and new approaches and applications. Thus, in designing the current symposium, effort was made to look intensively for state-of-the-art technology. For example great advances were seen in Europe where automated biomonitoring has been in place for at least a decade with outstanding results. Biosensor technology is also being developed there. Other innovative biomonitoring systems such as those described in the earlier volume (Butterworth et al., 1995) were also discovered there and in Mexico, Canada, and the US.

2.1. Automated Monitoring

2.1.1. On-Line Monitoring

The following seven chapters address on-line, automated monitoring. Automated biomonitoring or biological early warning systems are defined as systems that detect toxic conditions on a continuous basis in whole organisms, in this case, with behavioral endpoints. The examples described are currently employed monitoring environmental stresses in various rivers in Europe and Asia.

Gunatilaka and Diehl introduce the history of automated biomonitoring in the chapter, A Brief Review of Chemical and Biological Continuous Monitoring of Rivers in Europe and Asia. Here the point is clear: automated chemical and biomonitoring has had a profound effect on decreasing pollution and increasing river quality without much additional regulation.

Gunatilaka and colleagues present in the next chapter, The Evaluation of "Dynamic Daphnia Test" after a Decade of Use: Benefits and Constraints, describing the creation, development, and evaluation of a test that has been in use for the past 18 years discovering accidental spills and emissions into the Rhine and Danube Rivers. Particular attention has been made for the interpretation of upper and lower alarm levels in actual test situations.

Kramer and Foekema describe another invertebrate test, the Musselmonitor® in the chapter entitled The "Musselmonitor®" as Biological Early Warning System: The First Decade, which also detects escape behavior measured and recorded electronically. These latter two systems are widely used as indicated in Chapter 2.

The group led and introduced by Spieser (Quantitative Behavior Analysis—A New Approach to the Challenges of Environmental Toxicology) thoroughly describes in three chapters a fish behavior assay BehavioQuant®: Spieser et al. (An Introduction to Behavioral Monitoring—Effects of Nonylphenol and Ethinyl-Estradiol on Swimming Behavior of Juvenile Carp); Baganz et al. (How to Use Fish Behavior Analysis to Sensitively Assess the Hazard Potentials of Environmental Chemicals), and Blübaum-Gronau, et al. (Continuous Water Monitoring: Changes of Behavior Patterns as Indicators of Pollutants).

2.1.2. Automation Technologies

In addition, three chapters offer technologies that complement the above on-line, automated systems. Das et al. describe computer-controlled image recognition systems for microbiota in natural and waste waters in the chapter, Restoration and Classification of Water-Borne Microbial Images for Continuous Monitoring of Water Quality.

The Cowell et al. and Scully et al. chapters entitled, respectively: Screen-printed Disposable Biosensors for Environmental Pollution Monitoring; and Optical Sensors and Biosensors for Environmental Monitoring, describe technologies that can be free standing or become an adjunct to the whole-organism automata. Biosensor technologies offer intriguing possibilities for automated monitoring. Cowell et al. introduce a variety of enzymatic systems that can be exploited and also describe an LDH system under development in their laboratory. Scully et al. outline a dazzling array of optical sensors, techniques and applications complementing the Cowell chapter, both giving extensive reviews of the field.

2.2. Recombination and Recombinogen Detection

A seriously overlooked endpoint for biomonitoring is genetic recombination. Great effort has been focused on the detection of new mutations (errors in gene sequences). But there is a significant load of already-present mutations that cause disease and developmental defects. These genes, exposed by the recombination process, can wreak biological havoc. Most agents that cause mutations also cause increases in the rates of recombination, which in turn cause increases in the homozygosity of once heterozygous gene pairs. Thus, already present, recessive, deleterious mutations can be expressed in somatic tissues that are undergoing cell division, with the potential of causing far more damage to human health than mutation. The following chapters, which illustrate mouse, yeast, and fruit fly systems, explain the consequences of recombination and how to detect recombinogens. These presentations are aimed at stimulating interest in monitoring for these important toxins and in designing of even more sophisticated test systems. Although the tests involve somatic tissue, it is assumed that the mechanism, and thus the environmental effect on recombination, is the same as in germ-line cells, where "decks" of genes are "shuffled and dealt" to daughter gametes. The three *Drosophila* papers illustrate the sensitivity of this intricate system to detect genotoxins directly from the environment and to characterize genotoxic effects of complex mixtures.

Fahrig in the chapter, Recombination as Indicator for Genotoxic and "Non-genotoxic" Environmental Carcinogens, using a mouse and yeast test system, proposes a mechanism of recombination to explain the effects of combinations of chemicals.

Guzmán-Rincón et al. in their chapter, Somatic Mutation and Recombination Test in *Drosophila* Used for Biomonitoring of Environmental Pollutants, test the recombinogenic effects of environmental levels of nitrosamine in side-stream cigarette smoke employing diluted condensates.

McGowen et al. in the chapter Genotoxic Effects of Mixtures of a Polychlorinated Biphenyl, a Polyaromatic Hydrocarbon, and Arsenic, create a wide range of "complex" mixtures (in concentrations of chemicals as low as parts per trillion) that result in startlingly, unpredictable results and offer a new approach to study unknown mixtures.

Finally, Ramos et al. in their chapter, *Drosophila* is a Reliable Biomonitor of Water Pollution, indicate that their system is sensitive enough to detect toxins directly in municipal drinking water. Because McGowen et al. show arsenic to be a genotoxic potentiator at extremely low concentrations, the Ramos et al. chapter indicates the importance of biomonitoring public water supplies that naturally carry high levels of arsenic.

2.3. New Approaches and Applications of Established Approaches and Systems

2.3.1. Sentinel Systems

One of the most direct forms of monitoring is to exploit biological systems already present in the environment, called sentinel biomonitors. The following five chapters address new approaches to using sentinel monitoring.

Gerhardt in her chapter, A New Multispecies Freshwater Biomonitor for Ecological Relevant Control of Surface Waters, presents an ingenious system where the operator can choose a set of species chosen from reference sites depending on which species are present in the test site and what environmental conditions need to be monitored. Thus "sentinel monitors" can be employed in the test site. Behavioral/physiological endpoints lend potential to on-line monitoring, giving this system added interest.

Gonsebatt et al. in their chapter, Cytogenetic and Cytotoxic Damage in Exfoliated Cells as Indicators of Effects in Humans, describe the only direct human biomonitor in the book. Because most cancers are epithelial in origin, and the epithelia tested here have endpoints that are manifested relatively quickly, this test is extremely important.

Hussain and Harris in their chapter, *p53* Mutation Load: a Molecular Linkage to Carcinogen Exposure and Cancer, introduce us to the concept of using the frequency and type of mutations in the human *p53* gene as a dosimeter for environmental carcinogens. The authors use human cell cultures, a highly sensitive genotypic mutation assay and the *p53* gene as a target to identify an unknown mutagen or carcinogen in the environment.

The Lovett-Dousts present in their chapter, *Vallisneria americana* as an In Situ/Sentinel Monitor of Pollution in Natural Waters, a rooted, aqueous plant that can simultaneously distinguish toxicity in the sediment and water column. The authors propose this system to monitor the remediation of contaminated sites.

The chapter by Uribe-Alcocer and Diaz-Jaimes, Fish Chromosomes as Biomarkers of Genotoxic Damage and Proposal for the Use of Tropical Catfish Species for Short-term Screening of Genotoxic Agents, introduces a new sentinel system of various species peculiar to various tropical regions and habitats, employing cytogenetic endpoints for monitoring coastal and estuarine waters.

2.3.2. Laboratory-Based Biomonitors

Finally, there are four chapters on interesting laboratory-based biomonitors.

Jaffe describes in the chapter, Utility and Reliability of The *Tetramitus* Assay as a Monitoring Instrument for Toxicity Reduction and Risk Assessment, a set of protozoan species that evaluates drinking water quality through measuring the toxic effects of suspended particles on cell division. This test is the only one in the book that is in the process of being considered by the USEPA as a standard toxicity assay. It measures particle toxicity without extraction, under nonsterile conditions, and is 5–10X more sensitive than the USEPA Whole Effluent Toxicity assay.

In the Rico et al. chapter, The Use of Aquatic Invertebrate Toxicity Tests and Invertebrate Enzyme Biomarkers to Assess Toxicity in the States of Aguascalientes and Jalisco, Mexico, in addition to the standard *Daphnia* acute toxicity test, the chapter introduces newer and more sophisticated lab test systems: ingestion rates in *Daphnia* and cytofluorescence of enzyme activity in the rotifer, *Lacane* which have the potential for automation.

Gomez-Arroyo et al. in the chapter, Biomonitoring of Pesticides by Plant Metabolism: an Assay Based on the Induction of Sister-chromatid Exchanges in Human Lymphocyte Cultures by Promutagen Activation of *Vicia faba*, demonstrate how plant bioactivation of pesticides applied to crops may play a significant role in human (and animal) susceptibility. The paper provides an excellent review of the literature as well as experimental information.

Villalobos et al., in their comprehensive review, Genetic Monitoring of Airborne Particles, emphasize the critical need to monitor the particulates in the exposure medium. The paper provides an excellent platform for those involved in or planning studies in particulate monitoring.

2.4. By Abstract Only

The work of the following conference participants is presented by abstract only: Baganz et al., Bergeron and Crews, Brennan et al., Burton, Chappie et al., Chandy et al., Drouillard and Nordstrom, Edlund and Stoermer, Espinosa et al., Gerhardt and Janssens de Bisthoven, Gillis et al., Gunselmann et al., Hoffmann et al., Kaag and Scholten, Márquez et al., Muir, Munro et al., Ramírez-Victoria et al., Vahl, Vanderwall and Lovett-Doust, Yauk and Quinn. Addresses are given where possible for these authors (as well as the others) so that readers may follow up for more details. For example, the Munro group has amassed a wealth of chemical-emissions data collected over the past decade using automated chemical-monitoring technologies. Gunselmann and colleagues presented unique diagnostic technologies that have shown close correlations of environmental pollutants with human disease.

3. BIOMONITORING IN THE 21ST CENTURY

If our diminishing freshwater resources are to be protected, the development and deployment of biomonitoring technology will have to be a critical activity in the coming years. But progress in the recent past in North America is discouraging. The Introduction to the previous volume (Butterworth, 1995) outlines the history, existing databases, rationale and design, and benefits of biomonitoring. In the above reference and also Butterworth and Hough (1998) a survey of the Great Lakes, which hold 20% of the world's fresh water, biomonitoring was found to be inadequate. The biological systems in use were judged expensive, time consuming and often inaccurate.

Hopefully, the success of on-line, automated biomonitoring in Northern Europe will encourage other governments. This has been the case in several instances. Indonesia judged by many as a developing country has, because of a critical shortage of clean water in one of its industrial river basins, installed a multi-million-dollar, on-line, automated, monitoring system that includes state-of-the-art biomonitoring stations. The state of Tlaxcala in Mexico realizing agriculture is irrigating with polluted water has begun, as a result of the previous book (Butterworth, et al., 1995), a chemical and biomonitoring program. The First Nation in the St. Clair River Delta in Ontario threatened by industrial discharges upstream is, as a result of the first volume, on-line monitoring their drinking water. Perhaps, progress is only need driven.

In addition to the slow progress in deploying existing technology, there are other serious obstacles. Developing new technology and implementation of existing technology in governmental test regimes is needed. But resistance to change is a major obstacle. A second major obstacle is the "chemical-by-chemical-paradigm" approach to toxicology. Until a new paradigm that is biologically based is established, progress in protecting the environment will continue to be slow. It is hoped that this current volume in a small way will play a role in overcoming these obstacles.

4. REFERENCES

Butterworth, F.M. 1995. Introduction. In: *Biomonitors and Biomarkers as Indicators of Environmental Change: A Handbook.* Butterworth, F.M., L.A. Corkum, and J. Guzmán-Rincón, Eds., Plenum Press, NY.

Butterworth, F.M., L.A. Corkum, and J. Guzmán-Rincón, Eds. 1995. *Biomonitors and Biomarkers as Indicators of Environmental Change: A Handbook.* Volume 50, Environmental Science Research, Series Editor: H. S. Rosenkranz, Plenum Press, NY.

Butterworth, F.M., and R. Hough. 1998. Rehabilitation of the North American Great Lakes watershed: Past and future. *Landscape Ecology Volume III. Rehabilitation of Rivers*, L.D. Waal, A. Large, M. Wade, and G. Pinay, Eds., John Wiley, London.

Gerhardt, A., Ed. 1999. *Biomonitoring of Polluted Waters.* Trans Tech Publications, Switzerland.

Gruber D.S., and J.M. Diamond, Eds. 1988. *Automated Biomonitoring of Living Sensors as Environmental Monitors.* Ellis Howard Publishing, Chichester.

Huggett, R.J., R.A. Kimerle, P.M. Mehrle, Jr., and H.L. Bergman, Eds. 1992. *Biomarkers: Biochemical, Physiological, and Histological Markers of Anthropogenic Stress.* Lewis Publishers, Boca Raton.

Jeffrey, D.W., and B. Madden, Eds. 1991. *Bioindicators and Environmental Management.* Academic Press, London.

Kramer, K.J.M., Ed. 1994. *Biomonitoring in Coastal Waters and Estuaries.* CRC Press, Boca Raton.

Mathewson, P.R., and J.W. Finley, Eds. 1992. *Biosensor Design and Application.* American Chemical Society, Washington.

Waters, M.D., S.S. Sandhu, J.C. Huisingh, L. Claxton, and S. Nesnow, Eds. 1980. *Short-Term Bioassays in the Analysis of Environmental Complex Mixtures II.* Plenum Press, New York.

2

A BRIEF REVIEW OF CHEMICAL AND BIOLOGICAL CONTINUOUS MONITORING OF RIVERS IN EUROPE AND ASIA

AMARA GUNATILAKA[1] AND PETER DIEHL[2]

[1]Verbundplan Ltd., Engineers and Consultants, Parkring 12, A-1011 Vienna, Austria.
[2]State Office of Water Affairs, Rhine Water Control Station Worms, Am Rhein 1, D-67547 Worms, Germany

1. INTRODUCTION

The continuous chemical and biological monitoring in Europe dates back to the sixties and seventies. After three decades of development there are a number of reliable chemical monitoring systems for river monitoring, but biological monitoring is still in the stages of infancy (Baldwin and Kramer, 1994; Kramer and Botterweg, 1994; Gunatilaka et al., 2000). However it is a fast developing field with the availability of powerful tools from biochemistry, molecular biology and genetics. At the molecular level the developments are only two decades old. Research and development has stimulated development of a series of novel on-line sensors which are quite promising and different to conventional sensors (see Cowell, 2000; Scully et al., 2000; Girotti et al., 1993, 1992; Roda, 1989). Increased reliability requirements has stimulated research into non-invasive sensor techniques using laser technology and into optical sensors. The non-invasive laser sensor has the advantage of not being prone to fouling, freedom from electromagnetic interference and of being less susceptible to corrosion (Meredith, 1997). Optical sensors are promising and it is a fast emerging branch in the monitoring technology (Scully et al., 2000; Zhang et al., 1996; Bruno et al., 1995).

The major reasons for improvements in river monitoring in Europe are: (a) establishment of an international monitoring net along the main rivers and tributaries (e.g. Rhine; there are nine International Monitoring Stations from Switzerland to North Sea), (b) improvements in the regulatory policies of regional and national governments, and (c) strict regulations on industrial and municipal waste emissions. As a result by the end of eighties, off-line and on-line monitoring nets were established along almost all the major river systems in Europe. During the last two decades rivers have been investigated using sophisticated technologies, later extended from detailed chemical investigations of water (more than 300 parameters are routinely analyzed) to sediment, biota and to comprehensive biological surveys.

These developments were supported parallel by the impact of micro-electronic development in the last three decades. This led to the use of new instrumentation and their application in information gathering, data transmission, computation and process control through advancements in Information Technology (IT). Alarm and warning services became a cardinal aspect of river monitoring through the availability of real-time information (Gunatilaka and Dreher, 1996; Hendriks and Stouten, 1994). The expansion of the automatic control systems through micro-electronic development has been a major contributor to the increased use of instruments and their applications in river monitoring.

2. NEED FOR RIVER MONITORING

Since the well documented Sandoz incident in Basel, Switzerland on 1st November 1986 (LWA, 1986), where the river Rhine was polluted due to a fire in the industrial storage building, the development and installation of biological early warning systems has been seriously pushed forward in Europe especially in Germany, the Netherlands and the United Kingdom. During this accident, pesticides in large quantities (nearly 30 tons) were released into the Rhine which resulted in complete damage of a large portion of the river biotope. As a consequence fisheries suffered very badly; the eel population (specialized benthos feeder) was completely eradicated in the Upper Rhine. However, 500 km down stream, in North Rhine-Westphalia (NRW), the "Dynamic Daphnia Tests" in operation registered an alarm situation. This gave a signal to the environmental protection organizations for the importance of biological monitoring systems as "Biological Early Warning Systems (BEWS)". The basic idea of the use of automated biological sensor systems for water quality management was first proposed in the early seventies by Juhnke and Besch (1971) and Cairns et al. (1973a, 1973b) and similar systems were deployed in the Rhine in the late seventies (Poles, 1977; Knie, 1978).

Chemical monitoring was introduced in most of the large rivers of Europe during the last three decades, with varying amount of sophistication. Earliest monitoring was limited to easily measurable physico-chemical parameters such as temperature, pH, conductivity and oxygen. Today it has evolved into automated laboratories (Figures 1–4), that measure a large number of parameters including organics (Diehl et al., 1997; Gunatilaka and Dreher, 1996), which need elaborate sample preparation procedures. This is achieved through development of efficient on-line methodologies for screening of a large number of organic micropollutants which include halogenated hydrocarbons, herbicides and pesticides, nitro

FIGURE 1. Rhine Water Monitoring Station at Worms, Germany, located at river-km 443.3. The instruments for sampling, measuring and analyzing river water are located within the historic bridge tower and the laboratory (white) building. Samples are taken by means of diving motor pumps at four sampling sites across the river section for which two pumps are located at the opposite river banks, and the other two pumps are fixed to the central bridge piers. As the contamination plumes from the discharges are concentrated close to the river banks, the biomonitors (Dynamic Daphnia Test, DF-Algae Test) continuously monitor the left and right bank sampling sites (Photo: State Office for Water Affairs, Rhineland-Palatinate, Germany).

aromatics, phosphoric esters and other priority substances (screening on-line techniques using GC, GC-MS and HPLC). The results are obtained after a few hours of laboratory analyses (see Frintop, 1984; Albert and Willemsen, 1989; Brusske and Willemsen, 1990) in time to inform the down-stream water supplies. However, according to Hendriks et al. (1994), the above methods are still incapable of detecting all the organic micropollutants; about 1,000 individual compounds could be identified with GC-MS analyses. In the Rhine, since early seventies the chemical monitoring is supplemented with continuous biomonitoring; the organisms used at the beginning were fish (Hendriks and Stouten, 1994). Later a number of biomonitors were deployed in continuous monitoring either in flow-through systems or

FIGURE 2. The water from the four sampling sites (located across the river section) is led through 4 sets of on-line instruments where water temperature, pH, oxygen concentration, electrical conductivity, turbidity, spectral absorption coefficient (SAC, 254 nm) and fluorescence are measured. Also four sets of automatic samplers (2 for each sampling site) were installed for continuous sampling for routine (major and minor ions, metals, nutrients and organic micro-pollutants etc.) analyses. In addition to the chemical/physical on-line measurements, biomonitoring (algae, Dynamic Daphnia Test) and routine analysis, a GC/MS screening for organic micro-pollutants is made daily. The photo shows the on-line instruments and the automatic samplers which are integrated into the restored vaults of the historic bridge tower (Photo: State Office for Water Affairs, Rhineland-Palatinate, Germany).

in situ (e.g. *Daphnia*, algae, mussels, fishes and photo bacteria; see Kramer and Botterweg, 1991). Biomonitoring is an alternative to chemical tests, and it is capable of rapidly detecting acutely toxic conditions in river water. Because of this distinctive feature they are referred to as Biological Early Warning Systems (BEWS). The basic principle used is that of monitoring some function of physiology or behavior, in a test organism which is changed when exposed to a toxic substance at a sufficient concentration. BEWS are automated, continuous monitors which employ biological organisms or material as primary sensing element, and their uses are discussed in several recent publications (LAWA, 1998, 1996; Knie, 1994; Irmer, 1994). Puzicha (1994, 1995) has discussed the field testing of several biomonitors under the research project "*Wirkungstests Rhein*". Baldwin and Kramer (1994),

FIGURE 3. On-line monitoring station at Lobith (Netherlands) on the Dutch-German border at river km 862.3, right river bank. [At Kleve-Bimmen, River km 865.0 on the left bank, is the corresponding German continuous monitoring station. Both serve as International trans-boundary monitoring stations.] The ship anchored at the shore houses a well-equipped laboratory which analyzes a wide variety of pollutants in the river ranging from standard parameters, nutrients, trace elements to GC-MS and HPLC screening for organic micro-pollutants and biomonitoring (algae and Dynamic Daphnia Test) (Photo: Gunatilaka).

Kramer and Botterweg (1991) and Gerhardt (1999) provide exhaustive reviews on commercially available biomonitors.

Of all the biomonitoring methods that have been deployed over the last two decades, many have remained at a laboratory stage. Only a few tested under field conditions, even fewer are really commercially available. After intensive field testing (Puzicha 1994, 1995), the "Working Group of the Federal States on Water Problems", in Germany recommended a few reliable continuous biological test systems for river monitoring (LAWA, 1998) and they are listed in Table 1.

3. TRANS-BOUNDARY MONITORING

Water quality monitoring of international rivers has always been a difficult task in Europe, but great strides have been achieved through international commissions such as International Commission for the Protection of the Rhine (ICPR) and the recently established International Commission for the Protection of the Danube River Protection Convention (ICPDR). On-line monitoring has been introduced in most of the large rivers in Europe and they served as cornerstones in

TABLE 1. Continuous Biotest Systems for river monitoring—a synopsis of the recommendations of LAWA (1996, 1998; Working Group of the Federal States on Water Problems, Germany), based on biotest systems tested in the research project "*Wirkungstests Rhein*" and those implemented by the state authorities in Germany (modified and translated from German). Since the table was updated by the LAWA working committee on "Biomonitoring" in 1996, several more biomonitors have been developed and tried to be established on the market. None of these new biomonitors has been tested in such a systematical way as those mentioned in the table. Nevertheless some of the tentatively tested new systems seem to have a realistic chance to be well established in the next few years. The LAWA working committee observes the further developments critically and discusses them routinely during their meetings.

No.	Biotest system	Organisms	Principle of the measurement	Evaluation	Maintenance (hrs) Operation costs DM/year	Producer and Instrument cost	Dimensions (mm)
1	Dynamic Daphnia Test	*Daphnia magna*	swimming behavior by IR light sensors	dynamic alarm thresholds	4 hrs. per week 1.500 DM	Elektron GmbH, Magdeburger Str. 19, D-47800 Krefeld 47.000 DM	H 1,000 W 600 D 650
2	Koblenz Behavioral Fish Test*	golden ide (*Leucisus idus melanotus*)	behavioral parameters as video images	dynamic alarm thresholds	2 hrs. per week 900 DM	GSF Inst. f. Toxikologie Ingolstädter Landstr. 1 Oberschleißheim D-85758 67,000 DM	H 1,800 W 1,500 D 800
3	"Mossel-Monitor"	bivalves (*Dreissena polymorpha*)	opening/closing of valves, distance of valves	comparison of activity patterns, dynamic alarm thresholds	1 hr. per week 700 DM	Delta Consult B P.O. Box 71 NL-4420 AC Kapelle 36.000 DM	H 350 W 600 D 600
4	Dreissena-Monitor	bivalves (*Dreissena polymorpha*)	opening/closing of valves	comparison of activity patterns, dynamic alarm thresholds	2–5 hrs. per week 200 DM	Envicontrol Kapellenstr. 53 D-50226 Frechen 35.000 DM	H 700 W 1200 D 600

table 2 (cont.)

5	DF-Algae Test	algae (various species)	delayed fluorescence (DF),	comparison of fading curves, sample vs. control	5 hrs. per week 1.200 DM	V. Gerhardt, J. Putzger, Univ. Regensburg Universitätsstr. 21 D-93040 Regensburg 50.000 DM	H 1,000 W 900 D 900
6	FluOx-Algae Test*	algae (various species)	fluorescence and oxygen production	comparison of actual with former data	3 hrs. per week	Kolibri, Schwerte 110.000 DM	H 1,600 W 500 D 600
7	Biosens Algae Toximeter*	green algae (*Chlamydo monas reinhardtii*)	spontanuous, variable chlorophyll-a fluorescence	fluorescence parameter (ET-value) difference	3 hrs. per week 6.900 DM	Dr. Noack, Käthe-Paulus-Str. 1 D-31157 Saarstedt 86.000 DM	H 2,000 W 620 D 560
8	bbe-Algae Toximeter	algae (various species)	variable chlorophyll-a fluorescence	comparison sample vs. control, adaptive Hinkley detector	1 hr. per week 600 DM	bbe Moldaenke Schauenburger Str. 116 D-24118 Kiel, 57.000 DM	H 600 W 350 D 380
9	Regensburg Luminous Bacteria Test	*Photobacterium phosphoreum*	decline of luminescence	comparison sample vs. control	1 hr. per week 1.200 DM	V. Gerhardt, J. Putzger, Univ. Regensburg Universitätsstr. 21 D-93040 Regensburg 45.000 DM	H 600 W 500 D 500
10	BioLum Luminous Bacteria Test*	*Photobacterium phosphoreum*	decline of luminescence	comparison sample vs. control	2 hrs. per week	Kolibri, Schwerte 70.000 DM	H 700 W 500 D 550 (2 pieces)

* Contact the producer for further information.

FIGURE 4. Donaukraft—Groundwater on-line monitoring station along the Danube at Nussdorf (river km 1932.26). The station controls the water quality of the river bank filtrate that is recharged into the ground water aquifer in the second and twentieth districts of Vienna. The station also continuously monitors river water; facilities installed includes both chemical and biomonitoring (Dynamic Daphnia Test) (Photo: Gunatilaka).

establishment of trans-boundary monitoring between cross border countries (e.g. Rhine, between Switzerland/Germany and Germany/Netherlands).

In the nineties with establishment of the New Independent States in Europe (and together with East European countries) the number of international river boundaries increased rapidly. In 1992 in Helsinki, Water Framework Directive of the European Commission under the auspices of United Nations Economical Commission for Europe (UNECE), established the Convention on Protection and Use of Trans-boundary Watercourses and International Lakes. The convention was set up as a framework for management of nearly 115 major trans-boundary rivers and 20 international lakes. This came into effect in 1996 and under the UNECE Water Convention, the Task Force on Monitoring and Assessment has drafted "Guideline for Water-Quality Monitoring and Assessment of Trans-boundary Rivers". The implementation of the guidelines will be effected in a series of trans-boundary river basins in Central and Eastern Europe. On-line monitoring stations will be established on either sides of the boundaries for compliance and regulatory monitoring and as well as a control measure between the bordering countries with the aim of improving health and environmental impacts. As a result a large number of on-line monitoring stations will be established not only in the main rivers, but also along the tributaries of large river systems. For example the following Danube tributaries will be installed with on-line monitoring stations: Ipel/Ipoly (between Slovakia and Hungary); Mures/Maros (between Romania and Hungary); Latorytsya/Latorica and Uz/Uh (between Ukraine and Slovakia); Morava (between Czech Republic and Slovakia).

Trans-boundary water pollution is a widely recognized environmental problem in Central and Eastern Europe. A key objective of the EU environmental policy for the future is the protection of water resources (Hayward, 1999). The establishment of trans-boundary monitoring in Eastern Europe as well as New Independent States (NIS) will lead to improvement of the river systems through the control of "trans-boundary impacts" in river systems. This will include pollution control through emission limitations, action programs for improvements in waste water treatment, water quality improvement and water management in the basin and sub basins. This will eventually result in improvements in water quality in larger river systems.

Such improvements are already visible in the Rhine and the Danube, but monitoring efforts were initiated three decades back. Today an effective early warning system in both rivers can deal with short and sudden contamination, as in case of accidents spillage and malfunctioning of waste water treatment plants (Figures 5 and 6; Gunatilaka and Dreher, 1996; Rodda, 1994; Botterweg and Rodda, 1999; Pintér, 1999; IRC, 1987). In the Rhine, the pollution load from the nutrients, heavy metals and organic micropollutants have decreased nearly to one tenth and the

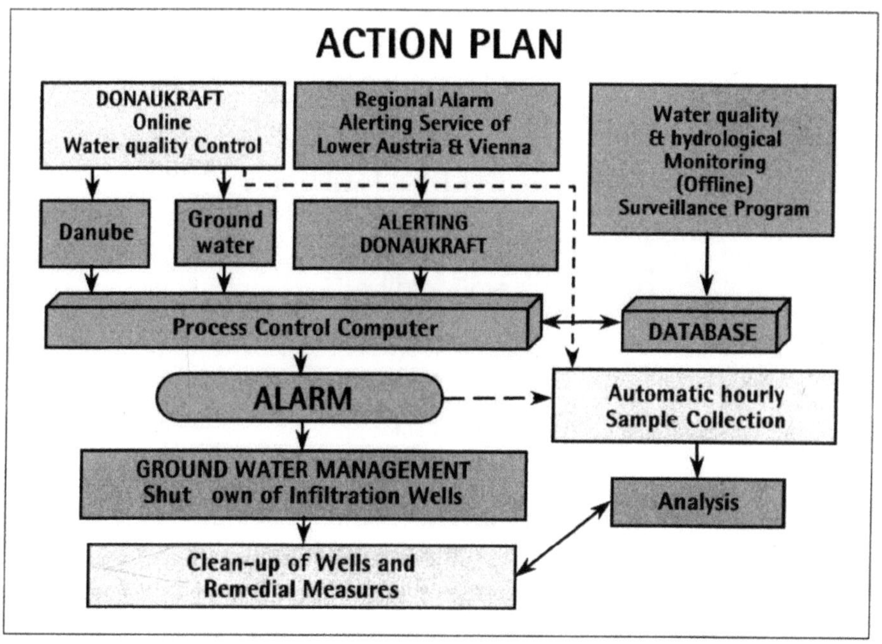

FIGURE 5. Alarm alerting through Danube Action Plan which is in operation in the Austrian stretch of the Danube (executed through Donaukraft). Hierarchical organization of the alarm plan couples the on-line as well as off-line information and stored in a central database which is used for decision making processes. In case of an alarm, groundwater recharging will be automatically stopped and remedial measures undertaken.

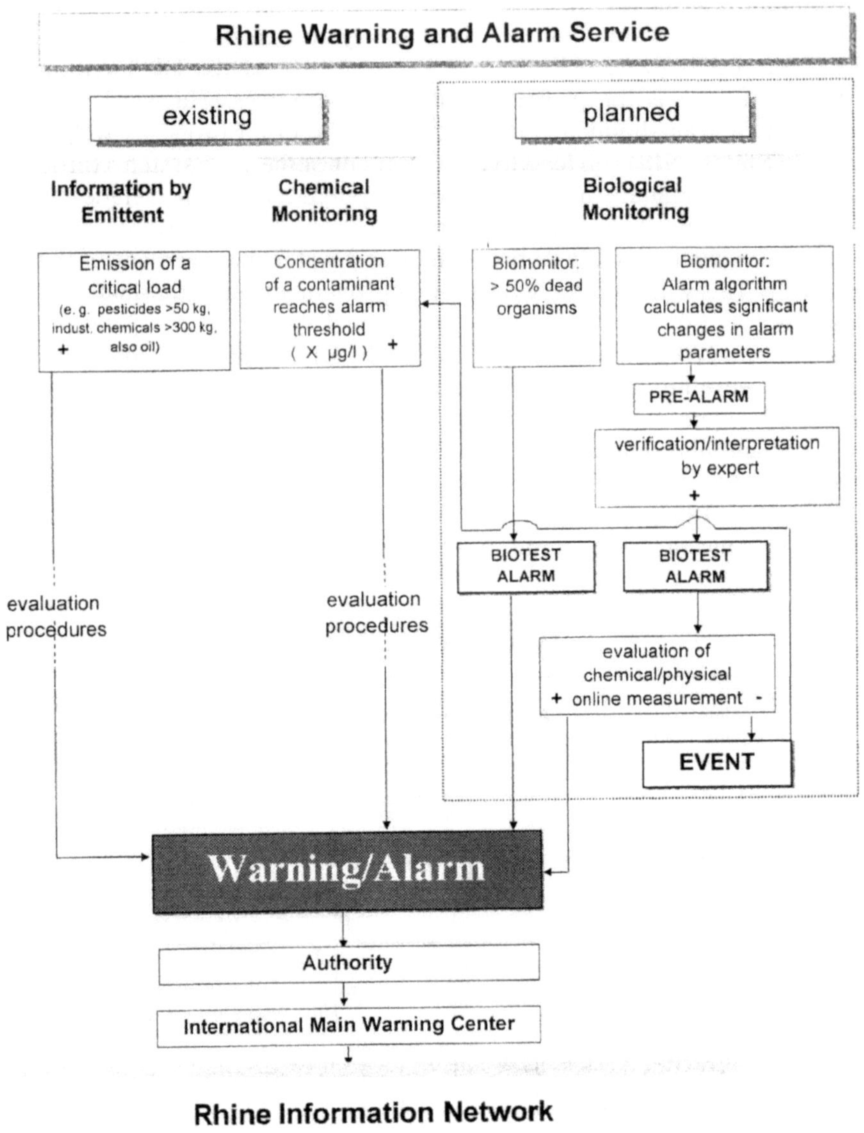

FIGURE 6. Proposed alarm warning plan for the Rhine. At present the Rhine Warning and Alarm Service relies on chemical measurements and analyses and on information about emissions. In the future information from both chemical as well as biological monitoring will be coupled for conformation of alarms. (Monitoring concept of the Rhine, LAWA working committee "Biomonitoring", Diehl et al., manuscript in preparation).

organic load to half that of the seventies. The International Rhine Action Program adopted in 1987 stresses the further reduction of pollution load to improve the Rhine ecosystem and to protect the North Sea (IRC, 1987). The effectiveness and successes of protection measures resulted from among others, a continuous water quality monitoring net built along the Rhine and its tributaries where trans-boundary monitoring, vigilance and peer-pressure has played an important role. Figure 7 (Panel A and B) show a comparison of water quality in the Rhine during 1969/1970 and 1994, in the State of North Rhine-Westphalia, Germany (the State which is adjoining The Netherlands; trans-boundary stations located at: Lobith / Netherlands; Kleve-Bimmen / Germany). Here there is a drastic improvement in the water quality as well as in the river environment; the biological assessment shows the transformation of the Rhine and the tributaries from *alphamesosaprobic* (Class III, heavily contaminated) to *betamesosaprobic* (Class II, moderately polluted) (see Figure 7 for the details of the classification). However, this transformation took nearly twenty five years.

In the early seventies due to high organic load in the Rhine through untreated sewage, the oxygen saturation was low. Gradually treatment plants were built for the treatment of industrial and municipal wastes in many cities and industrial complexes along the river. These measures resulted in improvements in oxygen levels in the river, and at present even under dry weather flow, oxygen concentrations are above $6\,mg\,L^{-1}$ (Irmer, 1996), which is also the critical limit for fish. As Rhine water was heavily polluted in the sixties and seventies, it caused acute deaths of Daphnia and trout embryos (Poles et al., 1980; Slooff, 1982, 1983; Sloof et al., 1983) but in early nineties, the toxicity of Rhine water in terms of acute mortality of *Daphnia magna* has markedly improved (see Hendriks et al., 1994).

Macrozoobenthos are good indicators of river water quality. With the reduction of the pollution load, macrozoobenthos populations have reacted immediately, their numbers started increasing since the mid seventies. However, in the lower Rhine the species composition has changed; some sensitive species have disappeared and several new species (neozoa) have colonized the vacant niches. This is seen with molluscs, chironomids, caddies flies, and flatworms; crustaceans became dominant in the nineties. In general immigrant species have shown a steady increase in the Rhine and this is true for macrozoobenthos as well for fish populations (van den Brink et al., 1990). The Ecological Master Plan under the Rhine Action Plan (IRC, 1987) is directed towards restoring the river as a habitat for sensitive migratory fish species ("Salmon 2000"). This includes the protection, preservation and improvement of ecologically important stretches of the Rhine and its tributaries (Irmer, 1996).

Both off-line as well as on-line monitoring programs executed in the upper Danube (from source in Germany to Austrian/Slovakian border; nearly 350 km stretch) in conjunction with stringent environmental regulations have resulted in improvement in the river water quality (Class II, *betamesosaprobic*). Also structural improvements in the main river as well as tributaries have contributed to improved habitat structure for the fish fauna. The fish fauna of the Austrian Danube is characterized by its large number of species (Jungwirth, 1984; Schiemer and Spindler, 1984). Fifty seven species have been recorded during the recent surveys,

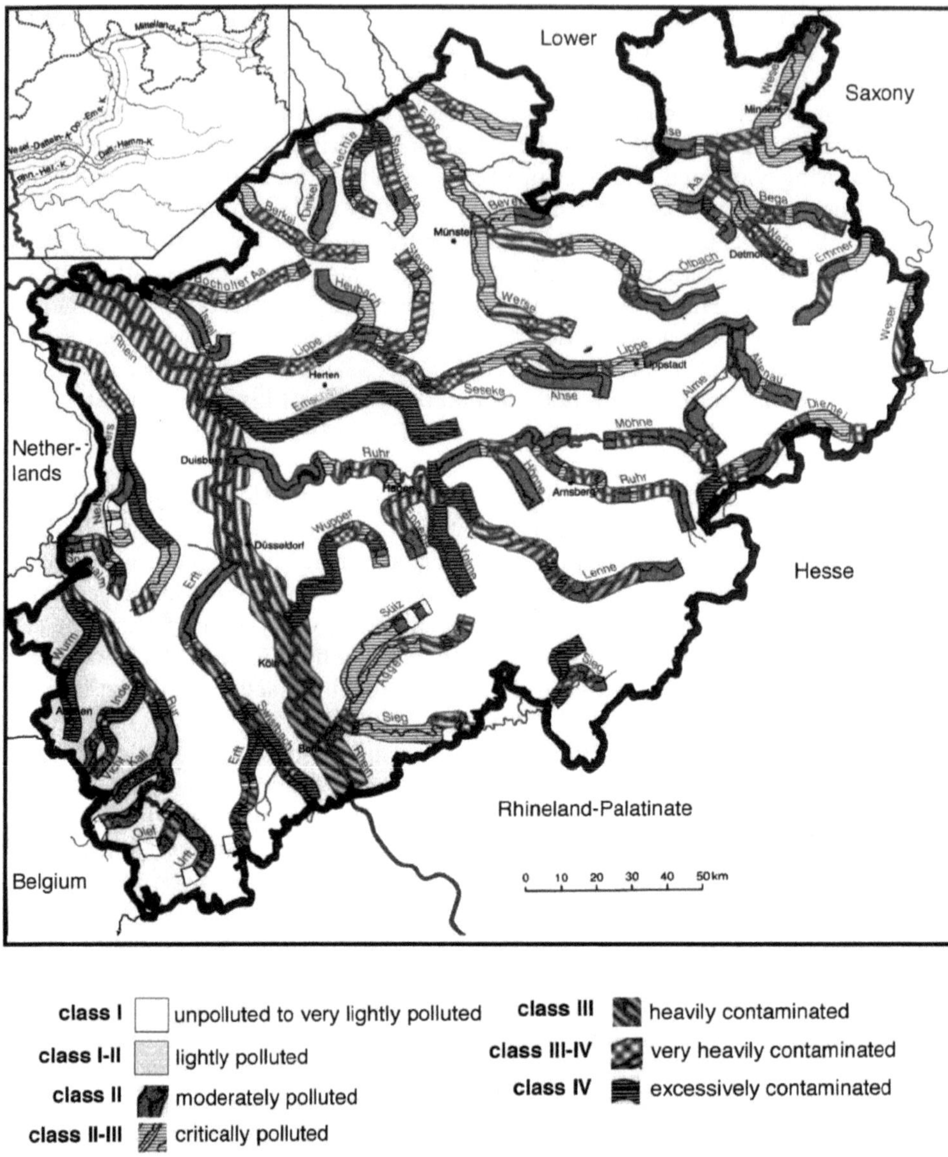

FIGURE 7. A comparison of the water quality of the Rhine River in the State of North Rhine-Westphalia, Germany, during 1969/1970 (Panel A) and 1994 (Panel B). Water quality is both chemically and biologically assessed. The water quality assessment shown in the map is presented according to the following classification:

Quality class I:	Unpolluted to very lightly polluted (*oligosaprobe*)
Quality class I–II:	Lightly polluted (*oligosaprobe* to *betamesosaprobe*)
Quality class II:	Moderately polluted (*betamesosaprobe*)
Quality class II–III:	Critically polluted (*betamesosaprobe* to *alphamesosaprobe*)
Quality class III:	Heavily contaminated (*alphamesosprobe*)
Quality class III–IV:	Very heavily contaminated (*alphamesosaprobe* to *polysaprobe*)
Quality class IV:	Excessively contaminated (*polysaprobe*)

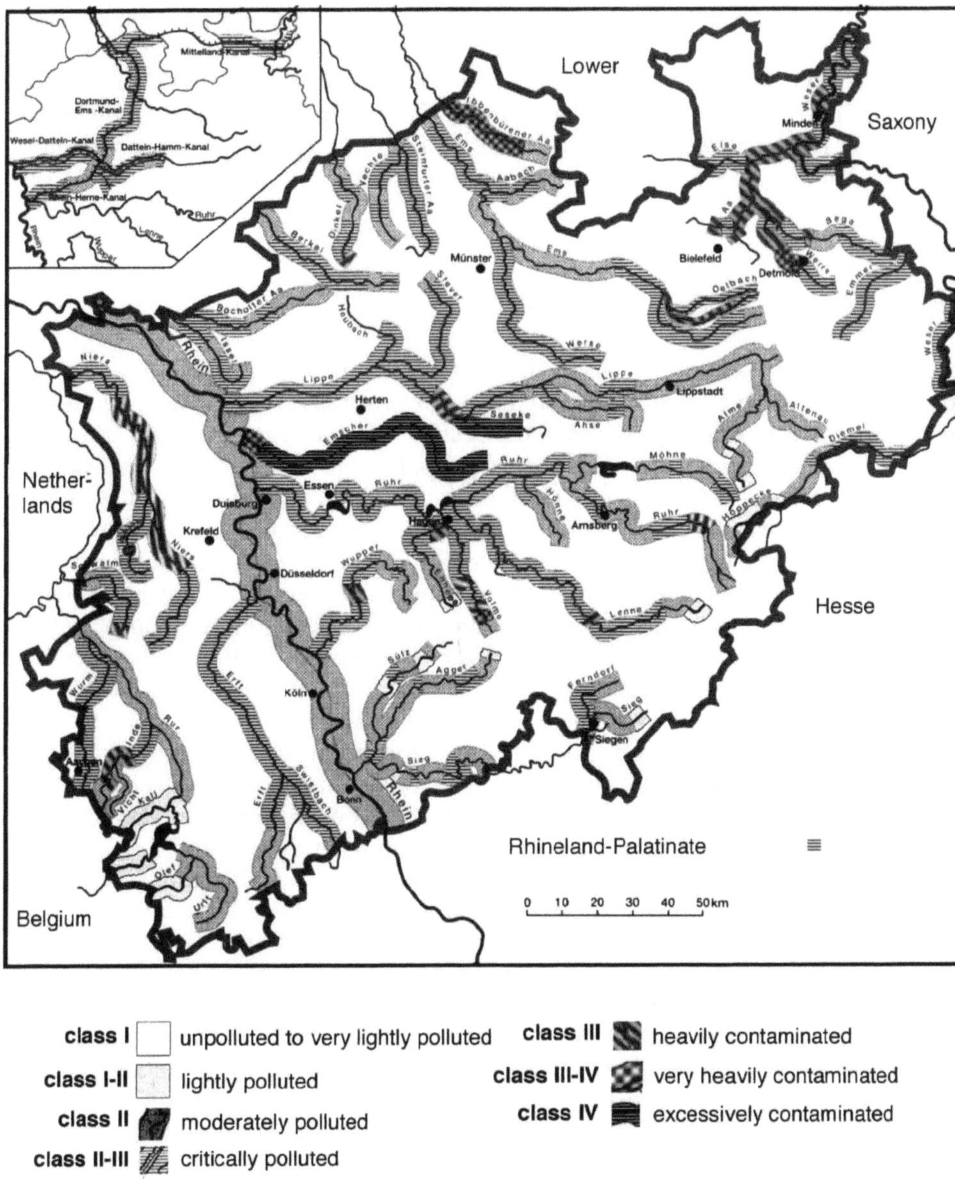

FIGURE 7. *Continued* Note the water quality change in the Rhine from heavily contaminated (Class III/*alphamesosaprobe*) in 1969/1970 to moderately polluted (Class II/*betamesosaprobe*) in 1994. Under *alphamesosaprobic* conditions, the river was heavily polluted with organic matter which consumed oxygen leading to low oxygen levels, and periodic fish deaths were common. Under these conditions sewage micro flora were dominant and macro organisms were limited to sub-oxic species adapted to live under low oxygen concentrations. Similar conditions prevailed in almost all the major tributaries. Due to strong national and international monitoring efforts, the water quality of the Rhine improved gradually and the 1994 inventory shows a change to *betamesosaprobic* conditions. This means that although the river is moderately polluted, there is no oxygen depletion. Due to the improved oxygen condition reduced pollution load, the missing macro invertebrates and fish assemblages have begun to return to the river along with new migrant species. Also similar changes are observed in the tributaries.

(Maps: Through courtesy of State Office for Water and Waste North Rhine-Westphalia, Essen, Germany; modified by the authors).

52 of which are autochthonous or long-term established elements (Schiemer and Waidbacher, 1992). These include rare salmonids like *Hucho hucho* which are very sensitive to poor water quality and oxygen concentrations; the other salmonids found commonly in the Austrian Danube are *Salmo trutta forma fario*, *Salvelinus fontinalis* and *Oncorhynchus mykiss*. In general fish communities are good indicators for habitat structure as well as for the ecological integrity of large river systems.

4. NEEDS FOR STANDARDIZATION

With the expansion of on-line monitoring, the need for standardization of commercially available measuring systems are increasing. It is needed by the users, law enforcing authorities to compare the results of measurements from different sources carried out by using similar instruments, but produced by different manufacturers. At present it is difficult to compare instruments from different vendors due to lack of proper documentation as well as test protocols of the instrumentation. Moreover, most of the on-line instruments which are sold, are not field calibrated and tested. Rivers are dynamic systems and the instruments may be exposed to completely different matrices than the laboratory calibration medium or physical conditions (e.g. floods, high sediment loads). The users as well as authorities are unable to test and compare different products before purchasing due to lack of the test protocols and some shortcomings described below.

During installation of on-line, water-quality monitoring stations along the Danube for river-water and ground-water control (Gunatilaka and Dreher, 1996), we had to find answers by ourselves for many questions in regard to the installed instrumentation. This included specific information such as rise and fall time, up time, dead (lag) time, drift, memory effects, trueness / bias and accuracy to mention a few. To our surprise some of the information supplied by the vendors on basic specifications such as linearity, lowest detectable change and limit of detection was either inaccurate or completely wrong. Common sensors used in monitoring like red-ox electrodes (oxidation reduction potential measurements) or analyzers used for dissolved organic carbon (DOC) or total organic carbon (TOC) were plagued with such problems. The red-ox sensors we installed had inherent problems such as platinum electrodes getting coated with reduced substances (common in ground water) which needed daily cleaning and calibration to obtain accurate results. Other than that, the instrument recorded a false redox value without subtracting the electrode potential of the reference electrode (see Zo Bell, 1949) and introduction of a correction was not possible. We tested five DOC analyzers from well-known instrument makers (2 from USA, 2 from Europe, 1 from Japan) before the final selection. The main criteria given for selecting an instrument was that they should stand a six months continuous field test with no break down and should have a capability of measuring low DOC concentrations (as low as 1 $mg L^{-1}$). The measuring range given in the technical specifications for all instruments were 1–10 $mg L^{-1}$. However, almost all the instruments tested were either inaccurate or incapable in measuring low concentrations (in the stations we are monitoring, ground water DOC concentrations are usually below 2 mg L^{-1}, and it is common to have values around 1.20 mg L^{-1}) and the precision and accuracy of the measurements were poor. Two instruments were withdrawn from the test during the first week; another two in the third week. With the one left, the test was continued up to the second month. As we needed an accurate measuring instrument for DOC for the monitoring stations, thereafter we personally collaborated with the technicians from the vendor in rectifying the faults in the instrument which in total took 18 months. The same company produced an on-line, TOC (total organic carbon) process analyzer which we installed for river water analyses. Again, it took another two years for modifications and adjustments to produce reproducible accurate results. The above examples show the dilemma users have to face, and the problems mentioned are solvable only by trained experts. In most cases, the users buy the instruments on the recommendation of a vendor without the consultation of a specialist.

On the other hand the same on-line sensors today are used in large quantities by the water industry, food and beverages industry, and waste water treatment plants. There is a large effort in Europe to upgrade the waste-water treatment plants (numbering more than 60,000) through process optimization using on-line monitoring and up to date information technology (IT). On-line measurements deliver "real time" data and serve as a link to the real world. However, inaccurate on-line measurements coupled to high computational power will not bring the expected results. Redox potential and oxygen in waste water treatment plants (WWTP) are measured using electrodes (the classical sensors used in waste water treatment industry) and they are coupled to activation of blowers for aeration. When the redox potential in treatment ponds drops close to zero mV (at the nitrate break point; parallel the oxygen concentration drops), the blowers are activated for re-oxygenation. However, faulty redox measurements naturally could lead to high energy consumption in the WWTP. The same is true for on-line nutrient measurements (ammonia, nitrate and orthophosphate) which provide basic information on nitrification and denitrification processes as well as the biological and chemical processes removing phosphorous. Based on calculations for Denmark (5 million population, 1.5 billion ECU investments, operational costs/year > 5%), with accurate on-line measurements, it is estimated that 15–25% could be saved on investments alone. If this is extrapolated to Europe (>250 millon population) savings for investments and operations will be high as 15 billon ECU and 1 billon ECU/year, respectively (Jakobsen, 1998). [1 ECU is approximately equivalent to 1 US $ in 1999].

Based on this information, to remedy this deficiency a European project for evaluation of on-line sensors was established in 1998 referred to as ETACS (European Testing and Assessment of Comparability of on-line Sensors). The scope of the project is to present: (a) a standardized description of on-line in-situ sensors, (b) establishment of standards for on-line sensors, and (c) development of an accurate test protocol useful for manufacturers as well as users (ETACS, 1999).

It is the intention that the findings of the ETACS project to be used as a basis for a European as well as a ISO-Standards for on-line, in situ analyzers and sensors. The standards will ensure the uniformity of the on-line water quality measurement.

This will bridge a serious gap in the monitoring and control technology and help to produce reliable data for real-time process control as well as for river monitoring. Additionally, it will be an advantage for environmental authorities for compliance monitoring as well as harmonization of databases.

5. CONTEMPORARY DEVELOPMENTS

The urgent need for drinking water, food and recreational areas on one hand, and dangers of increasing eutrophication and pollution (chemical and microbiological) on the other, are problems well known in the temperate zone. Although not yet serious as in many industrial, temperate countries, the latter problems may develop quickly and probably become much more serious in fast developing regions in the tropics, especially in South East Asia.

In the developing countries, problems concerning water quality have become aggravated during the last decades (World Bank, 1994). In industrialized countries, the problems of pollution through intensive agricultural practices, release of large amounts of industrial effluents and domestic wastes gradually developed during the last century, but fast developing countries are experiencing the same effects in much shorter time (WHO, 1989). Wide spread anthropomorphic pressures on aquatic ecosystems have been caused by rapid urbanization (organic and inorganic load), industrial expansion, and excessive sedimentation from land clearance and deforestation. The major river systems in most of the South East Asia today suffers from these anthropomorphic stresses to such an extent, that in the lower reaches of the rivers, the water quality was considered substandard [untreatable by conventional means; based on 80% samples taken; see Gunatilaka, 1999]. To avoid social chaos and environmental catastrophe, the management of existing water resources (as there are nearly no new resources to develop) should take priority; and hence the management of rivers and river basins is an important task. In time, it is hoped that the standardized, improved sensor techniques and automated control systems developed for river monitoring in Europe will become available for developing countries.

Similar to most of the developing countries in South East Asia, Indonesia has most of the contemporary water quality problems of the developed nations such as pollution, eutrophication, toxicity development, ecosystem dysfunction, acidification from air pollution now aggravated through long standing forest fires. Industrialized countries have faced most of these problems sequentially, but Indonesia is facing them simultaneously. Due to severely degraded water quality several developing countries are facing water shortages (United Nations, 1997) and it is true for Java. Even more grotesque is that there are a few new sources of water available in Java to be developed, which means that remediation of available sources will be a necessity in the near future. For example, in Indonesia, a country with 204 million population, the demand for bottled drinking water has increased from 323,000 liters in 1987 to 4 million cubic meters in 1998 (an increase of nearly 12,000 times). Therefore, development of river basin or sub-basin management programs should be given priority to avoid water scarcity in the next 25 years. The major problem the government has faced today is to improve the water quality of the river systems to meet the demand from different sectors and to improve the capacity of water supply for domestic, industry, agriculture, energy, tourism and other uses.

To achieve this goal the government of Indonesia emphasizes the importance of an integrated approach for water resources management based on river basins. The intention is to meet the conflicting demands through managing the available water resources using up-to-date, state-of-the-art technology and appropriate tools. A key to water resources management is availability of reliable data, which are lacking in most of the developing countries. The new water resources management activities introduced by the government will eliminate this deficiency. The proposed program is comprised of the following elements: a water resources data management system, a basin water allocation/operation system, a quality management system and a river management system. As a first step, an elaborate water quality and water quantity monitoring network comprising both on-line and off-line monitoring stations is implemented in one of the largest rivers in East Java (Brantas), along with other river basin management operations. If this program is successful, similar monitoring systems will be established in other river systems in West and Central Java (Gunatilaka, 1999).

6. CONCLUSIONS

The above mentioned developments stress once more the importance, as well as the need for the availability of accurate monitoring sensors and analyzers. The expansion of the continuous monitoring with the interests in obtaining accurate real-time data for process management as well as river monitoring has become a global need which is pushed forward through development of sophisticated sensors as well as rapid development of the information technology (IT). These processes will be facilitated through development of standardization for all type, on-line sensors and analyzers, and the economic impact of adopting such products has to be emphasized. As the sensors generate vast amounts of data, parallel development is necessary in data acquisition and telemetry, IT—concepts for data handling, database management and guidelines for efficient use of the generated information.

With respect to the development of new, on-line sensors the following requirements must be addressed: sensitivity, precision, reliability, repeatability, reproducibility, stability, plausibility of data. Also, self-cleaning, low maintenance requirements, auditable performance to a defined standard, availability (up-time) should be anticipated characteristics of on-line sensors. As analyses are carried out in remote locations, they require more rugged but reliable instrumentation than laboratory installations. Another proviso is that the instruments will be in operation round the clock (continuous, unsupervised operations), and hence, total automation of calibration of the instrumentation as well as simplicity in operation will be required.

With large numbers of new chemical substances entering river systems, continuous monitoring systems for their detection will become increasingly important with respect to environmental effects they produce, in addition to carcinogenic, endocrine or other toxic effects. Much effort has to be directed towards the on-line detection of such pollutants in rivers. Thus, the challenges to continuous chemical and biological monitoring will be immense. Finally the importance of integration of biological and chemical monitoring has to be emphasized. The on-going efforts for such integration by the International Commission for the Protection of the Rhine should serve as an example for other large river systems in Europe.

7. REFERENCES

Alberti, J., and Willemsen, H.G. 1989. Enrichment of organic substances in trace concentrations by solid-liquid-phase-adsorption as a special form of sampling. Forum Städte-Hygiene 4: 118–125.

Baldwin, I.G., and Kramer, K.J.M. 1994. Biological Early Warning Systems (BEWS). In: Biomonitoring of Coastal Waters and Estuaries. K.J.M. Kramer (ed.), pp. 1–23.

Botterweg, J., and Rodda, D.W. 1999. Danube River Basin: progress with the Environmental Programme, Wat. Sci. Tech. 40(10): 1–8.

Bruno, J.C.D., Lewis, E., Scully, P., and Edwards, R. 1995. A novel technique for optical fiber pH sensing based on Methylene blue adsorption. J. Lightwave Technol. 13(7): 1407–1414.

Brusske, A., and Willemsen, H.G. 1990. Short time monitoring of the Rhine-conception and the first results. Vom Wasser 74: 393–399.

Cairns, J. Jr., Hall, J.W., Morgan, E.L., Sparks, R.E., Waller, W.T., and Westlake, G.F. 1973a. The development of an automated biological monitoring system for water quality. Virginia Water Resources Research Center, Blacksburg, Bulletin 51.

Cairns, J.R., Sparks, R.E., and Waller, W.T. 1973b. The use of fish as sensors in industrial waste lines to prevent fish kills. Hydrobiologia 41: 151–167.

Cowell, D.C. 2000. Screen-printed Disposable Biosensors for Environmental Pollution Monitoring (this volume).

Diehl, P., Krauß-Kalweit, I., and Lüthje, S. 1997. The new Rhine Water Quality Station in Worms (in German), Wasser and Boden 49: 25–29.

ETACS, 1999. European Testing and Assessment of Comparability of On-line Sensors/analysers [SMT4-CT96-2131 (DG XII-SMT)]—ETACS Project, News letter No. 3, p. 9.

Frintop, P.C.M. 1984. Signalisierung erhöhter Gehalte an organischen Mikroverunreinigungen (SIVEGOM). Staatliches Wasserwirschaftsamt—staatliches Institut für Abwasserreinigung (RIZA), Lelystad, Juli 1984 (unpublished report).

Gerhardt, A. 1999. Recent trends in on-line biomonitoring for water quality control. In: A. Gerhard (ed.) Biomonitoring of Polluted Water. Environmental Science Forum, Trans. Tech. Pub. 96: 95–118.

Girotti, S., Ferri, E., Ghini, S., Rauch, P., Carrera, G., Bovara, R., Roda, A., Giosuè, M.A., Masotti, P., and Gangemi, G. 1993. Bioluminescent flow-sensing device for the determination of Magnesium (II). Analyst 118: 849–853.

Girotti, S., Ferri, E., Ghini, S., Rauch, P., Carrera, G., Bovara, R., Piazzi, S., Merighi, R., and Roda, A. 1992. L-Glutamate determination by bioluminescent flow sensor. Fresenius J. Anal. Chem. 343: 176–177.

Gunatilaka, A. 1999. What does the future hold for Brantas River (Indonesia)?—a common concern and commitment. Proc. Int. Symp. "Hydropower into the Next Century—III", Gmunden, Austria, Oct. 1999. Int. J. Hydropower and Dams, 189–199.

Gunatilaka, A., and Dreher, J. 1996. Use of early warning systems as a tool for surface and ground water quality monitoring. Proc. IAWQ and IWSA Symp. on Metropolitan areas and Rivers, Rome, May 1996. TSI—River quality surveying and monitoring methods 2: 200–211.

Gunatilaka, A., Diehl, P., and Puzicha, H. 2000. The evaluation of the "Dynamic Daphnia Test" after a decade of use: benefits and constriants (this Volume).

Hendriks, A.J., Maas-Diepeveen, J.L., Noordsij, A., and van der Gaag, M.A. 1994. Monitoring response of XAD-concentrated water in the Rhine delta: a major part of the toxic compounds remains unidentified. Wat. Res. 28: 581–598.

Hendriks, A.J., and Stouten, M. 1994. Monitoring response of flow-through *Daphnia magna* and *Leuciscus idus* assays to microcontaminants in the Rhine Delta: early warning as a useful supplement. Ecotoxicol. Envir. Safty. 26(3): 265–279.

Hayward, K. 1999. Europe's new frontiers. Water 21(4): 10–11.

IRC, 1987. Rhine Action Program (in French/German). International Rhine Committee, Strasbourg, France.

Irmer, U. (ed.) 1994. Continuous biotests for water monitoring of the river Rhine. Summary, recommendations, description of test methods. Umweltbundesamt Texte 58/94, Berlin, p. 30.

Irmer, U. 1996. Rhine water quality monitoring methods and rehabilitation. Proc. IAWQ and IWSA Symp. on Metropolitan areas and Rivers, Rome, May 1996. TSI—River quality surveying and monitoring methods 2: 41–47.

Jakobsen, H.S., 1998. Hans Savankjaer Jakobsen, VKI, DK. 1st ETACS Workshop, Brussels, 26-11-98, personal communication.

Jungwirth, M. 1984. "Einige Fragen zur Herkunft der Süßwasserfischfauna der europäisch-mediterranen Unterregion. Arch". Hydrobiol. 57: 16–134.

Juhnke, I., and Besch, W.K. 1971. A new method for acute toxic compounds early warning. Z. Wasser Abwasser Forsch. 11: 161–164 (in German).

Knie, J. 1994. Biomonitore zur kontinuierlichen Überwachung von Wasser und Abwasser in der Bundesrepublik. Spektrum der Wissenschaft 5: 94–98.

Kramer, K.J.M., and Botterweg, J. 1991. Aquatic biological early warning systems: an overview. In: Bioindicators and Environmental management, D.W. Jeffrey, and B. Madden (eds.), Academic Press, London, pp. 95–126.

LWA, 1986. Sonderbericht: Brand bei Sandoz und Folgen für den Rhein in NRW, November 1986.

LAWA, 1996. Empfehlungen zum Einsatz von kontinuierlichen Biotestverfahren für die Gewässerüberwachung. Kulturbuchverlag, Berlin, pp. 1–38.

LAWA, 1998. Recommendations on the deployment of continuous biomonitors for the monitoring of surface waters. Kulturbuchverlag Berlin, pp. 1–46.

Meredith, W.D. 1997. On-line monitoring of water quality, New World Water 4: 103–108.

Pintér, G.G. 1999. The Danube Accident Emergency Warning System. Wat. Sci. Tech. 40(10): 27–33.

Poels, C.L.M. 1977. An automatic system for rapid detection of acute high concentrations for toxic substances in surface water using trout. In: "Biological Monitoring of Water and Effluent Quality" (J. Cairns et al. Eds.), ASTM STP 607: 85–95.

Poels, C.L.M., van der Gaag, M.A., and van de Kerkhoff, J.F.J. 1980. An investigation into the long-term effects of Rainbow trout. Wat. Res. 14: 1029–1035.

Puzicha, H. 1994. Biomonitoring under field conditions at the rivers Rhine and Main and experiments with waste water (in German). Landesamt für Wasserwirtschaft, Rheinland-Pfalz, Mainz, Hessische Landesanstalt für Umweltschutz, Wiesbaden, p. 198.

Puzicha, H. 1995. Bewertung verschiedener Biomonitore für den Routineeinsatz zur Schadstoffüberwachung in Oberflächenwässern. PhD Thesis, Johannes-Gutenberg-Universität, Mainz, p. 174.

Roda, A., Girotti, S., Ghini, S., and Carrea, G. 1989. Coupled reactions for the determination of analytes and enzymes based on the use of luminescence. J. Biolumin. Chemilumin. 4: 423–436.

Rodda, D.W. 1994. The Environmental Programme for the Danube River Basin. Wat. Sci. Tech. 30(5): 135–145.

Schiemer, F., and Spindler, T. 1989. Endangered fish species of the Danube river in Austria, Regul. Riv. 4: 397–407.

Schiemer, F., and Waidbacher, H. 1992. Strategies for conservation of a Danubian fish fauna. In: "River conservation and Management", P.J. Boon, P. Calow, and G.E. Petts (Eds.), pp 363–382, John Wiley, NY.

Scully, P.J., Chandy, R.P., Edwards, R., Merchant, D.F., and Morgan, R. 2000. Optical sensors and Biosensors for Environmental Monitoring (this volume).

Sloof, W. 1983. Benthic microinvertebrates and water quality assessment: some toxicological considerations. Aquat. Toxicol. 4: 73–82.

Sloof, W. 1982. Skeletal anomalies in fish from polluted surface waters. Aquat. Toxicol. 3: 157.

Sloof, W., de Zwart, D., and van de Kerkhoff, J.F.J. 1983. Monitoring the rivers Rhine and Meuse in the Netherlands for toxicity. Aquat. Toxicol. 4: 189–198.

United Nations, 1997. Comprehensive assessment of the freshwater resources of the world. Report of the Secretary General, Commission on Sustainable Development, 5th Session, 7–25 April 1997. Document E/CN.17.

Van den Brink, F.W.B., Van der Velde, G., and Cazemier, W.G. 1990. The faunistic composition of the freshwater section of the River Rhine in The Netherlands: present state and changes since 1900. Limnologie Aktuell 1: 191–216.

World Bank, 1994. Indonesia—Environment and Development: challenges for the future. Report No. 12083-IND, Environment Unit, Country Department III, East Asia and Pacific Region. World Bank, Washington, 294pp.

WHO, 1989. Global environmental monitoring system, Global freshwater quality. Cambridge, Basil Blackwell, 293pp.

Zhang, F.H., Scully, P.J., and Lewis, E. A novel optical fibre sensor for turbidity measurement. Proc. Appl, Opt. Div. Conf., Reading, 16–19 Sept. 1996, IOP Pub., 370–373.

Zo Bell, C.E. 1946. Studies on redox potential of marine sediments. Bull. Amer. Ass. Petrol. Geol. 30: 477–513.

3

THE EVALUATION OF "DYNAMIC *DAPHNIA* TEST" AFTER A DECADE OF USE

Benefits and Constraints

AMARA GUNATILAKA,[1] PETER DIEHL,[2] AND HEIKE PUZICHA[3]

[1]Verbundplan Ltd., Engineers & Consultants, Parkring 12, A—1010 Vienna, Austria & Lab. of Ecotoxicology, Inst. of Medicinal Biology, Univ. of Vienna, Schwarzspanierstrassê, 17, A-1090 Vienna, Austria
[2]State Office of Water Affairs, Rhine Water Control Station Worms, Am Rhein 1, D-67547 Worms, Germany
[3]State Office of Water Affairs, P.O. Box 3024, D-55020 Mainz, Germany (Present address: Hildebrandstr. 31a, D-76227 Karlsruhe-Durlach, Germany

1. INTRODUCTION

The aim of biological monitoring is to determine the effects of as many pollutants as possible in shortest conceivable time. In contrast to chemical tests, biomonitoring is an alternative which is capable of rapidly detecting acutely toxic conditions in river water. Using standard GC-MS analytical procedures (which are comparatively time consuming), more than 1000 individual organic contaminants have been reported from the lower Rhine (Puyker et al., 1989; van Genderen and Noij, 1991). However, there are a large number of other toxic compounds that cannot be detected by these methods in river water. No consideration has been made of the potential effects of these unknown or chemically undetected substances. The overall consequences of toxic substances in river water, including their synergistic and antagonistic effects, can only be assessed by biological systems. Biotests can also detect pollutants with biological effects which are not included in analytical matrices, or have very high detection limits (LAWA, 1996, 1998). Taking these facts into consideration, and also because of the specific action of a multitude of pollutants, Hendriks and Stouten (1994) point out that more emphasis has to on biological test methods with more sensitive organisms.

Biomonitors and Biomarkers as Indicators of Environmental Change 2, Edited by Butterworth *et al.*
Kluwer Academic/Plenum Publishers, New York, 2000

Biotest methods have been used for several decades in a number of areas for the monitoring of water courses (Juhnke and Besch, 1971; Adema, 1978; Knie, 1978; Hellawell, 1978; Gruber and Diamond, 1988) and accordingly an attempt has been made to use them as a necessary supplement for physical-chemical monitoring (LAWA, 1996, 1998). The tests assess the effect of biologically harmful and/or stimulating substances on selected test organisms (Poles, 1977; Knie et al., 1983; Kramer and Botterweg, 1991; Borcherding, 1992; Baldwin and Kramer, 1994; Borcherding and Volpers, 1994; Gruber, et al., 1988, 1994; Hendriks and Stouten, 1994; Irmer, 1994; Knie, 1994). Mayer and Ellersieck (1986) in their database have included information on acute toxicity data of 410 chemicals to 66 species of freshwater animals.

In continuous biotests the test organisms are exposed to the test medium, for example river water, more or less continuously. During this exposure changes in metabolism or behavioral parameters induced by toxic or sublethal effects in the test animals are measured using automatic detection systems. However, there is very little data available on toxic effects of sublethal concentrations of many of the pollutants. The determination of acute toxicity levels as practiced now has little relevance in the estimation of ecological consequences. The designation "biomonitor" characterizes the main function of the system as a monitoring and early warning system for the identification of combined pollutant effects. In the monitoring of river water the tests are primarily deployed as warning systems which, for instance, indicate elevated concentrations caused by accidental discharges upstream above the usual background concentration of the water course. In conjunction with chemical analysis and more detailed biological examinations, the tests assist in securing evidence of illegal discharges and accidents (LAWA, 1996, 1998). As pollution events are liable to be rapid in onset and vanish almost quickly, thus if the monitoring is to be effective, it needs to be continuous (Cole et al., 1994; Gunatilaka and Dreher, 1996). This effectiveness could only be achieved, when biomonitors are added to physical and chemical sensors as they are more rapidly responsive to biocidal organic and inorganic pollutants giving a broad-band indication of anomalous water quality. It has to be emphasized that the management decisions will be more effective, if the information is generated rapidly.

The dynamic daphnia test has been used for nearly a decade as a standard biotest procedure in on-line river monitoring installations and proved to be useful in the detection of contamination resulting from accidental spilling or emissions into rivers. Swimming behavior of twenty young daphnia (in duplicate) exposed in continuous flow cells for a week to the test water is closely observed with infrared sensors and recorded as impulses per 10 minutes. Dynamic upper and lower alarm levels are statistically evaluated for each measuring cycle. A lapse in the whole exercise is the definition of the alarm and also distinguishing between true and false alarms. In addition, a largely neglected issue is the physiological status of the animals during the test. Availability of an adequate amount of food (preferably algae and bacteria) in test media influences the physiological status of the daphnia and therefore the threshold reaction levels for toxic substances. This would definitely influence the dynamic alarm limits during the test. In this paper are discussed field results from Daphnia test systems along the Rhine and Danube water-quality monitoring stations. Suggestions are made for distinguishing between false and true alarms, improvements in the algorithm used for alarm detection and for optimal location of such installations. Also clues are given for distinguishing between false and true alarms, proper maintenance of *Daphnia* and algal cultures for the test, and suggestions are made for the optimal location of test installations.

2. THE DYNAMIC *DAPHNIA* TEST

2.2. Background

One of the oldest continuous biotest systems is the "Dynamic *Daphnia* Test" (Knie 1978, 1988, 1994; Puzicha, 1992, 1994, 1995). It was developed at the State Office for Water and Waste (LWA) of North Rhine-Westphalia (Germany), and has been employed as an automatic biomonitor for continuous water monitoring since 1982, predominately in Germany (North Rhine-Westphalia, several other Federal States), Benelux and a couple of Middle European Countries. Commercially produced field instruments were available for this test during the last ten years from three different firms in Europe, but at present Elektron Ltd. (Germany) is the only supplier of this instrument. At present, the Dynamic Daphnia Test is run at approximately thirty different river sites (monitoring stations) in Germany, Austria, The Netherlands and Belgium. Now the test has been introduced in some east European and NIS countries (Newly Independent States) as well as in South Korea.

In the Dynamic Daphnia Test, *Daphnia magna* or *Daphnia pulex* (Crustacea; family Daphnidae) are the most commonly used test organisms. They are highly specialized crustaceans found in a variety of freshwater habitats such as ponds and lakes or slow flowing streams where they form the bulk of the zooplankton biomass.

Daphnidae (e.g. *Daphnia magna*, *Daphnia pulex* and *Ceriodaphnia reticulata*) are filter feeders and as low-order consumers, they take up ecologically important position in the food chain between destruents and primary producers on the one hand and higher-order consumers on the other. They are quite sensitive to environmental toxicants such as heavy metals, pesticides, herbicides and an array of organic toxic chemicals. The data bases AQUIRE (Pilli et al., 1989) and AQUATOX (BKH, 1989) have records on short-term toxicity of more than 800 chemical compounds to Daphnidae. They do not possess the capability to bioaccumulate nor elaborate mechanisms to excrete them. Therefore Daphnidae are ideal biomonitors (response indicators). Primarily they have been used for toxicity tests (lethality, reproduction) and recently for static and dynamic toxicity tests as well as continuous monitoring (behavior).

2.3. Principle of the Measurement

The test system currently in operation is comprised of a continuous flow system with two test chambers and a self-priming-pump system. A small stream of

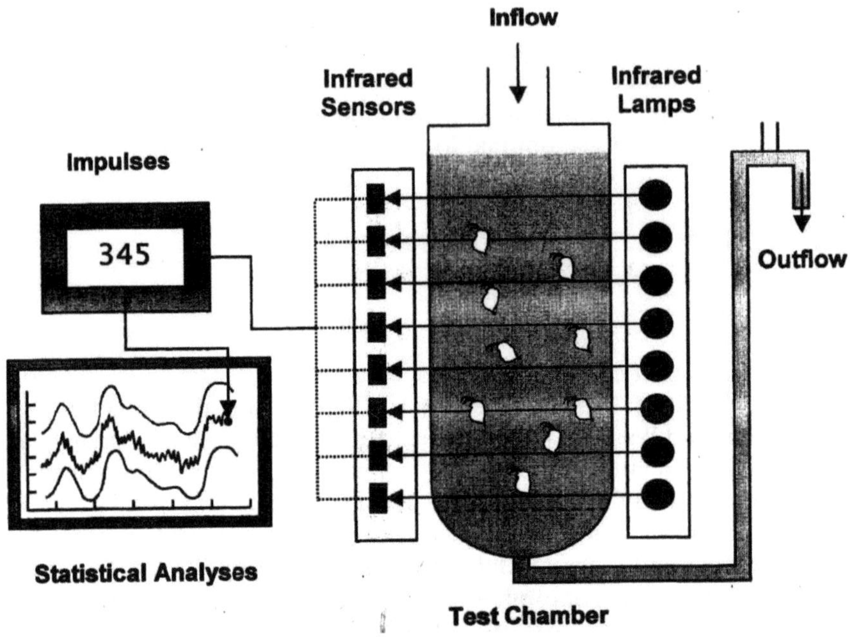

FIGURE 1. Dynamic *Daphnia* Test. The flow through system (note: only one test chamber is shown in the diagram: test system is equipped with two chambers). River water flows through both test chambers in which twenty *Daphnia* spp. juveniles in each (24–72 h old at the beginning of the test cycle) are kept at constant temperature (20 ± 1 °C) under subdued light. The swimming activity of the test animals is registered with the aid of infra red light sensors, (placed on each side of the test cuvette) and counted as pulses (each movement through the light barrier is counted as a pulse) per 10 or 15 minutes (e.g. test systems along the Rhine and the Danube respectively). The activity of the animals is continuously recorded in a computer system and evaluated.

test water (river water) with temperature adjusted (Peltier element) flows through two similar test cuvettes, where 20 test animals (juveniles) in each flow-through cell are kept under observation. The swimming behavior of these test animals is continuously observed and recorded with the aid of a battery of infra-red optical sensors (6–10 pairs, Figure 1) which act as light barriers. The swimming activity of the test animals is registered as an impulse when they cross the light barrier, and the impulses are summed up as counts per 10–15 min. If water quality changes occur in the test water through a contamination by a toxic substance, the animals react by changing their swimming behavior. They swim more slowly (retarded activity or death) or react with an increased activity (hyper-active). The changes in behavior for a specific monitoring cycle are measured as mentioned above and the "impulses per 10- (or 15-) minute cycle" of the two measuring chambers are transmitted continuously to a computer. A typical analysis of such a data set is shown in Figure 2. When a defined threshold limit is exceeded, an alarm is triggered, and this results in alerting of an early warning system. Validation of the results is done by comparing the plots of both left and right chambers. A basic criterion

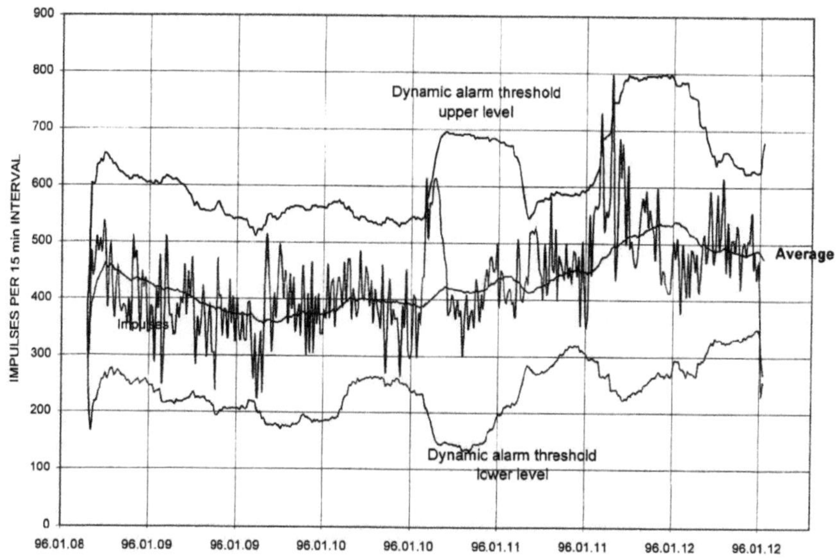

FIGURE 2. A typical activity plot of *D. magna* during a routine test from 8–12 Jan. 1996 at Danube on-line monitoring station at Nussdorf, Vienna (river km 1932.53; right test chamber). *Daphnia* activity is shown as impulses per 15 min period. The mean swimming response of the test animals as well as the upper and lower dynamic levels (±3σ, standard deviation of the mean) are shown in the diagram. The upper and lower alarm thresholds are defined by the operator, and when an "alarm" is triggered the controlling authority is alerted immediately to take up necessary remedial measures.

in confirming an alarm is that the plots should show a similar swimming behavior pattern.

For the test, water fleas used are cultured according to modified DIN (1990) and ISO (1989) standard procedures (see Appendix 2 for details). Laboratory cultures of Daphnia (mother cultures) and algal cultures for feeding them are required. Parthenogenetic adult females, born from a single animal are used in cultures so that the progeny will be genetically similar. Both *Daphnia* and algal cultures are maintained under controlled, optimal conditions. For *Daphnia* cultures older animals (up to 2 months old) are kept in large beakers (2 L, for 20 individuals) in a culture medium (ca. 1,300 ml M4- medium; after Elendt, 1990). The beakers are kept in a water bath or a climate chamber at 20 ± 1 °C, 12 h light and dark cycle but in subdued light. The young have to be separated every other day as well as the culture medium has to be refreshed, which is essential for maintaining healthy cultures. The algal cultures (*Scenedesmus, Chlorella, Chlamydomonas*; mono cultures or a mixture) used for feeding are grown in Chu-10 medium (Chu, 1942) under constant conditions and 20–25 ml aliquot is used (3–4 days old algal cultures, 2.0–3.0 mg C content). Sufficient number of adults have to be cultured to yield sufficient number of juveniles for the test. In practice, usually 6–24 h old neonates are used. However, to suit the field conditions and also according to requirements of the investigation, 2–3 days-old, lab-fed juveniles could be used.

3. PROBLEMS AND CONSTRAINTS

3.1. Practical Aspects

Practical know-how and experience with the Dynamic *Daphnia* Test have been available from monitoring stations in Germany, the Netherlands and Austria for the past 15 years. This allowed the operators to gather a wide range of positive and negative findings with regard to the commercially-available instrumentation for the test, type of test animals that should be used in field tests, and the required infrastructure and personnel. Some of these details are discussed in Puzicha (1994, 1995) and DOKW (1996). The collected information is summarized in two recent publications of the German Working Group of the Federal States on Water Problems (LAWA, 1996, 1998).

A major problem in conducting the test is occasional registering of false alarms. In the test when the alarm thresholds set by the operator are surpassed an alarm is registered. For the confirmation, in most cases the data set has to be analyzed by the operator. A computer program is available for the data analyses and it can be used for the rapid calculation of: (i) the mean impulse rate (10 or 15 min), (ii) the $\pm 3\sigma$ standard deviations which is used for the detection of upper and lower alarm levels and (iii) the final plotting of the results (Figure 2). The plot obtained shows the moving averages of the impulses with calculated standard deviations of the mean as dynamic upper and lower threshold values (Figure 2; see LUA, 1994). Based on this analysis it has to be decided that either the alarm is true or false. In case of a true alarm a number of simultaneous measures have to be taken such as (a) the checking the duration of the alarm, (b) informing the responsible persons/organization or authority immediately such as the early warning surveillance and the associated network to take necessary steps for the prevention of the spreading of a contamination plume, (c) diagnosis of the cause which may involve costly chemical analysis, and finally (d) law enforcement. All three last procedures mentioned here are bound with high costs and interactive processes which need personnel judgement. Therefore the alarms registered by the instrument have to be very accurate.

The sources of false alarms (see Appendix 1 for details) can be categorized into three groups; alarms arising through (i) the malfunctioning of the instrument and (ii) the poor or unpredictable physiological condition of the animals used in the test and (iii) the unsuitable adjustment of the alarm algorithm. Some of the practical problems according to their priority are discussed below. The solutions for the problems with the algorithms are discussed in section "Evaluation of the data".

Most common problems encountered with the instrument are:

1. Disturbance with suspended solids; a high load of suspended solids can lead to malfunctioning of the instrument. In most cases continuous feed-in of test water is not guaranteed due to high amount of suspended matter content which eventually can lead to blockage of the test cuvette sieves (placed at the bottom of the cuvettes; Figure 1) and result in overflowing of the test chambers. If the blockages interfere with the flow, despite the presence of flow monitors, the unit may register an artifact. As a solution

pre-filter systems were installed, but they were not very successful due to poor construction and other associated problems. Two or more gas wash bottles loosely packed with either glass beads or Teflon wool were arranged in series to serve as pre-filters to trap large particles and also to optimize the water flow. The Rhine water control station at Worms (river km 443.3) has had unique experience with Teflon wool filled into glass funnels to which the tubing is attached which leads the river water to the instrument. The funnel itself is attached in a 1 L tank with a renewal time of 2–3 min.

An inherent problem of such a system is the quick build up of a biofilm over the enhanced surfaces which may lead to adsorption and desorption problems connected with toxic substances (see Appendix 1) and therefore needs frequent cleaning or renewal. An alternative is to use a partitioned pre-sedimentation tank with a renewal time of 15–20 min. The river water flow into the tank is adjusted so that water cascades from one consecutive chamber (the number of partitions required depend on the location) to another allowing heavy particles to sediment quickly (e.g. sand) and at the same time retain large suspended matter. From the last chamber strained water is pumped into the instrument for the test and such a system is installed at the Danube monitoring station at Nussdorf (river km 1932.53).

2. Temperature: the temperature control installed in the equipment is not capable of guaranteeing a constant temperature of 20 °C under all conditions. The temperature changes are due to heat generated in the instrument itself (from the electronics and illumination). During hot summer days, especially when the equipment is run in a control station installed in a container without air conditioning, high temperature variations can result in false alarms. Temperature increases in the system can cause non-specific evolution of gases and the air bubbles may interrupt the test water flow (see Appendix 1).
3. To avoid false alarms all parts of the instrument have to be cleaned regularly and the time required for maintenance is 3–4 h per week (see Appendix 1 for details).

The problems arising with test animals could be prevented by proper maintenance of cultures (quality assurance; see Appendix 2) and by giving due consideration to their reproductive physiology. The suggested solutions are: firstly, juveniles used for the test should be from a healthy stock and care has to be taken during transfer not to damage them. Secondly, their age should be known at the onset of the test; this is a prerequisite to confirm that they do not produce young during the test and also a helps to decide the duration of the exposure. The age at which the daphnia are introduced into the system must be adjusted to the conditions of each location (e.g. lack of food in winter; abundance in summer). The age of the suitable animals for the test could be between 1 and 3 days. The fitness can be improved when the animals are well fed in the laboratory before they are introduced in to the test chambers. On the other hand the instrument runs more reliably when the animals are older. They have less chance to escape through the sieves (compared to 24 h-old animals), or adhering to the water surface; whereas the light barriers are

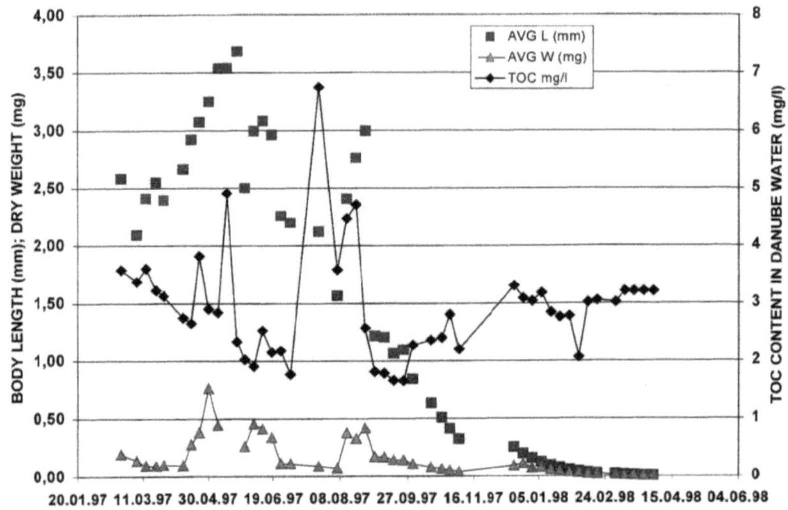

FIGURE 3. A comparison of total organic carbon content (TOC) dynamics in the Danube with growth of *D. magna* cohorts. The weekly increase in size of the animals (total body length and weight as growth indicating parameters) in the exposure period to the Dynamic *Daphnia* Test is defined here as growth. Animals were exposed to river water in continuous-flow cells, for a period of seven days. TOC of the river water plotted as weekly averages (n = 6/day) is used here as an indicator of food availability (however, it is not an absolute measure of food quality as part of the carbon is not assimilated). *D. magna*, 20 individuals placed in each test chamber, recovered after the exposure and 10 randomly selected individuals (from a total of 40) were used for dry weight and body length determinations; the weekly averages are plotted in the diagram. In general all three curves show corresponding similarities: at higher carbon levels, there is a parallel increase in body length and weight (late spring and summer) but a drop starting in the onset of autumn, reaching the lowest levels during the winter.

crossed more frequently (larger size, more active), leading to better statistics in activity plots (DVWK, 1997). Another advantage is that the older animals have a higher resistance against infections and predators.

Daphnia are very sensitive to temperature changes; an increase of temperature in the test cuvettes could lead to death of test animals, and this was observed in summer as well as in winter (heat gain in the instrument due failure). A deficiency in the present instrumentation is that there is no direct measurement of temperature in the test cuvettes. An improvement foreseen for the instrument is to introduce direct temperature measurements in flow-through cells and to record in parallel with the impulses of the left and right chambers. One of the other major concern is the condition of the test animals during the test, partly based on their physiological status and food availability which are treated in detail below.

3.2. Physiological Status of the Animals

Within the duration of the daphnia test (6–7 days), more attention has to be paid to the physiological condition of the test animals. The animals may show stunted growth due to lack of food (Figures 3 & 4) when subjected to long hunger

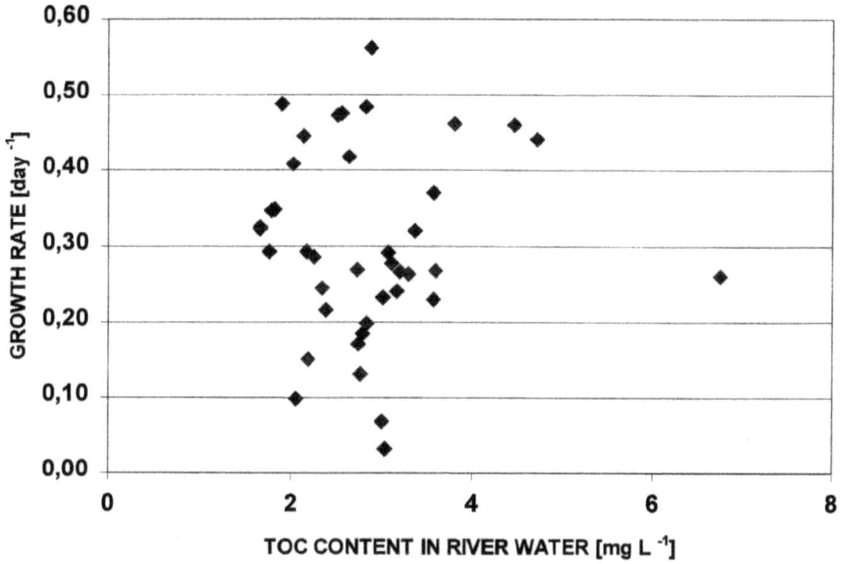

FIGURE 4. The accordance of daily growth rates of *D. magna* in the Dynamic *Daphnia* Test with food availability during the exposure (see Figure 3. for experimental details and comments on TOC). The mean growth rate per day is calculated from the difference between the initial weight of the exposed neonates and final weight of the animals after exposure for seven days in the test system, using the following equation: $g = (\ln W_t - \ln W_i)/t\ day^{-1}$, where g = growth rate; W_t = final weight after seven day exposure and W_i = initial weight (both as dry weight). Most of the observed TOC values lie between 1.8 and 3.8 mg C L^{-1} (92%) but even at low carbon levels animals can testify enhanced growth rates (0.3–0.5) showing that food quality (nutritional value) is an important factor. Correspondingly, the results show the occurrence of slow growing cohorts (40–50%) during the year, yielding smaller size animals with lower fitness.

periods due to either unavailability of food, or poor quality food. Thus the physiological status of the animals during the exposure is an important factor for the avoidance of false alarms in the Dynamic *Daphnia* Test.

It is a well known fact that food quality plays an important role in growth as well as physiological fitness of animals. Animal health plays an important role in how they react to toxic substances. Well fed animals are larger, stronger (higher fitness) and they react differently to an alarm situation whereas weak animals show exaggerated reactions. Experiments conducted by Matthias and Puzicha (1990) using the test system monitor swimming behavior of well fed and hungry *Daphnia* to DDT confirms this observation. Experiments on *Daphnia galeata* have shown (Repka, 1997a,b) that low quality food imparts a reduction in intrinsic growth rate in the animals. The availability of an adequate amount of food (preferably algae or bacteria) in river water influences the physiological status of *Daphnia* and therefore the threshold reaction levels for toxic substances; this would definitely influence the dynamic alarm limits during the test. Usually algal biomass in both Danube as well as in the Rhine is low and the availability is subjected to strong seasonal variations. With regard to food quality there are two major aspects: a) availability and b) nutritional value. In literature the food habits of daphnids are well documented (Nauwerck et al., 1979; Gliwicz, 1977; Bern, 1990a,b; Müller-Navarra and Lampert, 1996).

Daphnia filter-feed rather non-selectively on small sestonic particles (DeMott, 1986, 1988; Bern, 1990a), but selectively on large particles (Gliwicz, 1977; Bern, 1990b). According to Burns (1968), DeMott (1995) and DeMott and Müller-Navarra (1977) the upper size range for edible particles varies directly with body length, both between and within species of Daphnia (particle size: 18 μm optimal; 40 μm max. and <30 μm preferable; see Müller-Navarra and Lampert, 1996; Nauwerck et al., 1971). Their diet consists of a wide spectrum of particles, differing in quality, including bacteria, protozoa, detritus, nanoplankton, smaller net algae, and fragments of animal food. The morphology, digestibility, nutritional value and possible toxic compounds of these food items determine the food quality.

The literature also documents that the biochemical composition of algal food affects the production and growth responses of animal herbivores such as *Daphnia* (Gulati and DeMott, 1997). A variety of algal cellular constituents are important determinants of food quality for zooplankton: carbon (C), nitrogen (N), phosphorous (P), lipids, (Kilham et al., 1997a,b; Kreeger et al., 1997; Brett, 1993; Hessen, 1992; Sommer, 1992; Sterner et al., 1992; Urabe and Watanabe, 1992; Sterner and Hessen, 1994); essential fatty acids such as EPA (eicosapentaenoic acid) and DHA (docosahexaenoic acid), (Ahlgren, 1993; Müller-Navarra, 1995a,b; Müller-Navarra and Lampert, 1996; DeMott and Müller-Navarra, 1977); proteins, lipids and essential amino acids (Ahlgren et al., 1990; Kreeger et al., 1997). There are likely to be taxon-specific differences in responses to limiting resources in both algae and zooplankton (Sterner and Hessen, 1994). The observed variations among studies in biochemical composition of algae are discussed in Kilham et al. (1997b) and in Gulati and DeMott (1997). In the Danube, algal seston C:N:P composition during the summer months varies between 81:9:1 and 135:7:1, showing clear changeable N or P deficiencies which in turn influence the food quality and growth rate of filter feeders.

We have observed the growth of *Daphnia magna* during seven-day-exposure period to river water (Danube water quality monitoring station at Nussdorf, Vienna: Danube river km 1932.53, right river bank) and the data are summarized in Figure 3 and Figure 4. As particulate organic matter (particle size: 18 μm—optimal; 40 μm max. and <30 μm preferable; see Müller-Navarra and Lampert, 1996; Nauwerck et al., 1971) is the best predictor of food availability, the total carbon dynamics in river water (continuous, daily measurements of TOC) are compared with the growth of Daphnia cohorts during the seven day exposure period, using individual body length and dry weight as growth indicating parameters (Figure 3). The results show that there are seasonal fluctuations of the particulate organic carbon level in the river and at the same time the nutritional quality of the particles varies (dominated by abioseston, which is inert but <30 μm) during the year. This is in turn reflected in the growth pattern of the animals. This is further confirmed in Figure 4 where daily growth rate is compared to food availability. Most of the observed TOC values lie between 1.8–3.8 mg C L^{-1} (92%), but even at low carbon levels, animals can testify enhanced growth rate (0.3–0.5) showing that food quality (nutritional value) is an important factor. Correspondingly, the results show the occurrence of slow growing cohorts (40–50%) during the year, yielding smaller size animals with lower fitness. Variation in the distribution of genetic, physiologic and physical characteristics of individuals in a population together with the biogeochemical environment of the population determines the characteristics of the effects resulting from chemical exposure (Hallem et al., 1990). Therefore in the *Daphnia* test, the physiological status of the test animals should be carefully scrutinized to avoid false alarms. If inadequacies are found in the diet (Figure 3), alternative remedies have to be implemented (see Appendix 2).

3.3. Evaluation of the Data

In theory continuous biotests are a reliable method to estimate the effects that a contamination may have on aquatic organisms. In a biotest, if clear responses are registered as in Figure 2, the interpretation should be easy, and the rapid generation of information for the responsible water quality authority is possible, and the results can be delivered in the shortest possible time. With this the biotest accomplishes the function of an ideal early warning system.

In the beginning of the development of continuous biotest systems (biomonitors) it was hypothesized how an ideal system should function. The expected response of the biotest system was a change in the measured parameter (e.g. swimming behavior, animal movement) with an increase or an decrease in activity depending on the ambient concentration of the contaminant. On the contrary, if the dose is too high, it could retard the animals altogether or may result in ultimate death of the organism. Therefore for the use of the *Daphnia* test as an early warning system it was necessary to define an alarm threshold. The first generation test instruments used for the Dynamic Daphnia Test were equipped with software using static alarm thresholds that could be chosen by the operator (Puzicha, 1989; Knie, 1990). Subsequently as the static alarm thresholds were insufficient to define an alarm adequately, dynamic upper and lower limits were suggested for a statistical evaluation of the results (Puzicha and Blohm, 1991; LUA, 1994; see also section "Practical Aspects"). The algorithm used calculated moving averages and standard deviations from the observed data, and the alarm threshold for each actual value was defined as $\pm 3\sigma$ (standard deviation of the mean). The factor 3 was introduced for 99% probability which assured better reliability (LUA, 1994) and Figure 2 illustrates the dynamic alarm thresholds calculated with the $\pm 3\sigma$-algorithm. This algorithm gave better reliability for the test and it is widely used by operators of the test. Based on an evaluation done by LAWA in 1995 (LAWA working group on "Biomonitoring"), the Dynamic *Daphnia* Test emerged as one of the most reliable on-line biotests (LAWA 1998).

However, in practice the alarms rarely follow this ideal scheme. The Figures 5 through 7 give examples for plots of actual alarm situations in the Rhine (based on test results from the Worms monitoring station at River km 443.3) recorded from a Dynamic *Daphnia* Test system. The diagrams show the reactions of *Daphnia* to contaminants (2-chloropyridine, 3-nitrobenzene sulfonic acid) and as well as to high turbidity in river water. The dynamic alarm concept mentioned above (LUA, 1994) used to evaluate these results was unable to detect these alarms. The cause of the alarm was the elevated concentrations of the contaminants in river water and these examples are discussed in detail in section "Monitoring-.".

During many years of operation of the Dynamic *Daphnia* Test, this $\pm 3\sigma$-algorithm for determining the upper and lower alarm levels was found to be inadequate (DVWK, 1997). The assumption used for this concept was that the data recorded during the test have a normal (Gaussian) distribution. The test results show that the data deviate often from the normal distribution pattern which may lead to abnormalities such as failing to recognize alarm patterns and delayed reactions. In addition it is not possible for theoretical reasons to calculate a standard deviation from values that are not normally distributed and the latter is used to calculate the upper and lower dynamic threshold levels. Detailed analysis of records from several Daphnia Test installations in Germany demonstrated that only very small parts of the examined recordings consist of data that are normally distributed (DVWK, 1997). It is clear that though the 3σ-algorithm led to a much better reliability of the system than the static alarm threshold, it failed when there were drifts or physiological rhythms (e.g. growth, aging) overlapped with the activity (swimming behavior) plots. *Daphnia* which are exposed for seven days, steadily grow during the test period and are liable to show different reactions to toxic substances, depending on their age. Also, there could be drifts or jumps in the data due to changes in the instrument itself (see Appendix 1). These drifts as well as rhythms lead to non-Gaussian distribution of the continuously recorded data.

As a solution to the problem, an improved algorithm was developed based on the adaptive Hinkley detector (Hinkley, 1971; DVWK, 1997). It was first used as a non-linear filter algorithm for the detection of drifts and missing events in data series in biological systems (Draber et al., 1991, 1993; Schultze and Draber, 1993; Draber and Schultze, 1994) and subsequently adapted for dynamic data models (Schultze, 1992, 1993; DVWK, 1997). The algorithm first eliminates the drifts as well as jumps in a data series, then calculates a mean curve based on preceding twenty seven values (moving averages), predicts the subsequent values sequentially and evaluates how the real values deviate from the predicted values. If two subsequent values deviate, it is recognized as an alarm. The theory of the Hinkley detector is given in DVWK (1997), and "DYNSPRUNG" (Moldaenke, 1997) is a software program developed for the automation of Dynamic Daphnia Test using this algorithm. In using the adaptive Hinkley detector there are eight parameters that can be adjusted to define the exact alarm threshold (see Moldaenke, 1997; DVWK, 1997). To obtain optimal results the fine tuning of the computer program has to be done by the operator by defining them (see section "Monitoring-..", and Figure 5 for explanation). Because of this it became clear from all Dynamic Daphnia Test users that the test can only be carried out with the support of specially qualified personnel (LAWA, 1996, 1998). However, the adaptive Hinkley detector is not widely in use for alarm recognition in the daphnia test. More experience in practice is necessary before it could be recommended as an overall solution.

Though the application of the adaptive Hinkley detector for defining dynamic alarm thresholds seems to be the best at the moment, it has to be stated that it is

a)

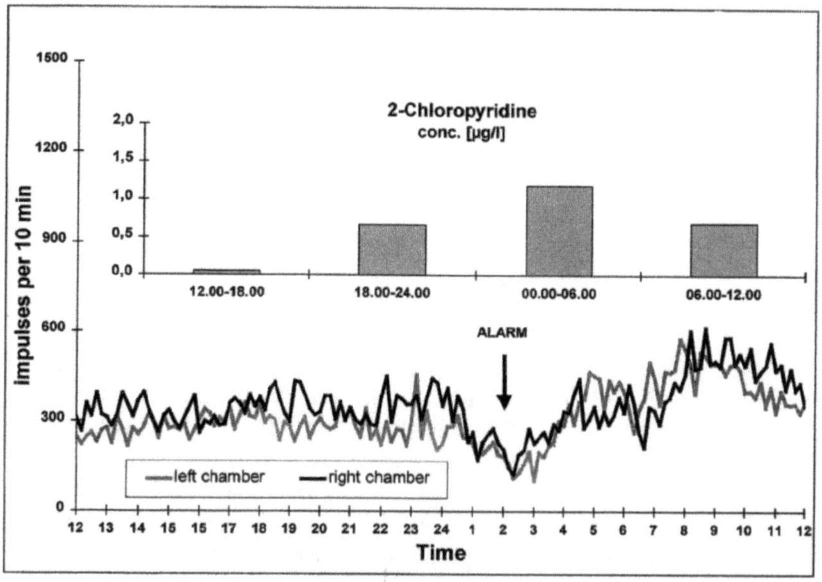

b)

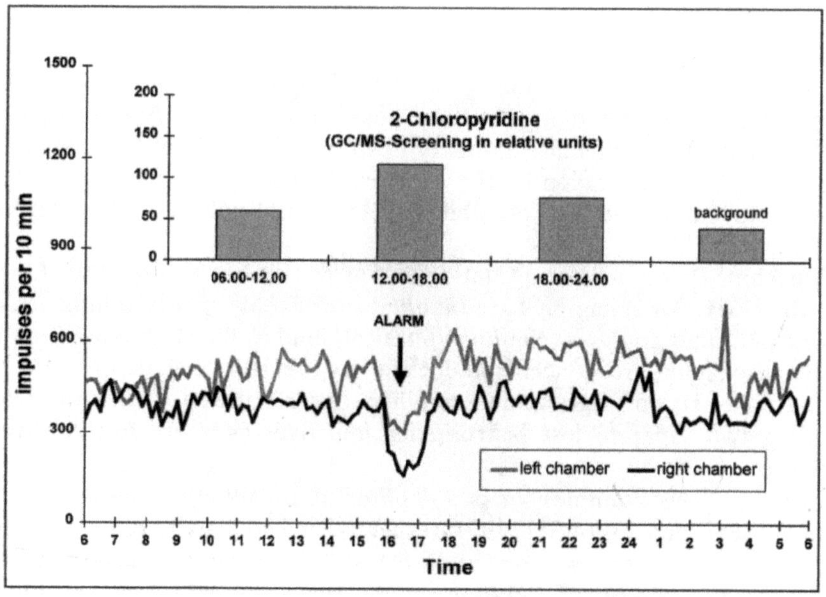

FIGURE 5. Two examples for *Daphnia* test responses recorded in the Rhine water monitoring station at Worms (Germany) on the right river bank (river km 443.3). The diagram shows the course of daphnia activity (recorded as impulses per 10 min.) over time (in hours). Insert: Concentration of 2-chloropyridine in river water, analyzed in 6 h composite samples: (a) Incident in February, 4–5, 1997: decrease of impulses in both test chambers of the Dynamic Daphnia Test; increase in concentration of 2-chloropyridine (in µg/L) parallel to the daphnia test response. (b) Incident in September, 13–14, 1997: decrease of impulses in both chambers of the Dynamic *Daphnia* Test; increase of concentration of 2-chloropyridine (in relative units, calculated from the peak areas in the GC/MS-screening chromatograms).

still inadequate to cover all types of alarm situations. For example, it cannot be used to detect laggard changes in activity which may result from slow increases of concentration of a contaminant (see section "Monitoring-.."). It is suggested that an additional algorithm that considers the ceiling of the activity curve may have to be used for an automatic alarm system (DVWK, 1997).

4. MONITORING—PRESENTING RHINE RIVER CASE AS AN EXAMPLE

The Rhine water monitoring station at Worms (river km 443.3) came into operation in 1995. A number of large scale industries and treatment plants which discharge directly into the river are situated approximately 10–15 km upstream of the monitoring station. One of the most prominent dischargers on the left bank is the waste water treatment plant of one of the largest chemical industrial complexes in Europe. The plant releases about 4.5 m^3/s of treated waste water into the river (1,400 m^3/s; mean flow of the river) with an organic load of 150,000 population equivalents (the untreated waste water entering the plant has 6,000,000 P.E.). On the right bank there are also conspicuous emissions. The Neckar River which receives effluents from a wide industrialized area empties into the Rhine approximately 15 km upstream. Other direct dischargers into the Rhine are: a waste water treatment plant of a cellulose factory in Mannheim; a large communal treatment plant which treats the industrial waste water of chemical and other factories; at least two treatment plants of a community and a smaller chemical industry site. A multitude of chemicals are found in the Rhine near Worms, primarily as a result of discharges from the large chemical industry upstream. Despite the improvements in security and precautions taken by the chemical industry since the notable Sandoz accident in November 1986, the possibility of accidental spillage in the river has not been eliminated.

The Rhine River monitoring station at Worms checks water quality of the river around the clock. Water samples are taken at four evenly-spaced points across the river for continuous analyses (chemical/physical), and as the contamination plumes from the discharges are concentrated close to the river banks, only the water taken from the left and right bank sampling sites are used for dynamic Daphnia tests (using two separate test instruments; mid-river sites are not continuously monitored).

Figure 5 shows examples for two daphnia alarm situations on the right river bank on 4–5 Feb. and 13–14 Sep. 1997, respectively. The activity plots of both right and left chambers of the test instrument are shown in the diagram. In both cases simultaneous decrease of the *Daphnia* activity was observed for several hours but the on-line-module of the adaptive Hinkley detector did not register an alarm. A subsequent calculation with the off-line-module showed that an alarm should have been released, if some of the parameters of the algorithm had been adjusted to a more sensitive mode.

For a further evaluation of the incident, storage samples from the automatic samplers were analyzed. As shown in Figure 5a simultaneous increases of

2-chloropyridine up to 1.5 µg/L were found and this increase was sufficient to register an alarm. In a subsequent incident where an alarm was recorded (Figure 5b) the concentration rose to about 3 to 10 times higher than the background concentration. Also the emitter could be detected; a communal waste water plant to which the producer of the substance was an indirect discharger. According to an administrative regulation of German Environmental Ministry classification of substances in water, 2-chloropyridine is classified as a hazardous chemical (VwVS, German Federal Ministry for the Environment, Nature Protection and Nuclear Safety).

The examples illustrated in Figures 6 and 7 show another difficulty in interpreting alarms in *Daphnia* tests, even if there is direct evidence of contamination. There could be parallel events (natural or man-made) which may lead to ambiguity in daphnia reactions which make interpretations difficult, as well as automation, a laborious task. In March 1998 the daphnia test system at the left river bank registered an alarm (Figure 6). Here the Hinkley algorithm is adjusted to a more sensitive alarm threshold than at the right bank where the discharges cause a larger risk for the Rhine water quality. At the time corresponding with the alarm, the important discharger had an accidental discharge of about 3 tons of 3-nitrobenzene sulfonic acid into the river. Figure 6 shows also the results of the laboratory analysis of stored samples from on-line automatic samplers at the Worms monitoring station. A parallel event was a heavy rain and an alarm was recorded in the *Daphnia* test system on the right bank (Figure 7). However, although there was a coincidence in both alarms, detailed analysis of the samples from automatic samplers confirmed that there was no 3-nitrobenzene sulfonic acid contamination at the right bank.

It was known from previous observations that heavy rains lead to increased responses in daphnia biotests. This is due to an increase in turbidity in river water and for a comparison, both turbidity and rainfall data are included in Figure 6 and 7. Although the parallel events here may produce a certain amount of ambiguity, the detailed analysis of the Daphnia plots in Figures 6 and 7 show that the different nature of the activity patterns. The huge discharge of 3-nitrobenzene sulfonic acid in to the river increased the concentration in water up to 200 µg/L, largely retarded the Daphnia activity (drop close to 500 impulses; Figure 6) which produced an alarm situation on the left bank. Contrary on the right bank, the heavy rains (and eventually turbidity) produced nearly a two fold increase in activity in Daphnia (up to 1,500 impulses; Figure 7) and that was the cause of the alarm.

All these examples show that in case of an alarm, before a final judgement is made, a careful scrutiny of the test results is important. The follow up actions of an alarm are expensive as it involves a series of protocols and expensive laboratory analysis of stored samples (e.g. International Rhine Alarm System). Therefore the interpretation of a biotest alarm should be very clear and accurate. On the other hand, the continuous biotests to function as early warning systems, the authorities have to be informed in time so that they could implement necessary measures.

With the several years experience of running the Dynamic *Daphnia* Test, it has to be emphasized that daphnia test plots cannot be interpreted as alarms, even if the mathematical algorithms confirm the situation. The problem in the practice

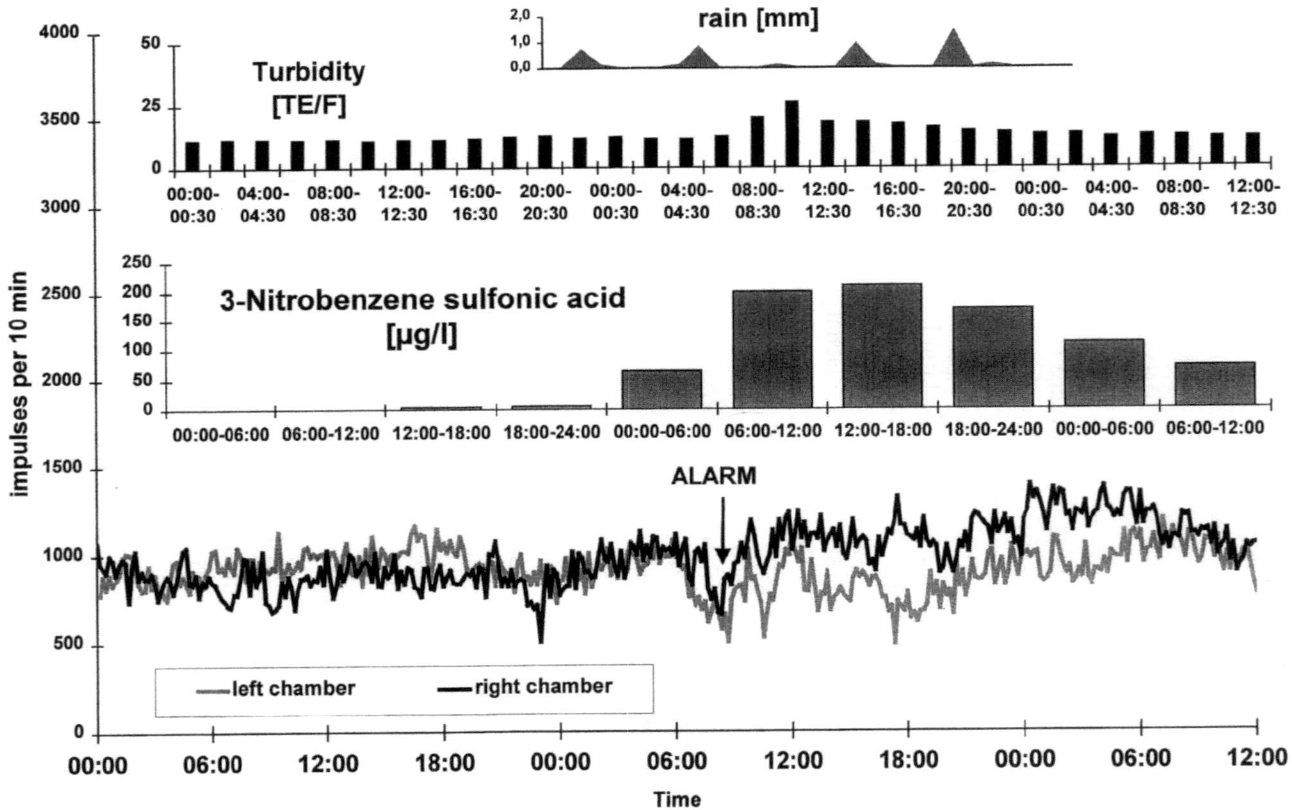

FIGURE 6. Example for *Daphnia* test responses documented in the Rhine water monitoring station at Worms (Germany) on the left river bank (March, 4–6, 1998) showing the course of the *Daphnia* activity (impulses per 10 min.) over time (in hrs.). Inserts (N.B. common x-axes): *upper*: amount of rainfall in mm at the time; *middle*: turbidity of the Rhine water (in TE/F), measured every 2 hrs. for 30 min.; *lower*: concentration of 3-nitrobenzene sulfonic acid in Rhine water, analyzed in 6 hr-composite samples. An alarm was released before the turbidity and the concentration of the contaminant reached their maximum.

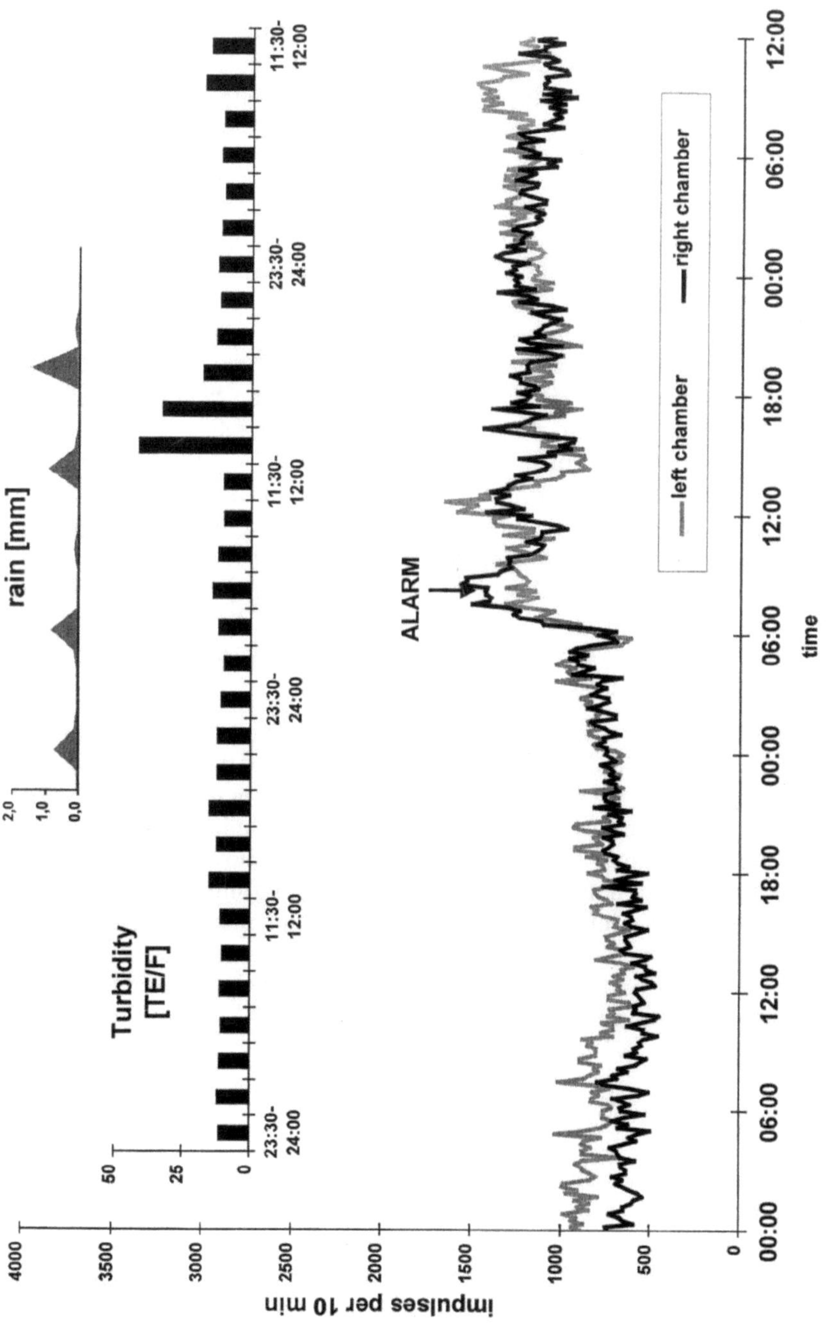

FIGURE 7. Example for *Daphnia* test response at the Rhine water monitoring station at Worms (Germany), right river bank (March, 4–6, 1998) showing the course of the *Daphnia* activity (impulses per 10min.) over time (in hrs.). Inserts (N.B. common x-axes): *upper:* amount of rainfall in mm at the time; *lower:* turbidity of the Rhine water in TE/F, measured every 2 hrs. for 30 min.; A *Daphnia* test alarm was released before the turbidity reached its maximum; 3-nitrobenzene sulfonic acid was not detected at this sampling site.

of river protection—(i.e. the protection of drinking water, the ecosystem etc.)—is whether and in which form information of daphnia test responses shall be submitted to the authorities. To overcome this hurdle the Rhine water monitoring station at Worms has devised a three step alarm concept to evaluate *Daphnia* Test results. This alarm classification uses the available supplementary information (e.g. on-line chemical and other biomonitoring data) to verify the alarms and the categories are as follows:

1. "Irregularity": there are only changes in the *Daphnia* activity pattern but no changes in water quality are observed; if the changes do not reach the threshold, the case is categorized as an "irregularity" (German "Auffälligkeit", Figure 8a).
2. "Event": When the threshold is reached (upper or lower) and a *Daphnia* alarm is released automatically, but no other indications are found (supplementary information), the case is registered as an "event" (German "Ereignis", Figure 8b) and further analyses are carried out on stored samples to confirm the results.
3. "Alarm": When the threshold is reached, an automatic *Daphnia* alarm is released and there are indications of changes in water quality (on-line chemical and other biomonitoring data) the incident is referred to as an "alarm" (Figure 8c). The authorities are informed promptly and detailed analysis of samples carried out immediately. (Note: If there are clear indications that the changes in water quality are caused by rain, the case is registered as a "rain event").

This classification is still in the test phase. The International Commission for the Protection of the Rhine (IKSR) plans to establish an international alarm system based on biotests, supplementary to the existing system which is based on chemical analysis and information given by the dischargers. Such an alarm system should work with a classification system comparable to that shown above. If the *Daphnia* test system is to be extended from river water quality monitoring (immission control) to direct discharge control (waste water treatment plant outlets, factory outlets etc.), a number of constraints have to be considered to ensure quality control. At present the German Working Group of the Federal States on Water Problems (LAWA) has focused attention on this issue.

5. SUMMARY AND RECOMMENDATIONS

Experiences have shown that the physiological fitness of the organisms used in continuous biomonitors is an essential feature for them to function as efficient early warning systems. This is true also for the Dynamic *Daphnia* Test. The main causes for lower fitness in test animals are a decrease in food availability or poor quality of food. There is a clear correlation between the food supply in the river and the growth rate of *Daphnia* in the test systems. For the optimal functioning of the test systems, the food supply has to be optimized or other alternatives have to be found. Another factor that affects the physiological state of the test animals is the

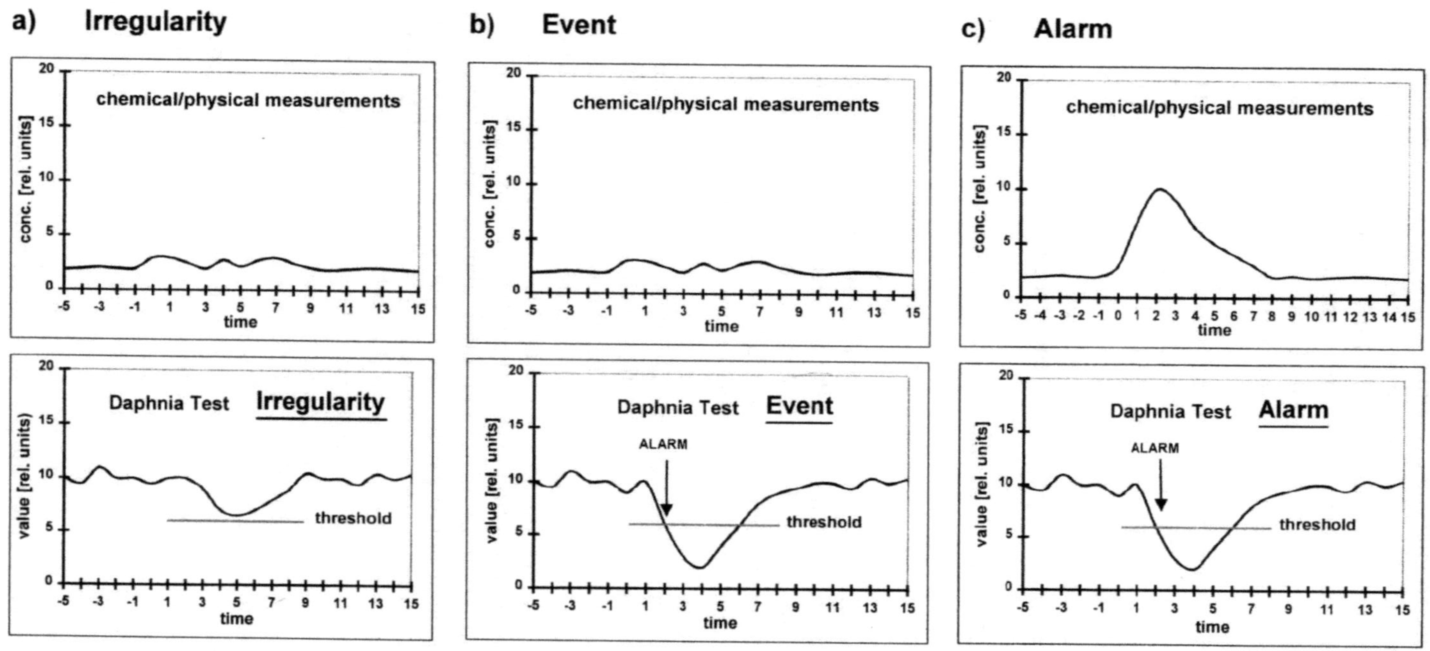

FIGURE 8. Alarm classes: a three-step alarm system proposed for the Rhine water quality monitoring station at Worms (Germany)—a schematic presentation. (a) "Irregularity": *Daphnia* activity pattern shows changes but does not reach a threshold; there are no supplementary changes in water quality, (b). "Event": *Daphnia* activity reaches a threshold but no changes in water quality, (c). "Alarm": *Daphnia* activity pattern with a threshold as well as accompanied by changes in water quality (supportive on-line chemical and other water biomonitoring data) [for details see text].

poor culture conditions. To guarantee a continuous supply of juveniles for the test, the cultures have to be maintained under optimal conditions, and the necessary precautions are listed in detail in Appendix—2. A further shortcoming in the test is the incidence of false alarms and their causes, and remedies to avoid them are given in Appendix—1. The other problems associated with evaluation of alarms are discussed here and suggestions are made for the improvements of the algorithms and alarm handling procedures.

Our results lead to some recommendations in regard to the operating of the Dynamic Daphnia Test systems. These should supplement the recommendations published elsewhere (Puzicha and Blohm, 1991; LUA, 1994; LAWA, 1996, 1998). In Germany recommendations for the deployment of continuous biomonitors for the monitoring of surface waters have been published recently (LAWA, 1996, 1998). The optimal conditions for the operation of the test systems are given below:

1. Adaptation of the test animals to local conditions is essential. Such adaptation includes for. e.g. addition of food for the test organisms in the event of a food shortage, or the adjustment of the sensitivity level of the alarm. However, addition of food during the test leads to a number of problems including false alarms. This can be avoided by deploying well fed older juveniles (2–3 days old) for the test, but then test, animals have to be replaced twice a week (see Appendix 1 & 2: *Daphnia* reproductive cycle). This should cause no additional problems as the daily availability of a sufficient number of neonates could be guaranteed by the proper maintenance of mother cultures (Appendix 2).
2. For the test water feed for the *Daphnia* test, pre-filtration should be avoided. Devices used in pre-filtration such as glass balls, Teflon shavings etc. are liable to be covered with a biofilm within a short time which tend to accumulate contaminants (adhesion onto surfaces; fine particles which get adhered on to surfaces are capable of adsorbing contaminants). Adsorption and desorption lead to the release of toxic substances from the biofilm resulting in artifacts and false alarms. Also, these effects are enhanced through microbial succession which lead to release extra doses of toxins. As a remedy we recommend pre-sedimentation tanks with laminated constructions, which separate the tank into different chambers. Here the construction is such the larger particles are sedimented or trapped first and the water overflows from one chamber to the other, leaving behind larger particles (*Daphnia* filter particles <30 µm as a source of food). Total turnover time for water in the tank should be 15–20 min.
3. The test system itself should be improved further. Future generations of instruments should be equipped with a better temperature control for both pre-conditioning of the sampling water in sample in the sample loop (Peltier element) as well in the test chambers. The actual temperature in the test chambers should be measured parallel with the *Daphnia* activity. The method of measuring the activity with the aid of light barriers can be improved by increasing their number and use of photo-detectors with a higher sensitivity. An increase in the number of light barriers should lead to a better statistical basis for the alarm algorithms (DVWK, 1997). Further recom-

mendations for the improvement of the test system can be found in Puzicha and Blohm (1991), LUA (1994) and LAWA (1996, 1998). It has to be ensured that the data records document changes in the activity of the daphnia and not artifacts of the instrument. Therefore, improvements in sensitivity of the instrument is a major contribution towards the reliability of the test.

4. To ensure proper operation (for routine care and maintenance) trained technical personnel are essential. They may have to seek expert advise in assessing difficult alarm situations (see section on "Monitoring"). This involves the interpretation of the biological effects, assessment of the level of potential damage to the ecosystem, and establishment of interfaces to chemical monitoring.
5. Continuous on-line transmission of the test data to the supervisory personnel, or to the alarm center, must be ensured for the proper functioning of the early-warning system. In addition, on-line measurement of supplementary environmental parameters such as water temperature, pH, conductivity, turbidity etc. should be measured as they are helpful in the final assessment of daphnia test alarms.
6. Running a computer-controlled sampling facility (e.g. self-emptying or semi automatic) parallel to the continuous biotests is necessary, in which composite samples are taken over short periods (e.g. 1–6 hrs.) and stored (4 °C) for further analysis.
7. The location of the Dynamic *Daphnia* Test installations should be carefully considered according to site-specific monitoring objectives (e.g. early warning system, water use, baseline data or data inventory, waste water discharge control, law enforcement, water quality modeling, net work development, etc.). Especially in large rivers, the dispersion of the contamination plumes depend on the location of point sources (right or left river bank) or accidental spills (e.g. middle reaches or the river banks). Taking all the possibilities into consideration, an optimal solution could be to place the monitors on both banks (e.g. Worms on Rhine). However, such an optimal solution will totally depend on the available funds.
8. The evaluation of alarms and treatment of the data need more careful analyses. In *Daphnia* tests, during the one week test cycle, the activity curve of the animals (Figures 2, 7 and 8) is influenced by their growth patterns results in a positive drift (section 3.3). This positive drift causes problems in using static alarm thresholds (Puzicha and Blohm, 1991) and for theoretical reasons does not allow to calculate a dynamic threshold based on standard deviation (DVWK, 1997). The proposed solution for calculating dynamic alarm thresholds is an algorithm based on the "adaptive Hinkley detector" (DVWK, 1997) but it is still under investigation. The algorithm should be implemented both for on-line and off-line evaluations, the latter used to recalculate alarm situations by varying the parameters of the algorithm. In practice, for conformation of an alarm, the results should be analyzed with both available algorithms. For the interpretation of alarm situations a three-step alarm classification is under consideration where alarms are categorized as an "irregularity", "event" or "alarm" (section 4). An "alarm" is confirmed only when accompanied with changes in water quality.

6. CONCLUSIONS

The Dynamic *Daphnia* Test is the most common continuous biomonitor used as an early warning system in Central Europe. In 1999, 17 water quality monitoring stations at several river locations in Germany, used the Dynamic *Daphnia* Test. At present there are more than 20 years of field experience in using this test system. It has been in continuous development and has received top rating by government environmental organizations as a reliable test (LAWA, 1996, 1998). The users of the test are in constant contact with each other which has led to a consistent advancement in the development of the instrument, as well as the optimal deployment of the system. As a result, at present, none of the newly developed systems could replace the Dynamic *Daphnia* Test completely. However, the experience showed that the Dynamic *Daphnia* Test has several constraints which have to be taken into account to obtain reliable results. Some of these constraints and suggestions for their solution are discussed in this paper and we have presented here the guidelines for more efficient use of the test system.

ACKNOWLEDGMENTS

A.G. thanks Heinz Jacksch for supervision and for making available his expertise on *Daphnia* cultures; Karl Wittmann for providing facilities at the Ecotoxicology Laboratory, Faculty of Medicine, University of Vienna; Nicola Dreher, Christian Foster and a number of graduate students for careful maintenance of cultures and conducting the *Daphnia* tests, Stefan Wimmer for the drawings and Donaukraft for financing the test operations on the Danube. P.D. thanks Ina Kolland for her meticulous operation of the Dynamic *Daphnia* Test at Worms Station and for her help in interpreting alarms; the members of the LAWA "biomonitoring" committee, especially Elke Blübaum-Gronau, Werner Blohm, Brigitte von Danwitz, Michael Lechelt and Michael Marten for their input for improvements of the test system and alarm interpretation; Christian Moldaenke for advice on the adaptive Hinkley detector; Sven Lüthje, Director of the State Office of Water Affairs Rhineland-Palatinate, granting permission for this publication; the Federal States Baden-Wurttemberg, Hesse and Rhineland-Palatinate for financing the project.

7. LITERATURE CITED

Adema, D.M.M. 1978. *Daphnia magna* as a test animal in acute and chronic toxicity tests. Hydrobiologia, 59: 125–134.

Ahlgren, G. 1993. Seasonal variation of fatty acid content in natural phytoplankton in two eutrophic lakes. A factor controlling zooplankton species? Verh. Int. Verein. Limnol. 25: 144–149.

Ahlgren, G., Lundstedt, L., Brett, M., and Forsberg, C. 1990. Lipid composition and food quality of some freshwater phytoplankton for cladoceran zooplankters. J. Plankton Res. 12: 809–818.

Baldwin, I.G., and Kramer, K.J.M. 1994. Biological early warning systems (BEWS). In: Biomonitoring of Coastal Water and Estuaries' Kramer, K.J.M. (ed.)., CRC Press, Inc., Boca Raton, USA. pp. 1–24.

Bern, L. 1990a. Size-related discrimination of nutritive and inert particles by freshwater zooplankton. J. Plankton Res. 12: 809–818.

Bern, L. 1990b. Postcapture particle size selection by *Daphnia cacullata* (Cladocera). Limnology and Oceanography, 35: 923–926.

BKH (1989). AQUATOX, an inventory on ecotoxicological data of 290 compounds (in Dutch). Institute for Inland Water Management and Waste Water Treatment, RIZA, Lelystad and BKH, The Hague, The Netherlands.

Boersma, M., de Meester, L., and Spaak, P. 1999. Environmental stress and local adaptation of Daphnia magna, Limnol. Oceanogr., 44: 393–402.

Borcherding, J. 1992. Another early warning system for the detection of toxic discharges in the aquatic environment based on valve movement of the freshwater mussel *Dreissena polymorpha*. Limnologie Aktuell 4: 127–146.

Borcherding, J., and Volpers, M. 1994. The "Dreissena-Monitor"—First results on the application of this biological early warning system in the continuous monitoring of water quality. Wat. Sci. Technol. 29: 199–201.

Brett, M. 1993. Comment on "Possibility of N and P limitation for planktonic cladocerans: an experimental test" (Urabe and Watanabe) and "Nutrient element limitation of zooplankton production" (Hessen). Limnology and Oceanography, 38: 1333–1337.

Burns, C.W. 1968. The relationship between body size of filter-feeding Cladocera and the maximum size of particle ingested. Limnology and Oceanography, 13: 675–678.

Cole, J.A., Norton, R.L., and Montgomery, H.A.C. 1994. Countering acute pollution events: procedures currently being adopted in the United Kingdom. Wat. Sci. Tech. 29: 203–205.

DeMott, W.R. 1986. The role of taste in food selection by freshwater zooplankton. Oecologia, 69: 334–340.

DeMott, W.R. 1988. Discrimination between algae and detritus by freshwater and marine zooplankton. Bulletin of Marine Science, 43: 486–499.

DeMott, W.R., and Müller-Navarra, D.C. 1977. The importance of highly unsaturated fatty acids in zooplankton nutrition: evidence from experiments with Daphnia, a cyanobacterium and lipid emulsions. Freshwater Biol. 38: 649–664.

DeMott, W.R., and Müller-Navarra, D.C. 1995. The influence of prey hardness on Daphnia's selectivity for large prey. Hydrobiologia, 307: 127–138.

Diamond, J., Collins, M., and Gruber, D. 1988. An overview of automated biomonitoring—past developments and future needs. In: D.S. Gruber, and J.M. Diamond (eds.), "Automated biomonitoring, living sensors as environmental monitors". Ellis Horwood Publ., Chichester, pp. 23–38.

DIN (1989). DIN 38412: Bestimmung der nicht akut giftigen Wirkung von Abwasser gegenüber Daphnien über Verdünnungsstufen (L30), Teil 30.

DOKW (1996).—Management of Daphnia and algal cultures for Dynamic Daphnien Test. Water quality control for ground water management—Hydro Power Plant—Freudenau, Vienna (Unpublished report in German).

Downing, J.A., and Rigler, F.H. 1984. A manual of methods for assessment of secondary production in fresh waters. IBP Handbook–17, Second Ed., Blackwell, Oxford. 428p.

Draber, S., Schultze, R., and Hansen, U.P. 1991. Patch-clamp studies on the anomolous mole fraction effect of the K^+ channel in cytoplasmic droplets of *Nitella*: an attempt to distinguish between a multiion single-file pore an enzyme kinetic model with lazy state. J. Membrane Biol. 123: 183–190.

Draber, S., Schultze, R., and Hansen, U.P. 1993. Co-operative behavior of K^+ channels in the tonoplast of *Chara corallina*. Biophys. J. 65: 1553–1559.

Draber, S., and Schultze, R. 1994. Correction for missed events based on a realistic model of a detector. Biophys. J. 66: 191–202.

DVWK (1997): Weiterentwicklung eines mathematischen Modells zur on-line-Erkennung von signifikanten Messwertveränderungen in dynamischen Biotestverfahren. Bonn, pp. 1–80.

Elendt, B.-P. 1990. Selenium deficiency in crustacea. An ultrastructural approach to antennal damage in *Daphnia magna* Straus. Protoplasma, 154: 25–33.

Gliwicz, Z.M. 1977. Food size selection and seasonal succession of filter feeding zooplankton in a eutrophic lake. Ekoligia Polska, 25: 179–225.

Gruber, D.S., and Diamond, J.M. (eds.), 1988. Automated biomonitoring: living sensors as environmental monitors. Ellis Horwood, Chichester, UK., 99–208.

Gruber, D., Frago, C.H., and Rasnake, W.J. 1994. Automated biomonitors—first line of defense. J. Aquat. Ecosyst. Health, 3: 87–92.

Gulati, R., and DeMott, W.R., 1997. The role of food quality for zooplankton: remarks on the state-of-the-art, perspectives and prorities. Freshwater Biol. 38: 753–768.

Gunatilaka, A., and Dreher, J. 1996. Use of early warning systems as a tool for surface and ground water quality monitoring. Proc. IAWQ and IWSA Symp. on Metropolitan areas and Rivers, Rome, May 1996. TSI—River quality surveying and monitoring methods 2: 200–211.

Hallem, T.G., Lassister, R.R., Li, J., and McKinney, W. 1990. Toxicant-induced mortality in models of *Daphnia* populations. Environ. Toxicol. Chem. 9: 597–621.

Hellawell, J.M. 1978. Biological surveillance of rivers. Water Research Centre, Medmenham, Stevenage, UK. p. 332.

Hendriks, A.J., Maas-Diepeveen, J.L., Noordsij, A., and van der Gaag, M.A. 1994. Monitoring response of XAD-concentrated water in the Rhine delta: a major part of the toxic compounds remains unidentifies. Wat. Res. 28: 581–598.

Hendriks, A.J., and Stouten, M. 1994. Monitoring response of flow-through *Daphnia magna* and *Leuciscus idus* assays to microcontaminants in the Rhine Delta: early warning as a useful supplement. Ecotoxicol. Envir. Safty. 26(3): 265–279.

Hessen, D. 1992. Nutrient element limitation of zooplankton production. American Naturalist, 140: 799–814.

Hessen, D. 1993. The role of mineral nutrients for zooplankton nutrition: Reply to the comment by Brett. Limnology and Oceanography, 38: 1340–1343.

Hinkley, D.V. 1971. Interferences about change points from cumulative sum test. Biometrica 57: 1–17.

Irmer, H. (ed.) 1994. Continuous biotests for water monitoring of the river Rhine. Summary, recommendations, description of test methods. Umweltbundesamt Texte 58/94, Berlin, p. 30.

Juhnke, I., and Besch, W.K. 1971. Eine neue Testmethode zur Früherkennung akut toxischer Inhaltstoffe im Wasser. Gewässer und Abwasser, 50/51: 107–114.

Kilham, S.S., Kreeger, D.A., Goulden, C.E., and Lynn, S.G. 1997a. Effects of nutrient limitation in biochemical constituents of *Ankistrodesmus falcatus*. Freshwater Biol. 38: 591–596.

Kilham, S.S., Kreeger, D.A., Goulden, C.E., and Lynn, S.G. 1997b. Effect of algal food quality on fecundity and population growth rates of *Daphnia*. Freshwater Biol. 38: 639–647.

Knie, J. 1978. Der dynamische Daphnientest—ein automatischer Biomonitor zur Überwachung von Gewässern und Abwässern. Wasser und Boden, 12: 310–312.

Knie, J. 1988. Der dynamische Daphnientest—praktische Erfahrungen bei der Gewässerüberwachung. Gewässerschutz Wasser Abwasser 102: 341–356.

Knie, J. 1990. Betriebserfahrungen mit biologischen Testanlagen. VDI-Umwelt 20: 517–519.

Knie, J. 1994. Biomonitore zur kontinuierlichen Überwachung von Wasser und Abwasser in der Bundesrepublik. Spektrum der Wissenschaft, 5: 94–98.

Knie, J., Hälke, A., Juhnke, I., and Schiller, W. 1983. Ergebnisse der Untersuchungen von chemischen Stoffen mit vier Biotests. Dtsch. Gewässerkundl. Mitt. 27: 77–79.

Kramer, K.J.M., and Botterweg, J. 1991. Aquatic biological early warning systems: An overview. In: Bioindicators and environmental management. D.W. Jeffrey, and Madden, B. (eds.), Academic Press, London, pp. 95–126.

Kreeger, D.A., Goulden, C.E., Kilham, S.S., Lynn, S.G., Datta, S., and Interlandi, S.J. 1997. Seasonal changes in the biochemistry of lake seston. Freshwater Biol. 38: 525–537.

Lampert, W., and Trubetskova, I. 1996. Juvenile growth rate as a measure of fitness in Daphnia. Ecology 10: 631–635.

LAWA 1996. Empfehlungen zum Einsatz von kontinuierlichen Biotestverfahren für die Gewässerüberwachung. Kulturbuchverlag Berlin, pp. 1–38

LAWA 1998. Recommendations on the deployment of continuous biomonitors for the monitoring of surface waters. Kulturbuchverlag Berlin, pp. 1–46.

LUA 1994. Der Dynamische Daphnientest—Erfahrungen und praktische Hinweise—Bericht der Bund/Ländergruppe "Wirkungstests Rhein", Landesumweltamt Nordrhein-Westfalen, Materialien Nr. 1, Essen, pp. 1–39.

Lürling, M. de Lange, H.J., and van Donk, E. 1997. Change in food quality of the green alga *Scenedesmus* induced by *Daphnia* infochemicles: biochemical composition and morphology. Freshwater Biol. 38: 619–628.

Matthias, U., and Puzicha, H. 1990. Erfahrungen mit dem Dynamischen Daphnientest—Einfluss von Pestiziden auf das Schwimmverhalten von *Daphnia magna* unter Labor- und Praxisbedingungen

(Experiences with the dynamic waterflea assays—influence of pesticides on the swimming activity of *Daphnia magna* under laboratory and field conditions). Z. Wasser & Abwasser Forsch. 23: 193–198.

Mayer, E.L., Jr., and Ellersieck, M.R. 1986. Manual of acute toxicity; interpretation and database for 410 chemicals and 66 species of freshwater animals. Fish and Wildlife Service, United States Dept. of the Interior, Washington D.C. Resource publication 160. 508 pp.

Moldaenke, C. 1997. DYNSPR—Programm zur dynamischen Sprungerkennung, Projekt zur Weiterentwicklung eines mathematischen Modells zur on-line-Erkennung von signifikanten Mewertveränderungen in dynamischen Biotestverfahren. Bbe, Kiel. Software Manual.

Müller-Navarra, D.C. 1995a. Biochemical versus mineral limitation n *Daphnia*. Limnol. Oceanogr. 40: 1209–1214.

Müller-Navarra, D.C. 1995b. Evidence that a highly unsaturated fatty acid limits *Daphnia* growth in nature. Archiv Hydrobiol. 132: 297–307.

Müller-Navarra. D.C., and Lampert, W. 1996. Seasonal patterns of food limitation in *Daphnia galeata*: separating food quantity and food quality effects. J. Plankton Res. 7: 1137–1157.

Nauwerck, A., Duncan, A., Hillbricht-Ilkowska, A., and Larsson, P. 1979. In Le Cren, D., and Lowe McConnel (Ed.) Functioning of freshwater ecosystems, IBP-22. Cambridge University Press, Cambridge, England. 588 pp.

Pilli, A., Carle, D.O., and Sheedy, B.R. 1989. Aquatic information retrieval data base AQUIRE, database and technical support document, Environment Research Laboratory; U.S. Environmental Protection Agency, Duluth, Minn.

Poles, C.L.M. 1977. An automatic system for rapid detection of acute high concentrations of toxic substances in surface water using trout. In: Biological Monitoring of Water and Effluent quality (J. Cirns et al., eds.), ASTM-STP, vol. 607: 85–95.

Puyker, L.M., and van Genderen, J. 1989. Organic micro contaminants in Rhine and Meuse in 1988; pesticides and mutagenicity. Report SWO 89.245, Netherlands Waterworks Testing and Research Institute, KIWA, Nieuwegein, The Netherlands (in Dutch).

Puzicha, H. 1989. Untersuchungen über die Auswirkungen verschiedener Pestizide auf die Schwimmfähigkeit von *Daphnia magna* Straus mit Hilfe des Dynamischen Daphnientests. Master thesis (Diplomarbeit), Technical University of Karlsruhe, Germany.

Puzicha, H. 1992. Der dynamische Daphnientest: Erfahrungen aus dem Messstationsbetrieb am Gewässer. Schr.-Reihe Verein WaBoLu, Gustav-Fischer Verlag, Stuttgart, 89: 349–359.

Puzicha, H. 1994. Biomonitoring under field conditions at the rivers Rhine and Main and experiments with waste water, (in German). Landesamt für Wasserwirtschaft, Rheinland-Pfalz, Mainz, Hessische Landesanstalt für Umweltschutz, Wiesbaden, p. 198.

Puzicha, H. 1995. Bewertung verschiedener Biomonitore für den Routineeinsatz zur Schadstoffüberwachung in Oberflächenwässern. PhD Thesis, Johannes-Gutenberg-Universität, Mainz, p. 174.

Puzicha, H., and Blohm, W. 1991. Ergebnisprotokoll vom 1. Anwendertreffen zum Austausch von Erfahrungen über den Einsatz des Dynamischen Daphnientests in Messstationen zur Gewässerüberwachung. Veranstaltung am 14./15. Februar 1991 in Hamburg. Landesamt für Wasserwirtschaft Rheinland-Pfalz, Mainz, 15 pp.

Repka, S. 1997a. Effects of food type on the life history of *Daphnia* clones from lakes differing in trophic state. I. *Daphnia galeata* feeding on *Scenedesmus* and *Oscillatoria*. Freshwater Biol., 38: 675–683.

Repka, S. 1997b. Effects of food types on the life history of *Daphnia* clones from lakes differing in trophic state. II. *Daphnia cucllata* feeding on mixed diets. Freshwater Biol., 38: 675–683.

Schultze, R. 1992. Robust identification for adaptive control: the dynamic Hinkley-detector. In: Proc. Fourth IFAC symposium on "Adaptive Systems in Control and Signal Processing", 1992, Grenoble, France. Pergamon Press, London. 23–28.

Schultze, R. 1993. Sprungerkennung für die Systemidentifikation von Schiffen und zur Analyse von Patch-Clamp-Daten. Dissertation, Inst. Angewandte Physik der CAU, Kiel.

Schultze, R., and Draber, S. 1993. A non linear filter algorithm for the detection of jumps in patch-clamp data. J. Membrane Biol. 132: 41–52.

Sloof, W., de Zwart, D., and van de Kerkhoff, J.F.J. 1983. Monitoring the rivers Rhine and Meuse in the Netherlands for toxicity. Aquat. Toxicol. 4: 189–198.

Sommer, U. 1992. Phosphorous limited *Daphnia*: Intraspecific facilitation instead of competition. Limnology and Oceanography, 37: 966–973.

Stein, J.R. (Ed.) 1973. Handbook of Phycological methods. Culture methods and growth measurements. Cambridge University Press, Cambridge 448 pp.

Sterner, R.W., and Hessen, D.O. 1994. Algal nutrient limitation and the nutrient of aquatic herbivores. Annual review of Ecology and Systematics, 25: 1–29.

Sterner, R.W., Elser, J.J., and Hessen, D.O. 1992. Stoichiometric relationships among producers, consumers and nutrient cycling in pelagic ecosystems. Biogeochemistry, 17: 49–67.

Urabe J., and Watanabe, Y. 1992. Possibility of N and P limitation for planktonic cladocerans: an experimental test. Limnology and Oceanography, 37: 244–251.

Van Genderen, J., and Noij, Th. H.M. 1991. Organic micro contaminants in Rhine, Meuse, Ijsselmeer & Haringvliet and drinking water prepared from these surface waters. Report SWO 91.201, Netherlands Waterworks Testing and Research Institute, KIWA NV, Nieuwegein, The Netherlands (in Dutch).

Xiu, R., Xu, Y., and Gao, S. 1989. Toxicity of the new pyrethroid insecticide, deltamithrin, to *Daphnia magna*. Hydrobiologia 188–189: 411–413.

8. APPENDIX 1. DYNAMIC *DAPHNIA* TEST—POSSIBLE SOURCES OF FALSE ALARM

I. The source of false alarms

The sources for the false alarm could be generally classified into three categories:

(a) alarm by the malfunctioning of the instrument
(b) alarm originating from the test animals
(c) alarm through inadequacies of the alarm detecting algorithm

(a) Instrument

— Breakdown of the temperature controls, which maintain the water entering the test chambers at 20 °C.
— Break down of the pumps (inflow/outflow).
— Blockage of the flow through system (e.g. mesh, placed at the bottom of the cuvettes) either through suspended solids (sediments), filamentous algae or building up of large gas bubbles under the sieve.
— Entry of large suspended solid particles through a defective sieve.
— Desorption of contaminants from tubing after an alarm situation (tubing in the instrument have to be completely replaced in order to avoid such a situation).
— The release of toxic substances from bacterial aggregates attached on to the inner surface of the tubing or joints in the test system (bacterial growth leading to the formation of biofilms). Note: As a rule the tubing and all joints in the instrument have to be replaced with PTFE (Teflon) (the instrument is sold with silicone tubing); if silicone tubing is used for joints, they should be replaced during weekly clean up of the instrument. Tubing of the peristaltic pumps should be cleaned at regular intervals.
— Defective infra red light sensors.
— False impulses through reflections from glass test chambers.
— Entry of copepods (or other large zooplankters) into the test chambers.

(b) Test animals

— Weak test animals (neonates) due to their poor physiological condition.
— Sublethal damage caused during transfer of neonates into test chambers.
— Appearance of young daphnia in the cuvettes during the test cycle (usually happens due to increased supply of food). On the other hand at the beginning of the test neonates should be under 24 h of age and they have a fairly constant reproductive cycle. (In *Daphnia magna*—reproductive cycle is seven days; *D. pulex* six days; based on laboratory and field observations of J. Jacksch). Timely replacement of test animals in the instrument should be a routine procedure.
— Reduction of the number of test animals through losses; the main causes for the losses are:
  - 1. Slipping of the juveniles through the mesh placed at the bottom of the flow-through chambers (in the case the neonates if they are too small to be used in the test, the improvement of the mother cultures are necessary. This could be caused by either low food quality or through too old parental cultures),
  - 2. Predation by invaders (copepods),
  - 3. Parasites: the animals in the test chambers can be infected by rotatoria or ciliates.

(c) Algorithm

— The conditions for the optimal adjustment of the alarm detecting algorithm (adaptive Hinkley detector) are described in detail in DVWK (1997). Here it has to be stated that the adjustment of the alarm parameters must be in accordance with the specific location and fitted to suite the conditions of each site. They depend e.g. on the distance from the discharge, the trophic state of the river, the design of the river alarm system etc. It is difficult to find a compromise between "detection of all incidents" and "false alarms". However, false alarms caused by sub-optimal adjustment of the algorithms can be avoided.

II. Suggested remedies

(a) Instrument

1. At present the temperature of the continuously-fed river water is not measured in the test chambers in most of the instruments. It is necessary to measure the temperature directly in both test chambers and record along with the swimming behavior of the animals. Also the functioning of the Peltier element has to be improved by introducing adjustable temperature controls as animals are sensitive to small temperature changes (±1 °C). The existing instruments can be upgraded with these improvements for better functioning of the systems will lead to reduction of false alarms.
2. The formation of gas bubbles depends on temperature (degassing with increase of temperature) or with high dissolved CO_2 or O_2 content (algal photosynthesis). Improvements could be achieved through changing the form and dimensions of the outlet tube of the test chambers and also the flow rate.
3. Regular controlling of the functioning of the pumps and infra red lamps.

4. All parts of the instrument that are coming into contact with river water regularly (tubing, connectors and pumps etc.) must be cleaned regularly (weekly; if demands at shorter intervals). As a practice weekly replacement of all tubing in the instrument with a new set (Teflon tubing is recyclable) is time saving rather than cleaning them at the site. The maintenance work could be streamlined depending on the skill of the personnel.

(b) Animals

1. To improve the fitness of the animals (physiological status) used in the test, it is better to use 2–3 days old cohorts, which are pre-fed in the laboratory similar to parent cultures. This guarantees uniformly fed, healthier animals but they should be withdrawn from the test timely (6th or 7th day); a convenient procedure is to introduce animals twice per week into test chambers (e.g. on the 2nd and 5th day).
2. The age of the juveniles that are taken for the test should be carefully controlled to prevent the appearance of neonates during the test (see above and also Appendix 2).
3. Proper care has to be taken in maintaining parent cultures and they should yield sufficient neonates continuously (see Appendix 2).

(c) Algorithms

A careful analysis according to DVWK (1997) it is necessary to find out the best adjustment of the alarm parameters at each operation site. In order to adjust the instrument a longer test phase is necessary; during this phase animals should be exposed to a multitude of situations with uncontaminated as well as contaminated river water. This allows the collection of comprehensive data records that could be used for a better definition of the alarm parameters. By comparison of the daphnia test plots with other information such as chemical/physical characteristics of the test water it is possible to find an optimal balance between the needs for detecting as many contamination incidents as possible but at the same time avoiding false alarms.

9. APPENDIX 2. *DAPHNIA* AND ALGAL CULTURES—GENERAL MAINTENANCE AND QUALITY CONTROL

Culture media	*Daphnia*—M4 medium (Elendt, 1990) Algae—CHU10 (Chu, 1942) Water quality: The water used for the preparation of M4 and CHU10 media should be free of chlorine, metal ions and preferably with a low DOC content. Especially Zn, Pb or Cu containing material (for e.g. water pipes) should be avoided and if metals are present they have to be removed by passing through a suitable metal removing resin (e.g. LAVATEX®). Avoid using plastic containers or tubing. Instead the use of glass or PTFE Teflon as substitutes is recommended. Chlorine could be removed with using an active charcoal filter.
Cultures	It is recommended to refer standard manuals for culture methods for daphnia as well as algae (see Downing and Rigler, 1984; Stein,

1973; Lobban et al., 1988). *Daphnia* and algal cultures should be maintained under constant/controlled optimal culture conditions. Mother cultures should be maintained in 2 L glass beakers in approximately 1,300 ml M4-medium (this volume is sufficient to keep 20 parthenogenetically produced adults). The cultures are kept under subdued light conditions (12 h light and dark cycle) either in a water bath or in a climatic chamber kept at 20 ± 1 °C. At least 80–100 adults should be kept in culture to assure the production of sufficient number of healthy neonates for the test. The culture media in the mother cultures should be replaced every two days (this removes faeces and sedimented algae).

Daphnia cultures: preferably the clone should be identified by genotyping; timely renewal of cultures are necessary (the stock mother culture should be renewed at least every two years regularly).

Reference culture: a reserve culture has to be maintained in parallel, to be used in case of breakdown of a culture.

Daphnia in cultures should be fed with constant amount of algae with 1–2 fasting days for the week (excess of food leads to unhealthy cultures). Animals in culture are fed with a daily ration of approximately 20–25 ml of this algal suspension (3–4 days old, algae are in the logarithmic growth phase; ca. 0.1–0.2 mg C/daphnia/day), approximately five times a week.

Physiological status of the algae and daphnia cultures should be checked regularly (at least once a week; see details below under quality assurance).

Algal cultures: sterile techniques should be used in maintaining cultures and if possible axenic culturing as they are usually required for unialgal culture of freshwater species, as contaminating algae are often air-borne.

Algal cultures are in CHU-10 medium to be maintained in continuous culture under sterile conditions in 1 L glass bottles. They are kept under a constant light regime (light intensity 140 µE $m^{-2}s^{-1}$), 16 h a day and temperature (18–20 °C). Aeration: To keep the algae in suspension, filtered air (0.2 µm membrane filter) is purled through the cultures using a sintered glass bubbling device (after Dreschel).

Personnel — Technical assistants: should be in a position to identify problems associated with cultures; e.g. sicknesses through parasites or fungal attacks; copepods and ciliates in cultures.

Quality assurance of daphnia cultures

- Males or ephippia should not appear in the culture. This leads to break down of the culture.
- The mortality in the culture during a period of two months should not exceed 10%.

However, the adults should be renewed (mother culture) every two months.

— At the beginning of the reproductive phase (ca. 6–7 days), the daily reproduction rate of 20 Daphnia should not be less than 60 neonates/day. In general the variation coefficient of the reproduction rate should be ≥25%.
— The oxygen concentration of the cultures should not fall below 3 mg/l.
— Changes in pH should not exceed more than one unit.
— The cultures should be regularly examined, microscopically controlled to remove damaged or sick animals (at least once a week). Common illnesses are caused through bacterial attacks (brood sacs, cranial and caudal carapax areas), fungal attacks or ectoparasites (mostly Rotatoria or Ciliates). Most of the above illnesses could be cured through maintaining the cultures at optimal food level with one to two fasting days for the week (pers. com. H. Jaksch). Good quality algal food in the right quantity is of utmost importance; unconsumed algae may sediment and eventually facilitate building up of a high bacterial biomass. Feeding daphnia with unialgal cultures may lead to deficiencies which can be overcome by using an algal mixture.
  In case of a Selenium deficiency in daphnia, the common symptoms are either fouling or falling of antennae (Elendt, 1990). Other indicators that have to be cautious of are the appearance of males in the cultures as well as formation of duration eggs; both these factors may lead to a break down of the parental cultures.

Quality controls for algal cultures

— The commonly used alga for feeding is *Scenedesmus* spp. The control parameters that have to be checked are the cell density in the algae culture during the logarithmic phase of growth and the cell dimensions which could be used calculate the algal cell volume.
  Carbon content in algal cells—during log growth phase should be ≥50 pg/cell; when it is <25 pg, the algae cells begin to sediment. All the above mentioned factors could be used to check the state of health of algae.
  Surrogate measures such as algal cell number or light absorbance (cf. carbon determination) could be used to determine the food ration level for the Daphnia.
— The algae used for feeding Daphnia should be from the log growth phase for two reasons:
  (a) the algae are in a physiologically robust condition, and
  (b) they do not sediment.
  Sedimentation of algae in culture vessels (aged, senile algae) leads to high bacterial growth. Once they are sedimented to the bottom Daphnids could not filter them any more, a situation that eventually results in putrefaction leading to lowering of oxygen level in the culture medium. Alternative is to use motile forms such as *Chlamydomonas* and *Chlorella* which could be kept in suspension for a comparatively longer time.
— Measures have to be taken to control contamination through bacteria and invasion of other microbes (cyanobacteria, nuisance algae and fungi).

4

THE "MUSSELMONITOR®" AS BIOLOGICAL EARLY WARNING SYSTEM

The First Decade

KEES J. M. KRAMER[1] AND EDWIN M. FOEKEMA[2]

[1]MERMAYDE, P.O. Box 109, NL—1860 AC Bergen, The Netherlands [2]TNO-MEP, Department of Ecological Risk Assessment, P.O. Box 57, NL—1780 AB Den Helder, The Netherlands

1. INTRODUCTION

Traditionally[1] the monitoring of effluents, surface and other types of water is performed using dedicated physical and chemical methods. They may measure directly with in situ sensors or samples need to be collected which are measured later in a specialized laboratory.

The sampling-analysis way of operation is usually characterized as a non-continuous process and the condition of the waters tested can only be ascertained at intervals: in most monitoring programs a sampling frequency of once per day is considered as rather high. A check on the water quality will become available only for a given time (and location), and there is a risk that sporadic contamination will not be identified. The early detection of poor water conditions (e.g. as a result of accidental spills) is minimal and measures to prevent serious damage may come too late.

Due to their dedicated nature, these methods provide qualitative and quantitative information on the variables c.q. compounds that are analyzed. If for example, analyses are performed on the elements Cd and Hg in the water, the concentration of these elements will become available to a high level of accuracy, information that can even be used as "hard" evidence in court cases. Information on all the other elements is, however, non-existent: "What you don't look for, you don't find".

Musselmonitor® is a registered trade mark of Delta Consult, P.O. Box 71, 4420 AC Kapelle, The Netherlands.

[1] Abbreviations: BEWS: biological early warning system.

In situ measurements have the advantage that they provide on line information on the quality of the water tested. The "sampling" frequency is usually much higher (semi-continuous or continuous) allowing analytical data to be reported well in time to send out a warning that a certain criterion has been exceeded and an alarm situation has occurred. Measures may be taken immediately in order to prevent further deterioration of the environment. Unfortunately, the spectrum of sensors suitable for water quality monitoring is rather small. Only relatively few parameters can at present be measured (semi)-continuously in effluents and (surface) waters under natural conditions. Most common are: temperature, salinity/conductivity, pH, dissolved oxygen, redox, turbidity, fluorescence (Colin and Quevauviller, 1998). Sensors for pollutants (trace metals, organic micro pollutants) are not readily available, but it is expected that this will change within the next decade, when dedicated (bio)sensors will become available (Arnold and Wangsa, 1991; Scully, 1998). So far, only few biosensors are available that are capable of detecting single components. Their application under field conditions is, however, so far very limited. It is expected that this will change in the future. The strengthening of regulations will induce the need for greater specificity.

In addition, to in situ sensors, flow-through (automated) monitors are available, e.g. for (few) trace metals, BOD/COD, TOC, nutrients. They may be housed in automated monitoring stations that have been designed to monitor a wider spectrum of constituents. Instead of a single sensor, they are rather complete small laboratories with a host of analytical equipment (see Gunatilaka and Diehl, 2000, this volume; Brinkman, 1995; Colin and Quevauviller, 1998).

Next to these physico-chemical methods, typical biological monitoring systems have been developed that use (whole) organisms as sensors (Gruber and Diamond, 1988; Kramer and Botterweg, 1991). Like the dogs or geese around a farm, or caged birds in a mine, these systems are used as "biological early warning system" (BEWS). The first instruments developed already in the 1970's (using fish), received much attention, but due to the lack of computerized data handling they proved to be quite labor intensive. Today a whole series of BEWS instruments has been introduced on the market, which employ fish, bivalves (mussels, clams, oysters), invertebrates (*Daphnia* spp.), algae or bacteria as sensor organisms (Baldwin and Kramer, 1994; Gruber et al., 1994). Even multi-species designs have been developed and tested (Gerhardt, 1996).

Biological early warning systems have explicitly been developed to provide a rapid warning of the occurrence of contaminants at concentrations which could be of immediate threat to living organisms. In order to allow such a rapid response, the BEWS can only be based on biological functions that have the possibility to change with time. Therefore, BEWS are based on the monitoring of either a physiological or a behavioral function. Of course, the function should change when the organism is exposed to one or more contaminants, and a defined cause-effect relation should be established.

A typical characteristic of BEWS devices is that the organisms do react to a multitude of compounds, and not to one or a few analytes only. Many toxicants will be detected; the sensitivity of the monitoring system will depend on the type of compound and on the organism used. A BEWS will be able to provide neither qualitative nor quantitative information on defined compounds (unless only one compound is to be expected, such as in simple industrial effluents). Their strength lies in the continuous operation, their rapid response, and the broad spectrum detection of pollutants, potentially even those that are not detected by routine chemical monitoring programs. There are examples of well-defined alarm conditions generated by a BEWS, where the analysis of the related water samples could not reveal the nature of the toxicant (de Zwart et al., 1995).

As organisms in a BEWS will respond to "unfavorable environmental conditions", the signal will be a response to a sum of all elements that affect its functioning (or at least the functioning that is monitored by the instrument). This involves changes (increases) of one or more pollutants, but also changes in the natural conditions such as in temperature, conductivity, or suspended particulate matter. In general, the gradient of the changes must be relatively steep before the BEWS will detect a distinct reaction. Information on such natural conditions, e.g. from other on-line sensors, will prevent the occurrence of a false alarm.

The success of any BEWS depends on several factors. Important are of course the sensitivity for various types of pollutants, but also the ease of use, the price of the complete system, the cost of maintenance (staff), and the level of development for application under field conditions (Gruber et al., 1994; von Danwitz et al., 1998).

So far, biological early warning systems have not been given any legal status. This is partly due to the fact that standard operating procedures have not been defined (except for the (draft) fish ventilation behavior test; Diamond and Shedd, 1995). The second, more important reason is that all BEWS systems are multi-compound semsors that may react differently to different constituents in the water, present as single agent as well as in combination. This makes it very difficult to provide "hard evidence" on the basis of a BEWS alone. We see the role for a BEWS more in close combination with chemical monitoring (the BEWS will induce the sampling), followed by a quantitative and qualitative analysis: a legally valid "chemical proof".

One of the more successful BEWS is the Musselmonitor® (Figure 1). This instrument was a logical development of research carried out on the valve movement behavior of bivalves (e.g. Salánki, 1964; Salánki and Varanka, 1976; 1978; Manley and Davenport, 1979; Kramer et al., 1989). The development of the system was based on the many years experience at three Dutch institutes (KEMA, RIVM, and TNO) with the valve movement response of freshwater and marine mussels, both under laboratory and (semi)field experiments (Sloof et al., 1983; de Zwart and Slooff, 1987; Jenner et al., 1989; 1992; Kramer et al., 1989). In 1988 in the first prototype series was constructed. The very robust system can be used either in situ (Figure 2a) or in a flow through system (Figure 2b). The system makes use of the fact that bivalves under stress close their shells as protection against the hostile environment. The valve movement behavior is continuously followed and interpreted by the monitor. Prolonged closure (e.g. >4 min) may be a reason for alarm. A detailed description of the Musselmonitor follows in the next section.

Also, other groups have developed more or less robust biological early warning systems based on the valve-movement behavior of bivalves. Robust is used in a sense, that the systems have evolved from a system that only works under

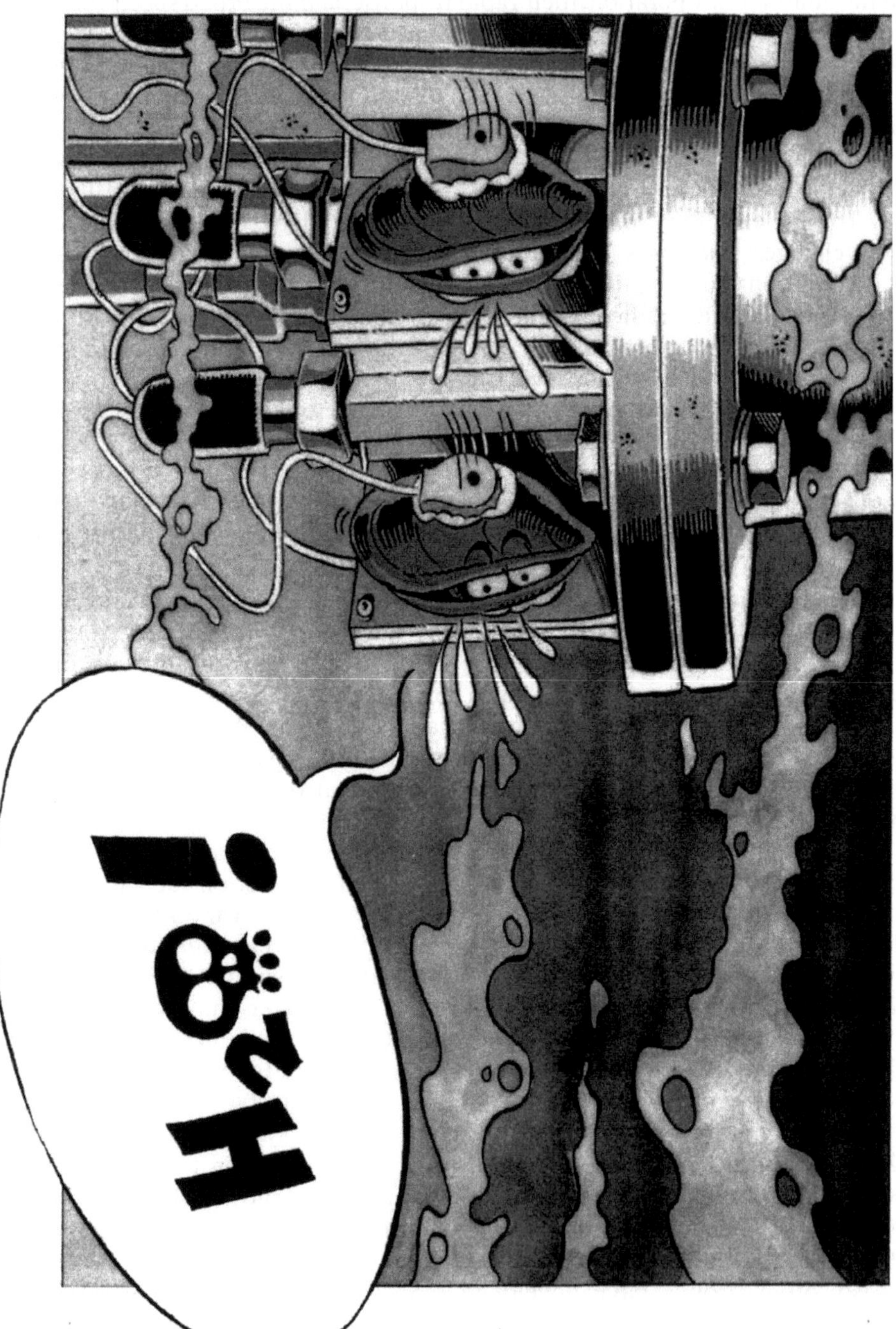

FIGURE 1. Biological early warning systems (BEWS) sound the alarm when e.g. accidental spills are detected. Here is an artist's impression of the functioning of the Musselmonitor®.

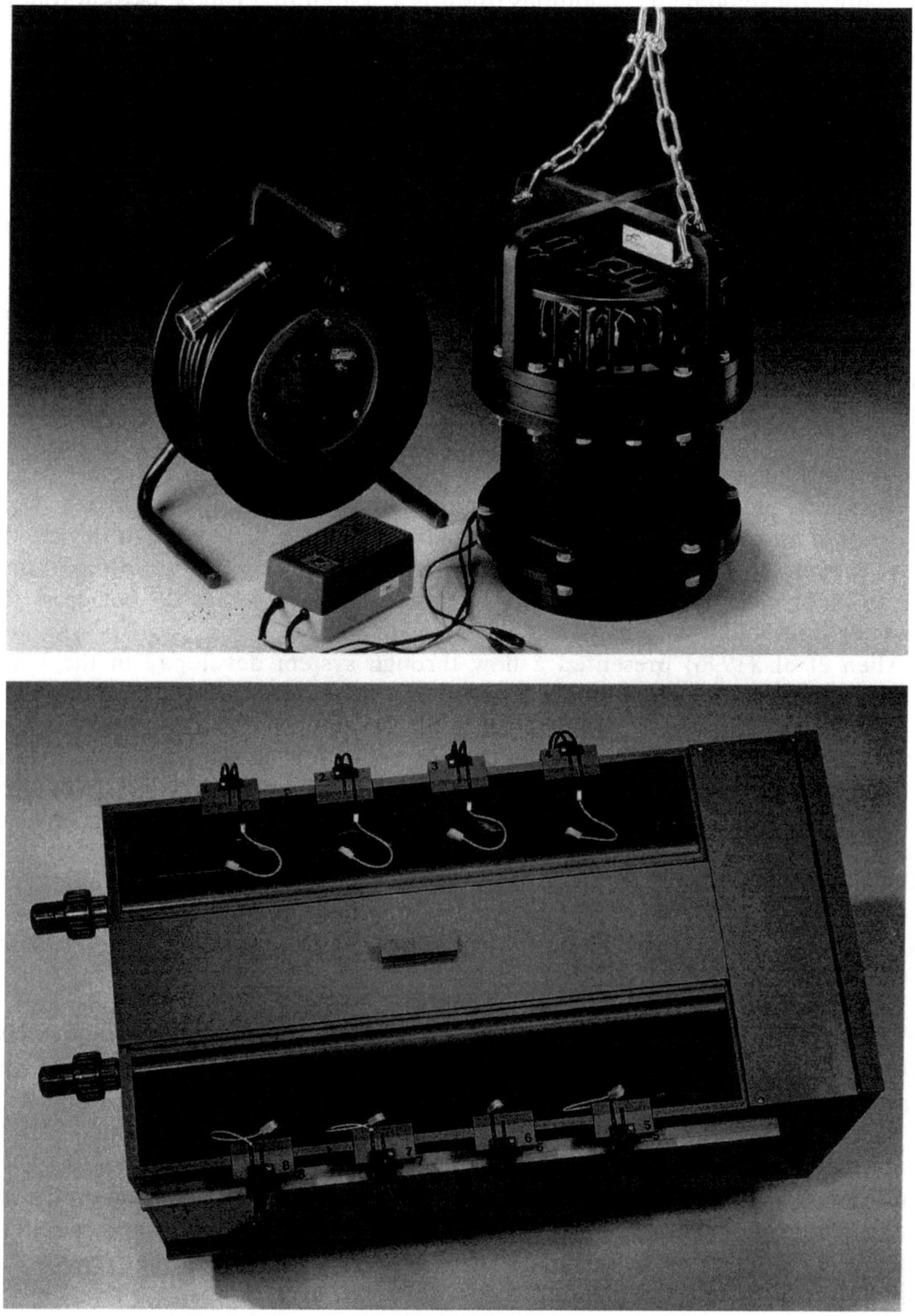

FIGURE 2. The Musselmonitor®. Shown the in situ version (top, with cable reel and battery charger) intended for direct use in the ambient water, and the flow-through system (bottom), that can be installed in any laboratory or monitoring station.

laboratory (development) conditions towards a system that can be used in routine (field) operation (Gruber et al., 1994). The following systems may be considered as BEWS:

• the "Dreissena Monitor", a German system, consists of two sets of 42 (freshwater) mussels (test system and control) in a flow-through installation. Each mussel, usually *Dreissena polymorpha*, is connected to a reed relay and a small magnet, which allows the detection of whether a given mussel is open or closed (binary information). No movement can be detected that does not cross the sensor threshold. The output signal is the averaged information of all 42 mussels (Borcherding, 1992);

• a Finnish design uses two switches made of silver wire that are glued to the bivalve shells; the tri-state output allows the detection of open, closed, and half-closed shells. The bivalves are not glued to a hard substrate, but can move freely. A typical set-up consists of 12 mussels (e.g. *Anodonta cygnea*). Since the system is operated at 100 Hz AC current, application is limited to freshwater environments. (Englund et al., 1994);

• Ham and Peterson (1994) described the use of a system using four Asiatic clams (*Corbicula fluminea*) in a floating system contained in a flow-through device; the valve movement triggers an electro-optical switch, enabling the recording of only the open or close situation. No movements are detected that do not cross the detector threshold;

• Allen et al. (1996) presented a flow through system developed in the USA, where a clam, *Corbicula fluminea*, is glued to a hard substrate. On the outer shell of each of the 15 clams a metal target is placed, opposite to a proximity sensor. This arrangement allows a semi-continuous recording of the shell gape. Alarm criteria are based on a simple time series analyses on the averaged data; to develop an alarm, the collective behavior of the unit of clams should change by, for example 15%, as compared to a previously collected value from the same set of clams;

• in France, the "Valvomètre" was made, but as this system copied the results of the Musselmonitor, it is hence not to be considered as a truly new design (Anon., 1994).

Several of these BEWS systems employing bivalves have been applied in routine monitoring networks (de Zwart et al., 1995; Borcherding and Volpers, 1994; Hoffmann et al., 1994; Irmer, 1994; Englund and Heino, 1994), usually in freshwater environments; others seem more suitable for laboratory investigations.

The first series of the Dutch Musselmonitor® was produced and tested in 1988. Since then this BEWS has, so far, been applied in six European countries, the USA and Australia. A wealth of practical information has thus been collected on the use of the system under various conditions, and on the day-to-day experiences in routine operation (stand alone time, maintenance, running costs). It will be obvious that also events of "strange" behavior of the bivalves has been observed, for example as influenced by the (changing) environmental characteristics. The causes for this behavior were either instrument-technical, physico-chemical or biological in nature. In general, when the causes were evaluated, precautions or adaptations in the application of the system could be suggested.

These "problems" will often not be limited to the use of the Musselmonitor only. The observations and solutions will be of interest to the users of this system and the other valve movement monitoring systems alike. In this chapter, after a description of the system, an overview is presented of these practical details that are connected to a decade of application of valve movement detection in monitoring of fresh- and marine waters. Topics that will be discussed include, for example, the influence of vibration, strong magnetic fields, light, temperature, suspended particulate matter, availability of food.

2. PRINCIPLE OF THE MUSSELMONITOR®

2.1. Biology

Bivalves are sedentary, hardy, commonly available organisms that are abundant in the freshwater as well as the marine environment. They have two shells (valves) that serve as outside "skeleton" and protection against predators. By closing the shells (escape behavior) they physically exclude the outside environment, and provided that the conditions do not go too bad, many species can thus survive for days.

There are numerous species of bivalves, but about 75 have been used in biological monitoring studies worldwide (O'Connor et al., 1994). Provided the shells are of suitable dimensions, potentially all bivalves (or even mono-valves like *Patella* spp.) can be used for studying the valve movement related to concentrations of toxic substances. Naturally, their sensitivity for given applications has to be validated. Species successfully used in the Musselmonitor include the zebra mussel (*Dreissena polymorpha*), swan mussel (*Anodonta cygnea*), the painter's mussel (*Unio pictorum*) and the dark false mussel (*Mytilopsis leucophaeta*). For the marine environment the blue mussel (*Mytilus edulis*) and the oyster (*Crassostrea gigas*) were applied. There is a slight preference for sedentary species that are attached with byssus threads to a hard substrate, but there is no proof that the normally free moving benthic species are affected by a fixed mounting in a BEWS system. As the BEWS will be used to provide information on the quality of the water, filter feeding species are preferred (*Dreissena* spp., *Mytilus* spp., *Perna* spp.). However, most benthic species are at least in part filter feeding as well. Although only a limited number of species were used so far (main reason their availability to the researchers), there are no *a priori* reasons why other endemic species would not be used, either mussels, clams, oysters or other bivalves. In any case it is highly preferable to use local species than to import new populations or—even worse—to import new (alien) species. Local species will be best adapted to the environmental conditions, provided of course that these specimens are not already adapted to polluted waters. It is expected that organisms collected from less polluted (pristine) sites will be more sensitive than those from polluted waters (see under adaptation below).

Under normal environmental conditions, mussels are submerged and their shells are open to allow for feeding and respiration. The typical behavior may vary from species to species, but almost all species tested close their shells only occasionally for a short period, e.g. to defecate. Three typical examples of "normal" behavior (no pollutants added) are shown for *Dreissena polymorpha* (Figure 3a,b), for *Mytilus edulis* (Figure 3c,d). The data were normalized to fully closed (set at 0%) to fully open (set at 100%), thus allowing similar scaling and easy comparison. The average valve opening of about 70–95% will be evident, as are the short periods of closure. The difference between the two graphs for the same species is the difference between the individual behavior of the two example mussels: one is more active (nervous) than the other, a quite normal situation when comparing individuals.

Closure of the shells for longer periods is to be considered as escape behavior. If one bivalve is closed for a prolonged time, e.g. for 5 min, this is not considered as unusual; if several mussels do this simultaneously, say 5 out of 8 mussels, this is highly unusual and a reason for alarm. Some species, like *Dreissena* and *Unio*, may show occasionally prolonged closure times (maximum 2–4 hours), obviously as part of their natural behavior. However, even in these cases, the simultaneous closure of (nearly) all bivalves remains highly unusual, and is a reason for alarm. A

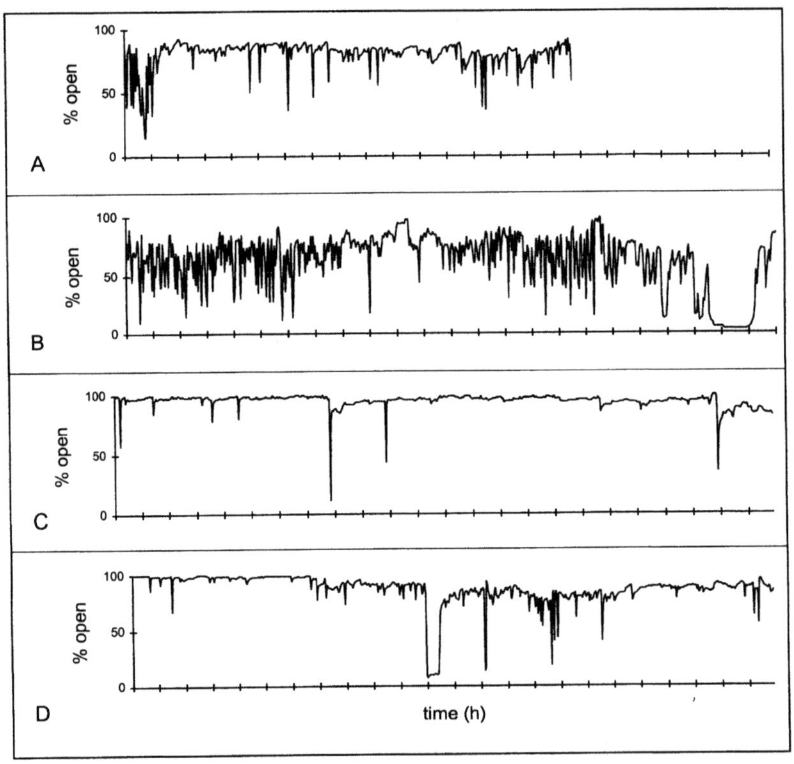

FIGURE 3. Typical registrations of the valve movement of individual bivalves in natural (sea)water (A, B) the fresh water zebra mussel (*Dreissena polymorpha*) and (C, D) the marine blue mussel (*Mytilus edulis*), each with a relaxed and a more nervous example (time scales are kept the same, data are normalised to closed (0%) and maximum open (100%)).

closure for much longer periods, even >24 h has been reported for zebra mussels employed in the "Dreissena Monitor" (Anon., 1998). In these cases no alarm was reported from other chemical or biological monitoring devices, leading to the assumption that the observations were not the result of poor environmental conditions but may result from the characteristics of the measurement system used (no instrumental adaptation to changes in the span of the valve opening, binary detection). In our experience we never observed such extremely long closure times when applying *Dreissena polymorpha* in the Musselmonitor.

Being closed for prolonged periods does not harm species like the zebra or blue mussel as they can change from an aerobic to an anaerobic metabolism (Eertman and de Zwaan, 1994). Unlike other species used in BEWS systems (such as fish, daphnia), these species have a distinct advantage in waters that may tend to become temporarily hypoxic or even anoxic. The closure of the shells will continue until the mussels find the water conditions acceptable. To check, they will open their shells for a short period to "test the water" (Figure 4a). The frequency of testing is highly dependent of the individual, and consequently the sensing abilities of the bivalves once they are forced to close, are limited. Opening of the shells is a guarantee that the water has been tested as OK, but the moment of opening may be (up to hours) later than the actual improvement water quality. Fortunately, this does in general not affect the "early warning" characteristics, as the detection of the alarm condition is usually considered far more important than the return of the improved water condition.

2.2. Technical Characteristics of the Musselmonitor®

The Musselmonitor was based on many years of experience with laboratory systems. From this experience it was clear that, provided the behavior of each mussel was compared with its own behavior, e.g. one hour previously, there was no need to middle out reactions by using relatively large numbers of specimens: natural variability is largely reduced when using the same individual as his own control. It was therefore essential that measurement and evaluation were to be carried out on each individual, and when averaged results were to be displayed this was only after full evaluation of the data. From experience with laboratory systems, it was considered sufficient to incorporate eight bivalves in this BEWS (a double set of 16 mussels was kept as optional).

When the system was designed, it was considered a preference to have a BEWS that could be immersed in the water to be tested: no flow through, no housing, no pumping system would be required. Thus, an in situ system was designed (Figure 2a). Surprisingly, many clients were more interested to have a flow through system, as they were using flow through for other sensors anyway. Therefore, in addition to the in situ design, a flow through system was added to the program (Figure 2b). Apart from the housing, the systems are identical.

The Musselmonitor (in situ design) consists of a water-tight casing (high-density polythene), which contains the rechargeable batteries and the electronic components for the actual measurement, data evaluation and data communication.

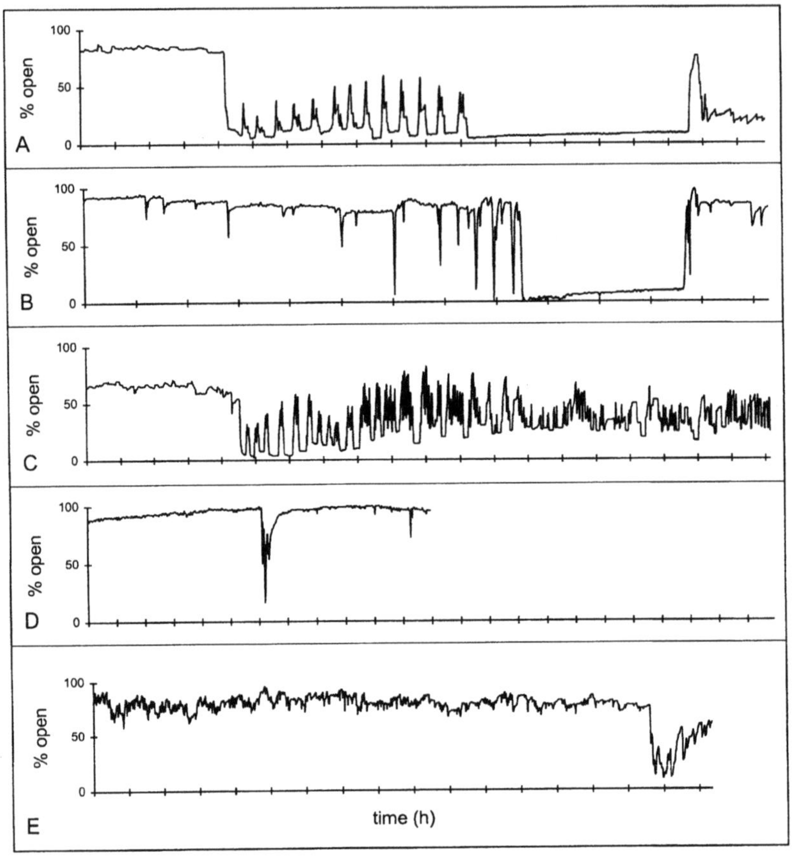

FIGURE 4. "Typical" behavior of the zebra mussel under various conditions: A) after closure, the water is repeatedly tested whether quality has improved; B) after some initial reaction a rapid closure reaction; C) after a closure reaction the activity increases considerably; D) a short spike caused by a sudden and short disturbance (average of 8 individual signals); E) the averaged signal of 8 individual mussels shows less closure (middled out), but near the end of the recorded period the entry of staff caused a closure alarm; note that the mussels get used to the vibrations and in the course of two hours slowly open.

On the upper part, eight mussels can be glued on plastic supports. We have good experience with a cement used in dentistry (Unifast); it is non-toxic, fills well, attaches perfectly to both support and mussel shell, and lasts at least several months. A protective cover eliminates the risk of debris hitting the mussels.

To each mussel a set of two sensors, consisting of tiny coils, is mounted. A high-frequency current generates in one coil a magnetic field which is received by the second coil. In this second coil the magnetic field induces a current, which is dependent on the distance between the coils (= shells). Within the range of operation, the signal is linear to the distance of the shells. The original low frequency electronic design of the measurement system (Schuring and Geense, 1972) has been updated by KEMA into a 500 kHz system (Jenner et al., 1989), and more recently by the producer Delta Consult, to a system of sequential measurements, thus minimizing between-sensor interferences.

The Musselmonitor is, when delivered, ready for use. Once the electronic cable is connected the system starts automatically. The built-in dedicated installation menu, accessible by PC, allows the tuning of the system for optimal settings of the measurement regime as well as the evaluation criteria (de Zwart et al., 1995).

The monitor measures (semi-)continuously the opening of each individual bivalve. It has been shown that measurement at intervals even up to 15 min does not affect the intrinsic value of the measurement (Kramer et al., 1989). Of course, a lower measurement frequency will increase the time needed for alarm generation, but this is only in the order of minutes. A commonly applied setting of measurement time is every 1–2 min. Such a frequency is a balance between the amount of data and the early warning character of the system.

The thickness of each individual mussel varies considerably. Hence, the output which is initially in an absolute value can vary considerably as well. Raw data of each mussel are normalized towards a span, with the historic minimum value set at 0% open, and the historic maximum value set at 100%. Because of this procedure, the Musselmonitor requires a few days installation time for this "learning process", but the advantage is that the results for each individual can be easily compared, and data evaluation becomes relatively simple. The results, when graphically presented, look like the examples in Figure 3.

The Musselmonitor will evaluate the data and compare results with preset criteria; when criteria are passed, an alarm will be generated. The system provides five different ways to evaluate the data (based on the observations for each individual):

- *closed alarm* (e.g. 5 out of 8 mussels are closed for 5 min, where closed is defined as <20%); this is the most definitive reaction; see Figure 4b;
- *decreased average alarm* (e.g. for 4 out of 8 mussels the average over the last 15 min is decreased by 25%, when compared to a block of 15 min 1 hour prior to this measurement); this behavior is often connected to the slowing down or stopping of pumping (Mouabad and Pihan, 1992);
- *activity alarm*; we observed for several species (*Mytilus* spp., *Dreissena* spp.) that the activity (open-close-open movement frequency) can increase dramatically when the species are exposed to toxic substances (organic solvents, free chlorine, and others); for evaluation the present frequency is compared with historic information collected 2 hours previously; an example is shown in Figure 4c (see other typical examples in Kramer et al., 1989; de Zwart et al., 1995); it has not yet been tested whether other species exhibit similar patterns in activity changes;
- *gaping alarm*; bivalves keep their shells closed because of the action of the adductor muscle; when they die, the muscle will loosen, and the shells will slowly open, but to a value much larger than the 100% defined under normal operation (gaping); gaping is a signal that the mussel is dead (or the sensor fell off the shell), and hence the data of this individual should no longer be used for evaluation;
- *battery alarm*: when the system operates without main power supply, the system will warn when a defined threshold has passed;

• "*system alive*"; in our early experiments we thought that no alarm meant that none of the alarm criteria mentioned above was triggered; in reality, however, the system could also have been down for the period; therefore we introduced a regular system check, which is documented in a log file.

All the variables given are to be seen as examples only, and can all be changed at will in the set-up procedure of the measurement program. In general, one may say that a high criterion, which is paired to non-sensitivity (8 out of 8 mussels must react) is more reliable than a low criterion (3 out of 8 mussels must react). There is an obvious balance between sensitivity and reliability. It is to the end user of the system which of the two parameters has the highest priority (in practice we find that users prefer a middle setting).

In addition to the standard evaluation contained in the Musselmonitor, Sluyts et al. (1995) developed a "split moving window" dynamic evaluation system (SMW), which is applicable on line as well to recorded data. This evaluation method samples over a given time window the frequency distribution of the averaged valve opening, and the individual activity. These data are compared with a similar but previous window, and tested for similarity, which is compared to an alarm criterion. A value below this threshold leads to alarm. SMW consists of a statistical method of data evaluation which helps to reduce the number of "false" alarms; in principle it may be applied to other BEWS as well.

In case an alarm is generated by the Musselmonitor, a potential free contact is closed which can trigger any action the user may want. The monitor contains a log buffer where the events, such as reset, alarms, system-alive, are stored with date and time. Measurement data are exported following a RS232 standard (new versions RS422); they are conveniently stored on a PC; a standard one week operation can easily be stored on a 1.4 Mb diskette.

3. A DECADE OF PRACTICAL EXPERIENCE

3.1. Feasibility Studies

Over the past decade we have performed many tests of and with the Musselmonitor. In the beginning the functionality of the instrument was the prime objective, later feasibility studies and demonstration projects allowed us to test the instrument under realistic conditions. In these situations it happened often that specific problems were encountered, which needed a solution. In the following sections a number of these items are discussed.

3.1.1. Construction and Installation

The prototype design consisted of a 1.5 cm thick 316 stainless steel housing, which was considered to be resistant to all media of concern. However, in waters with high loads of organic-rich particulate matter (effluents), deposited particles created an anoxic environment directly near the stainless steel surface. Within a few weeks of operation under these circumstances pit corrosion occurred, and the instrument was flooded (Jenner, pers. comm.). As a result the SS was replaced by high density polythene, which is also considered as potentially less toxic to the bivalves.

A second lesson learned was that automated long term application under environmental conditions is not to be underestimated, especially for the *in situ* instrument. Continuous movement leads to wear, and harp shackles must be protected against turning loose. For the same reason, the water-tight connector and cable have to be of durable quality.

Surprisingly, often basic requirements for operation, quite common to the laboratory environment, were lacking in the field. Power supply (mains) can easily be replaced by a solar panel connected to a 12 V truck battery as buffer. Depending on latitude (we tested at 53° N) and instrument settings (mainly the data transmission is an important power consumer) about 1 m^2 is generally sufficient.

Unless the BEWS is located in a specially designed monitoring station, where all facilities are present (flow through system, power supply, telephone lines, etc.) the transmission of data may be problematic in the field. To monitor remote locations, we used radio transmission successfully, enabling on-line data communication and read out on a PC-screen at the office desk. Since the measurements are carried out at about once per min, transmission speed is not important. The distance between the monitor and the receiver is limited only by the transmitter system used (antenna, output power); when using directional antennae, a distance of 20 km can be covered.

Sometimes theft (and vandalism) was of major concern. As the in situ model of the Musselmonitor was designed to be placed in the water unattended, but power supply and communication needs to be maintained, cables may attract fishermen, and others. Unfortunately, there is no general solution to this problem. Without communication there is no "early warning". The size of the monitor will make it relatively easy to construct a cage which can be mounted to a solid substrate. For applications where only retrospective data handling is required, the use of a water-tight data logger allows the entire system to be placed under water.

3.1.2. Limitations of the Bivalve Size

The size of the bivalves that can be used in the Musselmonitor may vary considerably. Although a rather constant distance between the sensors is preferred, some variation can be allowed. Interestingly, a 10 cm long painter's mussel or green mussel is not much thicker than a 2–3 cm long blue or zebra mussel. The tapered shape of the shells will allow an optimum setting between the sensors to be achieved. There are two physical limitations to the size of the bivalves, however. The species may be too small, so that one can not attach the sensors, and when this would be possible the movement of the sensor cables in running water will induce too much stress to the organisms. For this reason we failed, for example, to use pea mussels (*Pisidium* spp.) successfully. The minimum length is estimated to be 1.5 cm. The maximum length is in the *in situ* version of the instrument limited to about 10–12 cm by the protective cover. Without the cover, or in the flow-through system, the maximum size is less limiting. The optimum setting of the distance between the sensors can be adapted upon demand.

3.1.3. Optimization of Setting the Criteria

The settings of both the measurement and the data evaluation can be changed at will. This is a great advantage as it allows the setting to be optimized for the application. It was observed that the users felt insecure by the seemingly endless number of possibilities. A "standard" setting is provided, but this will usually not be the optimized situation. When a (demonstration) data set was available of sufficient length, like 2–3 months, a solution may be to perform "what-if" analyses. Although the data are collected at a user-defined setting, the data themselves are stored on diskette and thus allow retrospective re-evaluation. If, for example, during the measurement one of the alarm criteria was "5 out of 8 mussels must be closed for 6 min", we may in the evaluation re-run the data set and test the sensitivity and reliability (false alarms): if the criterion would have been "4 out of 8" or "3 out of 8", etc., or e.g. closed for 5, 4 or 3 minutes (each time resulting in a more sensitive, but less reliable situation), would that have created an alarm pattern that better suits my application? Table 1 provides an example of such a re-evaluation of the settings. A few "events" in Table 1 were caused by spiking of a toxicant to the flow-through system, allowing to estimate the connected response time.

TABLE 1. Example from a drinking water inlet for a "what if" analysis. Re-evaluation of a data set to examine the alarms generated in case the criterion for alarm was changed from 4 to 7 mussels (closed ≤20%; t ≥2 min).

date	time and number of mussels closed (out of 8)				
	4	5	6	7	spiking
26-03	14:04	14:04	14:04		copper
27-03	12:13	12:15	12:15		free chlorine
28:03	13:03				
	17:25	18:23			
08-04	17:02	17:08			NH_4Cl
09-04	18:11	18:45	19:03		
12-04	09:41	09:43	09:45		copper
13-04	01:34				
14-04	01:14				
	04:33	05:01			
	08:47	10:55	12:21	12:29	
17-04	02:11				
19:04	18:21	18:27	19:23	23:23	
30-04	00:41	00:49			
17-05	06:43	07:15			

3.2. Interferences

3.2.1. Adaptation?

In contrast to most other BEWS, such as those employing fish, daphnia or algae, the organisms used in the Musselmonitor need replacement only after months instead of after a week. This considerably cuts the cost of maintenance (staff). Since the bivalves do not show signs of malfunctioning and take the food and oxygen from the ambient water, there is no reason to replace the organisms. The growth rate is so small that we never experienced problems of loosening the attachment. In monitoring studies of river Rhine water employing 8 zebra mussels, the maximum exposure time of the same set of bivalves was up to 10 months, apparently without any obvious problems (de Zwart, pers. comm.). Borcherding and Volpers (1994) reported to apply the same zebra mussels for 6 months in the "Dreissena Monitor".

Adaptation to the environmental conditions, notably in slightly polluted situations, has been suggested to hamper the effectivity of this BEWS. The system may loose sensitivity, if the organisms are exposed to polluted waters for a longer time periods. We do not deny the existence of adaptation, but have—so far—not found any evidence of this effect. We have, however, not carried out well controlled experiments designed for the detection and quantification of possible adaptation. A reason that there is no proof may be that biological early warning systems usually react when relatively steep gradients (in e.g. the concentration of toxicants) are encountered (see also section 3.3.4). Steep gradients are typical for (accidental) spills, and their detection is one of the main objectives for application of any BEWS.

Although no effects of adaptation were reported, we advise to replace the mussels about every 2 months, just to be on the safer side concerning possible adaptation.

3.2.2. Light

Mussels are sensitive to light. Bennett (1954) already observed the modification of the valve movement behavior in the quahog (*Venus mercenaria*). A sudden change in the light level, e.g. when the light is switched on, or when the shadow of a person falls on the bivalves, a closure reaction may occur. We have the example, where at 3 AM all 8 eight mussels react with a sharp closure peak in a laboratory situation. It took some time until it was realized that at that time the night watch made his tour through the building, and had switched on the lights.

Also a diurnal cycle (see below) may be the result of light regime (Benedens and Hinz, 1980). We observed a distinct day-night rhythm when applying *M. edulis* in a open air mesocosm experiment in summertime, where the 8 mussels were exposed close to the water surface. In Figure 5a the average signal of 8 mussels is depicted, calculated for periods of 30 min. The dark bar indicates the hours of darkness. Obviously, the average valve opening is higher during nighttime. To investigate whether the light was of influence, we covered in a second experiment half of the

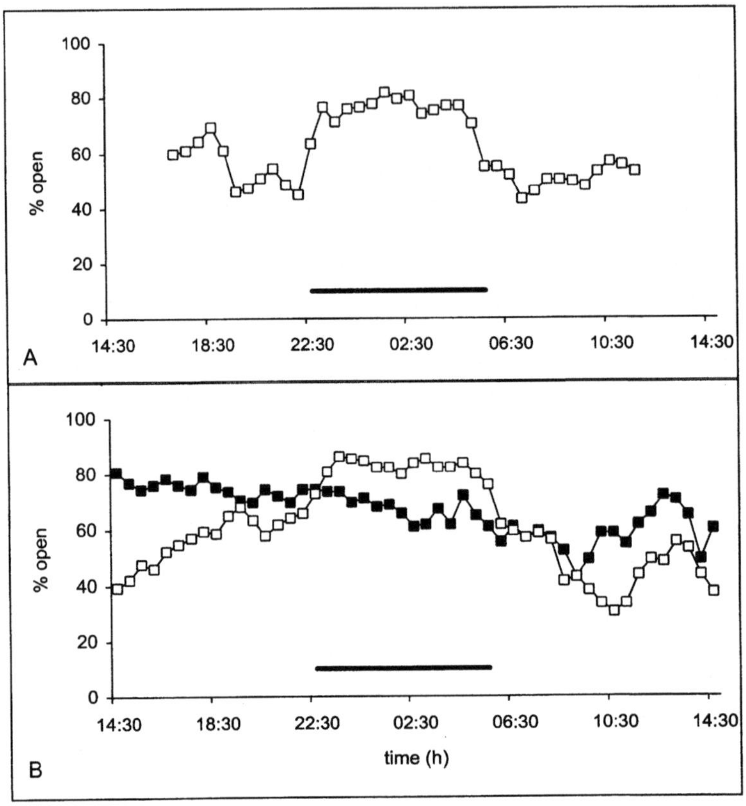

FIGURE 5. The effect of light upon the valve movement behavior of blue mussels. A) Average opening (n = 8) during day and night; B) a similar experiment where 4 mussels are exposed and 4 protected against sun light (recorded are the average values in time blocks of 30 min, open symbols = light, filled symbols = dark; the night period is indicated by the horizontal bar).

mussels by black plastic. From Figure 5b it becomes clear that indeed the mussels kept in the dark showed a more constant valve opening regime than the mussels exposed to sunlight. These appeared more closed during the day, and further open during night.

As a consequence, we believe that flow-through systems should be protected against direct light, either from artificial light sources or from the sun. Using opaque tanks and covers proved to be effective. For the in situ instrument the light absorptive effect of the water column will usually be sufficient to minimize the effects of light.

3.2.3. Temperature

Temperature is of major influence in many biological processes, whether it is an increase or decrease, and for bivalves this is not different (Gabbott and Baine, 1973; Spidle et al., 1995; Rajagopal et al., 1997). In general the activity will increase with an increase in temperature, as recorded for the filtration rate (Hinz and Scheil, 1972) or respiration rate (Schulz-Baldes, 1982). The valve movement activity (frequency) increases with temperature, as recorded for *Anodonta cygnea* (Salánki and Varanka, 1976) and for *Lamellidans corrianus* (Samant and Agarwal, 1977). Bivalves do react (closure) to drastic changes in temperature, as in the case of effluents, but for the monitoring of drinking water or river water the large heat capacity of the medium will prevent rapid changes. The Musselmonitor records data on the water temperature together with the valve movement data.

Seasonal differences in water temperature are reflected in the physiology of the organisms, and in wintertime at reduced temperature, the activity is expected to decrease. Our first fear that bivalves could not be used at temperatures below 10 °C, and hence there was no job for the Musselmonitor in wintertime, proved to be not justified. Indeed, mussels slow down when the temperature is lowered (e.g. observed for *Anodonta* by Salánki and Varanka, 1976), but they will still react to the sudden occurrence of a stress factor.

A change in the valve movement behavior as a function of temperature was coincidentally demonstrated during an experiment in a climate room. A lower temperature was desired for some other experiment, and the climate room was reprogramed from 18° to 8 °C). It took 10 h for the temperature to reach this lower level. During this forced decrease in temperature event, 8 mussels (*Mytilus edulis*) were contained in a tank. The valve opening dropped from the normal 80–90% towards about 65%, thus the span 0–100% changed over the period. This indicates that the maximum (absolute) values recorded at e.g. 20 °C will not be relevant at the lower temperatures. This has obvious implications for the setting of the various alarm criteria, and therefore a function had to be built in the Musselmonitor to update the span at regular intervals. Only then the criteria are correctly applied when there is a (seasonal) change in temperature. The closure for several days of zebra mussels as reported for the Dreissena Monitor (Anon., 1998) could well be due to this effect.

3.2.4. Vibration

Many BEWS, especially those that use fish as sensor organism, are sensitive to disturbances by vibration. The slamming of a door may easily cause a reaction. Bivalves are not different in this respect, but some interesting observations can be made.

A sudden vibration will cause a direct closure of the shells, usually of all mussels. However, when the vibration is only a short pulse, the bivalves will open almost immediately, and in most cases no alarm will be generated because the closure will not last sufficiently long. We have seen many data files where the entry of staff caused this type of "spikes" in the data (see Figure 4d for an example). The effect is, not surprisingly, often observed in the period following installation. When the system is new, staff wants to peek inside the tank to see how the mussels are, and many visitors will be shown the new tool. A system that was placed in a mobile housing showed this reaction every morning the staff came in to inspect the system or to exchange diskettes; the vibration transmitted through the wooden floor caused a closure reaction. Once the tank was placed on a firm support, the closure reaction no longer occurred.

Continuous vibration is something different. The bivalves "learn" that the vibration is not the forebode of danger, and they will slowly tend to open, and to continue with their normal behavior. In Figure 4e, the closure reaction was caused by people moving around and continuously causing vibration in a mobile laboratory. The zebra mussels do react but after some time they open, despite the continued activities in the room. In this case an alarm was generated as the adaptation period was in the order of hours rather than minutes. For example, drinking water industries prefer a BEWS near their water inlet. This is often the location where large pumps may generate vibration. We found that the often deafening noise of such equipment had no effect on the mussels employed in the monitor.

In conclusion, vibration is not considered a problem in the functioning of the Musselmonitor, provided abrupt disturbances are prevented.

3.2.5. Diurnal Variation

A number of investigators report on the occurrence of a biorhythm, such as a diurnal or day/night variations. For example, Brown et al. (1956) observed diurnal effects for oyster and quahog. Salánki (1964) discussed the effect for *Anodonta*, Benedens and Hinz (1980) for *Sphaerium*, and Ameyaw-Akumfi and Naylor (1987) for *Mytilus*.

The daily rhythm was discussed at a meeting of the "Dreissena Monitor" users. The diurnal variation was observed both for *Dreissena polymorpha* and for *Corbicula fluminea*, especially during summer and early spring; minima (highest number of mussels closed) seemed to occur early in the morning, the maxima (highest number considered open) late afternoon. In winter, no diurnal variation could be observed (Anon., 1998).

Interestingly, our experience has been that we, using the Musselmonitor (in situ version), hardly saw such diurnal behavior, using either zebra or blue mussel. It may be that specific surface water bodies induce such behavior, but it may also be the result of interferences from the system layout (effects caused by staff entering the station: lights on, vibration) or ambient daylight reaching the organisms (see section 3.2.2).

Since the effect is observed by several research groups, it remains a matter of concern. In cases where the diurnal rhythm occurs, the SMW data evaluation method described by Sluyts et al. (1995) may assist to decouple the effects, allowing a better evaluation and functioning of the BEWS.

3.2.6. Food Supply

For an in situ instrument it is essential that the organisms feed upon the material that is present in the water. Most bivalves are, at least in part, filter (or suspension) feeders, and thus rely for their food supply on the particulate matter (seston). In waters where no food is available, as e.g. in ground water, or water from a storage tank under laboratory testing conditions, additional feeding may be required.

For short term laboratory tests, such as for toxicological research, it will be easier not to feed the mussels, as they will show no effect within the first few days. For continuous monitoring this will not be an option, and the addition of a suspension of algae proves to be effective. We used successfully a continuous culture of *Phaeodactilum* spp. The system consisted of a bottle of medium which was connected to the culture via a peristaltic pump; the same pump was used via a second channel to transport the algal suspension to the exposure tank with the mussels. The refill of the medium bottle is a simple task that does not require a specialized skill.

When the food supply was switched off, it was noticed that the mussels detected this immediately. In order to evaluate the effect of intermittent feeding some tests were performed in a flow-through laboratory situation, employing blue mussels. In Figure 6a on the left side a typical "normal" behavior is displayed of mussels that have an abundant and constant food supply. If the feeding was stopped via a remote-controlled operation, almost immediately the mussels changed their behavior and the normally open shells were more closed than before (a reaction of 7 out of 7 mussels).

This new pattern (see also Figure 6b) is typical for mussels that find insufficient food in the water column for longer periods: they slow down the process of filtration and only once and again they test the water for availability of food, a matter of energy saving (Sprung and Rose, 1988; Bayne et al., 1993; Borcherding, 1995). We also observed this behavior in few field situations where no or insufficient food was available, e.g. in fast running streams where phytobenthos replaces phytoplankton. That the sensing for food abundance is rather continuous is reflected by

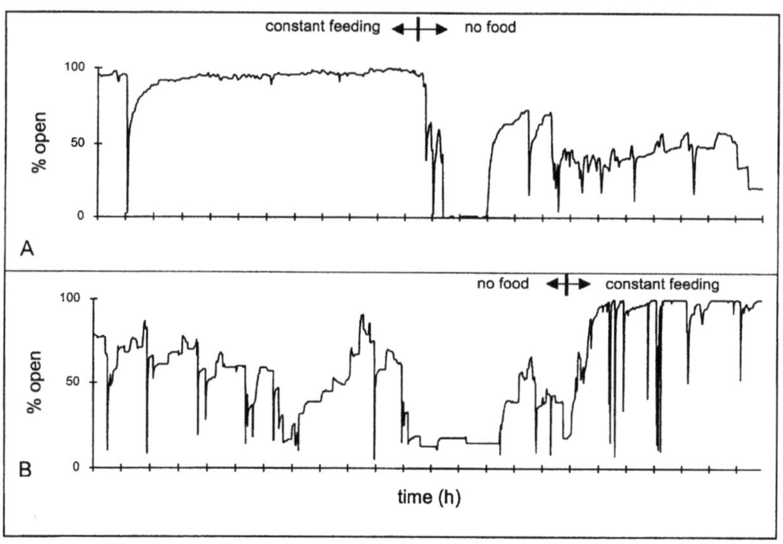

FIGURE 6. The effect of (interrupted) feeding with an algal suspension (*Phaeodactilum* spp.) in a laboratory exposure (see Figure 8) of the blue mussel: A) after days of constant feeding, the addition is stopped and B) turned on again. In between the two events, a typical "low food" pattern is recorded.

the return to a normal valve behavior the moment when the feeding is turned on again (Figure 6b). Although the blue mussels in this experiment may have been adapted to the feeding strategy ("trained"), this experiment stresses the need for a feeding pattern that is continuous in nature. Interrupted feeding will interfere with the valve movement response and hamper evaluation of alarm conditions. In most surface waters, however, the food supply is more or less continuous, and no problems should be expected. Well fed mussels will not show a starvation effect during the first 1–2 days, allowing their use in short term (laboratory) experiments without the addition of particulate matter (food).

3.2.7. Suspended Particulate Matter

Seston, or suspended particulate matter is composed of an organic and an inorganic fraction. The first consists of living and dead plankton cells, the second fraction of silt and clay particles. Although many inorganic particles have an organic coating, they will have a relatively poor nutritional value for bivalves.

In general the zebra mussel and the blue mussel can handle very well a large range of suspended matter. In Figure 7a an example is given of the fraction of time mussels (8 mussels, *Dreissena polymorpha*, averaged) are open (defined as >20%) as function of seston added to the exposure tank (mesocosm), concentrations ranging from 1 to 210 mg/l. The seston consisted of a variable mixture of natural

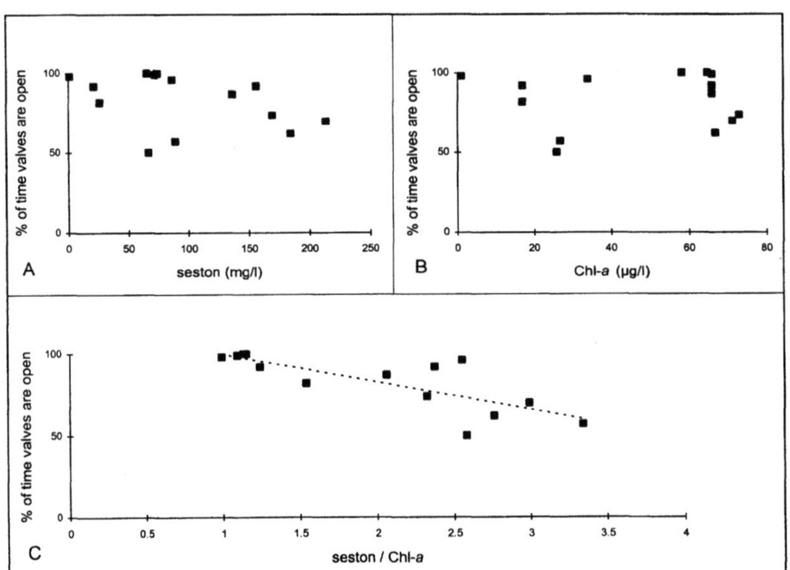

FIGURE 7. The effect of A) suspended matter and B) chlorophyll *a* concentrations upon the percentage of time the mussels are open (defined as >20%); C) only the ratio of seston/chlorophyll *a* (× 0.001) shows a distinct downward trend.

surface sediment from a clean lake and algae which was kept in suspension using a pump. There is no obvious change in behavior from very low to very high loads of suspended matter, or from its chlorophyll content (Figure 7b).

However, the ratio seston/chlorophyll showed a distinct trend, a reduced time the valves remain open with higher values. We believe that this demonstrates an energy-saving mechanism: if the seston is abundant, but the nutritional content low, it takes too much energy to get the organic matter sorted out.

In a study for a drinking water company, we faced the problem that despite the seston content was relatively high for natural conditions (average 20 mg/l, maximum 37 mg/l), the mussels (*D. polymorpha*) showed a distinct behavior of food limitation, often leading to long periods of closure (as the non-food conditions in Figure 6). Reduction of the seston load by placing a lamella filter, did not improve the situation. It appeared that the nutritional value of the seston was rather low (average 2.9 μg/l chlorophyll). This was found due to the fact that it mainly consisted of sediment that was resuspended by ship's movements and hence originated from deeper sediment layers.

3.2.8. Bio-Fouling

Once a biological problem occurred during a monitoring exercise of the river Rhine. In the summer of 1993 there was an outbreak of freshwater sponges, and serious fouling of the protective cover of the in situ Musselmonitor occurred. This fouling was that serious that the free access of the river water to the mussels was almost blocked, and the system had to be cleaned weekly during the event (de Zwart, pers. comm.).

3.2.9. Magnetic Fields

As part of the measurement, the sensors emit/receive a small magnetic field. We have never observed that this magnetic field would affect the valve movement behavior.

External magnetic fields may interfere with the measurement, however. In a laboratory experiment we once had unexplained sudden shifts in the output signal of the monitor. Its cause proved to be a large laboratory electrical oven used for ashing of organic matter, which was located in the next room(!). Even at a distance of about 6 m, the magnetic field generated by the coils in the oven could affect the sensitive measurement. Obviously, this problem does not occur when the system is applied directly in the field.

3.3. Examples of Applications

From the numerous experiments carried out with the Musselmonitor over the past decade, we already reported some example cases (e.g. de Zwart et al., 1995).

The overview of our experiences would not be complete without some additional case studies.

3.3.1. Laboratory Testing for Organic Compounds

In order to test the feasibility of application of the Musselmonitor as early warning system near an industrial effluent in the river Ems estuary (Netherlands), the sensitivity for three organic compounds was tested in a laboratory environment in 1992. The compounds were two commonly used solvents (dichloromethane, tetrachloromethane) and formaldehyde.

In toxicological experiments, where the lowest observed effect concentration (LOEC) is determined, one may slowly increase the concentration of the test compound until an effect is observed, as was demonstrated in de Zwart et al. (1995). As the objective of this study was to mimic the possible detection of an accidental spill, the test compounds were added as spikes. An effective mixing of the water ensured that the test concentration was reached almost instantly. The scheme of the flow through system is presented in Figure 8. The exposure tank was 10.7 L in size, the flow 12 L/h and the temperature 15 ± 2 °C. The common or blue mussel (*Mytilus edulis*) was used as test organism (8 individuals). Care was taken that the aeration did not take place in the exposure tank as it would strip part of the solvents from the water. The exposure vessel was blocked from light and in order to minimize vibration the test compounds were added by remote control during the night. No food was added. This all to prevent disturbances which could cause "false" results. For the testing of each compound a new set of mussels was installed. To allow recuperation of the organisms, additions were performed at daily intervals, starting with the lowest concentrations; in between experiments the tank was flushed with clean sea water.

The results of the study are given in Table 2. The following LOECs were observed (identified as: 4 out of 8 mussels react): tetrachloromethane (2.5 mg/L), dichloromethane (150 mg/L) and formaldehyde (10 mg/L), values that are in the same range as found chloroform, toluene, and trichloroethylene using the zebra mussel as test organism (de Zwart et al., 1995).

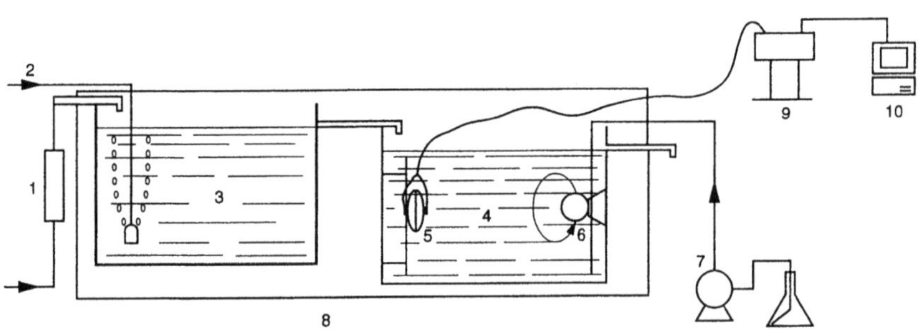

FIGURE 8. Schematic presentation of the laboratory exposure layout. 1) seawater line in with flow meter; 2) aeration; 3) tank 1; 4) tank 2; 5) mussels (8); 6) circulation pump; 7) remotely controlled peristaltic pump for addition of toxicants; 8) light tight container; 9) Musselmonitor®; 10) PC.

TABLE 2. Results for a dose-effect relation for the valve movement response of *Mytilus edulis* for three organic compounds (n = number of mussels).

compound	concentration (mg/L)	n positive response	n total number of mussels
tetrachloromethane			
	123	7	8
	25	7	7
	2.5	4	8
	2.5	3	7
	0.5	1	6
dichloromethane			
	400	7	8
	300	7	8
	200	8	8
	150	6	8
	100	3	8
	50	2	8
formaldehyde			
	10	8	8
	2	2	8

3.3.2. Monitoring of Drinking Water Intake

Near Diksmuide (Belgium) the drinking water company VMW has built an artificial storage basin. This basin is fed with river IJzer water. In order to test the feasibility of the Musselmonitor as BEWS at the river water intake, the system was operated on site for a period of 10 weeks (March–May 1991). To allow addition of toxic compounds (copper, ammonium chloride, chlorine), the in situ monitor was introduced in a polythene tank (200 L) with a flow through regime of 200 L/h). The zebra mussel was used as test organism. Of the 8 mussels, 7 completed the entire test period, one died after 3 weeks.

During the test period several alarm situations were observed, partly due to the intentional addition of toxicants. However, several very clear alarm situations were detected during the period which were due to the river water quality. As water sampling/analysis was not conducted at that moment, there is no indication of the nature of the causes. That several events occurred during the Friday or Saturday night may point in the direction of (illegal) spills. An example of an alarm event is presented in Figure 9. Of the 7 surviving zebra mussels, 6 clearly reacted, leading to a distinct drop in the averaged signal.

3.3.3. Monitoring of River Water

Over the past decade a number of performance tests have been carried out in the rivers Rhine and Meuse. Several typical river water monitoring examples with

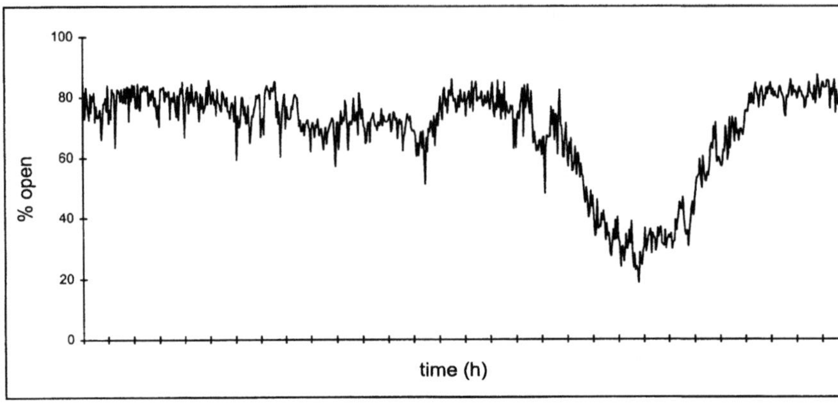

FIGURE 9. Field test of a drinking water inlet using the valve movement behavior of zebra mussels (averaged signal of 7 individuals), showing a distinct alarm situation.

the Musselmonitor® have been presented earlier (Jenner et al., 1991; de Zwart et al., 1995; and Irmer, 1994 for German river systems).

3.3.4. The Musselmonitor in Chemical Speciation Research for Cu

In natural seawater the free Cu^{2+} concentration is estimated to be in the order of 10^{-12} to 10^{-11} M. Adopting the expression $pCu = -\log[Cu^{2+}]$, thus a pCu of 11–12 (Sunda et al., 1990). In experiments where Cu-buffer solutions were used, constant free copper ion concentrations were maintained, a relation could be identified between this $[Cu^{2+}]$ and observed biological effects (Zamuda and Sunda, 1982; Sanders et al., 1983).

Despite the apparently low free copper concentration in natural waters, there is evidence that an increase by one or two orders of magnitude already leads to toxic effects in various groups of organisms, including cyano-bacteria (Moffet and Brand, 1996), phytoplankton (Sunda and Guillard, 1976; Gavis et al., 1981), zooplankton (Sunda et al., 1987; 1990), and several benthic species: the uptake of copper by bivalves was found to be directly related to the pCu (Creselius et al., 1982; Zamuda and Sunda, 1982) and not to the total Cu concentration (which is usually the basis for norms and quality criteria). There is an urgent need for methods that link the chemical speciation of copper (such as the determination of $[Cu^{2+}]$), or the "bio-available fraction", to biological effect studies (Allen and Hansen, 1996; Renner, 1997).

The study of a physiological parameter as indicator for possible sub-lethal toxic effects has the advantage that the reaction time is short, allowing a direct and fast coupling between chemical speciation and biological effects. We used the valve-movement response of the *M. edulis* in relation to changes in the pCu. The experimental layout as depicted in Figure 8 was used for the experiments. The Cu-buffer system consisted of a Cu-NTA buffer, based on chemical speciation calculations using the program "Titrator" (Cabaniss, 1987). In contrast to Cu-EDTA, Cu-NTA complex formation is relatively fast, allowing additions to be linked directly to the (non)-occurrence of a biological effect.

Seven mussels were allowed to become accustomed to the flow-through system and the seawater used. After the water flow was stopped, NTA was added to the exposure tank by remote operation. (spiking events are indicated by arrows in Figure 10), resulting in a concentration of 6×10^{-7}M/ NTA. No reaction was observed from the addition of NTA. Following this, Cu was added to increase step wise the pCu (equilibrium) concentration by spikes of Cu solution. Rapid mixing of the seawater allowed fast reaction times of the Cu-NTA complex, and the establishment of the pCu as predicted by the speciation modeling program: pCu = 11.8, 10.8, and 10.0, subsequently. In order to allow the mussels to recuperate, a final surplus of NTA reduced the pCu to >12. The valve movement was used as a signal for observed biological effects.

The results of this pCu speciation cause-effect experiment are provided in Figure 10 as the averaged signal. The first increase of free copper (to pCu 11.8) hardly had any effect; a small decrease is observed, but the average valve opening remains about 80%. This changes drastically after spike 3 (pCu = 10.8). All 7 mussels react by closing the shells. A slow recovery can be observed, but one hour later (after spike 4, pCu = 10) an even further closure can be seen. The recovery, shown by opening of all mussels rapidly follows the addition of an excess of NTA.

The conclusion that mussels react to a [Cu^{2+}] in the order of pCu = 10.8 confirms the observations for other groups of organisms (references above). Because of the fast response of the bivalves, the valve movement behavior in this type of experiment has been shown to be a very effective tool for the study of the relation between chemical speciation and biological effects.

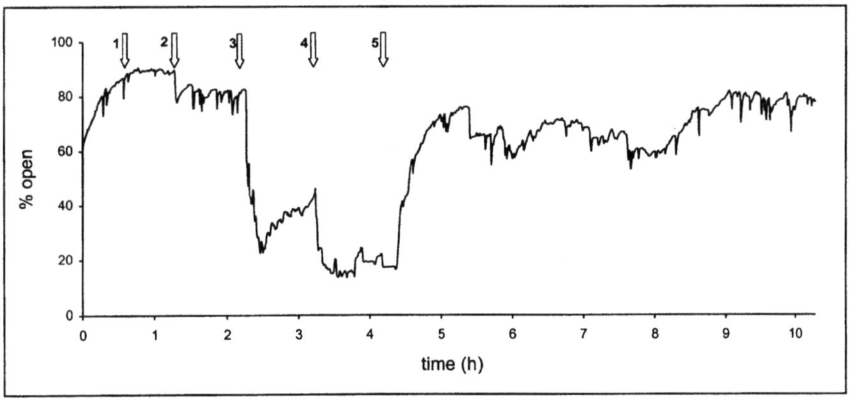

FIGURE 10. The reaction of the blue mussel to changing free copper ([Cu^{2+}]) concentrations in a controlled laboratory experiment (average signal of 6 mussels). 1) addition of NTA, resulting in 6×10^{-7}M NTA; 2) addition of Cu, leading to a (calculated equilibrium) pCu = 11.8; 3) Cu spike leading to pCu = 10.8; 4) Cu spike leading to pCu = 10.0; 5) addition of surplus NTA (1.4×10^{-6}M) to reduce the free Cu to pCu >12.

4. CONCLUSIONS

One of the more successful biological early warning systems (BEWS) is the Musselmonitor®. This monitor is based on the valve movement behavior of bivalves (mussels, clams, oysters) from freshwater or marine environments. When a change in behavior exceeds a predefined criterion, an alarm is generated.

In several applications apparent changes in the behavior of mussels have been encountered, that were obviously due to the specific application rather than the functioning of the instrument. Factors like adaptation, light, temperature, vibration, diurnal variation, suspended particulate matter and food supply may interfere with the measurements, but it is argued that with proper precautions a number of these interferences can be eliminated.

The Musselmonitor® has been successfully used for over a decade, often in very different environments and using a variety of bivalve organisms and study objectives. In this chapter, typical examples demonstrate the versatility of the instrument in laboratory and field applications, e.g., to determine the sensitivity to organic solvents, to monitor a drinking water inlet, and to test the sensitivity for cupric ion activity [Cu^{2+}] in chemical speciation studies.

Installation of the system requires only half a day, instruction how to operate it another one day. Tuning of the settings to the local situation and needs (also in terms of sensitivity, avoidance of false alarms) may require some additional time. Since the sensing organisms need replacement at relatively large intervals, maintenance is minimized, making the monitor rather cost-effective in operation. One should, however, never forget that the successful application of this (or of any other!) BEWS is largely determined by the understanding that the instrument is based on biological functioning. This may be difficult to accept for technically oriented engineers, but it is part of the system that allows us to detect almost instantly a multitude of toxic compounds in water bodies, effluents and drinking water inlets.

5. ACKNOWLEDGMENTS

Over the past decade many were involved in the development, research and applications of the Musselmonitor®. Thanks are due to students and colleagues, especially Johan Verburgh, Tim Sikking, Nándor Oertel, and Jaqueline Tamis. The cooperation with the co-designers of the system, Dick de Zwart (RIVM), Henk Jenner (KEMA), and the staff of Delta Consult is gratefully acknowledged.

6. REFERENCES

Allen, H.E., and D.J. Hansen, 1996. The importance of trace metal speciation to water quality criteria. Wat. Environ. Res. 68: 42–54.

Allen, H.J., W.T. Waller, M.F. Acevedo, E.L. Morgan, K.L. Dickson, and J.H. Kennedy, 1996. A minimally invasive technique to monitor valve-movement behavior in bivalves. Environ. Technol. 17: 501–507.

Ameyaw-Akumfi, C., and E. Naylor, 1987. Temporal patterns of shell-gape in *Mytilus edulis*. Mar. Biol. 95: 237–242.

Anon, 1994. Le Valvomètre. Leaflet, IFREMER/Micrel, Plouzane, France, p. 2.

Anon., 1998. Protokoll. In: Proceedings: Anwandertreffen "Dreissena-Monitor", 27–28 January 1998, LUA-NRW, Dusseldorf (D), p. 48.

Arnold, M.A., and J. Wangsa, 1991. Transduce based and intrinsic biosensors. In: Fiber optic chemical sensors and biosensors, Vol. II; O.S. Wolfbeis (ed). CRC Press, Boca Raton.

Baldwin, I.G., and K.J.M. Kramer, 1994. Biological early warning systems (BEWS). In: Biomonitoring of coastal waters and estuaries, K.J.M. Kramer (ed). CRC Press, Boca Raton, pp. 1–28.

Bayne, B.L., J.I.P. Iglesias, A.J.S. Hawkins, E. Navarro, M. Heral, and J.M. Deslous-Paoli, 1993. Feeding behaviour of the mussel, *Mytilus edulis*—responses to variations in quantity and organic content of the seston. J. Mar. Biol. Assn. UK 73: 813–829.

Benedens, H., and W. Hinz, 1980. Day time periodicity of the filtration performance of *Dreissena polymorpha* and *Sphaerium corneum* Bivalvia. Hydrobiol. 69: 45–48.

Bennett, M.F., 1954. The rhythmic activity of the quahog, *Venus mercenaria* and its modification by light. Biol. Bull. Mar. Biol. Lab. 107: 174–191.

Borcherding, J., 1992. Another early warning system for the detection of toxic discharges in the aquatic environment based on valve movements of the freshwater mussel *Dreissena polymorpha*. In: Zebra mussel *Dreissena polymorpha*, D. Neumann, and H.A. Jenner, (eds), Gustav Fischer Verlag, Stuttgart, pp. 127–146.

Borcherding, J., 1995. Laboratory experiments on the influence of food availability, temperature and photoperiod on gonad development in the freshwater mussel Dreissena polymorpha. Malacologia 36: 15–27.

Borcherding, J., and M. Volpers, 1994. The "Dreissena-Monitor"—First results on the application of this biological early warning system in the continuous monitoring of water quality. Wat. Sci. Technol. 29: 199–201.

Brinkman, U.A.T., 1995. On-line monitoring of aquatic samples. Environ. Sci. Technol. 29: A79–A84.

Brown, F.A. Jr., M.F. Bennett, H.M. Webb, and C.L. Ralph, 1956. Persistent daily, monthly and 27-day cycles of activity in the oyster and quahog. J. Exp. Zool. 131: 235–262.

Cabaniss, S.E., 1987. TITRATOR: an interactive program for aquatic equilibrium calculations. Environm. Sci. Technol. 21: 209–210.

Colin, F., and P. Quevauviller (eds), 1998. Monitoring of water quality. Elsevier, Amsterdam, p. 269.

Creselius, E.A., J.T. Hardy, C.I. Gibson, R.L. Schmidt, C.W. Apts, J.M. Gurtisen, and S.P. Joyce, 1982. Copper bioavailability to marine bivalves and shrimp: relationship to cupric ion activity. Mar. Environ. Res. 6: 13–26.

de Zwart, D., and W. Slooff, 1987. Continuous effluent biomonitoring with an early warning system. Bengston, Norberg-King, and Mount (Eds). Effluent and ambient toxicity testing in The Göta Älv and Viskan Rivers, Sweden. Naturvardsverket report 3275.

de Zwart, D., K.J.M. Kramer, and H.A. Jenner, 1995. Practical experiences with the biological early warning system "Mosselmonitor". Environ. Toxicol. Water Qual. 10: 237–247.

Diamond, J.M., and T.R. Shedd, 1995. Guide for ventilatory behavioral toxicology testing of freshwater fish. American Assoc. Testing Materials, ASTM draft guide, Philadelphia, p. 14.

Eertman, R.H.M., and A. de Zwaan, 1994. Survival of the fittest: resistance of mussels to aerial exposure. In: Biomonitoring of coastal waters and estuaries, K.J.M. Kramer (ed). CRC Press, Boca Raton, pp. 269–284.

Englund, V., and M. Heino, 1994. Valve movement of *Anodonta anatina* and *Unio tumidus* (bivalvia, unionidae) in a eutrophic lake. Ann. Zool. Fennici 31: 257–262.

Englund, V.P.M., M.P. Heino, and G. Melas, 1994. Field method for monitoring valve movements of bivalved molluscs. Wat. Res. 28: 2219–2221.

Gabbott, P.A., and B.L. Bayne, 1973. Biochemical effects of temperature and nutritive stress on *Mytilus edulis*. J. Mar. Biol. Ass. UK 53: 269–286.

Gavis, J., R.R.L. Guillard, and B.L. Woodward, 1981. Cupric ion activity and the growth of phytoplankton clones isolated from different marine environments. J. Mar. Res. 39: 315–333.

Gerhardt, A., 1996. Behavioural early warning responses to polluted water—performance of *Gammarus pulex* L (crustacea) and *Hydropsyche angustipennis* (Curtis) (insecta) to a complex industrial effluent. Environ. Sci. Pollut. Res. 3: 63–70.

Gruber, D.S., and J.M. Diamond (eds), 1988. Automated biomonitoring: living sensors as environmental monitors. Ellis Horwood, Chichester, p. 208.

Gruber, D., C.H. Frago, and W.J. Rasnake, 1994. Automated biomonitors—first line of defence. J. Aquat. Ecosyst. Health, 3: 87–92.

Ham, K.D., and M.J. Peterson, 1994. Effect of fluctuating low-level chlorine concentrations on valve-movement behavior of the Asiatic clam (*Corbicula fluminea*). Environ. Toxicol. Chem. 13: 493–498.

Hinz, W., and H.G. Scheil, 1972. Filtration rate of *Dreissena* and *Pisidium* Eulamellibranchiata. Oecologica, 11: 45–54.

Hoffmann, M., E. Blübaum-Gronau, and F. Krebs, 1994. Die Schalenbewegungen von Muscheln als Indicator von Schadstoffen in der Gewässerüberwachung. Schr.-Reihe Verein WaBoLu 93. Gustav Fischer Verlag, Stuttgart.

Irmer, U. (ed), 1994. Continuous biotests for water monitoring of the river Rhine. Summary, recommendations, description of test methods.. Umweltbundesamt Texte 58/94, Berlin, p. 30.

Jenner, H.A., F. Noppert, and T. Sikking, 1989. A new system for the detection of valve movement response of bivalves. Kema Sci. Techn. Rep. 7: 91–98.

Jenner, H.A., D. de Zwart, and K.J.M. Kramer, 1991. Monitoring water quality with bivalves. Poster presentation. Proceedings of the conference "Bioindikation: Ein wirksames Instrument der Umweltkontrolle", Sept. 24–26, 1991, Vienna.

Jenner, H.A., G.H.F.M. Van Aerssen, and J. Terwoert, 1992. Valve movement behaviour of the mussel *Dreissena polymorpha* and the clam *Unio pictorum* for use in an early warning system. In: Zebra mussel *Dreissena polymorpha*, D. Neumann, and H.A. Jenner, (eds), Gustav Fischer Verlag, Stuttgart, pp. 115–126.

Kramer, K.J.M., H.A. Jenner, and D. de Zwart, 1989. The valve movement response of mussels: A tool in biological monitoring. Hydrobiol. 188/189: 433–443.

Kramer, K.J.M., and J. Botterweg, 1991. Aquatic biological early warning systems: An overview. In: Bioindicators and environmental management. D.W. Jeffrey, and B. Madden (eds), Academic Press, London, pp. 95–126.

Manley, A.R., and J. Davenport, 1979. Behavioural responses of some marine bivalves to heightened seawater copper concentrations. Bull. Environ. Contam. Toxicol. 22: 739–744.

Moffet, J.W., and L.E. Brand, 1996. Production of strong, extracellular Cu chelators by marine cyanobacteria in response to Cu stress. Limnol. Oceanogr. 41: 388–395.

Mouabad, A., and J.C. Pihan, 1992. The pumping behaviour response of *Dreissena polymorpha* to pollutants—a method for toxicity screening. In: Zebra mussel *Dreissena polymorpha*, Neumann, D., and H.A. Jenner (eds). Gustav Fischer Verlag, Stuttgart, pp. 147–154.

O'Connor, T.P., A.Y. Cantillo, and G.G. Lauenstein, 1994. Monitoring of temporal trends in chemical contamination by NOAA National Status and Trends Mussel Watch Project. In: Biomonitoring of coastal waters and estuaries, K.J.M. Kramer (ed). CRC Press, Boca Raton, pp. 29–50.

Rajagopal, S., G. van der Velde, and H.A. Jenner, 1997. Response of zebra mussel, *Dreissena polymorpha*, to elevated temperatures in The Netherlands. In: Zebra mussels and aquatic nuisance species, D'Itri, F.M. (ed); Ann Arbor Press, Chelsea MI, pp. 257–273.

Renner, R., 1997. Rethinking water quality standards for metals toxicity. Environ. Sci. Technol. 31: 466–468.

Salánki, J., and L. Viránka, 1976. Effect of copper and lead compounds on the activity of the fresh-water mussel. Annal. Biol. Tihany, 43: 21–27.

Salánki, J., and L. Varánka, 1978. Effect of some insecticides on the periodic activity of the fresh-water mussel (*Anodonta cygnea* L). Acta. Biol. Acad. Sci. Hung. 29: 173–180.

Salánki, J., 1964. Contributions to the problem of daily rhythm in the activity of the fresh water mussel *Anodonta cygnea* L. Annal. Biol. Tihany, 31: 109–116.

Samant, S., and R.A. Agarwal, 1978. Effect of some environmental factors on survival and activity of fresh water bivalve *Lamellidens corrianus*. Indian J. Exp. Biol. 16: 26–28.

Sanders, B.M., K.D. Jenkins, W.G. Sunda, and J.D. Costlow, 1983. Free cupric ion activity in seawater: effects on metallothionein and growth in crab larvae. Science 222: 53–55.

Schulz-Baldes, M., 1982. Tiere als Monitoringorganismen für Schwermetalle im Meer: Ein Überblick. Decheniana Beihefte (Bonn), 26: 43–45.

Schuring, B.J., and M.J. Geense, 1972. Een electronische schakeling voor het registreren van openingshoek van de mossel Mytilus edulis L. TNO-Rapport CL 72/47; TNO, Delft.

Scully, P., 1998. Optical techniques for water quality monitoring. In: Monitoring of water quality, F. Colin, and P. Quevauviller (eds). Elsevier, Amsterdam, pp. 15–35.

Slooff, W., D. de Zwart, and J.M. Marquenie, 1983. Detection limits of a biological monitoring system for chemical water pollution based on mussel activity. Bull. Environm. Contam. Toxicol. 30: 400–405.

Sluyts, H., F. van Hoof, A. Cornet, and J. Paulussen, 1995. A dynamic new alarm system for use in biological early warning systems. Environ. Toxicol. Chem. 15: 1317–1323.

Spindle, A.P., E.L. Mills, and B. May, 1995. Limits to tolerance of temperature and salinity in the quagga mussel (*Dreissena bugensis*) and the zebra mussel (*Dreissena polymorpha*). Can. J. Fish. Aquat. Sci. 52: 2108–2119.

Sprung, M., and U. Rose, 1988. Influence of food size and food quantity on the feeding of the mussel *Dreissena polymorpha*. Oecologia, 77: 526–532.

Sunda, W.G., P.A. Tester, and S.A. Huntsman, 1987. Effects of cupric and zinc ion activities on the survival and reproduction of marine copepods. Mar. Biol. 94: 203–210.

Sunda, W., and R.R.L. Guillard, 1976. The relationship between cupric ion activity and the toxicity of copper to phytoplankton. J. Mar. Res. 34: 511–529.

Sunda, W.G., P.A. Tester, and S.A. Huntsman, 1990. Toxicity of trace metals to *Acartia tonsa* in the Elizabeth river and southern Chesapeake Bay. Estuar. Coast. Shelf. Sci. 30: 207–221.

Von Danwitz, B., E. Blübaum-Gronau, P. Diehl, K.-W. Digel, V. Herbst, L. Höhne, F. Krebs, L. Küchler, M. Marten, M. Pfeiffer, and B. Rechenberg, 1998. Recommendation on the deployment of continuous biomonitors for the monitoring of surface waters; Report Landerarbeitsgemeinschaft Wasser (LAWA), Berlin, p. 46.

Zamuda, C.D., and W.G. Sunda, 1982. Bioavailability of dissolved copper to the American oyster *Crassostrea virginica*. I. Importance of chemical speciation. Mar. Biol. 66: 77–82.

5

QUANTITATIVE BEHAVIOR ANALYSIS—A NEW APPROACH TO THE CHALLENGES OF ENVIRONMENTAL TOXICOLOGY

O. HUNRICH SPIESER

GSF-National Research Center for Environment and Health, Institute of Toxicology, D-85758 Neuherberg, Germany

Serious degradative changes in ecosystems are reported worldwide. Therefore, the main goal of environmental toxicology is to detect the reasons of those changes as early as possible—as far as they are in a causal connection with chemical substances. The first effects of chemical agents in an ecosystem are influences on physiological processes in some species. The first signs of inconspicuous effects on physiological processes can be detected by quantitative changes in behavior. In both cases, whether these effects are beneficial or adverse to a species, the interspecific equilibrium can be influenced. In the long-term a polluted ecosystem might drift to a new equilibrium or it might perish in the worst case.

Quantitative behavior analysis is very suitable as a bioindicator. A general introduction to behavioral monitoring is given in the chapter from Spieser et al. in this book. Nowadays, the most sophisticated methods of behavior analysis are made by image processing. It is an additional advantage that this method is non-invasive. The following three chapters deal with different kinds of experiments done with the system BehavioQuant® (patents: Spieser and Frisch, 1987, 1988, 1991, 1992, 1996).

The chapters by Baganz et al. and part of the one by Spieser et al. deal with the observation of long-term effects on fish caused by low-level concentrations of chemicals in standardized flow-through test systems. The substances were added to the test water (Baganz et al.: microcystin-LR; Spieser et al.: nonylphenol) or by injection (Spieser et al.: ethinylestradiol). The expected results are concentration-effect relationships of known chemicals. The chapter by Blübaum-Gronau et al. reports continuous water monitoring. Here, the presence of primary

Biomonitors and Biomarkers as Indicators of Environmental Change 2, Edited by Butterworth *et al.*
Kluwer Academic/Plenum Publishers, New York, 2000

unknown pollutants in the river water were detected by short-term effects on behavior.

Baganz et al. used adult zebrafish as the test species. The investigations showed that the cyanotoxin microcystin-LR, the product of cyanobacteria, is a stressor on fish. The most sensitive indicator was behavior. The locomotion activity (average swimming speed per two minutes) has been analyzed mainly chronobiologically.

Spieser et al. worked on the effect of two man-made chemicals with endocrine disrupting abilities, nonylphenol and ethinylestradiol. As the paper indicates on juvenile carp, the test conditions were different compared to the test conditions of Baganz et al. The behavior analysis has been made with the original BehavioQuant® software. Here a special technique is used to extract information, which was hidden in the object traces, by transferring selected behavior qualities to standard frequency distributions. This kind of behavior analysis had been proven to be very sensitive to chemicals (Steinberg et al., 1994; Spieser et al., 1994; Lorenz et al., 1995, Steinberg et al., 1995). As an example, the locomotion activity has been analyzed as standard distribution frequencies of the swimming speeds. It has been calculated in a time raster with a period length of 10 seconds, taken from the traces of each fish, distributed to 8 size classes. This characterizes also the inconstancy of swimming speed. Since these patterns of motility showed very characteristic diurnal changes in the controls during the whole exposure time of 70 days, a chronobiological effect has been proved. The statistical comparison of the standard frequency distributions had been made in a non-parametric way as it is described in the article of Blübaum-Gronau et al. in this book. In this way also other behavior parameters like swimming height and habitat preferences and even distance behavior can be analyzed. A further method of distance behavior calculation, which is derived from the method published in the patents of Spieser and Frisch and in the paper of Spieser and Yediler, 1986, is described in the chapter of Spieser et al. in this book. Further analysis of combined behavior data like the standard frequency distributions, also regarding the behavior development during long term exposure, can be made with the semiquantitative Hasse diagram technique (Lorenz et al., 1996). Grillitsch et al. (1999) and Vogl et al. (1999) worked on short-term sublethal toxicity tests with fish using BehavioQuant®.

The chapter by Blübaum-Gronau et al. deals with the adaptation of the BehavioQuant® technique to the monitoring of surface waters with a combination of a continuous fish test and a continuous daphnia test. It can be applied also for drinking water quality control and effluent monitoring. Blübaum-Gronau et al. report about the test chambers for fish, which had been developed especially for the "*Koblenz behavioral fish test with BehavioQuant®*", which makes continuous monitoring possible even in dirty river water. The daphnia are observed in special test chambers from the *metacom® multispecies monitor®*. The whole test system is described. The purpose of this application is to detect poisonous chemicals in the water as early as possible. Therefore, during continuous monitoring the data analysis as described before is done automatically for at least five behavior parameters. The respective data are statistically compared referring to five different behavior parameters with the behavior data of a previous reference time of 6 hours. If they differ from the measurements made before, the BehavioQuant® system assumes a possible exposure to contaminants. The BehavioQuant® has sophisticated alarm-generating software which minimizes the possibility of false alarms. The sensitivity of this test system has been proved by some reported experiments in which defined doses of chemicals were added to the river water during the automated observation. The test species were the golden ide (*Leuciscus idus melanotus*) and *Daphnia magna*.

The three given examples prove quantitative behavior analysis to be a very sensitive tool for bioindication. Since the BehavioQuant® method can be used with nearly all aquatic or terrestrial organisms which are able to move, the reader may get some ideas for his own research.

1. REFERENCES

Grillitsch, B., Vogl, C., and Wytek, R. (1999): Qualification of spontaneous unidirected locomotor behavior of fish for sublethal toxicity testing. Part II. Variability of measurement parameters under toxicant induced stress. Environmental Toxicology and Chemistry 18(12). 2743–2750.

Lorenz, R., Spieser, O.H., and Steinberg, C. (1995): Neue Wege in die Ökotoxikologie: Quantitative Verhaltensmessungen an Fischen als Toxizitätsendpunkt.—New ways to ecotoxicology: quantitative recording of behaviour of fish as Toxicity Endpoint. Acta hydrochimica et hydrobiologica. Acta hydrochim. hydrobiol. 23(5), 197–201.

Spieser, H., and Frisch, H. (1988): Method of measuring the types of motion and configuration of biological and non-biological objects. U.S. Patent 4,780,907. Oct. 25, 1988.

Spieser, H., and Frisch, H. (1986/1991): Procéde pour la mésure des modes de déplacement et des configurations d'objects, biologiques ou non, notament pour le déplacement des bancs de poissons. French Patent 8613588.

Spieser, H., and Frisch, H. (1996): Japanese Patent 2.109.008.

Spieser, H., and Frisch, H. (1992): Canadian Patent No 1294047.

Spieser, O.H., and Scholz, W. (1992): Verfahren zur quantitativen Bewegungsanalyse von mehreren Objekten im selben Medium. Deutsche Patentschrift P. 4224750.0.

Spieser, O.H., Scholz, W., Blübaum-Gronau, E., Hoffmann, M., Grillitsch, B., and Vogl, C. (1994): Das System BehavioQuant® zur Bioindikation anhand des Verhaltens von Fischen und von anderen aquatischen Organismen. in: *Eco-Informa-'94, Band 5, Umweltmonitoring und Bioindikation*. Hsg. K. Alef, H. Fiedler, O. Hutzinger. Umweltbundesamt Wien. Seite 429–448.

Spieser, O.H., and Yediler, A. (1986): *Empfindliche Parameter bei der Entwicklung von Langzeittests an Fischen*. Umweltforschungsplan des Bundes (10603030). Institut für Toxikologie der Gesellschaft für Strahlen- und Umweltforschung, München.

Steinberg, C.E.W., Mayr, C., Lorenz, R., Spieser, O.H., and Kettrup, A. (1994): Dissolved Humic material amplifies irritant effects of terbutylazine (Triacine herbicide) on fish. Naturwissenschaften 81: 225–227.

Lorenz, R., Brüggemann, R., Steinberg, C.E.W., and Spieser, O.H. (1996): Humic material changes effects of terbutylazine on behavior of zebrafish (Brachydanio rerio). Chemosphere 33(11), 2145–2158.

Steinberg, C., Lorenz, R., and Spieser, O.H. (1995): Effects of atrazine on swimming behaviour of zebrafish, *Brachydanio rerio*. Water Research 29, 981–985.

Vogl, C., Grillitsch, B., Wytek, R., Spieser, O.H., and Scholz, W. (1999): Qualification of spontaneous unidirected locomotor behavior of fish for sublethal toxicity testing. Part I. Variability of test parameters under general test conditions. Environmental Toxicology and Chemistry, 18(12). 2736–2742.

6

AN INTRODUCTION TO BEHAVIORAL MONITORING—EFFECTS OF NONYLPHENOL AND ETHINYLESTRADIOL ON SWIMMING BEHAVIOR OF JUVENILE CARP

O. HUNRICH SPIESER,[1] JULIA SCHWAIGER,[2] HERMANN FERLING,[2] AND ROLF-DIETER NEGELE[2]

[1]GSF-National Research Center for Environment and Health, Institute of Toxicology, D-85758 Neuherberg, Germany. [2]Bavarian State Office for Water Resources Management, Institute of Water Research, D-82407 Wielenbach, Germany. (Dedicated to the memory of Wilfried Scholz. Without his ingenious contributions and untiring enthusiasm the BehavioQuant® system described here would not exist)

1. BIOLOGICAL FUNCTIONS OF BEHAVIOR

Before entering the complex field of behavioral monitoring one has to consider biological functions of behavior. Behavior has to do with ecological adaptation of organisms to their environment, and with the intra- and extraspecific interactions of individuals. There are many different purposes of behavior, which is determined by evolution. The most organisms have innate behavior mechanisms which allow them to adapt to almost every situation occurring in their life and environment. The more subtle these mechanisms are differentiated, connected with a higher flexibility of reactions, the more the animal will be successful in its struggle for life, even under changing environmental conditions. These innate behavior mechanisms are permanent under evolutionary selection. The differences between "fixed action patterns" in the sense of Lorenz (1950) and "modes of action patterns" in the sense of Marler and Hamilton (1966) will not be discussed here. Barlow (1968) argues: "If the fixed action pattern is to be used to generate models of use to the neurophysiologist, we must examine it closely to see if it qualifies as a basic unit of behavior". One of his criteria to answer this question is to analyze if there is a variability in the pattern. The behavior patterns we can use for behavioral monitoring have all a certain range of variability. So we like to follow the suggestion of Barlow: "It might be wise to drop the term fixed from the action pattern."

Behavior can also be regarded as an expression of the inner state of an organism, since many behavior patterns have a signal function for the communication with conspecifics and also with members of other species.

Human beings, too, have to follow inborn behavior mechanisms, and also in man these mechanisms can be affected by chemical substances. For example, humans will keep an interindividual distance no less than 2 and no more than 5 feet if they meet in the open and communicate, unless they are lovers or enemies of each other. Those who doubt the existence of this phenomenon may conduct a very simple experiment: maintain a distance of less than 1 foot or more than 7 feet while speaking to a person: if the reason for such behavior has not been announced before, the other person will feel uncomfortable in either case. A "strange feeling" will be perceived at any time a person's behavior differs from subconscious expectations. However, this "strange feeling" will not be experienced by a person under the influence of ethanol: this chemical has the ability to modify the inborn respect for minimal distance.

Shoaling fish do not maintain interindividual distances as strictly as schooling fish do, but they keep distances. The main function of this behavior may be to protect against predators. They swim as a loose unit at distances allowing for the search of food or courtship display without interfering with each other. Appearance of a predator or any other alarming event will cause the fish to reduce interindividual distances. Evidently this behavioral mechanism must have a high adaptive value, since it can be observed with many species from vastly different phyla. Much speculation has been focused on the potential reasons how this might serve the preservation of species (see Shaw, 1970), yet despite the presence of extensive selectional pressure it might not contribute too much to the survival of the species, but it helps for the survival of the individual: A fish being chased by a predator is safest if it hides between a large number of its conspecifics, since this effectively lowers the individual probability of being eaten. This behavioral pattern can also be elicited by permanent exposure to chemicals, e.g. the pesticide atrazine, in a controlled experimental setting, although in the absence of any alarming event. Under these conditions the interindividual distances will be reduced in a concentration-dependent fashion for several days or even a week. Thereafter, the interindividual distances return to normal, probably due to habituation. At high concentrations of atrazine the fish will completely cease to take notice of each other.

In case of a lack of oxygen supply freshwater fish will move to deeper regions of the water and swim very slowly. This helps to save energy; in addition the water temperature there is normally lower and the content of oxygen higher. If the lack of oxygen becomes critical, the fish will move to the surface of the water, and some fish species will begin to move quickly. Movement of the surface will prompt dissolution of oxygen from the air into the water. Before the situation becomes critical, there is a very typical situation of ambivalent behavior to be observed: some fish are at the bottom, some at the surface. In the rest of the waterbody, there are no fish; with the only exception that some fish from time to time swim quickly and straight from the surface to the bottom or from the bottom to the surface. This adaptive reaction can also be observed under the influence of toxicants even if they have nothing to do with a lack of oxygen. One example would be the action of 3,4-dichloroaniline on zebrafish (*Brachydanio rerio*) which, at concentrations of 8 µg/L to 524 µg/l, will stay increasingly close to the bottom of waters, whereas at 2,048 µg/L they show the ambivalent behavior as described above. They partly prefer the bottom, but much more the surface of the waters, and there is nearly no presence in the rest of the aquarium (Spieser et al., 1994). This behavior had been registered automatically with the video processing apparatus BehavioQuant®, which is described in the second part of this chapter. Since the adaptive behavior of fish can be influenced temporarily by many external factors, it is very important, that this kind of observations are made under strictly defined experimental conditions and without any disturbance by the presence of persons.

With exposure to chemicals many species of fish will change their preferences for light or dark habitats if the bottom half and the adjacent sidewall of the test aquarium are underlayed with a white and the other side with a black background. The frontside and the backside of the aquarium are not underlayed in that way since the observation is done from the frontside of the aquarium. Zebrafish (*Brachydanio rerio* Hamilton-Buchanan) prefer the dark underlayed part of the test aquarium when exposed to atrazine (Steinberg et al., 1995) or simazine (Spieser et al., 1994) or terbutylazine (Lorenz et al., 1996) compared to the control groups.

The habitat selection of fish is influenced by water quality e.g. temperature, pH. Steinhörster et al. (1997) provided a two chamber aquarium with a lock, allowing fish to choose test conditions such as different pH values or ambient temperatures, or combinations of both. They used the BehavioQuant® test system in an arrangement monitoring both test chambers simultaneously with one camera. The behavior of two month-old nase fry (*Chondrostoma nasus* L.) had been assessed for both test chambers separately, with respect to frequency, duration of stay, swimming speed, and number of turnings. The fish preferred pH 9 and even pH 10 in comparison with pH 8 and temperatures of 16 °C/18 °C/19 °C in comparison with 10 °C/12 °C/13 °C. At pH 10 the swimming speed and the number of turns was reduced. These results are surprising in that one would expect pH 10 water to be more detrimental to fish then pH 8 water. Nase fry have a pronounced termophilous behavior covering the effect of pH-value. The authors concluded, that this may explain why nase fry are mostly found in shallow waters tolerating even bad environmental conditions such as high pH values or high oxygen supersaturation, although there are better conditions in adjacent areas. It can be assumed, that this way of behavior had been developed in an environmental situation before the man-made eutrophication of the rivers had been started.

Avoidance behavior and preference behavior are sensitive indicators for many chemicals and suitable to find lowest observed effect concentrations. But it cannot be judged that avoidance indicates that the tested chemical is harmful and preference indicates that the chemical is beneficial for the tested species. McNicol and

Scherer (1991) reported behavioral responses of the lake whitefish (*Coregonus clupeaformis*) to cadmium in concentrations from 0.2 to 256 µg/L during preference-avoidance testing. At over half of the concentrations, responses at any one concentration were dichotomous among individuals; whereas most fish showed avoidance, a significant number displayed preference. It is difficult to find an explanation for this counter intuitive result. We may comment on it that way, that there was no possibility in evolution to develop an avoidance reaction to cadmium. The fish react to something unknown, for most specimens it seems to be repulsive, for some of them attractive.

Fish maintain their position in flowing water with the help of two sensory systems: i) perception of water flow through the lateral line system, which activates the rheotactic reflex to counter the flow, and ii) an additional optomotoric reaction causing a follow-up swimming which has been studied using the striped cylinder apparatus (Horstmann, 1959, Shaw and Tucker, 1965). Most vertebrates respond rapidly to changes of the optical perception of their surroundings. For example, a frog placed horizontally on the palm of a hand will turn the head with a turn of the hand. This corrective movement restores the original perception of the surroundings and the angle at which light hits the eye. The combined optomotoric and rheotactic reactions of fish facilitate preservation of their position in the presence of water flow. Strong water movement even forces fish to swim against the flow. From the early sixties until about ten years ago rheotaxis had been used as an indicator in surface water monitoring. The rheotactic reflex movement is a very strong reaction. It still functions if a fish is damaged and very weak. Therefore, it is only good as an indicator for lethal concentrations. The optomotoric reaction of fish in the striped cylinder apparatus might be a more sensitive indicator, since it is not a very strong reaction. But here the problem is the habituation to the stimulus (see Shaw and Tucker, 1965). The simplest striped cylinder apparatus consists of an upright standing glass vessel which is surrounded by a movable cylinder which is at the inside striped in vertical direction black and white. If the cylinder is turned by hand or by a machine with a constant velocity, fish follow the moving pattern for several minutes. If the turning velocity changes, the reaction of the fish continues.

This chapter will not give a complete overview on trends, criteria and methods in behavior toxicology, but it will, through examples and ideas, give some suggestions to categorize the used test parameters and to understand the connection to the biology of these test species. It may be helpful for planning new experiments, and it will introduce some ideas suitable for the assessment of conventional and unconventional test methods. Background on behavioral responses as indicators of environmental alterations can be found in Scherer (1992) and information about learning ability as a bioindicator can be found in Marcuella and Abramson (1978).

Trends and thought patterns in science are products of historic developments. Still at the end of the twentieth century it can happen that a scientist who speaks about the feeling ability of animals is prejudged to give anthropomorphic interpretations of animal behavior. That means he is said to be no scientist at all despite a genuine effort for objectivity by the behaviorists. The animal brain is considered by the behaviorists as a black box. It was said that it is impossible to look inside on principle. Working hypotheses thus became axioms. But nowadays we come to quite different conclusions in analyzing the experimental conditions and the thinking categories of behaviorism. In all learning experiments the behaviorists have shown that animals respond to reward and punishment. Which properties require the "black box" if it reacts on punishment and reward?—If we accept that there is a feeling ability in animals, we can understand better how low concentrations of chemicals have an effect on behavior. It is still a working hypothesis, but a well-founded one, that many chemicals which affect physiological processes influence the well-being of animals.

As said above, behavior can also be regarded as an expression of the inner state of an organism. In that way we might get a new aspect to understand spontaneous locomotion activity as a bioindicator. Spontaneous activity follows endogenous rhythms, that can be triggered by the daily light period for example. All what is said below about influences on the "motility patterns" is also valid for locomotion activity in general.

Spontaneous locomotion activity, or motility, is a good behavioral indicator of environmental changes in waters. During the last 30 years a lot of methods for measuring motility were invented. The most important of them are mentioned by Scherer (1992). Nowadays, video processing methods such as BehavioQuant®, are the most suitable. With that method it is possible to analyze the behavior for discovering more sensitive indicators.

Specifically, the inconstancy of swimming speed is a very sensitive indicator for the biological activity of chemicals in water at very low concentrations. Fish do not swim all the time at a constant speed, but display short periods of high activity followed by periods of slower movements, or even pause swimming. This inconstancy of fish swimming speed can be quantified by the image-processing-behavior analyzer BehavioQuant®. From the traces of the objects it is categorized in the form of a standard frequency distribution of swimming velocities. Normally—as in the example presented below—tracings of swimming fish are partitioned into periods of 10 seconds each. Average swimming velocities within these 10 second periods are grouped into 8 categories, and the frequency of incidence per category is evaluated. This frequency distribution is transformed by the simple division of all values by the total number into a standard frequency distribution. By this division the sum of all values becomes 1.

We call these standard frequencies, derived from swimming speed, "motility patterns". They are not only species-specific. They also reflect environmental conditions such as water temperature, oxygen content, pH value, weather conditions, size and gender of the fish. They are also influenced by social components such as group behavior, isolation, hormone levels, or courtship behavior. These motility patterns, as well as overall activity, are subject to a reproducible diurnal cycle (cf. the chapter of Baganz et al. in this volume), a fact which has to be considered carefully when assessing the influence of chemicals on motility (see below).

Animals kept in isolation tend to display abnormal behavior. Sometimes the motility curves are not reproducible. Social deprivation can cause disturbed behavior in fish, and therefore it is not a good practice to use individual animals for biomonitoring unless one deals with a species which prefers a solitary life. Franck (1964) even reports that in his experiments the most of his young Xiphophorus which were kept solitary since birth died before reaching maturity. Allee (1931) found that poisons were less detrimental to fish in groups than to fish on isolation. Alekseeva (1963) found a higher oxygen consumption rate when individual fish were visually isolated from conspecifics, but when individual fish were not visually isolated, he obtained comparable rates in individuals and groups. The social necessities of animals are legally recognized: In some countries the use of isolated animals in test batteries may result in a conflict with the animal protection law. Animals are to be kept under conditions that are appropriate for the species.

"Aggregations based upon biotaxic responses may provide the setting for the reciprocal stimulation of the collected animals" (Tobach and Schneirla, 1968). Nakamura (1952) has presented striking proof for the existence of social attraction in fish. He used an upright glass cylindrical aquarium where a single fish moved freely. Surrounding it was a ring-shaped container subdivided into four compartments. Two of the four compartments each contained three conspecific fish. Monitoring the residence of the single fish at given times allowed conclusions toward its social attraction. A fish spending more of its time in the part of the test chamber where the empty tanks are, displays social avoidance behavior. The tested fish stayed mostly near to the compartments containing the conspecifics. This supports the hypothesis given by Tobach and Schneirla (1968) about biosocial attraction and withdrawal. Since sick fish prefer to be isolated, Nakamura's method could be used for biomonitoring, and with the help of BehavioQuant® it could even be automated.

Courtship behavior will necessarily affect all behavior parameters in a given experimental setting. This is true for a variety of behavior patterns; examples are the violent dashing and chasing of spawning zebrafish (*Brachydanio rerio*), aggressive approaching, defending their territory, and digging of courting cichlids (*Cichlidae*), or courtship movements of guppies (*Poeciliidae*), that all changes the motility patterns, the horizontal and vertical positions in the test chamber and the distance behavior.

Courtship behavior of fish itself is a suitable indicator for low level chemical effects, particularly in species with permanent courtship display, e.g. mouthbreeding cichlids or viviparous guppies or sword-tails. Here constant behavioral patterns can be observed over extended periods of time which underlay diurnal rhythms.

Assessments of behavior must be conducted under standardized conditions. This includes for example the surroundings, form and size of test chambers, light intensity, length of daily photoperiod, ambient temperature, pH value and oxygen content of the water as well as the physiological condition of the test animals. With respect to the latter it is important to use fish of the same age and provenience. The test animals shall be genetically uniform. But if highly inbred strains are used, it shall be considered that homozygous animals often have a loss of fitness. The advantage of the uniformity of the test animals goes on the costs of the disadvantage of less comparability to normal conditions. If extremely uniform test animals are needed, F1-hybrids between two different inbred strains can be used, but normally this is not necessary. Using *Daphnia*, it is recommended to work with cloned animals. Since *Daphnia* are temporally parthenogenetic, the *Daphnia* clones are homogeneous but not homozygous.

Previous exposure to chemicals is strictly to be avoided. The best is to breed and to keep the test animals under standardized clean and disease- and parasite-controlled conditions, where medical treatments are not necessary. Also, the quality of the food and the water shall be supervised. Animals which have been previously exposed to a chemical may retain permanent damage, or induction of cytochrome P450 enzymes may alter their sensitivity to all chemicals which are metabolized with these enzymes.

Another important consideration is to give animals a sufficient period of adaptation to the experimental conditions. Diurnal activity patterns are influenced at least for several days by changed conditions such as onset and duration of light-dark cycles, feeding schedules or isolated aggravating events like a shock caused by sudden loud noise or unexpected gross handling. Also, the length of the light period itself has an important influence. This may be exemplified by the spawning behavior of zebrafish: under a 12/12 hour light/dark cycle they will spawn during the first two hours of the light cycle, whereas under a 16 hour light, 8 hour dark cycle they may spawn sometimes as late as during the second half of the light cycle. Hisaoka and Battle (1958) have successfully synchronized the time of spawning to within one-half hour precision by turning an additional light source on for half an hour always at the same time of day.

Various kinds of endogenous and exogenous influences can modify the behavior. This means that experimental animals have to be protected from any kind of exogenous stimuli, if reproducible measurements are to be obtained.

Finally it is to be kept in mind that the different qualities of behavior which are to be measured are not fully independent from each other. The whole organism's integrative behavioral response is very complex, and different chemicals cause different answers depending on the physiological processes which they affect. Often a chemical affects more than one organ. The dose-effect-relationships on the different affected mechanisms in the same organism are different. Especially regarding long-time experiments, the obtained behavior data are to be considerated as time-dependent processes including the effects of toxikokinetics, accumulation, and metabolization of chemicals, and sometimes habituation of the tested organism to a harmful exposure. Such habituation effects might cover low level effects, if the time processes are disregarded.

2. THE VIDEO-PROCESSING BEHAVIOQUANT® SYSTEM

BehavioQuant® has been designed to record and analyze automatically behavioral parameters of various moving objects (e.g. fish, *Daphnia*, mice, rats, protozoa, tubifex, leaves of plants like *Phaseolus* or *Mimosa*). It is to be used as well for research and for applied continuous monitoring. BehavioQuant® is able to register and track up to 30 untagged objects (up to 200 with some restrictions). The BehavioQuant® system consists of an specially-equipped personal computer, an additional VGA-Monitor, extensive software, and at least one video-source (Figure 1).

The video signal (on-line from CCD, off-line from VCR) is processed by a digitizer board at a frequency of 25 or 30 scans per second. The object recognition is

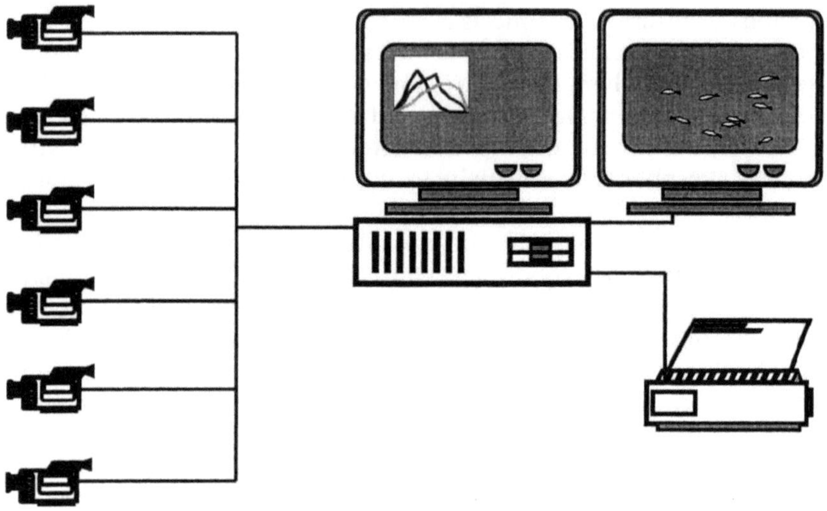

FIGURE 1. The video-processing BehavioQuant® system for automated observation and behavior analysis.

done in real-time, the x/y-positions and dimensions for each object are written to a hard disk together with the time and some other information for each scan. The system can process up to 16 video sources sequentially, and can be set to run measurements automatically around the clock. If for example five cameras are connected and each camera is always active for 2 minutes, the first camera will be active again after 10 min. Track reconstruction, behavior parameter extraction and statistical calculations are performed (on line) in between two subsequent measurements, or parallel using an additional CPU.

The object recognition occurs in principle by comparison of the actual image with the background image, 25 respectively 30 times per sec, depending on the circuit frequency of the electric mains. The background image can be generated from time to time automatically or by menu selection. It is calculated as the average image from 64 single pictures taken at a user-defined or preset time interval which corresponds to the kind of animals to be observed. In the average background picture the moving objects are not present, because they are in the single pictures at different places. Therefore, slower moving species need a longer background image generating time. There are many automated procedures inbuilt to prevent artifacts. But there are also a lot of possibilities for the user to change the presets for special experimental situations.

There is the possibility to select the objects to be traced by size, for example to exclude the new born *Daphnia* and to calculate only the behavior of the adults, or reversed, or to compare both groups in the same sample. The objects traces contain all information of behavior like locomotion activity, inconstancy of swimming speed, number of turns per time, shoaling behavior, swimming height, habitat preferences, phototaxis, preference of dark and light background, energy preserving behavior (in case of lack of oxygen), and flight reactions.

Standard frequency distributions similar to those described above for the inconstancy of motility, can also be used for characterizing the different kinds of habitat selection and even distance behavior. If the diurnal changes of these behavior patterns are taken in consideration, the influences of chemicals on behavior can be evaluated statistically by comparison of these standard frequency distributions of the different behavior parameters. This can be done by a simple nonparamatric method, (see the chapter of Blübaum-Gronau et al. in this book). Another way of analysis, using the motility data, number of turns per time etc., without using the described standard frequency distributions, can be seen in the article of Baganz et al. in this book. All these possibilities are included in the BehavioQuant® software.

The distance behavior analysis can also be done in another way, as described by Spieser et al. (1988) and modified in Blübaum-Gronau et al. (1994). Since there is usually only one video-source per test chamber, the information about the real interindividual distances is lost. But it is possible to estimate changes of the real distances using the information of a big number of screens by the BehavioQuant® GINA ("Group Inrange Normation Arithmetics") program. To understand it's function one should imagine as a theoretical model two fishes which move in the three dimensions with the constant distance AB. The field projection A′B′ of this distance AB moves between the length of AB and zero. If an arbitrarily selected distance A′C is smaller than A′B′ and if enough screens are analyzed, the frequency, that A′B′ is smaller than A′C, is constant, depending on the sine function. These facts are used by GINA: There are four concentric circles calculated around each object screen by screen. Then, the numbers of objects inside the calculated circles are counted and listed, sorted by size, separately for each circle size. The result are four sequences of descending numbers for each screen. These listings are averaged range by range (all first positions, all second positions . . .) for each measuring period. The results are four average sequences of falling numbers per measuring period. With the increasing number of analyzed screens the standard deviations of the means become small. That means that the four average sequences of numbers represent the distance behavior very well. The differences between such sequences of numbers from different measurements are estimated by a varied Euclidean distance formula. The statistical evaluation is similar to the method used for the standard frequency distributions as described above.

Special calculations are made with user definable areas/points of interest. The definitions are to be made easily by moving the cursor over the screen. All previous mentioned calculations can be done separately for each area without losing the tracing information over all areas. Such an application has been used by Steinhörster et al. (1988), see above. Some more data are to be gained for each area, like length of stay, time until entering (from start) and so forth. There is also the possibility to define non-interesting areas, for example the running numbers of a video timer in a tape can be excluded. All that settings can be carried out easily with the inbuilt dialogue and help system.

Further information concerning the applications of the BehavioQuant® system and the possibilities to obtain BehavioQuant®, *Daphnia* test chambers, the fish behavior test equipment IchtyObserve, and the Metacom® Multispecies Monitor®, can be obtained from the first author of this chapter.

3. APPLICATION OF BEHAVIOQUANT® TO THE STUDY OF LONG-TERM EXPOSURE OF JUVENILE CARP TO NONYLPHENOL OR ETHINYLESTRADIOL

Nonylphenol is suspected to cause a number of feminizing effects (Soto, et al., 1991; Jobling and Sumpter, 1993; Jobling et al., 1996; Gimeno et al., 1996; Gray and Metcalfe, 1997; Christiansen et al., 1998). An ongoing major research endeavor, including the present results, involves long-term studies with juvenile carp (4 months old at the start of the experiment) exposed for 10 weeks to 1, 5, 10 or 15 µg/L nonylphenol in order to be comparable to other studies with trout under different conditions. These concentrations are unusual in that they represent more an arithmetic rather than a geometric series. In the experiment reported here the tested animals were also studied after the end of the behavior measurements in histopathology, biochemistry and hematology.

The present investigations were carried out in a combined research program which predominantely focused on endocrinological endpoints. The treatment with the synthetic estrogen ethinylestradiol was performed, to compare the estrogenic potency of nonylphenol with that from a known, highly potent estrogen such as ethinylestradiol. The behavior study was made with the question whether the two compounds, which are found in low concentrations in the environment, have the potential to influence the behavior of fish, that means under a toxicological aspect. But it should not be excluded, that behavioral changes might also be a consequence of endocrine disorders.

Juvenile fish were obtained from the institute's breeding stock. They were bred and kept under standardized and disease- and parasite-controlled conditions.

The experiments were performed in a basement laboratory with artificial lighting, lights on from 7:00 am to 10:30 pm. Neither noise nor undesired light leakage could influence the behavior of the fish. Nobody was allowed to enter the experimental room between 15 min. before onset and 15 min. after termination of the measurements to prevent any influence of the presence of humans on the fishes' behavior. Entrance of animal care personnel or feeding just before or just after the experiment would have caused an undesired activity burst at the beginning or at the end of the measurement series at morning, at noontime or at evening. Fish were fed with BioMar A/S (Brande, Denmark) once per day (1.5% of body weight) during the experiment.

Both chemicals, ethinyl-estradiol and technical nonylphenol, consisting of 90% 4-nonylphenol, 10% 2-nonylphenol, and 2% dinonylphenol, were obtained from Sigma Aldrich (Deisenhofen, Germany).

The test water was continuously renewed at a rate of 9 L/h. The water supply was spring water, which was regularly checked for pH (7.6), oxygen saturation (60%), hardness (21.2°dGH), conductivity (730 µS/cm). The temperature in the test aquaria was kept constant at 15 °C.

After an acclimatization period of 10 days, test substances were added to the water supply of the aquaria, beginning at the evening before the start of the behavior measurements. The required amount of a nonylphenol stock solution was added with the aid of a Gilson® dosing pump via a mixing chamber.

Throughout the experimental period, NP concentrations within the test waters were measured once a week. After several steps, including extraction and purification processes, nonylphenol was quantified by gas chromatography (Hewlett Packard 5970 A). According to the analytical data obtained, the differences between the real concentrations of nonylphenol and the nominal concentrations did not exceed 20%.

For reasons of workplace safety ethinylestradiol was injected rather then dosed via water. An effective high dose of 500 μg/kg body weight was administered i.m. every three weeks. Sham treatment to the other groups was not given since no substance exists where we could be sure that it would have no side effects using it for sham treatment. This decision was proven right later on, since data obtained on the day after injection never showed any deviation from data obtained at other time points during this experiment. Evidently the injection procedure by itself did not have an effect on the behavior patterns measured here.

Behavior measurements were conducted Monday through Friday between 8:00 and 12:00 am and between 4:00 pm and 10:00 pm. The avoidance of Saturday and Sunday observations had a practical reason: The possibility for 2 daily automatic starts of test series was not jet implemented im the BehavioQuant® system when these experiments were done. Six test aquaria were used, and the software was set to switch to the next camera every two minutes. Thus five measurements per hour were obtained for each test group.

No measurements were conducted during the lights-off phase. Although BehavioQuant® permits this type of work (cf. Baganz et al., this volume), no infrared lighting was available in the test facility. The prolonged break at noon time was introduced to allow personnel to access the test lab, feed the fish, clean tanks, check equipment and draw water samples for analysis of dosing levels, and to allow fish to recover from these disturbances before the afternoon measurements were started.

Each aquarium was stocked with 22 fish per 80 L of water. This is a comparatively high loading, but there were two important reasons for it: i) the histopathological, hematological, etc., follow-up investigations required this large number of fish, and ii) a bigger tank volume would have produced serious logistical problems like the daily supply of large volumes of stock solutions. For that reason, the test aquaria did not have a size where the fish would have place enough to show a distance-behavior as in free water, and to select preferred positions in the test tank. It would not have been useful to calculate distance behavior and habitat selections (for example, horizontal and swimming distributions) of the fish under this condition. Only the diurnal changing motility patterns (shown as standard frequency distributions of swimming speed per 10 sec in 8 size classes, see above) were analyzed.

In order to obtain an overview of the development of behavior changes during the exposure time of 70 days, the *average* standard frequency distributions of the patterns of motility deriving from the daily 50 individual measurements for each treatment group were calculated separately for each day. Since these results were very difficult to be interpreted, it was suspected that there was a chronobiological effect of the substances involved.

Therefore, the motility data over one-hour periods were averaged and those hourly data over one week's time were pooled. Thus the averaged standard distributions for each treatment group were obtained for Monday through Friday, 8:00 through 9:00, 9:00 through 10:00, and so on. Now it was possible to elucidate the time courses of changes of behavior, both with respect to changes with diurnal cycles as well as over the duration of the experiment.

During the course of a day, the patterns of motility change in a characteristic way. This is highlighted by a comparison of motility data from the control group between first and the second week of treatment, respectively (compare the dotted lines of figures 2a–j with figures 3a–j).

In the morning of the *first week* motility patterns of the treated animals were relatively similar to those of the controls. They differed statistically from the controls only in up to two from eight size classes. The only exceptions were found in 5 µg/L nonylphenol (NP) between 8 and 9 am and ethinylestradiol (EE2) between 9 and 10 am (Figures 2a–d). But during the afternoon hours particularly the ethinylestradiol-treated group showed a strong decrease in velocity (Figures 2e–h), statistically proven by up to 6 size classes difference to the controls. In evening the NP groups (with the exception of the 10 µg/L NP group, see below) displayed a decrease of motility. The strongest effect was between 8 and 9 pm (Figure 2i). During the *second week* of treatment the slow-down effect in EE2 was observed instead during the morning hours (Figures 3a–d). In the afternoon of the *2nd week* the group treated with 15 µg/L nonylphenol displayed a rapid increase of fast movements in the afternoon (Figures 3 f–i), and a similar effect was observed during evening hours in the ethinylestradiol group (Figures 3h–j). During the *weeks 3 to 5* the motility patterns of the nonylphenol-exposed fish resembled mostly control patterns, whereas in the ethinylestradiol group a marked deceleration of movement was observed for most of the time (Figures 4a–d). The form of the morning patterns of EE2 remind to afternoon patterns of the controls (compare figures 4a–c with 3e–f). During this period, lasting until the *8th week* in the 15 µg/L nonylphenol group, the swimming activity predominantly was increased in the evening hours. The animals exposed to 10 µg/L nonylphenol began to swim slower in the *8th week* of treatment from the morning hours until 6 pm. In contrast the 5 µg/L NP group displayed more quick motions in the morning until noontime, and the 1 µg/L NP group had more activity at noon time (Figures 5a–d). Toward the end of the experiment, at the *10th week* of exposure, all exposed fish showed a drastic increase of slow motions during the evening hours (Figures 6e–j). This effect was concentration-dependent for nonylphenol and, even in the lowest exposure group of NP, exceeded values found in the ethinylestradiol-exposed group. In contrast, during the morning hours of the *10th week* only the ethinylestradiol group and the groups exposed to 10 or 15 µg/L nonylphenol displayed reduced swimming velocities (Figures 6a–d).

There was an interesting temporal exception found in the concentration-effect relationship: In the *first week* there is nearly no effect in the 10 µg/L NP group. In the *second week* there is no effect in the 5 µg/L NP group. The observation that there

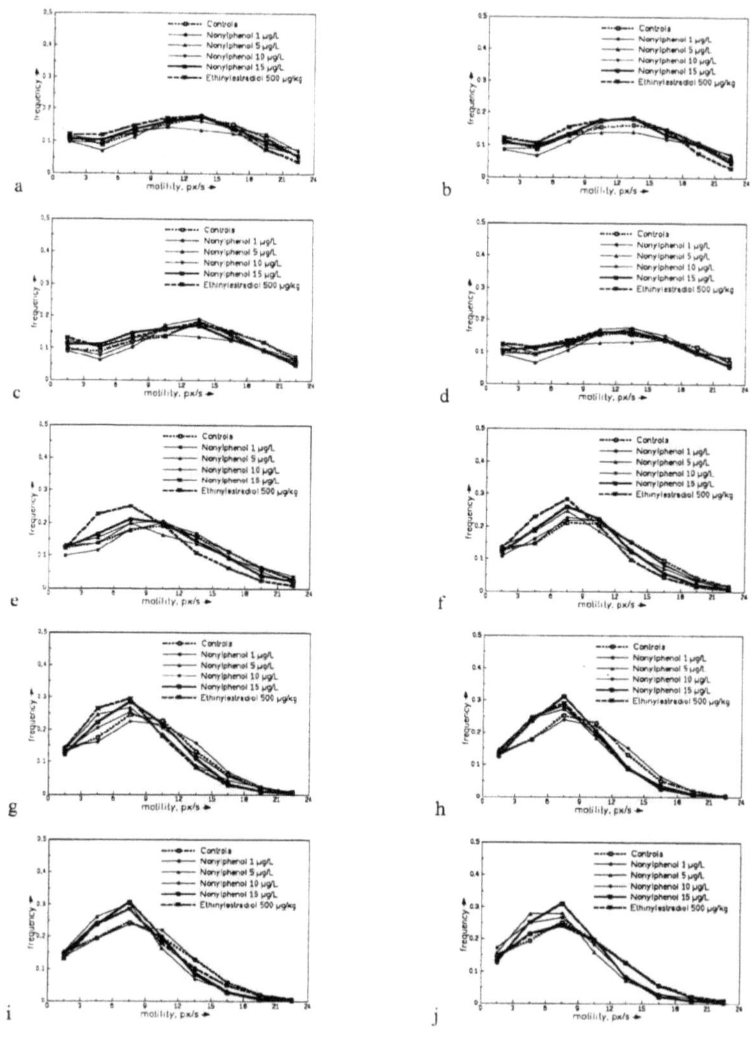

1st week	8-9 am	9-10 am	10-11 am	11-12 am	4-5 pm	5-6 pm	6-7 pm	7-8 pm	8-9 pm	9-10 pm
1 µg	1	2	1	2	0	0	0	2	4	2
5 µg	3	1	2	2	1	2	3	4	5	4
10 µg	1	1	0	0	0	0	0	0	0	0
15 µg	0	0	2	0	1	5	5	5	6	5
EE2	1	3	1	1	6	6	6	3	0	0

FIGURE 2. Standard frequency distributions of the short time swimming speeds, "motility patterns", during the first week of treatment, averaged from Monday to Friday. a) from 8:00 am to 9:00 am, b) from 9:00 am to 10:00 am, c) from 10:00 am to 11:00 am, d) from 11:00 am to 12:00 am, e) from 4:00 pm to 5:00 pm, f) from 5:00 pm to 6:00 pm, g) from 6:00 pm to 7:00 pm, h) from 7:00 pm to 8:00 pm, i) from 8:00 pm to 9:00 pm, j) from 9:00 pm to 10:00 pm. x- axis: motility per 10 seconds in 8 size classes, y-axis: frequency.

Number of size classes with significant differences to the controls at the level of 1% error probability, double-sided questioning, calculated with the U-test according to Wilcoxon, Mann and Whitney.

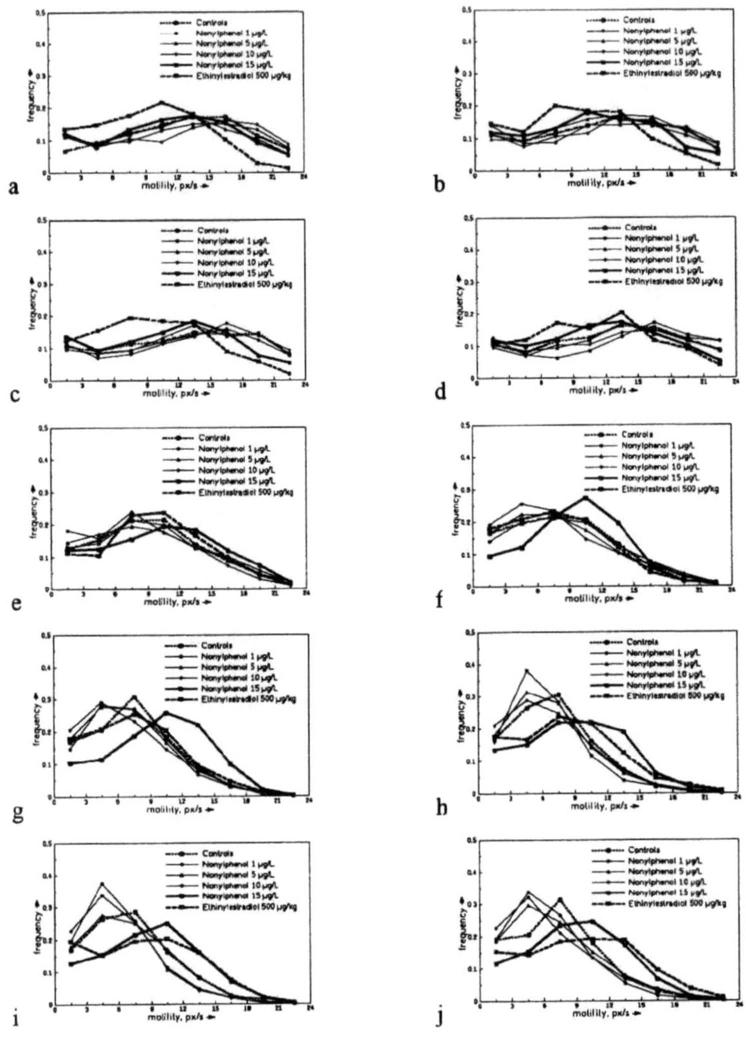

2nd week	8-9 am	9-10 am	10-11 am	11-12 am	4-5 pm	5-6 pm	6-7 pm	7-8 pm	8-9 pm	9-10 pm
1 µg	0	1	0	0	0	1	0	1	0	1
5 µg	0	0	0	0	0	0	0	0	0	0
10 µg	1	0	0	0	0	1	2	1	0	3
15 µg	0	1	1	0	0	4	4	4	3	1
EE2	2	3	2	1	0	0	1	2	2	3

FIGURE 3. Standard frequency distributions of the short time swimming speeds, "motility patterns", during the second week of treatment, averaged from Monday to Friday. a) from 8:00 am to 9:00 am, b) from 9:00 am to 10:00 am, c) from 10:00 am to 11:00 am, d) from 11:00 am to 12:00 am, e) from 4:00 pm to 5:00 pm, f) from 5:00 pm to 6:00 pm, g) from 6:00 pm to 7:00 pm, h) from 7:00 pm to 8:00 pm, i) from 8:00 pm to 9:00 pm, j) from 9:00 pm to 10:00 pm. x- axis: motility per 10 seconds in 8 size classes, y-axis: frequency.

Number of size classes with significant differences to the controls at the level of 1% error probability, double-sided questioning, calculated with the U-test according to Wilcoxon, Mann and Whitney.

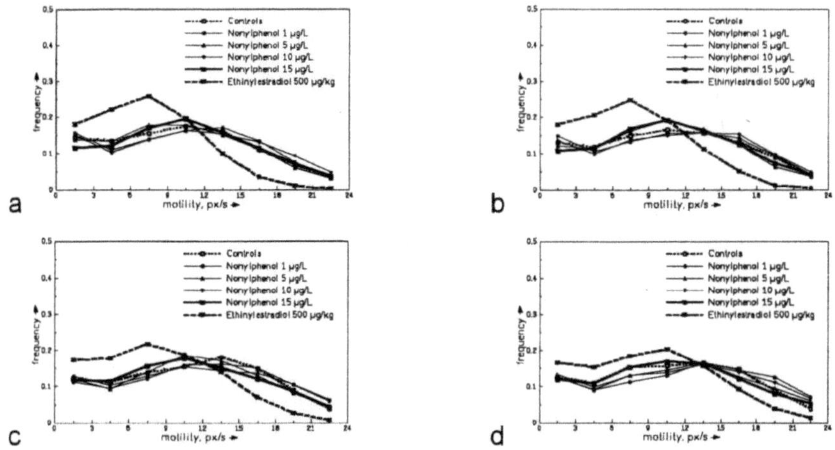

4th week	8-9 am	9-10 am	10-11 am	11-12 am
1 µg	1	0	0	1
5 µg	0	1	0	0
10 µg	1	0	1	0
15 µg	0	0	0	1
EE2	6	7	8	6

FIGURE 4. Standard frequency distributions of the short time swimming speeds, "motility patterns", during the fourth week of treatment, averaged from Monday to Friday. a) from 8:00 am to 9:00 am, b) from 9:00 am to 10:00 am, c) from 10:00 am to 11:00 am, d) from 11:00 am to 12:00 am. x- axis: motility per 10 seconds in 8 size classes, y-axis: frequency.

Number of size classes with significant differences to the controls at the level of 1% error probability, double-sided questioning, calculated with the U-test according to Wilcoxon, Mann and Whitney.

was for a longer time no effect in a certain concentration in between others with an effect, was made by the authors in several long-term experiments in low level concentrations with different chemicals. It might have to do with a basic phenomenon like hormesis (Luckey, 1980 and 1982; Wachsmann, 1987) and needs further research for explanation.

During the 10 week duration of the experiment, the fish increased in size, which caused a natural slowing of their movements. But this effect is small compared with the slow down caused by the applied substances. This example shows how important it is to monitor controls at all times in a toxicological behavior study.

Thus, nonylphenol affected the swimming behavior of young carp in a concentration-dependent manner. The treatment did not affect swimming velocity in general, but only during specific times of the day, it changed with the duration of the experiment, and it was both increased and decreased, respectively. These complex temporal patterns of behavior changes represent superimposition of

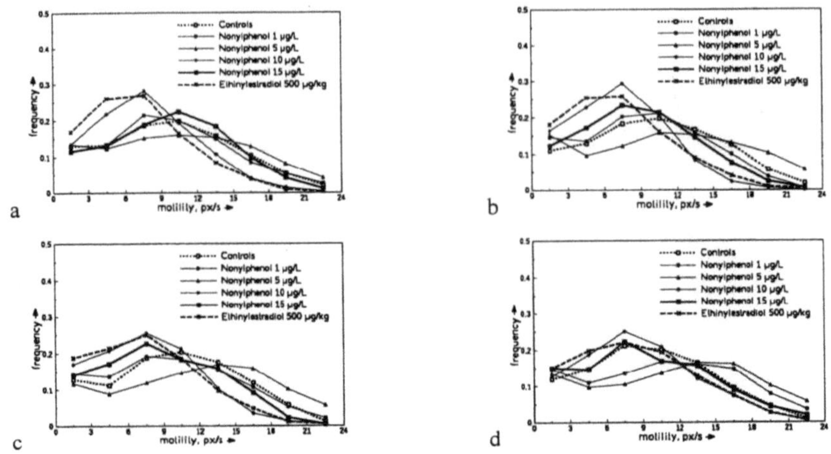

8th week	8-9 am	9-10 am	10-11 am	11-12 am
1 µg	0	0	0	4
5 µg	1	5	3	6
10 µg	3	7	4	0
15 µg	2	2	0	0
EE2	4	4	0	0

FIGURE 5. Standard frequency distributions of the short time swimming speeds, "motility patterns", during the eighth week of treatment, averaged from Monday to Friday. a) from 8:00 am to 9:00 am, b) from 9:00 am to 10:00 am, c) from 10:00 am to 11:00 am, d) from 11:00 am to 12:00 am. x- axis: motility per 10 seconds in 8 size classes, y-axis: frequency.

Number of size classes with significant differences to the controls at the level of 1% error probability, double-sided questioning, calculated with the U-test according to Wilcoxon, Mann and Whitney.

several phenomena. For example, it has been observed that several chemicals will increase or decrease motility, depending on the level of exposure. The most well-known examples for this phenomenon are d-amphetamine and ethanol. Low doses stimulate, high doses let the motility decrease. This effect could be explained in terms of different threshold levels regarding different functions of the nervous system and in other organs. The temporal shifts in diurnal activity—the maxima and minima of activity were moved especially in the 15 µg/L nonylphenol group and in ethinylestradiol—indicate that both substances cause also a chronobiological effect.

Hematological investigations revealed a distinct anemia in fish exposed to the higher test concentrations of nonylphenol and in ethinylestradiol-treated individuals. Severe histopathological alterations however, were restricted to the group treated with ethinyl-estradiol (Schwaiger et al., 2000). These results, especially the anemia found at the end of the experiment in the groups treated with ethinyl-

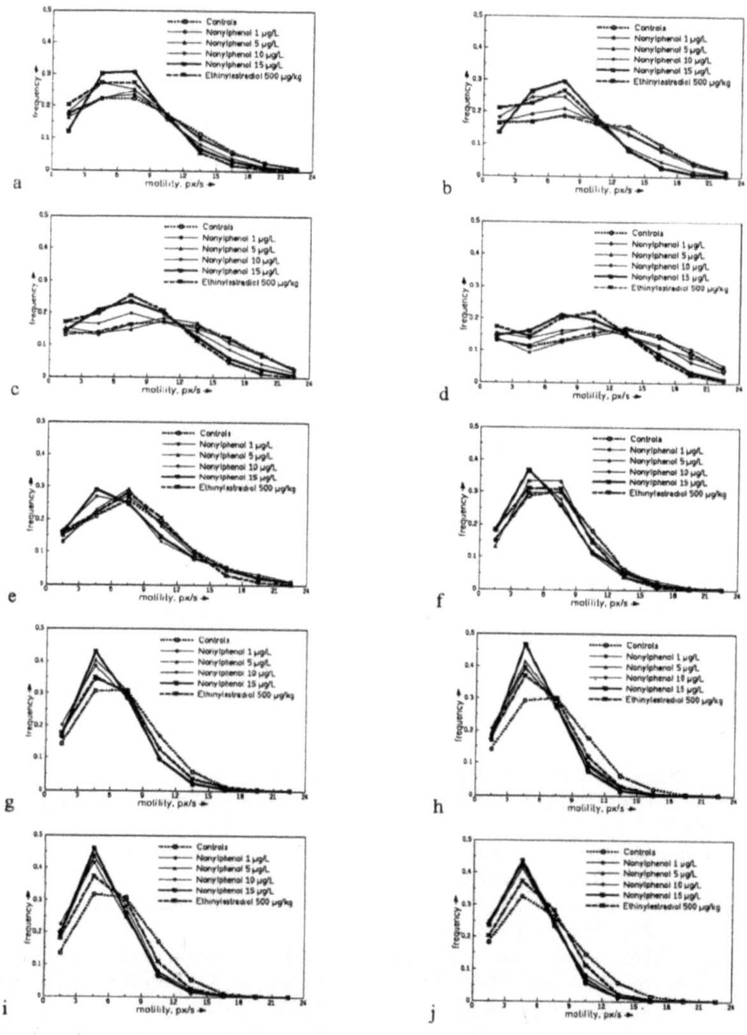

10th week	8-9 am	9-10 am	10-11 am	11-12 am	4-5 pm	5-6 pm	6-7 pm	7-8 pm	8-9 pm	9-10 pm
1 µg	0	0	0	0	0	0	1	5	5	4
5 µg	0	0	0	4	0	0	3	5	6	5
10 µg	0	6	1	0	1	3	4	5	6	4
15 µg	7	6	5	6	1	3	4	5	6	5
EE2	6	7	8	7	0	0	3	5	5	3

FIGURE 6. Standard frequency distributions of the short time swimming speeds, "motility patterns", during the tenth week of treatment, averaged from Monday to Friday. a) from 8:00 am to 9:00 am, b) from 9:00 am to 10:00 am, c) from 10:00 am to 11:00 am, d) from 11:00 am to 12:00 am, e) from 4:00 pm to 5:00 pm, f) from 5:00 pm to 6:00 pm, g) from 6:00 pm to 7:00 pm, h) from 7:00 pm to 8:00 pm, i) from 8:00 pm to 9:00 pm, j) from 9:00 pm to 10:00 pm. x- axis: motility per 10 seconds in 8 size classes, y-axis: frequency.

Number of size classes with significant differences to the controls at the level of 1% error probability, double-sided questioning, calculated with the U-test according to Wilcoxon, Mann and Whitney.

estradiol and with the higher concentrations of nonylphenol, gives a convincing explanation for the slow down of locomotion activity down to 10 µg/L. But this does not explain the strong slow down effect in the lower concentrations during evening.

Nonylphenol had been detected in some European surface waters at concentrations in the range of <0.2—12 µg/L (in polluted British river waters, Blackburn and Waldock, 1995), and in the range of <0.3—45 µg/L in the river Glatt in Switzerland, with only one result above 10 µg/L (Ahel et al., 1994).

For the hazard assessment of nonylphenol it is important to notice that after the exposure of ten weeks in the evening the motions of juvenile carp are slowed down drastically, even in the concentration of 1 µg/L. Although it is not proved by these experiments, it can be expected there would be an effect on predation if the locomotion activity of a species in an ecosystem is slowed down. In this concentration no other effects were found in the juvenile carp. As often to be observed with chemical substances, regarding nonylphenol the quantitative behavior analysis with BehavioQuant® has been proved, compared to other methods, to be the most sensitive.

4. REFERENCES

Ahel, M., W. Giger, and C. Schaffner (1994): Behaviour of alkylphenol polyethoxylate surfactants in the aquatic environment—II. Occurence and transformation in rivers. Wat. Res. 28(5), 1143–1152.

Alee, W.C. (1931): *Animal aggregations*. Chicago (cited from Barlow, 1968).

Alekseeva, K.D. (1963): The significance of vision in the "group effect" in a few Black Sea fishes. VOP Ikhtiol. 3(4), 726–733.

Barlow, G.W. (1968): Ethological units of behavior. In: Ingle, D. (Ed.): *The Central Nervous System and Fish Behavior*. The University of Chicago Press, Chicago and London, 217–232.

Blackburn, M.A., and M.J. Waldock (1995): Concentrations of alkylphenols in rivers and estuaries in England and Wales. Wat. Res. 29(7), 1623–1629.

Blübaum-Gronau, E., O.H. Spieser, and F. Krebs (1992): Bewertungskriterien für einen Verhaltensfischtest zur kontinuierlichen Gewässerüberwachung. Schr.-Reihe Verein Wasser-, Boden- und Lufthygiene 89, Gustav-Fischer-Verlag, Stuttgart, 333–348.

Blübaum-Gronau, E., M. Hoffmann, O.H. Spieser, and F. Krebs (1994): Der Koblenzer Verhaltensfischtest, ein auf dem Meßsystem BehavioQuant® beruhender Biomonitor zur Gewässerüberwachung. Schriftenreihe des Vereins für Wasser-, Boden- and Lufthygiene 93, 87–117.

Blübaum-Gronau, E., B. Grillitsch, and O.H. Spieser (1994): Quantitative Verhaltensmessungen mit BehavioQuant® als Instrument der kontinuierlichen Gewässerüberwachung. In: *Biomonitoring—ein Instrument zur kontinuierlichen Überwachung von Gewässern*. Bundesanstalt für Wassergüte, Wien, ÖVGW, ÖWAV, Bundesministerium für Land- und Forstwirtschaft, Wien. Seite 23–33.

Christiansen, T., B. Korsgaard, and Å. Jespersen (1998): Effects of nonylphenol and 17β-oestradiol on vitellogenin synthesis, testicular structure and cytology in male eelput *Zoarces viviparus*. J. Exp. Biol. 201, 179–192.

Franck, D. (1964): Versuche zur Schwertfunktion bei *Xiphophorus helleri*.—Zool. Anz. 173, 315–325.

Gimeno, S., A. Gerritsen, T. Bowmer, and H. Komen (1996): Feminization of male carp. Nature 384, 221–222.

Gray, M.A., and C.D. Metcalfe (1997): Induction of testis-ova in Japanese medaka (*Oryzias latipes*) exposed to p-nonylphenol. Environ. Toxicol. Chem. 16(5), 1082–1086.

Hisaoka, K.K., and H.I. Battle (1958): The normal developmental stages of the zebrafish, (*Brachydanio rerio* Hamilton-Buchanan). Jour. Morph., 102, 311–321.

Horstmann, E. (1959): Schwarmstudien unter Ausnutzung einer optomotorischen Reaktion bei *Mugil cephalus* (Cuv.). Pubbl. Staz. Zool. Napoli 15, 143–158.

Jobling, S., and J.P. Sumpter (1993): Detergent components in sewage effluent are weakly oestrogenic to fish: An *in vitro* study using rainbow trout (*Oncorhynchus mykiss*) hepatocytes. Aquat. Toxicol. 27, 361–372.

Jobling, S., D. Sheahan, J.A. Osborne, P. Matthiessen, and J.P. Sumpter (1996): Inhibition of testicular growth in rainbow trout (*Oncorhynchus mykiss*) exposed to estrogenic alkylphenolic chemicals. Environ. Toxicol. Chem. 15(2), 194–202.

Lorenz, K.Z. (1950): The comparative method in studying innate behaviour patterns. Symps. Soc. Exp. Biol. 4, 221–268.

Lorenz, R., R. Brüggemann, C.E.W. Steinberg, and O.H. Steinberg (1996): Humic material changes effects of terbutylazine on behavior of zebrafish (Brachydanio rerio). Chemosphere 33(11), 2145–2158.

Lorenz, R., O.H. Spieser, and C. Steinberg (1995): Neue Wege in die Ökotoxikologie: Quantitative Verhaltensmessuneg an Fischen als Toxizitätsendpunkt.—New Ways to Ecotoxicology: Quantitative recording of Behaviour of fish as Toxicity Endpoint. Acta hydrochimica et hydrobiologica. Acta hydrochim. hydrobiol. 23(5), 197–201.

Luckey, T.D. (1980): *Hormesis with Ionizing Radiation*. CRC Press, Raton/Florida.

Luckey, T.D. (1986): Ionizing radiation promotes protozoan reproduction. Radiation Res. 108.

Marcuella, H., and C.I. Abramson (1978): Behavioral Toxicology and Teleost Fish. In: Mostofsky, D.I. (Ed.): *The Behavior of Fish and Other Aquatic Animals*. Academic Press, New York, San Francisco, London. pp. 33–77.

Marler, P., and W.J. Hamilton, III. (1966): Mechanisms of animal behavior. New York: Wiley.

McNicol, R.E., and E. Scherer (1991): Behavioral responses of lake whitefish (*Coregonus clupeaformis*) to cadmium during preference- avoidance testing. Environmental Toxicology and Chemistry, 10, 225–234.

Nakamura, Y. (1952): Some experiments on the shoaling reactions in *Oryzias latipes* (Temminck et Schlegel). Bull. Jap. Soc. Sci. Fish. 18, 93–101.

Scherer, E. (1992): Behavioural responses as indicators of environmental alterations: approaches, results, developments. J. Appl. Ichthyol. 8, 122–131.

Shaw, E., and A. Tucker (1965): The optomotor reaktion of schooling Carangid Fishes. Animal Behavior 18, 330–341.

Schwaiger J., O.H. Spieser, C. Baner, H. Ferling, V. Mallow, W. Kalbfus, and R.D. Negele (2000): Chronic toxicity of nonylphenol and ethinylestradiol: haematological and histopathological effects in juvenile common carp (*Cyprinus carpio*). Aquatic Toxicology, in press.

Shaw, E. (1970): Schooling in Fishes: Critique and Review. In: Aronson, L.R., E. Tobach, D.S. Lehrman, and J.S. Rosenblatt (Eds.): *Development and evolution of behavior*. W.H. Freeman and Company, San Francisco, 452–480.

Soto, A.M., H. Justicia, J.W. Wray, and C. Sonnenschein (1991): p-Nonylphenol: An estrogenic xenobiotic released from "modified" polystyrene. Environ. Health Perspect. 92, 167–172.

Spieser, H., and H. Frisch (1988): Method of measuring the types of motion and configuration of biological and non-biological objects. U.S. Patent 4,780,907. Oct. 25, 1988.

Spieser, O.H., W. Scholz, E. Blübaum-Gronau, M. Hoffmann, B. Grillitsch, and C. Vogl (1994): Das System BehavioQuant® zur Bioindikation anhand des Verhaltens von Fischen und von anderen aquatischen Organismen. in: *Eco-Informa-'94, Band 5, Umweltmonitoring und Bioindikation*. Hsg. K. Alef, H. Fiedler, O. Hutzinger. Umweltbundesamt Wien. Seite 429–448.

Steinberg, C.E.W., C. Mayr, R. Lorenz, O.H. Spieser, and A. Kettrup (1994): Dissolved Humic material amplifies irritant effects of terbutylazine (Triacine herbicide) on fish. Naturwissenschaften 81, 225–227.

Steinberg, C., R. Lorenz, and O.H. Spieser (1995): Effects of atrazine on swimming behaviour of zebrafish, *Brachydanio rerio*. Water Research 29, 981–985.

Steinhörster, U., K. Starick, M. Schubert, and H. Stein (1997): Studies on behaviour of nase fry (*Chondostroma nasus* L.) at different temperature conditions and pH-values. In: Alef, K.,

J. Brandt, H. Fiedler, W. Hauthal, O. Hutzinger, D. Mackay, M. Matthies, K. Morgan, L. Newland, H. Robitaille, M. Schlummer, G. Schüürmann, and K. Voigt (Eds.): *Eco-informa'97*, Vol. 1.

Tobach, E., and T.C. Schneirla (1968): The biopsychology of social behavior in animals. In: Cooke, R.E. (Ed.): *The biologic basis of pedriatic practice*. McGraw-Hill, New York. pp. 68–82.

Wachsmann, F. (1987): Are small doses really so dangerous? Electromedica 55(3), 86–90.

7

HOW TO USE FISH BEHAVIOR ANALYSIS TO SENSITIVELY ASSESS THE HAZARD POTENTIALS OF ENVIRONMENTAL CHEMICALS

DANIELA BAGANZ,[1,3] GEORG STAAKS,[1] O. HUNRICH SPIESER,[2] AND CHRISTIAN E.W. STEINBERG[1]

[1]Institute of Freshwater Ecology and Inland Fisheries, Müggelseedamm 310, D-12587 Berlin, Germany. [2]GSF-National Research Center for Environment and Health, Institute of Toxicolgy, P.O. Box 1129, D-85758 Oberschleissheim, Germany. [3]Author to whom correspondence should be addressed

1. INTRODUCTION

During[1] their phylogenesis, all species have adapted to their distinct habitats and to naturally occurring environmental changes. By means of specific long-term self-regulation processes, species are able to react to these environmental changes and therefore maintain their physiological and ecological balance. If these changes take place within biogenetically short periods, species can not develop new strategies of adaptation and the self-regulating mechanism will fail.

Environmental factors become stressors, if they may cause harmful effects on organisms. Anthropogenic activities develop, and introduce, new substances such as xenobiotic chemicals into the environment that act as stressors. Furthermore, anthropogenic activity may extend the range of natural stressors, such as increasing UV-irradiation or secondary effects of eutrophication (exposure to algal toxins, such as cyanotoxins). Biological systems respond to chemical stresses on various levels of aggregation, from molecule to ecosystem.

Behavioral tests addressing whole organism-level effects, provide primary and very sensitive information about the hazard potential of environmental

---

[1] Abbreviations Microcystin-LR = MC-LR.

Biomonitors and Biomarkers as Indicators of Environmental Change 2, Edited by Butterworth *et al.*
Kluwer Academic/Plenum Publishers, New York, 2000

chemicals for organisms. From the evolutionary point of view the diversity of mechanisms which control the behavior of animals is a response to the diversity of ecological problems. There are regulative mechanisms which directly react to changes in the physical and/or social environment by changing the behavioral disposition of an animal in order to allow adaptability (Alcock, 1996). The use of behavioral parameters as biomarkers is based on the fact that every species needs, for its behavior, those environmental conditions, which correspond to its specific requirements.

In evaluating the impact of stressors in freshwater systems, fish have a special importance because they are situated at the end of the aquatic food chain, and thus may also indicate the contamination of their prey. In fish behavior analysis, different parameters respond non-specifically but very sensitively to stressors. The assessment and prediction of long-term sublethal effects of environmental chemicals on the behavior require a quantitative method with high sensitivity. By using the automated measuring system BehavioQuant®, several test parameters can be recorded and quantified: motility, turns, swimming height, places of sojourn, and distances between each single fish.

The *motility* (swimming velocity in video-pixels per second) depends on the spontaneous locomotor activity. The *turns* are the number of changes of the direction in a given period. Along with motility, turns characterise the swimming activity. The swimming activity, in general, is a sensitive indicator for environmental changes (Scherer, 1992; Tembrock, 1992), and can be significantly altered by pollution (i.e. Boujard and Leatherland, 1992; Al-Kahlem et al., 1994; Steinberg et al., 1995; Saglio et al., 1996; Paul and Simonin, 1996). Tembrock (1992) describes an increased activity as a changed spatial orientation for an active behavioral acclimatization to variable environmental conditions. This increase of activity after stress conditions, takes place only in a relatively short period of exposure, due to the energetic costs for the organisms. In contrast, long-term sublethal effects of stressors cause decreases in activity. Therefore, the recording of motility can also be used to estimate the energetic balance of behavioral reactions (Staaks, 1996). For indication of toxic exposures, the swimming activity during the spawning period and at the feeding times is of particular interest. Furthermore, some toxicological studies observed disturbances of the developing and maturing of gonads and the reproduction behavior, respectively (Braunbeck et al., 1990; Weber, 1993; Jones and Reynolds, 1997).

The *swimming height* often differs in the course of the day. Changes may indicate an exogenous or endogenous deficit of oxygen. *Places of sojourn* can show some aspects of the preference behavior in respect to a habitat. Alterations in the preference behavior are also often documented as being directly influenced by chemical substances (Smith and Bailey, 1990; Morgan et al., 1991; Scherer et al., 1991; Steinberg et al., 1995).

The *distance between each single fish* is an important property of a shoal, giving a scale for the quality of a shoal. It may clearly be modified by chemical exposures (Mikhaylova, 1991; Biermann, 1992).

All of the above mentioned parameters are time dependent. Endogenous rhythms are, under natural conditions, synchronised with environmental cycles by time triggers, e.g. light is the most important and best described environmental signal synchronising the endogenous rhythms to the 24-hour solar day. Disturbances of the synchronization between biological rhythms and their triggers may be used as biomarkers. A suitable parameter for analysing the biological rhythms is the motility. As a component of swimming activity, motility may show distinct diurnal differences, thus chronbiological aspects have to be necessarily considered.

As above mentioned, the test substance used as a stressor was the cyanobacterial toxin microcystin-LR (MC-LR). MC-LR is a cyclic peptide hepatoxin produced by the cyanobacterium *Microcystis aeruginosa* and other cyanobacteria. In contrast to acute toxicity studies (e.g. Bury et al., 1995; Oberemm et al., 1997), the ecological impacts and consequences of blooms of cyanobacteria associated with high concentrations of algal toxins are poorly known. Therefore, sublethal effects of MC-LR on the behavior of fishes were investigated.

2. MATERIALS AND METHODS

2.1. Measuring System

The BehavioQuant® system was applied to record and quantify major behavioral parameters of zebrafish (*Danio rerio*). It is an image processing system which employs pattern analysis algorithms (Figure 1). The movements of each individual fish are recorded by video imaging. The traces are numerically digitized by a real time object recognition process. With the BehavioQuant® system, it is possible to observe up to sixteen aquaria simultaneously, with one camera in the front of each aquarium. To ensure the continuous observation of individual fish, and to register the movements during periods of particular interest, such as dawn, dusk and at night, the system can be connected with cameras which are able to make use of normal as well as infrared light. The recording of the fish behavior takes place sequentially, i.e. two minutes for each aquarium in turn. The raw data are pre-processed and converted into tables, which contains the behavioral parameter values, i.e. motility. These tests generate a large amount of data. For instance, the subsequently described experiment produced 60. . . . 80 MB recording data, so 3–4 GigaByte raw data could be expected during an entire test. It is useful to save the data on CD or tape.

2.2. General Test Conditions

The following conditions should be adapted in behavior tests with fish:

1. Use a separate part of the laboratory with a very low level of optical and acoustical disturbances. It should be dark, without natural light, because its intensity cannot be controlled. If necessary a black curtain may be used as a screen.
2. Feeding, temperature changes and artificial light/dark rhythms should be constant and automatically controlled.

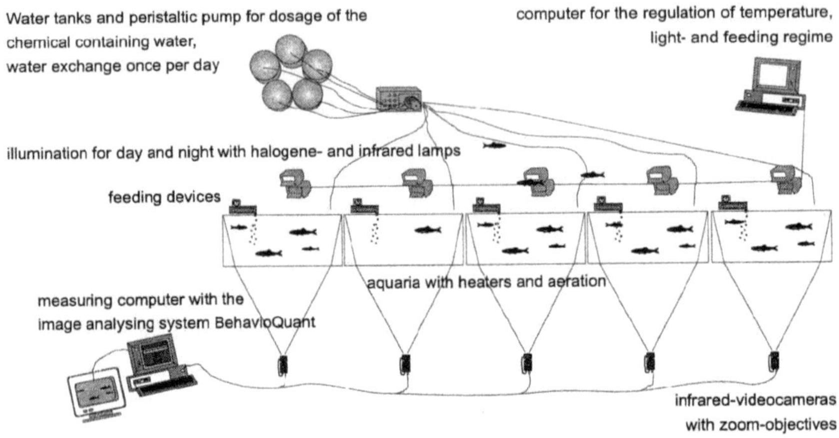

FIGURE 1. Scheme of the equipment.

3. The number of aquaria required depends on the number of different exposure concentrations, adding the replicates.
4. It is useful to carry out the tests in a flow-through system with a continuous water and chemical discharge from storage tanks.
5. The experiment can be divided in three phases. If the reversibility of behavioral changes should be studied, a fourth phase could be added.
   a) *Acclimatisation period* without any chemical exposure and without recording of behavior. Recommended are 2–4 weeks.
   b) *Observation period* to record the behavior without any chemical exposure over a period of 1–3 weeks. This period is necessary to recognise any significant differences between the various test groups. Normally no significant differences between the groups should arise, if there was an adequate period for acclimatization for the test organisms and if the external test conditions like feeding time, light and dark rhythm and the physico-chemical parameters, were equal in the various aquaria.
   c) *Test period* to record the behavior under chemical exposure. The concentration should be constant for each test group during the test period. If sublethal effects are to be investigated, the length of this phase should last 1–3 weeks.
   d) *Recovery period* without any chemical exposure. This phase is essential to determine the reversibility of the observed effects.

2.3. Specific Test Conditions for the Study of Microcystin-LR Effects

Zebrafish (*Danio rerio*) of 0.25 g mean weight were held in groups of 7 individuals (3 females, 4 males) in each of five 15-litre glass aquaria. The fish were kept under controlled conditions according to OECD guidelines (1992). Tests were carried out in a flow-through system with a continuous discharge of 10 litres aerated tap-water per day from storage tanks. Fish were exposed to a 12:12h light-dark rhythm. Temperature of 26 ± 0.5 °C and a pH of 7.6. . . . 7.7 were kept constant. Fish were fed on TetraMin® flakes at a ration of 3% body mass per day and food was dispensed automatically at 11 am and 3 pm. Data were recorded for 23 hours per day from 3 pm to 2 pm with a cleaning break of one hour.

2.4. Toxin Application

The aim of our experiment was to test the influence of the cyanotoxin MC-LR on zebrafish behavior. After an acclimatisation period of 3 weeks, there followed an observation period of 25 days during which time the test parameters were recorded without any chemical exposure. In the subsequent test period four groups were exposed to contaminated water and one group served as a control.

The storage tank water was renewed daily and in the four test groups, MC-LR was added at nominal concentrations of 0.5, 5 and 15 $\mu g L^{-1}$ for a period of 25 days and 50 $\mu g L^{-1}$ for a period of 6 days. The continuous toxin application reduced biodegradation process of MC-LR described by Cousins et al. (1996) and Welker and Steinberg (1999).

2.5. Statistics and Calculations

Exemplary values for the motility (swimming velocity in video-pixels per second) where mathematically extracted from the raw data. The computer program SPSS 6.1 was used for the subsequent statistical calculations of analysis of variance, Levenes test for equality of variances and Tukey-HSD multiple range tests.

3. RESULTS

A comparison of the motility, between the control and test groups showed a clear dose-effect relationship between fish activity and concentration of MC-LR (Figure 2). The daytime (8 am to 12 am) motility was significantly increased at the two lower exposure concentrations (0.5 and 5 $\mu g L^{-1}$), but decreased at elevated exposures compared to control. The mean daytime motility decreased significantly at the highest test concentration (50 $\mu g L^{-1}$). Exposure in the nighttime (from 9 pm to 4 am) showed reciprocal behavioral patterns compared to the daylight period. The mean motility during the nighttime decreased significantly at the lowest concentration, 0.5 $\mu g L^{-1}$. In contrast, all three higher exposure concentration groups exhibited significantly increased motility (Figure 2) as compared to the control group with a maximum motility at 15 $\mu g L^{-1}$ MC-LR.

The daily swimming activity curves of the highest test concentration (50 $\mu g L^{-1}$) and the control show clear behavioral changes caused by the exposure to MC-LR (Figure 3). Under undisturbed conditions the maximum of daily activity was recorded at 8 am. At this time the mating and spawning behavior occurred every

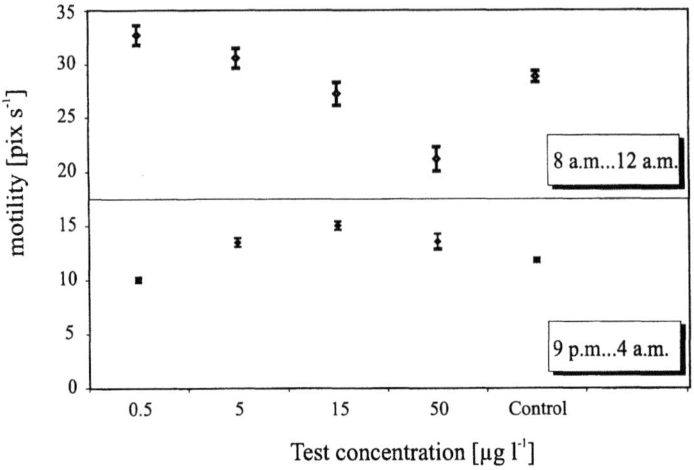

FIGURE 2. Average motility of zebrafish during the exposure to different concentrations of microcystin-LR, open squares indicate the average motility between 8 am and 12 am, solid squares indicate the average motility between 9 pm and 4 am. Means and 95% confidence intervals are shown.

day (see also Westerfield, 1989). More activity peaks were observed at the feeding times (11 am, 3 pm) and in the last hour before dusk. The activity decreased during the first hour after the light was off. Thereafter, fairly constant low activity levels were recorded for the rest of the night. The analysis of daily rhythms clearly showed day-active behavior of the fish.

In contrast, a dramatic reduction of activity was observed at the exposure to MC-LR ($50\,\mu g\,L^{-1}$) during the first hour after switching the light on (8 am). Under undisturbed conditions this is the mating and spawning period. A crucial consequence of MC-LR exposure, was that no spawned eggs were found at this exposure

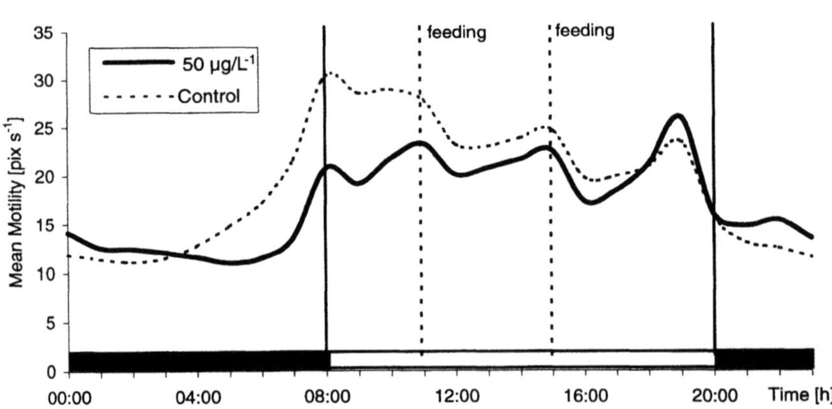

FIGURE 3. Smoothed curve of average motility of all days for the test group with $50\,\mu g\,litre^{-1}$ microcystin-LR in comparison with the control group.

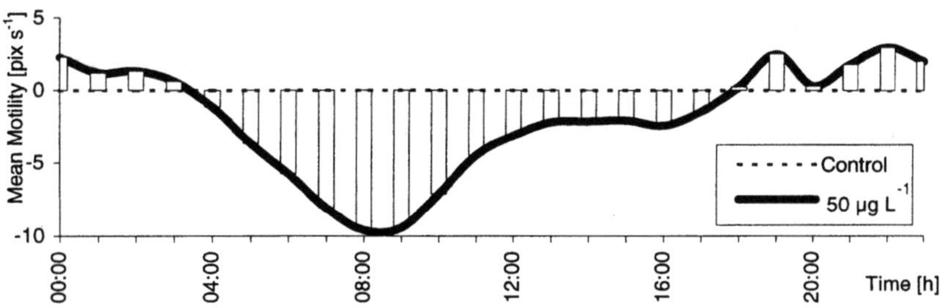

FIGURE 4. Smoothed curve of the difference of average motility of all days between the test group with 50 µg litre^{-1} microcystin-LR and the control group.

concentration. The maximum activity shifted 11 hours, from 8 am (without MC-LR exposure) to 7 pm at the highest concentration level of 50 µg L^{-1} MC-LR. This implied a stressor-induced phase shift of activity. Also, the reaction on feeding (11 am, 13 pm) was reduced at the highest test concentration of MC-LR (50 µg L^{-1}) as compared to the control group.

The absolute differences of the exposure (50 µg L^{-1}) and control groups is shown in Figure 4. Herein, the greatest difference with 10 pixel s^{-1} is found at 8 am, that was a reduction to two thirds of the motility under non-exposed conditions.

4. DISCUSSION

The observed effects on behavior of zebrafish clearly demonstrate that the cyanotoxin microcystin-LR (MC-LR) is a stressor for fish. The reduced motility, the increased rate of nighttime activity compared with daytime activity, the reduced activity during the spawning period and at feeding time, are indications of the hazard potential of MC-LR. In contrast to behavior, MC-LR diluted in water did not produce any obvious histopathological effect in fish (Krüger pers. comm.). Bury et al. (1995) registered an inhibited growth and disturbance of the ion balance in fish exposed to MC-LR at concentrations of 41–57 µg L^{-1}. It is obvious that the effects on the behavior as a toxicity endpoint are rather sensitive: the lowest observed effect concentration of MC-LR (LOEC) appears to be $\leq$ 0.5 µg L^{-1}. During studies on the effects of atrazine on the spontaneous locomotor activity of fish, Steinberg et al. (1995) also confirmed the high sensitivity of behavior as toxicological biomarker: The lowest observed effect of atrazine on behavior occurred at a concentration of three orders of magnitude below the acute toxicity.

A short summary of the conditions of the method follows, in order to answer the question what bioindication can expect from behavior analysis.

4.1. Advantages of the Method Itself

The method is automated, and it is possible to record multiple behavioral parameter descriptions from a single input source (the digitized fish movements). The automated test environment and the applied recording system BehavioQuant allow continuous observation of environmental chemicals and natural toxins as demonstrated with MC-LR. The test method may be applied to other test organisms, as well. A benefit for biomonitoring applications is the sensitive and non-impact measuring method of toxical events.

4.2. Feasibility for Standard Testing

The test system is useful for the detection of discontinuities and slight alterations in the normal behavior, such as shifts in the chronobiological activity (see above). It successfully uses behavior as a sensitive biomarker. Furthermore, the test system is also suitable for substance screening. Provided test conditions are standardized, critical concentrations on a very low level can be measured and potential dose-effect-relationships can be registered.

The test is suitable for long-term and continuous on-line monitoring (Blübaum-Gronau et al., 1992). Especially, in combination with chronobiological aspects, we propose a test, based on this method, has the potential to become a standard. As an early warning system, the test can be used to control water quality. Samples of environmental chemicals may be automatically collected on specific behavior signals and may be analysed afterwards.

4.3. The Integration of Pollution Research and Behavioral Ecology

The integrative behavior reaction on the level of the organism as a whole is an important part of ecological basic research (Scherer, 1992). Decreased activities during spawning and mating time, or during the feeding period, have clear disadvantages for the population dynamics. The registered chronobiological changes also indicate some adverse consequences in the precise temporal fit of the fish into its habitat. The overall effect of reproduction failure is simple to forecast: it may be the extinction of the sensitive part of a population, of a complete population or even a sensitive species.

Behavioral measurements allow the determination of the ecological relevant lowest observed effect concentrations (LOEC) and the no observed effect concentration (NOEC). These threshold values obtained under laboratory conditions may be transformed into a critical load concentration that are quantitative estimates of average expositions of one or more polluting agents, underneath that no significant harmful effects are known on the individual or populations level. Furthermore, behavioral tests could be performed with multiple exposures. This is relevant and environmentally realistic, because most of the anthropogenically caused contamination of the ecosphere is often complex and interferes with biogeochemical matrices. For example, humic material can significantly alter environmental and toxic properties of xenobiotic chemicals, with several instances of increased bioconcentration and toxicity (Haitzer et al., 1998; Steinberg et al., 1992; Steinberg et al., 1994).

The quantifying behavior analysis is an appropriate approach to survey sublethal damages and is suitable to answer ecotoxicological questions (with particular emphasis on ***eco***), i.e. to classify the noxiousness of environmental chemicals in the sublethal range.

5. REFERENCES

Al-Kahem, H.F., Z. Ahmed, and I. Alomer, 1994. A study on the behavioural response and haematological profile of carp, *Cyprinus carpio*, exposed to exogenous urea. Z. Angew. Zool., 80: 97–106.

Alcock, J., 1996. Das Verhalten der Tiere aus evolutionsbiologischer Sicht (Behavior of Animals from the Evolutionary Point of View). Gustav Fischer Stuttgart Jena New York.

Biermann, K., 1992. Circadiane Verhaltensparameter von Cypriniden als Indikatoren ökotoxikologischer Belastungen. Ph.D.-Dissertation, Humboldt-University, Berlin.

Blübaum-Gronau, E., O.H. Spieser, and F. Krebs, 1992. Bewertungskriterien für einen Verhaltensfischtest zur kontinuierlichen Gewässerüberwachung. Schriften-Reihe des Verein des Instituts für Wasser-, Boden- und Luft-Hygiene 89: 333–347, Gustav-Fischer Verlag, Stuttgart.

Boujard, T., and J.F. Leatherland, 1992. Circadian rhythms and feeding time in fishes. Environ. Biol. Fish 35: 109–131.

Braunbeck, T., G. Goerge, V. Storch, and R. Nagel, 1990. Hepatic steatosis in zebra fish (*Brachydanio rerio*) induced by long-term exposure to g-hexachlorocyclohexane. Ecotoxicol. Environ. Saf. 19: 355–374.

Bury, N.R., F.B. Eddy, and G.A. Codd, 1995. The effects of the cyanobacterium *Microcystis aeruginosa*, the cyanobacterial hepatoxin microcystin-LR, and ammonia on growth rate and ionic regulation of brown trout. J. Fish Biol. 46: 1042–1054.

Cousins, I.T., D.J. Bealing, H.A. James, and A. Sutton, 1996. Biodegradation of microcystin-LR by indigenous mixed bacterial populations. Wat. Res. 30: 481–485.

Haitzer, M., S. Höss, W. Traunspurger, and C. Steinberg, 1998. Effects of dissolved organic matter (DOM) on the bioconcentration of organic chemicals in aquatic organisms—a review. Chemosphere 37: 1335–1362.

Jones, J.C., and J.D. Reynolds, 1997. Effects of pollution on reproductive behaviour of fishes. Rev. Fish. Biol. Fish. 7: 463–491.

Mikhaylova, M.V., 1991. Effect of water-soluble fraction of Ust-Balyk petroleum on the early ontogeny of the sterlet (*Acipenser ruthenus*). Hydrobiol. J. 27: 1–10.

Morgan, J.D., G.A. Vigers, A.P. Farrell, D.M. Janz, and J.F. Manville, 1991. Acute avoidance reactions and behavioral responses of juvenile rainbow trout (*Oncorhynchus mykiss*) to Garlon 4), Garlon 3A) and Vision) herbicides. Environ. Toxicol. Chem. 10: 73–79.

Oberemm, A., J. Fastner, and C. Steinberg, 1997. Effects of microcystin-LR and cyanobacterial crude extracts on embryo-larval development of zebrafish (*Danio rerio*). Wat. Res. 31: 2918–2921.

Oberemm, A., J. Becker, G.A. Codd, and C. Steinberg, 1999. Effects of cyanobacterial toxins and aqueous crude extracts of cyanobacteria on the development of fish and amphibians. Environ. Toxicol. Water. Qual. (in press).

OECD Guideline for testing chemicals 210, 17.7.1992. Fish, early life stage toxicity test.

Paul, E.A., and H.A. Simonin, 1996. Effects of naled, synergized, and non-synergized resmethrin on the swimming performance of young trout. Bull. Environ. Contam. Toxicol. 57: 495–502.

Saglio, P., S. Trijasse, and D. Azam, 1996. Behavioral effects of waterborne carbofuran in goldfish. Arch. Environ. Contam. Toxicol. 31: 232–238.

Scherer, E., 1991. Behavioral responses of lake whitefish (Coregonus clupeaformis) to cadmium during preference-avoidance testing Environ. Toxicol. Chem. 10(2): 225–234.

Scherer, E., 1992. Behavioral responses as indicators of environmental alterations, approaches, results, developments. J. Appl. Ichthyiol. 8: 122–131.

Smith, E.H., and H.C. Bailey, 1990. Preference/avoidance testing of waste discharges on anadromous fish. Environ. Toxicol. Chem. 9: 77–86.

Spieser, O.H., and W. Scholz, 1992. Verfahren zur quantitativen Bewegungsanalyse von mehreren Objekten im selben Medium. (A Method of Quantitative Movement Analysis of Multiple Objects in the Same Medium.) German Patent P 4224750.0.

Staaks, G., 1996. Experimental studies on temperature preference behaviour of juvenile cyprinids. Limnologica 26: 165–177.

Steinberg, C.E.W., A. Sturm, J. Kelbel, S.K. Lee, L. Hertkorn, D. Freitag, and A. Kettrup, 1992. Changes of acute toxicity of organic chemicals to *Daphnia magna* in the presence of dissolved humic material (DHM). Acta Hydrochim. Hydrobiol. 20: 326–332.

Steinberg, C.E.W., C. Mayr, R. Lorenz, O.H. Spieser, and A. Kettrup, 1994. Dissolved humic material amplifies irritant effects of terbutylazine (triazine herbicide) on fish. Naturwissenschaften 81: 225–227.

Steinberg, C.E.W., R. Lorenz, and O.H. Spieser, 1995. Effects of atrazine on swimming behaviour of zebrafish, *Brachydanio rerio*. Wat. Res. 29: 981–985.

Tembrock, G., 1992. Verhaltensbiologie 2 (Behavioral Biology 2). Revised Edition. Jena Fischer Verlag.

Weber, D.N. 1993. Exposure to sublethal levels of waterborne lead alters reproductive behavior patterns in fathead minnows (*Pimephales promelas*). Neurotoxicology 14: 347–358.

Weber, D.N., A. Russo, D.B. Seale, and R.E. Spieler, 1991. Waterborne lead affects feeding abilities and neurotransmitter levels of juvenile fathead minnows (*Pimephales promelas*). Aquat. Toxicol. 21: 71–80.

Welker, M., and C. Steinberg, 1999. Indirect photolysis of cyanotoxins: one possible mechanism of their low persistence. Wat. Res. (in press).

Westerfield, M., 1989. The Zebrafish Book; A Guide for the Laboratory Use of Zebrafish (*Brachydanio rerio*). University of Oregon Press, Eugene.

8

CONTINUOUS WATER MONITORING

Changes of Behavior Patterns as Indicators of Pollutants

ELKE BLÜBAUM-GRONAU,[1] MICHAEL HOFFMANN,[1] O. HUNRICH SPIESER,[2] AND WILFRED SCHOLZ[†]

[1]German Federal Institute of Hydrology, Kaiserin-Augusta-Anlagen 15–17, 56068 Koblenz, Germany. [2]GSF-National Research Center for the Environment and Health, Institute of Toxicology, Neuherberg, 85758 Oberschleißheim, Germany. [†]In memorial to Wilfried Scholz. Without his great personal engagement the BehavioQuant® system had never reached the actual state of development.

1. BIOMONITORS IN WATER MONITORING

The detection of sudden contamination pulses in surface waters has gained increasing importance over the past few years with view to drinking water production and the protection of aquatic life communities. An efficient monitoring system for practical applications cannot be based alone on physico-chemical analyses because, on the one hand, suitable analytical methods are not always available and, on the other, continuous analyses of all relevant parameters are hardly practicable for reasons of time and economy (Juhnke and Besch, 1971; van Hoof, 1980; Nusch, 1993). Moreover, measured concentrations do not provide any information about the biological availability of the detected substance. Neither does one know which effects the compound has on aquatic organisms in combination with other substances to which they may be exposed simultaneously. Estimates about the unnatural compounds which occur in the River Rhine range in an order of magnitude between 30,000 to 50,000 (Botterweg, 1988), of which only about 150 to 200 substances are covered by routine analyses (Botterweg et al., 1989). Furthermore, the interpretation of the results of chemical routine analyses by reference to available toxicity data is problematic, since toxicity data that were measured under standardized laboratory conditions are hardly relevant for field conditions, because they leave the influence of the matrix of the water body out of account. For instance, humic acids, which occur ubiquitously, may alter the biological effectivity of contaminants by some orders of magnitude (Mayr, 1992).

For these reasons continuous biotests are a reasonable supplementation to chemical monitoring systems, although the signal from a biotest system is only an indication of the presence of toxicity in the water. The use of organisms is also supported by the fact that most chemical analyses take a relatively long time (hours to days), while organisms often respond within minutes (Kramer and Botterweg, 1991). The reliability of the summative indication of health-impairing substances in water can be achieved—according to Mäckle and Stabel (1989)—by using in the biotest organisms which belong to different systematic classifications and consequently have different tolerance ranges (such as bacteria, algae, *Daphnia*, mussels and fish). Life forms that differ fundamentally in physiological terms (bacteria, plants, animals) can detect a larger number of different effects of potentially present environmental chemicals (Peichl and Schmidt-Bleek, 1986).

In principle, all characteristics of normal behavior of an organism may be harnessed to monitor environmental impacts. In order to detect as many different chemical-induced effects in the same organism as possible, it is reasonable to choose such test criteria which depend on a wide spectrum of physiological processes. Behavioral changes, for instance, integrate many cellular processes, and they are essential for the life and survival of the organism, the population and the ecosystem. Thus, the observation of behavioral parameters offers the unique toxicological perspective to register sublethal effects. Behavioral changes are consequently a link between biochemical and ecological consequences of contaminant releases into the environment.

The behavioral response of aquatic organisms has been frequently employed as an indicator of contaminants (Warner et al., 1966; Anderson, 1971; Olla, 1974; Dicks, 1976; Larrick et al., 1978). Four principal lines of inquiry can be distinguished in the analysis of the behavior of aquatic organisms (Giattina and Garton, 1983): First, the locomotor responses, primarily preference-avoidance reactions and activity changes. Secondly, ventilatory responses (primarily opercular movements and coughs). Thirdly, behavioral changes which are based on predator-prey interactions, and fourth, toxicant-induced changes in conditioned behavior. The compilation by Beitinger and Freemann (1983), which considers 75 chemicals and 23 fish species, notes that behavioral responses were observed with 61.3% of the chemicals. Of these behavioral responses 37.3% could be characterized as constant avoidance reactions, with the majority of thresholds for avoidance reactions clearly remaining below the lethal concentration (LC_{50} 24 hh or 96 hh). These low thresholds may also be explained by the fact that fish perceive numerous chemicals with their very sensitive chemical sense. Because they taste contaminants already in concentrations which do not cause any physiological damage, these avoidance responses are a highly effective survival strategy.

A detailed analysis of behavior patterns used to be a very labor-intensive and thus ineffective approach. Moreover, the quantification of the data was problematic. A new approach is the use of video systems as they were described by Lubinski et al. (1977) as well as Miller et al. (1982), Smith and Bailey (1988) and Baillieul and Scheunders (1998). However, only the presently available efficient and fast PCs facilitate an automatic 24-hour observation of aquatic organisms. Such a method for objective quantification of the behavior of organisms, which are not exposed to any physical or psychological stress, is offered by the system BehavioQuant® that was introduced by Spieser and Yediler (1986) (Spieser and Frisch, 1985; 1988; Spieser and Scholz, 1992; Scholz, 1994).

2. BIOMONITORS WITH THE IMAGE-PROCESSING SYSTEM BEHAVIOQUANT®

This system uses an image processing unit for simultaneously analysing both locomotor responses and intraspecific relations. The fact that single behavioral parameters may be influenced in varying degrees by a substance, like the herbicide atrazine, had already been proved by Spieser and Freitag in 1984. BehavioQuant® can control up to 16 video cameras via a multiplexer. This feature can monitor a battery of test organisms with one measuring system. At the German Federal Institute of Hydrology, the measuring system BehavioQuant® was adapted to continous water monitoring with fish and *Daphnia* as test organisms.

2.1. Function of the Image-processing System BehavioQuant®

BehavioQuant® is a computer-controlled, image-processing data-acquisition system. The acquisition of data—as well as the interpretation and evaluation of the stored position coordinates of moving objects—is executed by special software, so that the fully automatic operation of the behavioral test is possible. The behavior of the test animals (e.g. fish and *Daphnia*) in several measuring chambers can be observed and computed in cycles over a freely chosen period of time. The duration of the monitoring period is defined by the user when he stops the system for the regular inspection and maintenance (e.g. every 7 days). Without such external intervention, data recording, evaluation and alarming is going on continuously. In the automatic operation mode, a maximum of 100 measurements per 24 hrs are possible, what is equal to a cycle length of 14.4 mins. During this period, the data of all measuring chambers are recorded and evaluated. When the maximum 16 chambers of such a unit are polled sequentially, information like image background, image brightness, contrast, level of object recognition, size and number of observed objects, positive or negative image (dark objects against a light background or light objects against a dark background) is also recorded for each camera separately.

The system is able to quantify a multitude of unlabelled objects in one and the same observation chamber. The principle of object recognition is a pixel-by-pixel comparison of successive video images with a reference background image. This reference background image is generated by averaging a larger number of digitized single images from a previous measurement. Through the great number of relocations of the moving objects, they occupy the pixels only for short moments, so that the average values of these pixels are nearly equal in brightness to the background.

This method generates a reference background image on which the moving objects are not visible, although they were present in the observation chamber throughout the preparatory measurement.

This procedure prevents an integration of the observed objects into the image background. In practice, this means that the intervals between the single pictures can be prolonged, when animals are observed who hardly change their position, e.g. in winter, when the mobility of organisms is noticeably reduced. Generating the reference background image regularly anew is done to avoid that the dirt particles, which may form in the intervals between maintenance inspections, disturbing the measurement as phantom objects or are registered as objects. Moreover, it prevents an impairment of image recognition by the aufwuchs that develops on the aquarium walls in the course of one week. In the evaluation of the video images, the positions and dimensions of the objects are stored in form of X, Y coordinates in a grid overlaying the whole frame. The assignment of objects to the momentary position coordinates is based on plausibility criteria of their locomotion. The dimensional coordinates of the objects are checked for the criterion whether these are objects within the given size range. The size of objects to be registered is defined by the operator with a view to the size of the test organisms. Other particles, e.g. fish feces, are thus eliminated. Averaging the initial and final coordinates gives the center point of the objects.

According to the underlying plausibility criteria and because of the rapid image sequence of 40 msec, it can be assumed that it is possible to assign to a pair of coordinates in the first image, the nearest coordinates pair of the second image. The vector that connects the two points indicates a direction and velocity of movement. Together with the positions in the second image, these vectors provide the basis for the assignments in the third image. If two objects superimpose for a short time, it is assumed that the direction and velocity of the movement of the object before and after crossing the other object are similar. In combination with some more criteria, a reconstruction of the locomotor behavior can be derived. The computed tracks, however, are inaccurate due to the shortening of distances in the plane projection. This error can be minimized by keeping the space depth of the measuring chamber short.

During the measuring cycles, five behavioral parameters are computed from the X, Y coordinates of the reconstructed tracks of all test organisms.

2.2. Definition of the Behavioral Parameters

The observed parameters of behavior can be related to the different qualities of activity, preferred locations in the tank and the intraspecific relations within the group. The activity is characterized by the parameters motility, number of turnings and inconstancy of movement; the preferred locations are expressed in form of frequency distributions of the swimming height and the positions in the dimension X.

Motility is computed for each object on the basis of the path on which it moved per unit of time (coordinate distances per second). The variations of motility which are shown indicate whether the observed object was equally active throughout the measuring interval or whether longer interruptions occurred between phases of increased activity. However, in routine monitoring this information is left out of consideration, it is more relevant in research applications of the system. To minimize the error through plane projection, the **number of turns** of the test organisms is additionally recorded as changes in the X direction. For further statistical computations the data are available as mean values with standard deviation. Then, the **inconstancy of movement** is obtained by computing from the reconstructed tracks of all objects for each two-minute measurement the swimming velocity of each object for successive 10 sec intervals. These data are normalized for all objects together in frequency distributions with the X axis "gridded swimming velocity" and the Y axis "frequency" within a grid of 10 sec swimming velocity.

For determining the **swimming height** the three-dimensional swimming space is projected onto a plane and covered with a coordinates grid. The Y axis is split into ten classes which designate the swimming height, with the swimming height "0" being directly below the water surface. For further statistical computations, the swimming heights of all objects are transferred into frequency distributions with the X axis "swimming height" and the Y axis "frequency" according to the procedure given for the constancy of swimming velocity.

The mean swimming height of a single object results from its averaged positions in the Y axis. Also in the case of the determination of the **horizontal place of preference** of the test organisms, the coordinate points of the X axis are subdivided into ten classes. In analogy to the computation of the swimming height behavior, frequency distributions are established for the horizontal positions.

The mean horizontal place of preference is found by averaging the positions of all animals along the X axis.

2.3. Statistical Alarm Evaluation

The statistical processing in the continuous water monitoring routine is performed automatically after each measurement. It pursues the strategy to check whether the measured data of a few recent single measurements differ from those measurements made before a possible exposure to contaminants.

The statistical evaluation is based on the following reflections: it cannot be assumed that the values computed for the behavioral parameters follow a normal distribution curve. This is particularly true for swimming height, where (depending on the character and intensity of a chemical effect) deviations have to be expected upwards **and** downwards or upwards **or** downwards. Additionally, a certain appetitive behavior at interfaces and local preferences as well as interactions between the animals influence their horizontal and vertical distribution in space.

A statistical evalution of behavioral responses is possible for instance by the Wilcoxon, Mann and Whitney test, also called U-test (Sachs, 1984). As a distribution-free statistical test for changes of the behavior patterns considered here, the measuring system BehavioQuant® executes the two-tailed U-test for the frequency distribution of the velocity of movement and for the horizontal and vertical positions. Additionally, the mean values of motility, swimming height and of the horizontal position are subjected to the U-test.

For defining the alarm, the possible differences are statistically determined between the current and the previous control measurements, the U-test according to every class of the frequency distribution of the individual behavioral parameters is used.

Thus, a single U-value is formed for each class. This approach is reasonable, because changes in the frequency distribution of the velocity of movement become first manifest in the increase or decrease of single classes. The analysis of the vertical and horizontal position uses the same procedure.

As false alarms must be strictly avoided, the comparative computations are of such a kind that external influences on behavior are of no consequence. Such external influences include among others the seasonal variations of the physical parameters of the water body, such as temperature. Here, a reference period of six hours, which corresponds to 24 measurements in 15-minute intervals, before the actual measurement is appropriate. As a safeguard against false alarms, the two previous measurements are included in the statistical comparison with the actual measurement, so that three current sets of measured data stand against the 24 sets of reference values. In order to detect slow continuous intoxication, which induces only slight, insignificant changes of behavioral parameters, during routine evaluation a distance of two measuring cycles is allowed between the 24 reference measurements and the current measurements.

The statistical comparisons are performed independently from each other for all five behavioral parameters. Only when the standards are exceeded simultaneously by several behavioral parameters registered, are alarm signals triggered, which are graded depending on the degree of deviation from the standard values, and transmitted e.g. to a central computer in a monitoring station. The transmission of a pre-alarm signal or a main alarm depends on the number of computed significant differences between the data of the three recent measurements and the measurements, which are added up separately for each parameter. Then, in consideration of the comparisons made for the respective parameter (e.g. the number of considered classes in the inconstancy of movement) these sums are weighted, so that the equal treatment of the different parameters is ensured irrespective of the size of the grid used for the computed classes. The weighted significances are then added up, and the result is compared with three alarm thresholds, which represent the pre-alarm and the main alarm. The alarm thresholds can be modified by the operator. If the highest threshold is exceeded, a main alarm is triggered directly (e.g. total disruption of any behavioral activity, while a permanent self-test of the system avoids false alarms here). Exceeding the mean threshold triggers a pre-alarm. If the two immediately preceding computations have already reached the pre-alarm level, the main alarm is triggered, if necessary, directly after computing the third measurement. If the lowest alarm threshold is exceeded, the event will be merely recorded. If this happens three times in sequence, a pre-alarm is released.

2.4. The Koblenz Behavioral Fish Test with BehavioQuant®

The Koblenz Behavioral Fish Test with the image processing system BehavioQuant® (Figure 1) is used in routine operations at the German Federal Institute of Hydrology with the fish golden ide *Leuciscus idus melanotus*, because the same species is employed also in the static biotest according the DIN Standard 38412 L15. The advantages are that for comparing the sensitivity of the behavioral test, the EC or LC values from the static test could be used and the availability of animals of appropriate size throughout the year. Another important argument is the temperature tolerance of this fish, which allows monitoring the water at all times of the year. In principle, this continuous fish test may employ any fish species that is native in the water body to be monitored. The native species acts as a representative of the ecosystem.

The observation tanks for the Koblenz Behavioral Fish Test were designed to accommodate a test-fish shoal of six (5–8 cm in size). The swimming space has a width of 50 cm, a height of 40 cm, and a depth of 10 cm. These aquaria were especially adapted for the requirements of continuous water monitoring. The shallow depth of 10 cm makes it possible to record the fish by video camera, even when concentrations of suspended matter are high.

To make sure that the dimensions of the tanks do not influence the behavior of the animals, a previous test series with reductions of the swimming space was executed (Blübaum-Gronau et al., 1992).

Moreover, the design of the tanks ensures that the flow conditions that are established in the tanks do not disturb the swimming behavior of the test fish. In previous tests it was found that the developing flow conditions does not induce any significant behavioral changes. In routine monitoring application of the Koblenz Behavioral Fish Test with BehavioQuant® two tanks are operated in parallel.

2.5. Daphnia Test with BehavioQuant®

For continuous monitoring of the water-flea *Daphnia magna*, the BehavioQuant® system was supplemented by another measuring unit. This unit contains measuring chambers that were developed at the GSF-National Research Center for the Environment and Health (Figure 2). The daphina are observed via a mirror that is mounted at an angle of 45° in front of the chambers, so that the space requirements of the measuring system are kept as low as possible (Figure 3).

The system is completed by a second slide-in unit which holds four pumps to control the inflow and outflow of the test water. In routine operation every 15 minutes video images are recorded of the ten daphnia in each chamber over 2 minutes and immediately interpreted and assessed automatically (see sections **Funtion of the image-processing system BehavioQuant®** and **Definition of the behavioral parameters** above).

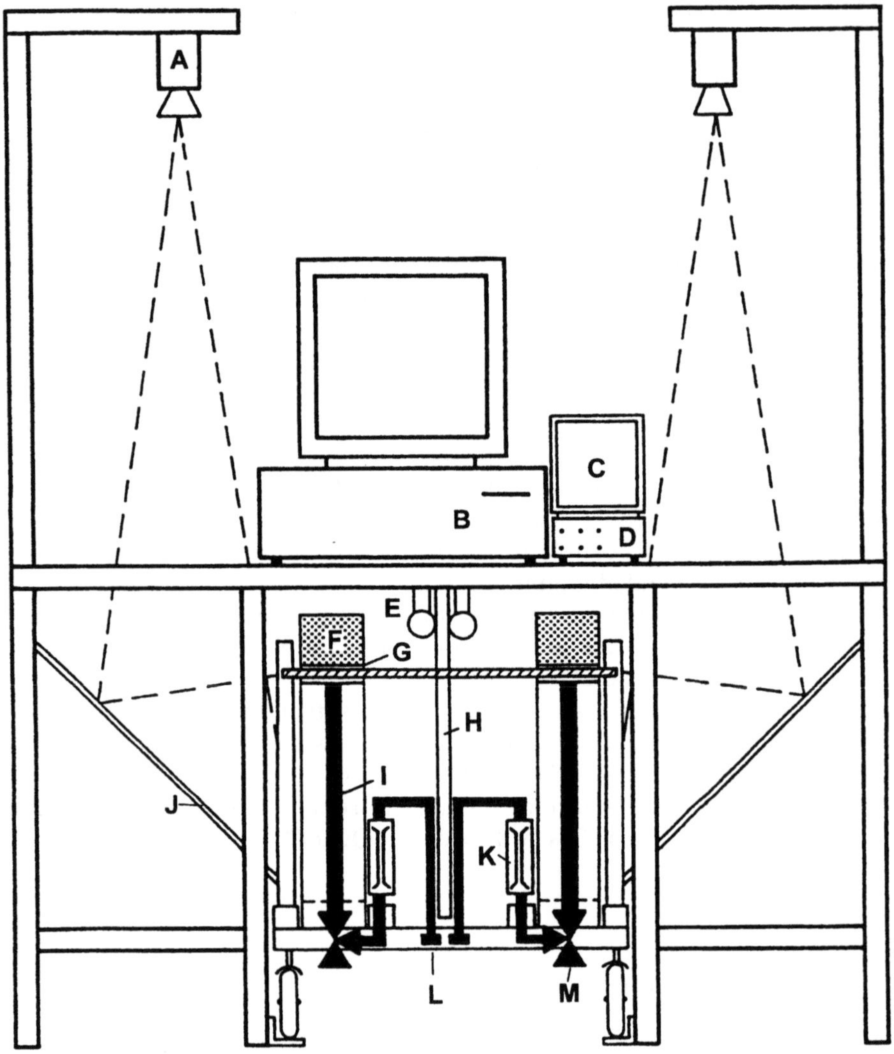

FIGURE 1. Schematic view of the Koblenz Behavioral Fish Test on the basis of the measuring system BehavioQuant® (A) video camera, (B) image processing system, (C) control monitor, (D) multiplexer, (E) light, (F) test tank (including the six Golden Ide), (G) overflow, (H) diving wall, (I) outlet, (J) mirror, (K) fluid-flow measurement, (L) water supply, (M) 3-way valve.

The fish are monitored by one video camera per tank suspended above mirrors that are mounted at an angle of 45° in front of the tank. The water, that is fed through a central opening in the tank bottom, is discharged again over the shortened walls on both sides. A perforated high-quality steel plate (perforation diameter 5 mm), that is loosely suspended about 5 cm above the tank bottom, functions as an equalizer. The inflow of test water is regulated at 5 L/min by means of flow meters.

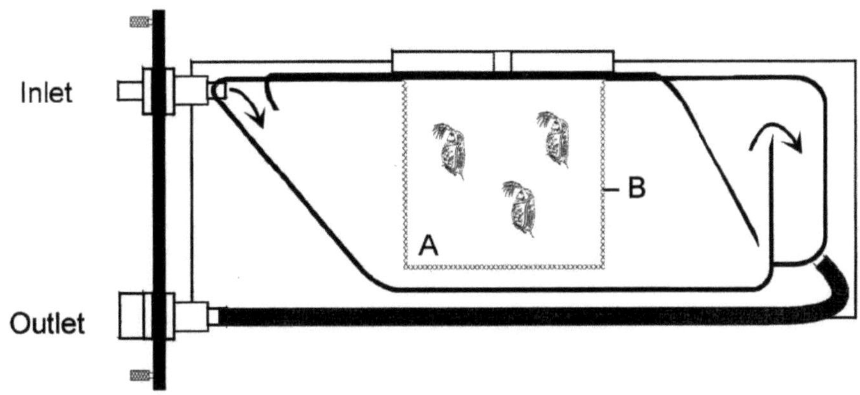

FIGURE 2. Side view of the measuring chamber for the observation of *Daphnia*. The main feature of the chambers is a nearly laminar flow which does not influence the swimming behavior of *Daphnia*. The actual swimming space (A) with dimension of 7.5 × 6 × 1 cm is separated from the total volume (125 mL) of these chambers by nets of high-quality steel (B). Water flow is set at 0.5 l/hr.

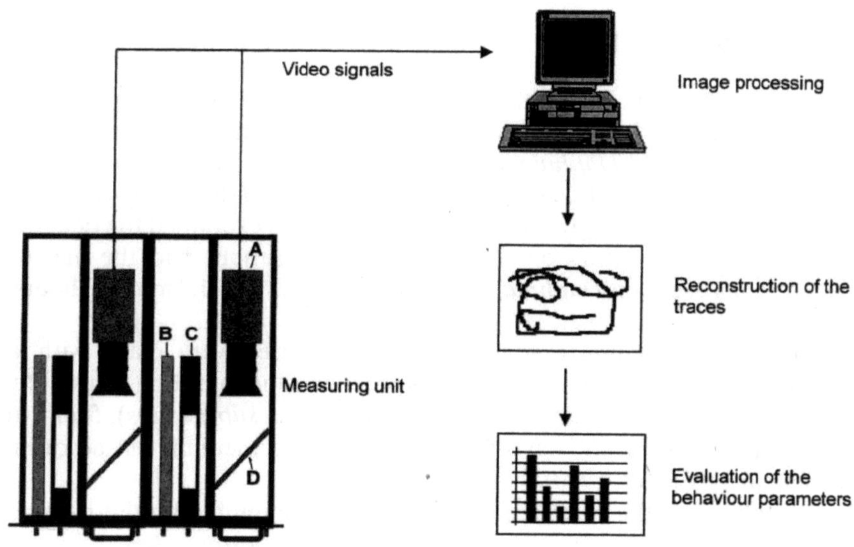

FIGURE 3. Scheme of a BehavioQuant® system with a measuring unit for *Daphnia* (A) video camera, (B) light, (C) test chamber, (D) mirror.

The chambers are integrated into a 19-inch slide-in unit which accommodates two more modules per measuring chamber which contain a electro-luminescence film for lighting and a video camera.

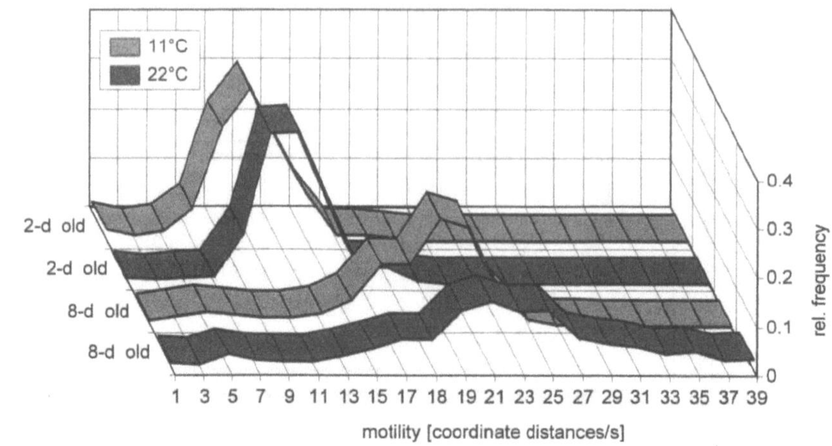

FIGURE 4. Relative frequency of the velocity of movement of 2 and 8 days old daphnia at different temperatures in water from the River Rhine. It is clearly noticeable that with growing age (2 versus 8 days old) the most frequent swimming velocity of 9 to 13 coordinate distances per second (2.7 to 3.9 mm/s) increases to 19 to 25 coordinate distances per second (5.7 to 7.5 mm/s). Moreover, the curves show that the maximum of the frequency distributions at a higher water temperature shift towards higher swimming velocity.

2.5.1. Efficiency of the Daphnia Test with BehavioQuant®

Before the Daphnia Test with BehavioQuant® could be used for continuous water monitoring, it had to be ensured that the measuring system could reliably identify 24 to 48-hour old *Daphnia* as objects. Moreover, it had to be demonstrated that the daphnia responded to changing environmental conditions by characteristic behavioral changes. For that purpose, different test series examined the dependence of the behavioral parameters on temperature, food supply, and the age of the daphnia. The tests used ten *Daphnia* of sizes between 1.1 and 1.4 mm (48 hours old) per measuring chamber.

Except for the feeding experiments, all test series were made with water from the River Rhine at determined temperature and aeration status. In the feeding experiments, an amount of algae (*Scenedesmus subspicatus*), 5 or 20 µg/L of chlorophyll-a, respectively, was added to the culture medium M4 according to Elendt-Schneider (Elendt, 1990).

In the following, the behavioral changes are exemplified by the parameters: inconstancy of movement, mean motility and number of turns.

Figure 4 shows the relative frequency of the velocity of movement of *Daphnia* of different ages. It is clearly noticeable that with growing age the most frequent swimming velocity increases. Moreover, it shows that irrespective of the age of the animals, the maximums of the frequency distributions at a water temperature of 22 °C shift towards higher swimming velocity. In a comparison of the respective distribution curves of young and old *Daphnia* it is striking that the curve for the older animals is broader, i.e. the range of observed velocity is markedly wider.

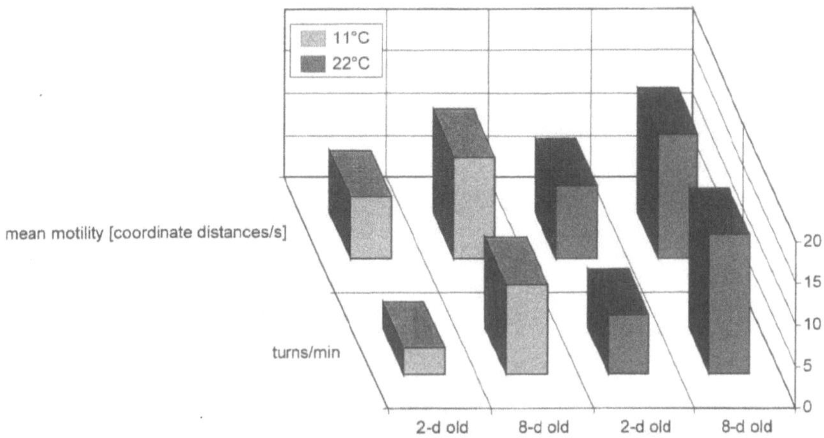

FIGURE 5. Mean motility and number of turns of 2- and 8-days-old *Daphnia* at different temperatures in water from the River Rhine. At a water temperature of 11 °C the mean motility increases from 7 coordinate distances per second (2-days old) to 12 coordinate distances per second for animals aged 8 days. The number of turns per minute rises from 3 (2-day-old) to 11 (8-days old). At a water temperature of 22 °C the increase of the swimming velocity is comparable with the result of 11 °C, but a much lower increase can be noted in the number of turns of older *Daphnia*.

Figure 5 presents the behavioral parameters mean motility and number of turns. These parameters also illustrate the different activity spectrum of younger and older daphnia. At a water temperature of 11 °C the mean motility of older daphnia (8-day old) is 42% higher than the motility of younger daphnia (2-day old). In percentages, a much higher increase (70%) can be noted in the number of turns. A comparable increase of the behavioral parameters motility and turns per second can also be noted at a water temperature of 22 °C. While the swimming velocity rose as at the lower temperature by about 40% with older animals, the number of turns increased by a mere 58%. If one considers the temperature dependence (temperature rise from 11 to 22 °C) of the parameters motility and number of turns separately for *Daphnia* aged 2 days and those aged 8 days, it shows that with a doubling of the surrounding temperature the increase in motility of 15% (2-days old) and 17% (8-days old) remains in the same order of magnitude, while the number of turns of 2-days old *Daphnia* increased by 56% and that of the 8-days-olds rose only by 36%.

If one considers the change in swimming velocity of 2 and 8 days-old *Daphnia* at 22 °C depending on food supply, then one finds in Figure 6 that with 5 µg as well as with 20 µg of chlorophyll-a per litre the frequency distributions have similar shapes to those from the tests with water from the River Rhine (at 22 °C). The maximums of the curves are also in the same order of magnitude. It is striking that with older animals a lower food supply causes comparable changes in the frequency distribution, as observed with increased temperature and a constant food supply. An explanation of this phenomenon in the two test arrangements can be traced to food shortage or an imbalance between food supply and water temperature, since higher temperatures induce intensified metabolic activity and consequently food demand.

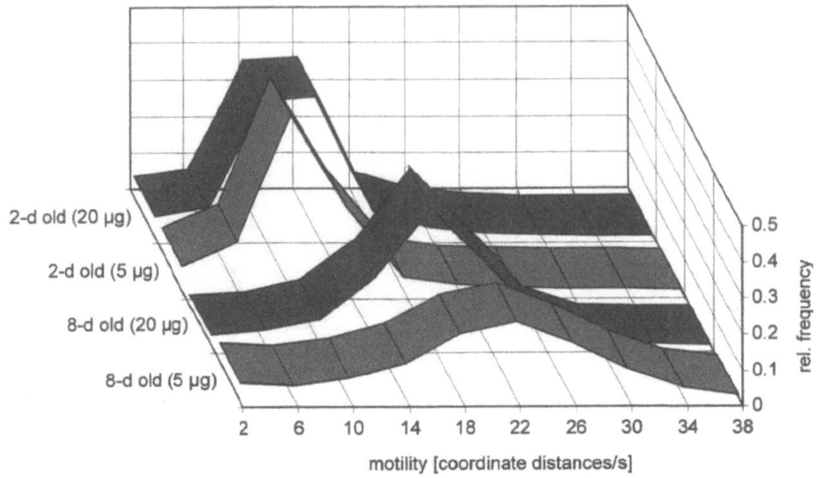

FIGURE 6. Relative frequency of the velocity of movement of 2- and 8-days old *Daphnia* with different food supplies (5µg or 20µg chlorophyll-a per L). The temperature of the medium was 22 °C. It is clearly noticeable that with growing age (2-versus 8-days old) the *Daphnia* needs more food. The greater number of motility classes being placed by 8-days old animals indicates a different behavior of the single *Daphnia*. On the other side, the frequency distribution of 2-days old *Daphnia* shows at 5 and 20µg chlorophyll-a nearly the same maximum.

3. SPIKING EXPERIMENTS WITH SODIUM PENTACHLOROPHENOLATE, α-ENDOSULFAN AND LINDANE

The sensitivity of these new continuous biotests with BehavioQuant® was determined by means of so-called spiking experiments (increasing concentration series). Immediately before the beginning of the experiment, water from the River Rhine was collected in a separate basin and mixed with the test substance in increasing concentrations and then circulated through the systems.

3.1. Results of the Spiking Experiment with the Koblenz Behavioral Fish Test with BehavioQuant®

The sensitivity of the Koblenz Behavioral Fish Test was determined with the reference chemicals sodium pentachlorophenolate (Blübaum-Gronau et al., 1994) and α-endosulfan (Ziegler, 1994).

These experiments showed that it is possible to detect in surface waters substances in sublethal concentrations by means of behavioral changes, especially changed locomotor behavior and preference-avoidance reactions. After recording reference values (pre-test: 2h in 15-min intervals), the spiked river water cirwlated for about 30min through the system. As changes in the frequency distribution of the considered behavioral parameter manifested themselves first in increases and

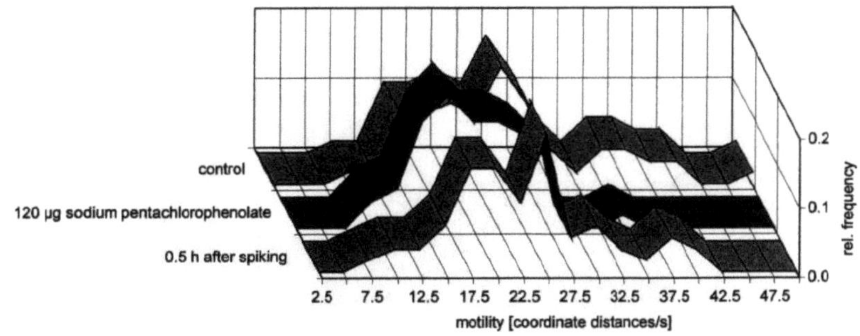

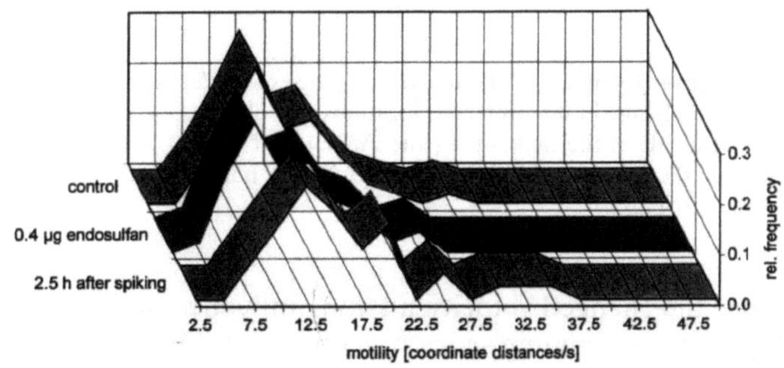

FIGURE 7. Relative frequency of velocity of movement of fish before, during and after the spiking of 120 μg/L sodium pentachlorophenolate (top) and 0.4 μg/L α-endosulfan (bottom). The two tests series are carried out at different seasons with different shoal of fish. Both tests showed that the **velocity of movement** of the fish was reduced. Under the impact of 120 μg/L sodium pentachlorophenolate, the maximum of the frequency distribution of motility shifts from class 10 to class 7. This corresponds to a decrease in the velocity from 23.8 coordinate distances (4.7 cm/s) to 16.3 coordinate distances (3.2 cm/s). Endosulfan concentrations of 0.4, 0.8 and 1.6 μg/L also caused a decrease in velocity by nearly 30%.

decreases of a single class, the data of the individual classes are subjected to the U-test for statistical evaluation.

The tests with sodium pentachlorophenolate and α-endosulfan showed both that the **velocity of movement** of the fish was reduced. The frequency distribution induced by sodium pentachlorophenolate shows a decrease of the maximum from 25 to 20 coordinate distances per second. A similar effect was observed under the influence of 0.4 μg/L endosulfan, i.e. the frequency distributions (consisting of 20 classes) shift toward the left-hand side (Figure 7).

The duration of the changed behavior depended both on the concentration and the substance. While in the experiment with 120 μg/L sodium pentachlorophenolate the behavior of the fish had returned to normal after 30 minutes, the observed effects disappeared only after 60 min at a concentration of 240 μg/L. The effects induced by the highly fish-toxic α-endosulfan (0.4 μg/L) began to reverse after 2.5 hrs. As a rule, recovery of the fish regarding their velocity of movement was reached only after 5 hrs.

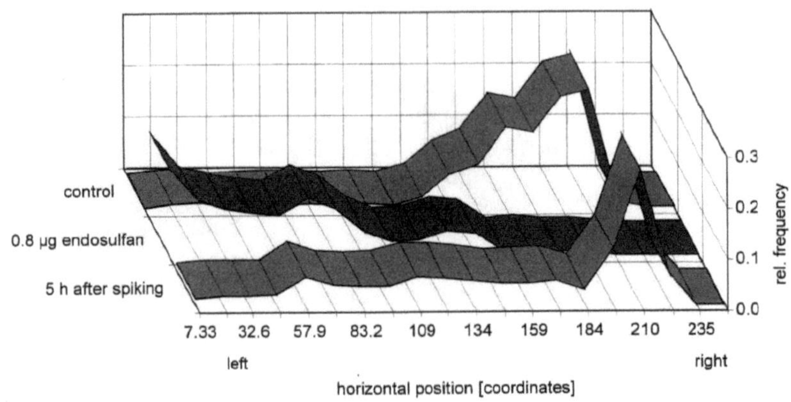

FIGURE 8. Relative frequency of the horizontal place of preference of fish with 0.8 µg/L (α-endosulfan. A pronounced change in the preferred locations was observed. While in the control sample the fish preferred the right side of the tank, under the influence of 0.8 µg/L endosulfan, the frequency distribution indicates that the fish changed to the left side. After 5 hours the fish have been changed back to the right side of the test tank.

Besides the changed velocity, a strong influence on the locations preferred by the fish within the swimming space was observed (Figure 8). Regarding the behavioral parameter **horizontal place of preference**, pre-studies had revealed that the animals always prefer a certain half of the swimming space depending on the test tanks used. This can be explained by the fact that the flow in the tanks is not exactly laminar, so that the water reaches certain zones earlier. These are the zones preferred by the fish. However, if the contaminant concentration in the water increases, the more highly contaminated water also reaches these zones first. The fish respond by changing their locations (avoidance behavior) towards those zones in the swimming space where the contaminant concentration increases only with some minutes delay.

This pronounced change in the preferred locations after addition of chemicals was observed equally with both reference chemicals. The duration of the induced change in behavior was proved to depend also on concentration and substance. While the effects of sodium pentachlorophenolate had faded away after 30 to 60 mins, the fish were observed in the originally preferred zone in the swimming space only nearly 5 hrs after the application of endosulfan.

In summary, the spiking experiments with river water from the River Rhine yielded the following result: the Koblenz Behavioral Fish Test records contaminants in sublethal concentrations within 30 mins, and the alarm is triggered at a time when the test fish have not yet been harmed.

3.2. Results of the Spiking Experiment with the Daphnia Test with BehavioQuant®

The sensitivity of the system was tested in a spiking experiment with the pesticide lindane, in order to find indications for the suitability of this continuous

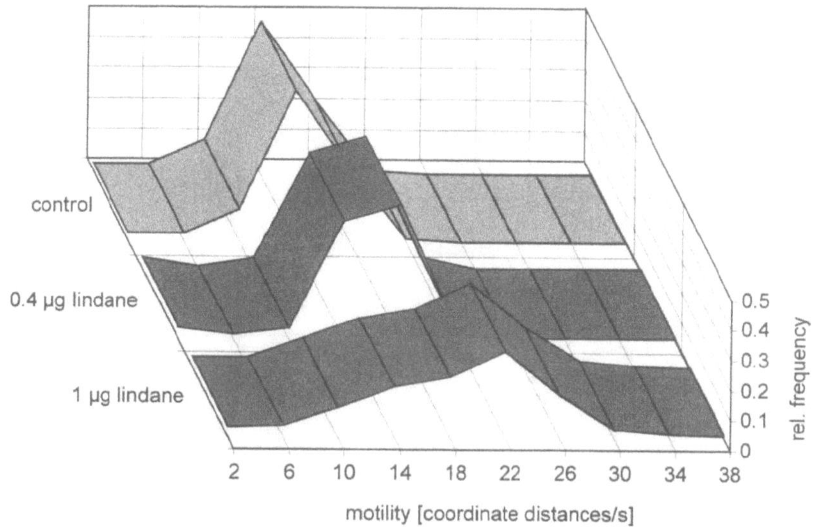

FIGURE 9. Relative frequency of the velocity of movement of 6-days old daphnia at 0.4 and 1 μg/L lindane in water from the River Rhine (11 °C). At a concentration of 0.4 μg/L, the maximum of the frequency distribution shifts towards higher velocity, and the number of classes covered on the whole increases. The spectrum of mean velocity reaches under the influence of 0.4 μg lindane per litre from 2 to 26 coordinate distances/s, while in the control sample velocity were only between 6 and 22 coordinate distances/s. This effect becomes more obvious at a concentration of 1 μg of lindane per litre of river water.

biotest for surface water monitoring. The River Rhine water at 11 °C was spiked with 1 and 0.4 μg lindane/L, respectively. The behavior of 6-days old *Daphnia* was registered over a period of two hrs in 10 mins intervals. After this pre-test, lindane-containing water was fed into the system. Figure 9 shows the change in the inconstancy of movement under the influence of this pesticide. Already at a concentration of 0.4 μg/L, the maximum of the frequency distribution shifts towards higher velocity, and the number of classes covered on the whole increases. This effect becomes more obvious at a concentration of 1 μg of lindane per litre of river water. In this case the maximum of the frequency distribution shifts from the class 10 to 14 coordinate distances per second in the reference test arrangement to the class 18 to 22 coordinate distances per second. Moreover, the spectrum of the mean velocity again widened, and up to 38 co-ordinate distances/s were registered. This widening of the frequency distribution is a clear indication that the individual daphnia respond individually to changes in their environment.

Figure 10 also illustrates the concentration-dependent changes in the mean motility and the number of turns. Particularly striking here is the unproportionally high increase in the number of turns. This suggests that the measuring system BehavioQuant® enables the recognition of the typical behavior of daphnia, namely the turning over of the head, even under poor environmental conditions.

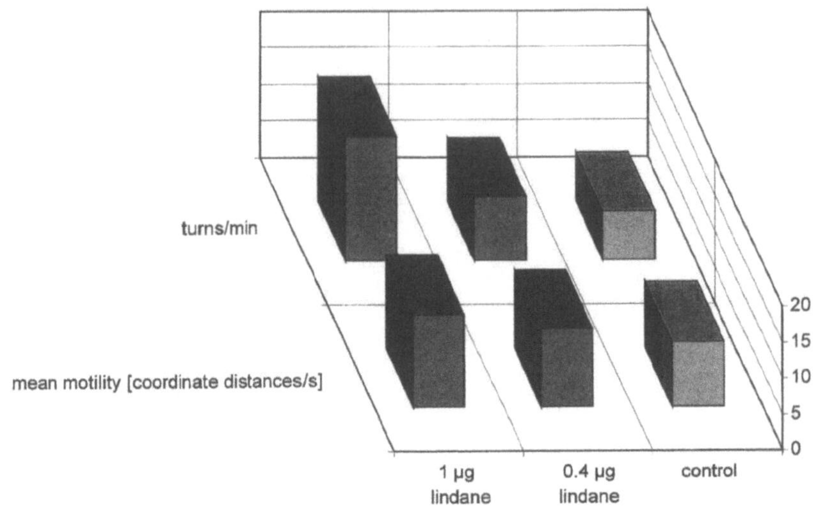

FIGURE 10. Mean motility and number of turns of 6-days old *Daphnia* at 0.4 and 1 µg/L lindane in water from the River Rhine (11 °C). The unproportionally high increase in the number of turns from 6.7 to 17.2 turns/min (1 µg/L) suggests that the measuring system BehavioQuant® enables the recognition of the typical turning over the head of the *Daphnia*. The increase of the mean motility is only 30% under the influence of 1 µg lindane/L.

4. CONCLUSIONS

This paper shows that the behavioral measuring system BehavioQuant® can be used in the continuous monitoring of surface waters. The system meets the requirement for a continuous biotest, namely to give a rapid response (alarm) to the occurrence of toxic pollutants within a period of a few minutes to half an hour.

With the new test system BehavioQuant® it is possible to detect changes in the behavior of several test organisms in one observation chamber in continuous and mainly automated modus. To date, the most common type of fish test in Germany is the rheotaxis monitor. The rheotaxis-behavior types of automated biotests typically assess the ability of a fish to maintain its position in a moving stream of water (e.g. Poels, 1977 and Ermisch and Juhnke, 1973). The ability of a fish to remain upstream is believed to be affected by the presence of toxicants, causing swimming impairment. These systems show a response (alarm) to the occurrence of toxic pollutants only when the animals become weakened and paralyzed.

Much more sensitive parameters are behavioral responses—such as changes in motility—of organisms. The motility is an important component of the functions of organisms that can provide important information concerning their physiology and behavior. It is therefore often suggested or used as an indicator of sublethal toxicity in toxicological or ecotoxicological research (Little and Finger, 1990; Williams and Dusenbery, 1990; Baatrup and Bayley, 1993).

Thus, the intergrative behavioral response at the level of the whole organism as an indicator for the effects of chemicals or environmental changes was considered.

The Koblenz Behavioral Fish Test with BehavioQuant® offers through use of the image processing system the possibility to register behavioral changes continuously and automatically. Comparative tests proved that the evaluation of behavioral parameters achieved a considerable increase in sensitivity against the rheotaxis tests. While in the Koblenz Behavioral Fish Test velocity of movement changed significantly in spiking experiments with sodium pentachlorophenolate already at a concentration of 120 μg per liter of water from the River Rhine, parallel experiments with rheotaxis tests could not record any responses even at the highest concentration (480 μg/L Rhine water) (WIR, 1995).

Another advantage of the behavioral measuring system BehavioQuant® is its application as multi-species test. By means of the multiplexer, which is integrated in this measuring system, 16 video cameras could be controlled, i.e. theoretically one measuring system can monitor the behavior of eight different species of organisms (two measuring chambers per organism).

In the German Federal Institute of Hydrology, BehavioQuant® is used as a combination of a continuous fish test and a daphnia test. It could be proven that the BehavioQuant® system is able to recognize not only six fish of 5–8 cm but even ten daphnia of 1.2 mm size and to keep track of them.

Thus, it also becomes possible to register differentiated behavioral changes with daphnia from an age of 48 hours. The efficiency of the system—with a view to recognizing small objects—could be demonstrated by induced behavioral changes by modifying abiotic factors (such as temperature) or food supply. Spiking experiments with the pesticide lindane showed that BehavioQuant® is also able to significantly recognize minor behavioral changes of daphnia due to the presence of xenobiotics in surface waters.

The data of the behavioral studies with BehavioQuant® illustrate that already slight changes in the environment of organisms lead to changes in behavior.

Consequently, the present study confirms the statement by Mello (1975): "The behavior of the organism is the endpoint of the functional integration of the nervous system ecompassing sensory, motor, and cognitive aspects. The functional capacity of the central nervous system cannot be determined by histological or even physiological studies independent of behavioral analysis."

5. ACKNOWLEDGMENT

We are grateful to Rowena Dooley for revision of the English text.

6. REFERENCES

Anderson, J.M., 1971. Assessment of the effects of pollutants on physiology and behavior. Proc. R. Soc. Lond. B 177: 307–320.

Baatrup, E., and M. Bayley, 1993. Effects of the pyrethroid insectcide cypermethrin on the locomotor activity of the wolf spider Pardosa amantata: Quantitative analysis empoying computer-automated video tracking. Ecotoxicol. Environ. Saf. 26: 138–152.

Baillieul, M., and P. Scheunders, 1998. On-line determination of the velocity of simultaneously moving organisms by image analysis for the detection of sublethal toxicity. Water Research 32: 1027–1034.

Beitinger, T.L., and L. Freeman, 1983. Behavioral avoidance and selection responses of fishes to chemicals. Residue Reviews 90: 35–55.

Blübaum-Gronau, E., M. Hoffmann, O.H. Spieser, and F. Krebs, 1994. Der Koblenzer Verhaltensfischtest, ein auf dem Meßsystem BehavioQuant® beruhender Biomonitor zur Gewässerüberwachung. Schr.-Reihe Verein Wasser-, Boden- und Lufthygiene 93, Gustav-Fischer Verlag, Stuttgart: 87–117.

Blübaum-Gronau, E., O.H. Spieser, and F. Krebs, 1992. Bewertungskriterien für einen Verhaltensfischtest zur kontinuierlichen Gewässerüberwachung. Schr.-Reihe Verein Wasser-, Boden- und Lufthygiene 89, Gustav-Fischer Verlag, Stuttgart: 333–348.

Botterweg, J., 1988. Continue signalering van toxische stoffen in het aquatisch milieu met behulp van biologische bewakingssystemen—literatuurstudie. In: Publikaties en rapporten van het project "Ecologisch Herstel Rijn". Publikatie no.5.

Botterweg, J., C. van de Guchte, and L.W.C.A. van Breemen, 1989. Bio-alarmsystemen: een aanvulling op de traditionele bewaking van de waterkwaliteit. H2O 22: 788–794.

Dicks, B., 1976. The applicability of the Milford Haven experience for new oil terminals. In: Marine ecology and oil pollution. Ed. Baker, J.M., Applied Science Publishers, Barking, Essex (UK): 67–87.

DIN 38412 Teil 15, 1982. Testverfahren mit Wasserorganismen (Gruppe L). Bestimmung der Wirkung von Wasserinhaltsstoffen auf Fische—Fischtest (L15). Deutsche Einheitsverfahren zur Wasser-, Abwasser- und Schlammuntersuchung.

Elendt, B.-P., 1990. Selenium deficiency in Crustacea. An ultrastructural approach to antennal damage in *Daphnia magna* Strau. Protoplasma 154: 25–33.

Ermisch, R., and I. Juhnke, 1973. Automatische Nachweisvorrichtung für akut toxische Einwirkungen im Strömungsfischtest. Gewässer und Abwässer 52: 16–23.

Giattina, J.D., and R.R. Garton, 1983. A review of the preference-avoidance responses of fishes to aquatic contaminants. Residue Reviews 87: 43–45.

Juhnke, I., and W.K. Besch, 1971. Eine neue Testmethode zur Früherkennung akut toxischer Inhaltsstoffe im Wasser. Gewässer und Abwässer 50/51: 107–114.

Kramer, K.J.M., and J. Botterweg, 1991. Aquatic biological early warning systems: an overview. In: Bioindicators and Environmental Management. Eds.: D.W. Jeffrey, and B. Madden, Acad. Press, London: 95–126.

Larrick, S.R., K.L. Dickson, D.S. Cherry, and Jr.J. Cairns, 1978. Determining fish avoidance of polluted water. Hydrobiologia 61: 257–265.

Little, E.E., and S.E. Finger, 1990. Swimming behavior as an indicator of sublethal toxicity in fish. Environ. Toxicol. Chem. 9: 13–19.

Lubinski, K.S., K.L. Dickson, and Jr.J. Cairns, 1977. Microprocessor-based interface converts video signals for object tracking. Computer Design/Dec. 16: 81–87.

Mäckle, H., and H.-H. Stabel, 1989. Bioteste—Einsatz und Auswirkungen im Wasserwerksbetrieb. In: Qualitätsüberwachung von Roh- und Trinkwasser—Messung, Analyse und Bewertung. Berichte aus Wassergütewirtschaft und Gesundheitsingenieurwesen, Technische Universität München 175–188.

Mayr, C., 1992. Kombinationswirkung von Terbutylazin und Humussäure—Untersuchung biologischer Wirkungen mittels quantitativer Verhaltensmessung und der Akkumultation am Zebrabärbling (Brachydanio rerio). Diplomarbeit an der Fakultät für Biologie der Ludwig-Maximilian-Universität, München.

Mello, N.K., 1975. Behavioral toxicology: A developing discipline. Fed. Proc. 34 No 9: 1832–1834.

Miller, D.C., W.H. Lang, J.O.B. Greaves, and R.S. Wilson, 1982. Investigations in aquatic behavioral toxicology using a computerized video quantification system. In: Aquatic Toxicology and Hazard Assessment. Eds.: Pearson, J.G., R.B. Foster, and W.E. Bishop, American Society for Testing and Materials (ASTM) Special Technical Publication (STP) 766, Philadelphia: 206–220.

Nusch, E., 1993. Biologische Testverfahren—Aussagekraft und Grenzen der Übertragbarkeit. UWSF—Z. Umweltchem. Ökotox. 5: 155–161.

Olla, B.L., 1974. Behavioral biotests—Behavioral measures of enviromental stress. In: Proceedings of a workshop on Marine Bioassays. Ed.: Olla, B.L., Marine Technology Society, Washington D.C. 24–31.

Peichl, L., and F. Schmidt-Bleek, 1986. Biosonden zum Früherkennen von Umweltschäden. Umwelt. Z. des Vereins Deutscher Ingenieure für Immissionsschutz, Abfall, Gewässerschutz (Düsseldorf). Sonderausgabe Biotechnologie 4: 285–288.

Poels, C.L.M., 1977. An automatic system for rapid detection of acute high concentrations of toxic substances in surface waters using trout. In: Biological monitoring of water and effluent quality. Ed.: Cairns, J., American Society for Testing and Materials (ASTM) Special Technical Publication (STP) 607: 85–95.

Sachs, L., 1984. Angewandte Statistik (6. Auflage). Springer-Verlag, Berlin, Heidelberg, New York, Tokyo: 230–238.

Scholz, W., 1994. Verhalten richtig analysiert. Elektronik Journal 3: 50–53.

Smith, E.H., and H.C. Bailey, 1988. Development of a system for continuous biomonitoring of a domestic water source for early warning of contaminants. In: Automated Biomonitoring: Living Sensors as Environmental Monitors. Eds.: Gruber, D., and J. Diamond, John Wiley & Sons, New York, Chichester, Brisbane, Toronto: 182–205.

Spieser, O.H., and D. Freitag, 1984. 28-Day Fish Test. In: Überprüfung der Durchführbarkeit von Prüfvorschriften und der Aussagekraft der Stufe I und II des E. Chem. G. Umweltforschungsplan des Bundesministers des Inneren (10604011/02). Institut für Toxikologie der Gesellschaft für Strahlen- und Umweltforschung, München: 132–169.

Spieser, O.H., and H.P. Frisch, 1985. Verfahren zur Messung der Bewegungsweisen und Konfigurationen von biologischen und nicht biologischen Objekten. Deutsche Patentschrift P 3543515.1.

Spieser, O.H., and H.P. Frisch, 1988. Method of measuring the types of motion and configuration of biological and non biological objects. United States Patent No. 4,780,907.

Spieser, O.H., and W. Scholz, 1992. Verfahren zur quantitativen Bewegungsanalyse von mehreren Objekten im selben Medium. Deutsche Patentschrift P. 4224750.0.

Spieser, O.H., and A. Yediler, 1986. Empfindliche Parameter bei der Entwicklung von Langzeittests an Fischen. Umweltforschungsplan des Bundesministers des Inneren (10603030). Institut für Toxikologie der Gesellschaft für Strahlen- und Umweltforschung, München.

Van Hoof, F., 1980. Evaluation of an automatic system for detection of toxic substances in surface water using trout. Bull. Envirnm. Contam. Toxicol. 25: 221–225.

Warner, R.E., K.K. Peterson, and L. Borgman, 1966. Behavioural pathology in fish: a quantitative study of sub-lethal pesticide toxication. In: Pesticides in the Environment and their effects on wildlife. Ed.: Moore, N.W., Journal of Applied Ecology. 3, Suppl.: 223–247.

Williams, P.L., and D.B. Dusenbery, 1990. A promising indicator of neurobehavioral toxicity using the nematode Caenorhabditis elegans and computer tracking. Toxicol. Indust. Health 6: 425–440.

WIR (Bund-Länder-Projektgruppe "Wirkungstest Rhein") 1995. Kontinuierliche Biotestverfahren zur Überwachung des Rheins. Ed.: Umweltbundesamt, UBA-Bericht 1/95, Berlin.

Ziegler, S., 1994. Der Koblenzer Verhaltensfischtest—eine Studie zum Schwimm- und Bewegungsverhalten von Leuciscus idus melanotus L. unter dem Einfluß des Insektizids Endosulfan. Diplomarbeit am Institut für Tierphysiologie der Justus-Liebig-Universität, Giessen.

9

RESTORATION AND CLASSIFICATION OF WATER-BORNE MICROBIAL IMAGES FOR CONTINUOUS MONITORING OF WATER QUALITY

MANOHAR DAS[1] AND FRANK M. BUTTERWORTH[2]

[1]Department of Electrical and Systems Engineering, Oakland University, Rochester, Michigan 48309, USA. [2]Institute for River Research International, 920 Ironwood Dr., Suite 344; Rochester, Michigan 48307, USA

1. INTRODUCTION

The past decade has witnessed phenomenal growth in the fields of image processing and pattern recognition. To a large extent, these have been made possible because of the advent of low cost microcomputers and microelectronic imaging devices, and development of sophisticated image processing and pattern recognition algorithms that are capable of replacing human experts.

With the advent of image processing hardware and software new arenas have emerged where such tools and techniques can be useful and, along with them, new problems that need to be addressed. One such arena, which holds a tremendous potential for real-world application, encompasses automatic evaluation of water samples for microbiota. Some of its obvious applications include: wastewater treatment analysis and management; pathogen detection in domestic drinking water, and the study of microbiotal dynamics and ecology in natural waters.

The above applications are important for obvious reasons, but the technical challenges faced today are overwhelming. For instance, wastewater treatment needs far better efficiency and round the clock monitoring, pathogen detection must be monitored rapidly enough to avoid human contact, and the vast unknown microbiotal system (s) in natural waters begs to be explored but this application requires sampling vast volumes of water by highly trained analysts experienced in identifying microbiotal species. Researchers have experimented with a number of new technologies to solve some of these problems. Two technologies have been found to be promising: particle counters, and flow cytometry.

The particle counters basically rely on the size of the organism, i.e., these are particle-size machines. These machines have serious limitations, because they can only distinguish ranges of sizes of particles and they cannot distinguish inanimate particles from organisms or the type of organisms within the size range (Sprules et al., 1992). The other technology, flow cytometry, utilizes fluorescence to identify the organisms either through autofluorescence the wavelength of which would be specific for the organism or through specific fluorochrome dyes that are linked to antibodies that are organism specific (Brailsford and Gatley, 1993). With this system specific identification can be made, but only if the organism autofluoresces at a species-specific wave length. If it does not autofluoresce, fluorochrome-antibody complexes have to be made for a specific organism. Thus the flow cytometry system is very limited: limited to the organisms that emit specific wavelengths, if they autofluoresce and limited to the necessity to make specific antibodies to a very large variety and range of organisms to anticipate the available organisms.

To overcome some of the drawbacks, a combination of the two technologies (looking at fluorescence of the particles) has been tried (Porter et al., 1995). But coupled immunofluorescent flow cytometry and cell sorting is still a particle counter that makes identification indirectly, which requires great expertise and cost to operate, and does not operate in real time.

In view of above, phase contrast microscopy seems to be one of the best approaches available today. Together with a video camera, an image digitizer, and an image processing unit, such a system allows continuous real-time automated monitoring of the actual microbial images. We explore this approach here and present some interesting experimental results.

2. MICROBIOTA CATEGORIZATION AND CREATION OF IMAGE DATABASE

To create a microbiota image database, water samples were collected from a local lake, stored in the laboratory for an appropriate period of time, and examined periodically by Dr. Butterworth. Using a phase contrast microscope, a CCD camera (equipped with microscope attachment), and a VCR, video tapes of microbiota in the samples were made. Next, appropriate frames from the video images of selected microorganisms were digitized and archived in an image database designed to facilitate retrieval and analysis. All the images captured for this study are monochrome pictures of size 480×512 and they are digitized to 256 gray levels.

The repertoire of the microorganisms chosen in this study include diatoms (centric- and pennate-shaped), flagellates, ciliates, bacteria (cocci-, bacilli-, vibrio-, and spirillum), algae, amoeba, and other creatures. Although some of them are either motion-less or relatively slow-moving, others are often discernible by their characteristic rapid movement patterns. For the stationary or slow-moving objects, we were able to gather multiple snapshots of the same objects, whereas for the rapidly moving ones, we could only capture a sequence of moving image frames.

Although motion constitutes an important attribute for the problem that we are concerned with, this study is focussed on the problem of recognition of stationary/slow-moving objects only. An extension of this work to detection of fast-moving objects is planned for the near future.

3. IMAGE ENHANCEMENT

The raw microbiota image samples are often very noisy and cluttered with unwanted objects. Thus, the enhancement of these images becomes a necessary pre-processing operation before one can proceed with segmentation and edge detection.

The noise in microbiotal images is contributed by several factors: i) relatively strong background noise arising from lights scattered by colloidal particles and/or other objects, ii) presence of various extraneous and/or foreign objects in the scene, and iii) lack of sharpness in the images of desired objects resulting from either low reflectivity or semi-transparent nature of some of these objects. Figure (1a) shows an example of such an image. The object in the center is the picture of a diatom, which happens to be the object of our interest. In the absence of any background noise, any commonly-used edge detection operator could have found the edges of the diatom. However, the presence of background noise makes the task of segmentation and edge detection quite difficult. This is evidenced from Figure (2a), which shows the edges found from the raw image by the so-called Sobel operator (Gonzalez, 1992; Jain, 1989). As can be clearly seen, the edges found are far from satisfactory. As is expected, better edge detection is possible using modified algorithms. These are briefly discussed below.

One of the potential problems in using conventional edge operators with noisy images is that many spurious objects and edges materialize due to the presence of the background noise. A simple solution to this problem would be utilization of a two-step process; namely, i) first, reduce the background noise using an optimal filter, such as a Wiener filter (Gonzalez, 1992; Jain, 1989), and ii) then segment a filtered image and detect its edges. Unfortunately, this simple idea does not always work very well, because Wiener filters (being low-pass in nature) tend to smear the edges in a picture. As a result, some of the not-so-prominent objects and their edges may remain undetected. Figures (1b) and (2b) depict the Wiener filtered image and its edges, respectively.

The above problem can be solved by employing a three-step strategy (Anand and Das, 1997; Anand, 1998); namely, i) first, use a robust edge detector to locate the candidate edge pixels of the raw image, ii) then, restore the raw image using a Wiener filter that is applied to the non-edge pixels only, and iii) finally, restore the original values of the edge pixels. The overall scheme is called edge-preserving Wiener filtering, because it preserves the sharpness of all the edges while removing most of the background noise. An example of edge-preserving Wiener filtering is

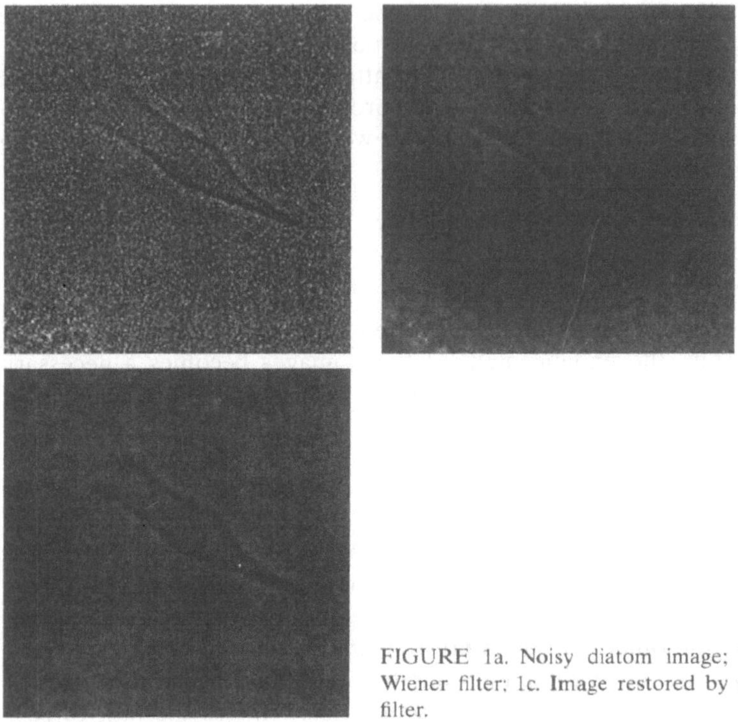

a b c

FIGURE 1a. Noisy diatom image; 1b. Image restored by Wiener filter; 1c. Image restored by edge-preserving Wiener filter.

shown in Figure (1c), which depicts the restored version of the noisy diatom image (which appears in Figure (1a)). Also, its edges are shown in Figure (2c).

The next tasks involve: thresholding, segmentation, and edge detection. These tasks are briefly discussed below.

4. GRAY-LEVEL ADJUSTMENT, SEGMENTATION, AND BOUNDARY EXTRACTION

The filtered image is often somewhat dull in contrast and requires some kind of gray level adjustment before segmentation and boundary extraction. The following gray-level adjustment algorithm was found to be quite useful in our experiments:

$$g(m,n) = af(m,n), \tag{1a}$$

$$a = \begin{cases} \text{a constant } (>1), \text{ if } f(m,n) \in [m_f - s_f, m_f + s_f] \\ 0, \text{ otherwise} \end{cases} \tag{1b}$$

where f(m,n) denotes the original pixel intensity, g(m,n) is the transformed value, and m_f and s_f denote the mean and variance, respectively, of f(m,n). The contrast-

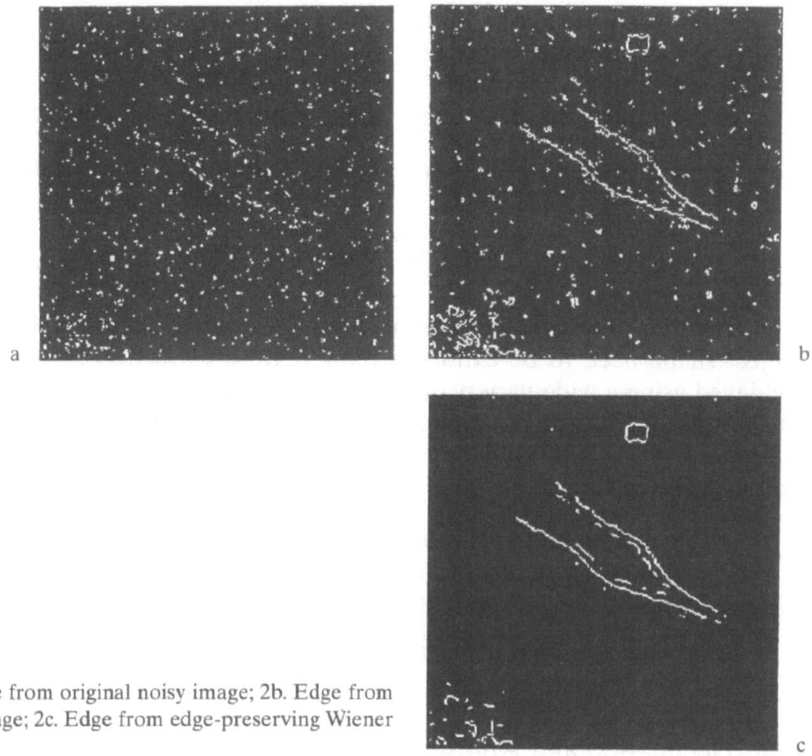

FIGURE 2a. Edge from original noisy image; 2b. Edge from Wiener filtered image; 2c. Edge from edge-preserving Wiener filtered image.

enhanced image is then binarized using a simple statistical algorithm based on bimodal distribution (Jain, 1989).

A microbiota image frame usually contains multiple objects pertaining to various object classes. These, therefore, need to be labelled appropriately so that the pattern classifier can analyze each object separately. This is achieved using an image segmentation algorithm. In our preliminary investigation, the segmentation problem was not addressed because the frames were chosen carefully to contain single objects only. However, a segmentation software based on connected component labelling strategy (Jain, 1989) can be easily developed.

Next, for each object in the field, we need to detect edges and boundary points so that its characteristics can be analyzed. Since our raw images were first restored using an edge-preserving Wiener filter, a simple Sobel edge detector (Gonzalez, 1992; Jain, 1989) was found to deliver adequate performance. Then, the boundary points were extracted using a boundary-following algorithm (Duda and Hart, 1973; Dubois and Glanz, 1986), outlined below.

The boundary follower is a simple level-change detection algorithm that follows the boundary in a clockwise direction. For simplicity, assume that the input image consists of a single black object in a white background. The operation of the boundary follower can then be summarized as follows.

- find a starting point on the object boundary;
- if the current pixel value is black, take a step toward left;
- else, if it is white, take a step toward right;
- continue, until the follower is back to the starting point.

The above algorithm is very easy to program and works quite well.

5. BOUNDARY MODELING AND FEATURE EXTRACTION

In order to classify different object shapes, the distinguishable features of each object shape need to be extracted using feature extraction software. This can be achieved using a wide variety of techniques (Ballard and Brown, 1982; Jain, 1989). In our preliminary investigation, we used a statistical technique called circular autoregressive (CAR) modeling (Kashyap and Chellappa, 1981). This consists of the following main steps:

- decimation and approximation of the original boundary sequence;
- statistical modeling of the approximate boundary sequence; and
- extraction of feature sets.

These are briefly discussed below.

5.1. Decimation and Approximate Representation of the Original Boundary Sequences

To start with, the boundary coordinates obtained from an image (or shape) can be viewed as a bivariate sequence, $\{x(k),y(k), 1 \leq k \leq L\}$, where x(k) and y(k) denote x and y coordinates values of the kth boundary sample. Decimation and approximation of the original sequences becomes necessary, because L may be very large for a typical image and consists of too many redundant samples. Thus, a given boundary shape is first approximated by a subset of N samples of its original boundary sequence, where N is chosen to be large enough to adequately represent the important information in the boundary. If the power spectrum estimates (Jain, 1989), $P_x(\omega)$ and $P_y(\omega)$, of the boundary coordinate sequences, $\{x(k)\}$ and $\{y(k)\}$, are calculated, one can determine the approximate cut-off frequency and thereby obtain the Nyquist sampling rate (Jain, 1989) for the boundary signal. After analyzing all the representative boundary shapes in this fashion, one can easily determine a suitable value of N.

Having determined N, as outlined above, the next task one faces is the choice of a sampling strategy. A variety of methods is available for this purpose (Kashayap and Chellappa, 1981): i) *equal-segment sampling*, which extracts samples in such a way that the lengths of the resulting boundary segments are equal to each other, ii) *equal-angle sampling*, which basically consists of drawing N radii from the centroid (at angular separations of $2\pi/N$) and marking off N points where the radii intersect the original boundary, iii) *equal cord length sampling*, where samples are selected such that the cord lengths between adjacent points are equal, and iv) *polygonal approximation*, where samples are placed at the vertices of a polygonal approximation of the shape. These are illustrated in Figure 4 (Paulik, 1989).

Each of the above sampling schemes has its advantages and disadvantages. The *equal segment sampling* tends to place more samples in areas of higher boundary complexity, because a greater portion of the total perimeter is concentrated in areas of high boundary variation. On the other hand, the *equal angle sampling* scheme tends to place more points on portions of the boundary closer to the centroid. Similarly, the *equal cord sampling* tends to produce samples that are placed at non-uniform intervals along the boundary curve, which is also true of the *polygonal approximation* method.

Two major advantages of the *equal angle sampling* technique are: i) unlike the other three, it produces a one-dimensional boundary sequence, and ii) the resulting boundary sequence is invariant of object rotation. These make the subsequent tasks of feature extraction and pattern classification somewhat easier. Therefore, for the sake of simplicity, we choose the *equal angle sampling* in this investigation. A detailed discussion of another technique, which concerns feature extraction from the two-dimensional boundary sequences, can be found in (Das, Paulik, and Loh, 1992).

5.2. Statistical Modeling of the Approximate Boundary Sequence

In one-dimensional (1-D) CAR modeling, one assumes that a given boundary can be represented approximately by a 1-D radii sequence, $\{r(k), 1 \le k \le N\}$, which consists of the radial distances of the N marked-off boundary points from the shape's centroid. This vector series is then modeled by an autoregressive model of the form:

$$y(k) = \sum_{j=1}^{p} a_j y(k-j) + \sqrt{\beta} w(k), \tag{2a}$$

$$r(k) = y(k) + \propto, \tag{2b}$$

where $\propto$ denotes the mean value of $\{r(k)\}$, $\{y(k)\}$ is the zero-mean sequence, a_j, $1 \le j \le p$ denote the AR coefficients of the zero-mean process $y(k)$, and $w(k)$ is a white noise sequence with variance equal to β. The parameters of the model (2) are estimated using the standard least squares technique details of which can be found in (Dubois and Glanz, 1986; Das, Paulik, and Loh, 1992).

5.3. Extraction of Feature Sets

The pattern recognition approach pursued here falls in the class of template matching, wherein an unknown item is compared to a set of prototype templates (or, feature vectors) corresponding to the available classes, and the most similar one is chosen as the destination class. The similarity is based on some matching criterion or distance measure.

In our case, the feature elements are derived from the autoregressive model parameters obtained for each object class. The construction of a feature vector which is invariant of translation, rotation and scaling can be easily done using the same basic principle as illustrated in (Kashyap and Chellappa, 1981). The feature vector, f, for pattern classification is chosen to be:

$$f = [a_1, a_2, \ldots a_p, \propto/\sqrt{\beta}]. \tag{3}$$

6. PATTERN RECOGNITION

Finally, we need to choose a pattern recognition scheme capable of classifying different objects based on their feature sets. In our preliminary investigation, we have used two minimum-distance pattern classifiers known as the feature weighting (FW) and rotated coordinate system (RCS) methods (Sebestyen, 1962; Dubois and Glanz, 1985). Also, we have experimented with a simple neural network classifier.

The FW method is basically simplified version of the Mahalanobis distance classifier (Jain, 1989), whereas the RCS technique is essentially akin to the Hotelling transform (Jain, 1989; Gonzalez, 1992). As is well known, given two prototype reference feature templates, q and s, a minimum distance classifier assigns an unlabeled feature sample, x, to the class which corresponds to the nearest of the two prototypes. Although many different minimum distance classifiers are available, FW and RCS are chosen here because: i) FW is a representative of the methods that require very little computation and deliver moderately good performance, whereas ii) RCS is a representative of the methods that require more computation, but deliver superior performance.

The feature weighting method attempts to minimize the mean squared distance between members of the same class by applying a metric which emphasizes the features that are closely clustered, and de-emphasizes those features that vary widely. Sebestyen (1962) has shown that the optimal feature weights for a particular class are inversely proportional to the standard deviations of each of their associated feature vector elements. Let $\{q_i, 1 \le i \le I\}$ denote the set of feature vectors pertaining to an object class, Q. The weighted cumulative distance, d(x,Q), between an unlabeled feature vector, x, and the feature class Q is given by:

$$d(x, Q) = \left[\prod_{j=1}^{J} s_j\right]^{2/J} [y^T y + J], \tag{4a}$$

where

$$y = [(x_1 - m_{1q})/s_1, (x_2 - m_{2q})/s_2, \ldots, (x_J - m_{Jq})/s_J]^T, \tag{4b}$$

J denotes the dimension of x, x_i is the ith element of x, and m_{iq} denotes the ith element of the mean of the pattern set $\{q_k\}$.

Notice that the FW method does not incorporate any measures to make sure that the feature elements are uncorrelated with each other, i.e., the coordinate directions associated with the feature elements are orthogonal to each other. The rotated coordinate system (RCS) method assures this by first rotating the coordinate system to form a set of uncorrelated features, and then weights these features optimally as in the FW method. Essentially it uses a Hotelling-like transformation to construct a set of uncorrelated features. The detailed information about this method can be found in (Sebestyen, 1962; Dubois and Glanz, 1985).

Finally, the neural net classifier (Haykin, 1994) tested here is a simple two-layer backpropagation neural network, consisting of a $(p + 1) \times 1$ input layer, a hidden layer consisting of several neurons, and a $(m \times 1)$ output layer (with sigmoid transfer functions). Here, p denotes the order of the autoregeressive models, whereas m denotes the number of object classes. The choice of number of neurons in the hidden layer was made empirically after several computer simulations.

7. EXPERIMENTAL RESULTS

This study focuses on seven different microbial shapes shown in Figures (3a)–(3g). The classifier was trained using fifteen images pertaining to each shape class, and classification accuracy was tested on five test images of each class. Also, care was taken to ensure that none of the images belonging to a class resemble any other pertaining to that class, i.e. they differed from one another due to varying amounts of translation, rotation, or scaling.

The experimental results are summarized in Tables 1–3. It is clearly seen that even a simple classifier like FW provides a reasonably good classification accuracy, and a better method like RCS works quite good. On closer examination, it was found that most of the misclassifications occurred among classes A, C, and F, because depending on the orientation and scaling many of these images looked quite similar.

Finally, the preliminary results from the neural net classifier is quite encouraging. However, much more data needs to be gathered before any concrete conclusion can be drawn. For instance, i) the training and test samples need to be much larger in size, and ii) the choice of the number of neurons in the hidden layer (which

TABLE 1. Classification results using FW method.

Model Order	No. of Misclassifications	Percentage Accuracy
2	10	71.4
3	6	82.9
4	5	85.7

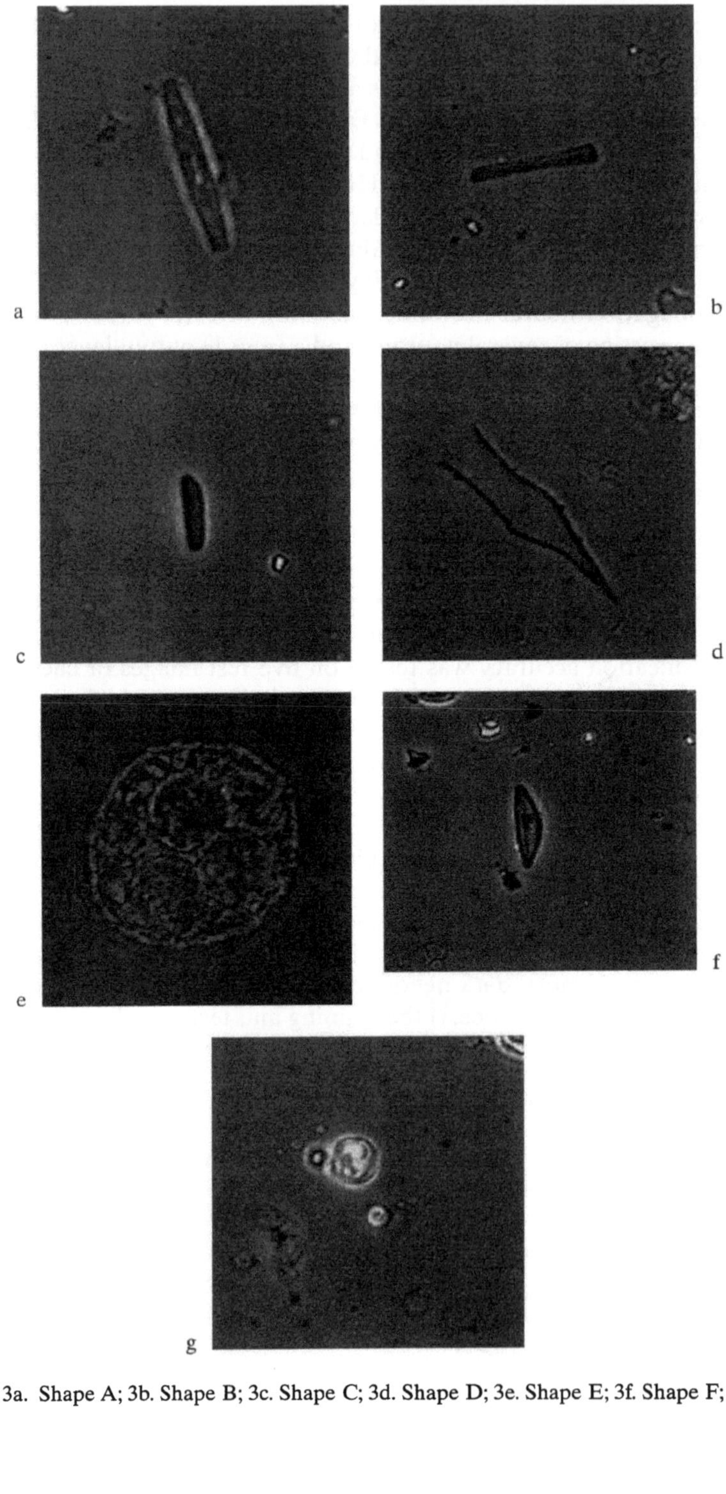

FIGURE 3a. Shape A; 3b. Shape B; 3c. Shape C; 3d. Shape D; 3e. Shape E; 3f. Shape F; 3g. Shape G.

TABLE 2. Classification results using RCS method.

Model Order	No. of Misclassifications	Percentage Accuracy
2	3	91.4
3	3	91.4
4	4	88.6

TABLE 3. Classification results using neural net classifier.

Model Order	No. of Misclassifications	Percentage Accuracy
3	0	(Almost) 100

is chosen to be six here) must be done carefully after conducting a more thorough analysis. The attainable classification accuracy of nearly 100%, as indicated in Tables 1 and 2, should be interpreted in the above light.

8. POTENTIAL APPLICATIONS

The image capturing, processing, and classification scheme described above can be useful in a variety of applications. Two potential applications encompassing *wastewater treatment* and *pathogen imaging* are briefly discussed below.

8.1. Wastewater Treatment

Typically the wastewater treatment operator aliquots small samples from the batch reactors, places a drop of the sample under the microscope and counts the various classes of organisms observed. The proportion of each class of organisms represents a stage of sewage digestion. For example, say there are eight classes of organisms: sarcodinia, bacteria, holophytic flagellates, holozoic flagellates, suctoria, free swimming ciliates, stalked ciliates and rotifers. Each class grows in number in response to the types of organic wastes (food) are present. Thus, at any given hour of digestion, the proportions of organisms will change. Thus, the operator will know at what stage of digestion the batch is in based on the proportions of classes present. This is very time consuming and inconvenient because the process continues on a 24 hr schedule. If anything happens to upset this schedule and the growth rate of these classes is altered, incomplete digestion will occur and the operator is faced with a very costly and hazardous problem of starting a new batch reactor while simultaneously trying to deal with the damaged batch. Knowing the status of this digestion continuously will provide a great savings in personnel time and avoid the problems of releasing incompletely digested and sometimes hazardous materials into the environment.

8.2. Pathogen Imaging

Great interest in waterborne pathogen imaging is resulting in research. However, the difficulties of identifying organisms such as Giardia or Cryptosporidium are enormous despite the development of specific, antibody-coupled stains. The general opinion is that accurate identification is still technician based. Vesey et al. (1994) and Hoffman et al. (1997) attempt to solve these shortcomings, but despite these improvements, the procedure is still slow and costly which is not good news to the thousands of municipal water facilities across the Nation. Thus, to date successful identification requires DIC microscopy and a microscope technician. In our opinion, the image capturing, processing, and classification scheme described above can provide a valuable tool for automating the above process.

9. REFERENCES

Anand, J., 1997. Image restoration and compression using two-dimensional predictive models. Ph. D. dissertation, Oakland University, Rochester, Michigan, USA.

Anand, J., and Das, M., 1997. Modeling and restoration of noisy images using a cascade structured 2-d multiplicative autoregressive Wiener filter. Technical Report, TR-97-ESE-07-01, Oakland University, Rochester, Michigan, USA. Also, presented at the 18th IFIP TC7 conference on Systems Modeling and Optimization, held in Detroit, Michigan, from July 22–25, 1997.

Ballard, D.H., and Brown, C.M., 1982. *Computer Vision*. Prentice Hall, New Jersey, USA.

Brailsford, M., and Gatley, S., 1993. Rapid analysis of microorganisms using flow cytometry. In: Lloyd, D. ed. 1993. *Flow Cytometry in Microbiology*. Springer Verlag, London.

Das, M., and Anand, J., 1995. Robust Edge Detection in Noisy Images Using an Adaptive Stochastic Gradient Technique. *Proc. 1995 IEEE International Conf. on Image Processing*.

Das, M., Paulik, M.J., and Loh, N.K., 1990. A Bivariate Autoregressive Modeling Technique for Analysis and Classification of Planar Shapes. *IEEE Trans. on Pattern Analysis and Machine Intelligence*, Vol. 12, No. 1, pp. 97–103.

Dubois, S.R., and Glanz, F.H., 1986. An Autoregressive Model Approach to Two-Dimensional Shape Classification. *IEEE Trans. on Pattern Analysis and Machine Intelligence*, January 1986, pp. 55–66.

Duda, R.O., and Hart, P.E., 1973. *Pattern Classification and Scene Analysis*, John Wiley & Sons, New York.

Haykin, S., 1994. *Neural Networks*, IEEE Press, New Jersey, USA.

Hoffman, R.M., Standridge, Jon H., and Bernhardt, Mat., 1997. Using flow cytometry to detect protozoa. For routine detection of Cryptosporidium and Giardia, flow cytometry with cell sorting bests immunofluorescence assat. *Journal of the american water works association*. 89(n 9): 104.

Jain, A.K., 1989. *Fundamentals of Digital Image Processing*. Prentice Hall, New Jersey.

Jeng, F.C., and Woods, J.W., 1988. Inhomogeneous Gaussian Image Models for Estimation and Restoration. *IEEE Trans. on Acoustics, Speech and Signal Processing*, Vol. 36, No. 8, pp. 1315–1312.

Kashyap, R.L., and Chellappa, R., 1981. Stochastic Models for Closed Boundary Analysis: Representation and Reconstruction. *IEEE Trans. on Information Theory*, September 1988, pp. 627–637.

Porter, J., Robinson, J., and Edwards, C., 1995. Recovery of a bacterial sub-population from sewage using immunofluorescent flow cytometry and cell sorting. *FEMS microbiology letters*. 133(n 1/2): 195.

Sebestyen, G.S., 1962. *Decision-Making Processes in Pattern Recognition*. The McMIllan Company, New York.

Sprules, W.G., Bergstrom, B., Cyr, H., Hargreaves, B.R., Kilham, S.S., MacIsaac, H.J., Matsushita, K., Stemberger, R.S., and Williams, R., 1992. Non-video optical instruments for studying zooplankton distribution and abundance. *Arch. Hydrobiol. Beih.*, 36: 45–58.

Vesey, Graham, Hutton, Primrose, Champion and Alan, 1994. Application of Flow Cytometric Methods for the Routine Detection of Cryptoporidium and Giardia in Water. *Cytometry*. 16(n 1): 1.

10

SCREEN-PRINTED DISPOSABLE BIOSENSORS FOR ENVIRONMENTAL POLLUTION MONITORING

DAVID C. COWELL,[1] ABDUL K. ABASS,[2]
ANTONY A. DOWMAN,[2] JOHN P. HART,[2] ROY M. PEMBERTON,[2]
AND SARAH J. YOUNG[1]

[1]Department of Environmental Health & Science, Faculty of Applied Sciences, University of the West of England, UK, BS16 1QY. [2]Department of Chemical & Physical Sciences, Faculty of Applied Sciences, University of the West of England, UK, BS16 1QY

1. INTRODUCTION

A biosensor has been defined as "a biologically sensitive material immobilised in intimate contact with a suitable transducing system which converts the biochemical signal into a quantifiable and processable electrical signal" (Lowe, 1985). Commercialisation of biosensors has seen the development of a disposable technology that offers the operator a reagentless analytical system whereby calibration and reagent addition are controlled at manufacture. This approach has been successfully developed for monitoring blood glucose in diabetic patients. The system, because of its simplicity, allows patients to monitor their own blood glucose levels at regular time intervals and consequently control their diabetes (Green and Hillditch, 1991). This device was specifically designed for a very large clinical market where tens of millions of sensors per month are currently being manufactured. For these types of devices to be employed for environmental pollution monitoring there is a requirement to research and fabricate generic devices, that will allow the mass production of the transducer which can then be modified simply for the detection of particular environmental analytes at low cost.

1.1. Generic Electrochemical Screen Printed Carbon Electrode Biosensors

The construction of a biosensor can be envisaged schematically (Figure 1), where the biorecognition element or biologically sensitive material may be an enzyme, antibody, antibody fragment, whole cell, cell organelle or tissue. The transducer is also variable, including ion selective electrodes measuring changes in ion activity or carbon dioxide levels, amperometric electrodes quantifying changes in oxygen levels or redox reactions and piezo electric crystals measuring mass changes. The transducing device thus converts the biochemical reaction into a quantifiable electrical signal which may be further processed to produce a relevant output depending on the application. This output may, in the simplest of application, be a "present" or "absent" LED, a quantitative analytical result or the decision of artificial neural net software. The latter relies on the device being "taught" to recognise particular signal outputs as representing one of many possible analytical answers.

Electrochemical screen printed carbon electrodes (SPCEs) are devices that operate by transducing the oxidation or reduction (redox) of an analyte of interest into an electrical signal. The redox reaction takes place at the surface of the SPCE, the potential of which can be controlled relative to that of a reference electrode. In addition to the direct sensing of inherently electroactive analytes, SPCEs can operate by measuring the electron transfer processes resulting from enzyme catalysed redox reactions. In relation to the definition of a biosensor, the biologically sensitive material in this case is an oxido-reductase enzyme, and the transducer is an SPCE, with a measuring system that quantifies the current flow between the SPCE and the reference electrode (Figure 2). Consequently the total system can be regarded as simplistic in nature.

To enable the electron transfer reactions to be monitored, a constant voltage is applied between the SPCE and the reference electrode. Electrocatalysts or mediators may be incorporated into the screen printing ink to reduce the applied voltage

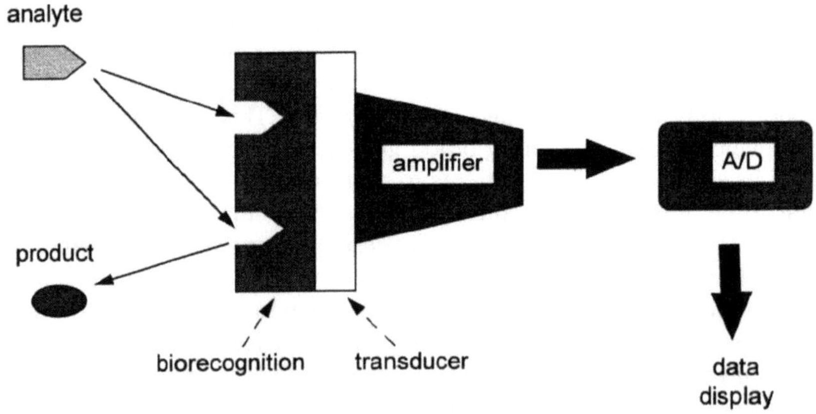

FIGURE 1. A schematic view of biosensor construction.

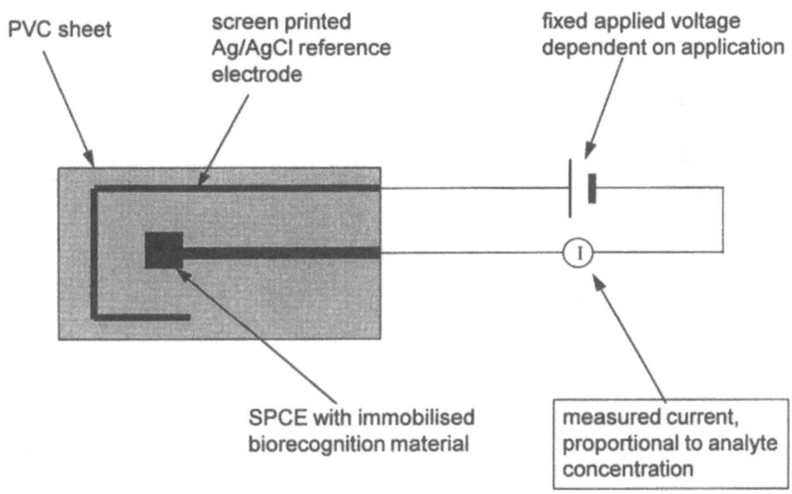

FIGURE 2. Schematic view of the measuring system of an SPCE biosensor.

by effectively lowering the operating potential of the system. The possibilities of interferent, electrochemically active, substances creating spurious currents are reduced in this way and selectivity improved. The technology and application of SPCEs has been extensively reviewed by Hart and Wring (1997).

A generic approach that has been adopted is to develop two screen printing inks, one for the oxidase group of enzymes (Gilmartin et al., 1994) and a second for the dehydrogenase group of enzymes (Sprules et al., 1994) providing an opportunity to develop biosensors based on over four hundred commercially available oxido-reductase enzymes. The approach for the oxidase enzymes has been to incorporate, within the carbon ink, the electrocatalyst, cobalt phthalocyanine (CoPC), which allows the detection of hydrogen peroxide produced by this group of enzymes at +0.4 V versus a silver chloride reference electrode. This system allows, for example, the enzyme, glucose oxidase (GOD), to transfer electrons from glucose via hydrogen peroxide and the CoPC electrocatalyst to the SPCE resulting in a measurable current (Figure 3), directly related to the concentration of the glucose present in the sample (Gilmartin et al., 1995).

The incorporation of the electrocatalyst Meldola's Blue (MB) into the screen printing ink has enabled the monitoring of the reduced form of the co-factor nicotinamide adenosine dinucleotide (NADH), which acts as the redox centre for the dehydrogenase group of enzymes. This system has been applied to ammonia detection in river waters, an analyte which is an indicator of environmental pollution from farm run off (Figure 4). The enzyme glutamate dehydrogenase (GLDH) oxidises the NADH dependant upon the ammonia content of the river water. The reduction in concentration of the reduced co-factor, NADH, is then monitored by a redox reaction involving the MB in "shuttling" electrons to the SPCE which has an applied voltage of +0.05 V. This has enabled the detection of ammonia in river waters down to 20 ppb (Hart et al., 1998).

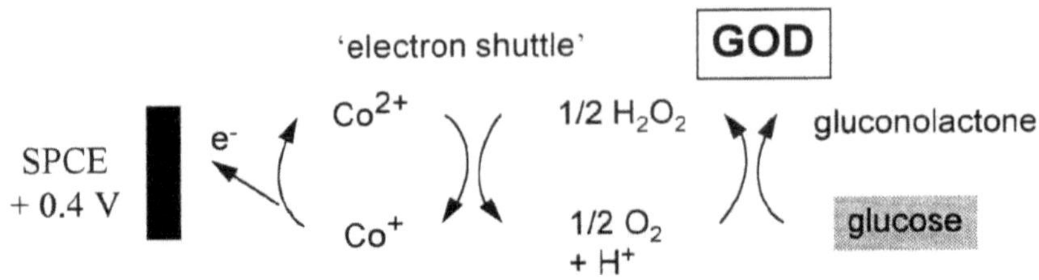

FIGURE 3. Electron flow in a glucose biosensor utilising glucose oxidase (GOD). The applied voltage to the SPCE is +0.4 V (Gilmartin et al., 1995).

These SPCEs have also been applied to enzyme systems that are not within the oxido-reductase group of enzymes, but have substrates or products that can be electrochemically oxidised or reduced directly at the electrode surface. An environmental application is in the detection of organophosphate pesticides. These pesticides inhibit the enzyme acetylcholinesterase (ACE) which can convert the substrate acetylthiocholine to thiocholine with the latter product being electrochemically oxidised at the SPCE containing CoPC (Figure 5). In this case a reduction in current output from the biosensor is seen when the organophosphate is present (Hartley and Hart, 1994). This analytical system has been applied to the detection of the organophosphate pesticide, paroxon, down to levels of 10^{-11} M in river waters (Rippeth et al., 1997).

The direct determination of substrates, as shown for the examples for glucose and ammonia above, are generally limited to a detection of around 10^{-6} M. For analytes that require much lower detection limits, for example some pesticides, the use of SPCE immunosensors is a possibility. In this case a monoclonal antibody to the analyte of interest is required which is immobilised onto the surface of the SPCE. A competitive immunoassay can then be employed, whereby the sample containing the analyte is mixed with a known quantity of the analyte labelled with an

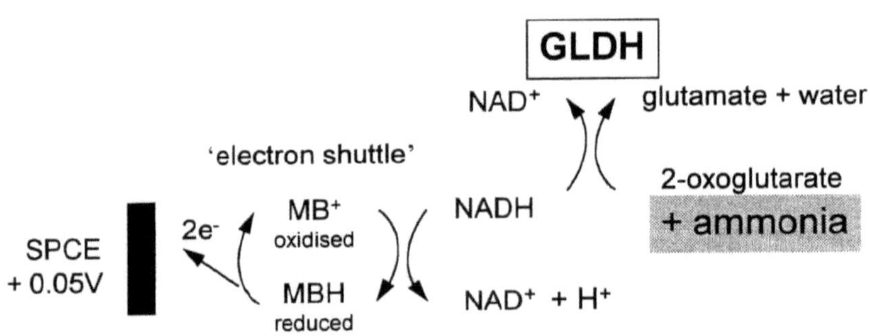

FIGURE 4. An ammonia biosensor using a "competitive" approach, with the oxidation of the co-factor NADH being catalysed by the enzyme glutamate dehydrogenase (GLDH). The reaction is dependant on the concentration of ammonia present, which is monitored via a Meldola's Blue (MB) electron "shuttle" to the screen printed carbon electrode. The applied voltage to the SPCE is +0.05 V (Hart et al., 1998).

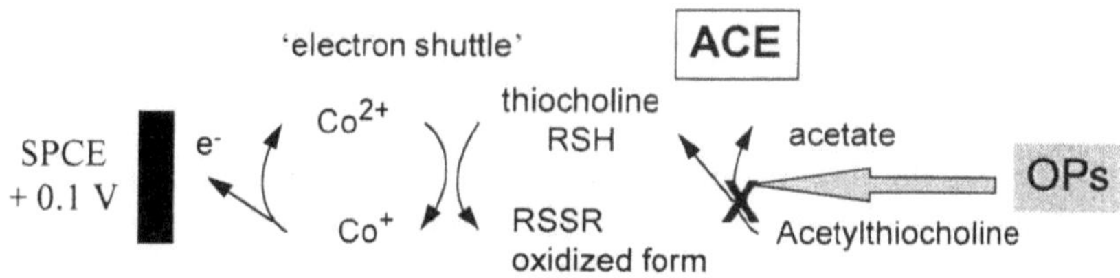

FIGURE 5. A biosensor for the detection of organophosphate (OP) insecticides based on the inhibition of the activity of the enzyme acetylcholinesterase (ACE) in the presence of an OP. The applied voltage to the SPCE is +0.1 V (Hartley and Hart, 1994).

enzyme e.g., alkaline phosphatase (AP). Competition then takes place between the labelled analyte and the sample analyte for the binding sites on the antibody immobilised to the SPCE (Figure 6).

When binding has taken place, a solution substrate for the enzyme label, is added to the electrode, which is converted to an electrochemically-active compound. In the case of the enzyme label, AP, aminophenol phosphate is converted to aminophenol which is electrochemically oxidised at the SPCE at +0.2 V. This type of device has been demonstrated for the hormone progesterone in cow's milk but would be applicable in the environmental area (Pemberton, Hart, and Foulkes, 1998).

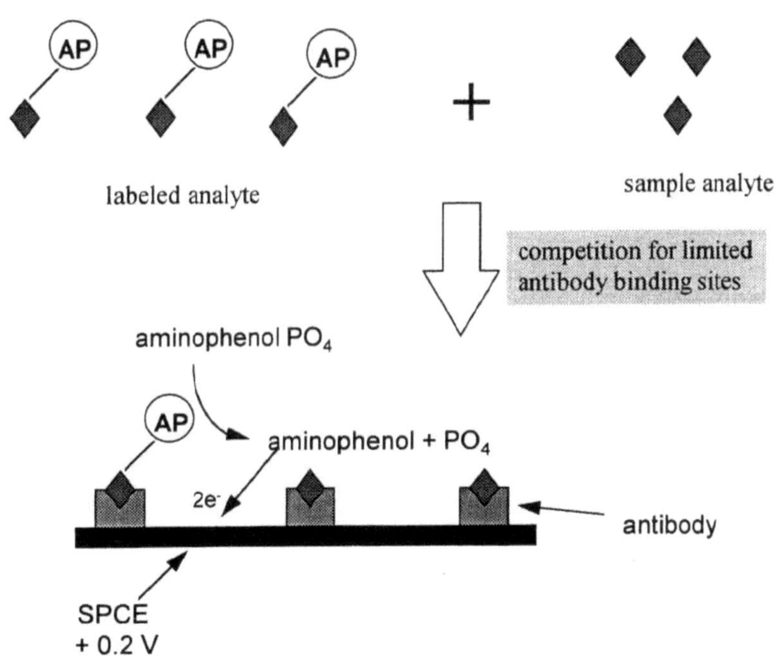

FIGURE 6. Schematic view of an immunosensor, utilising competition between the sample and labelled analyte in the solution phase for the limited antibody binding sites, employing an SPCE and the enzyme label alkaline phosphatase (AP). The applied voltage to the SPCE is +0.2 V (Pemberton, et al., 1998).

1.2. The Application of Non-specific Enzyme Inhibition to Environmental—Pollution Detection

To this point, the discussion has concentrated on the detection of specific analytes or pollutants. While this type of device may be extremely useful in the monitoring of a known pollutant, it offers little advantage over conventional chemical assays other than it uses a biological system and may be more readily deployed in the field. To detect known and unknown types of pollutants, a different approach must be adopted to the development of biosensors. If we consider the effect of pollutants on biological systems, they can be briefly summarised as occurring in three areas:

- they disrupt the integrity of cell membranes
- they cause a reduction in DNA, RNA replication and/or function
- they inhibit metabolic processes

Considering this last point in more detail, the immediate question that arises is one of whether the inhibition of enzymes within a metabolic pathway is specific or non-specific. A review of the literature related to enzyme inhibition (Zollner, 1993) indicates that the concept of specific inhibition of enzymes is unrealistic. We can conclude that this inhibition process, by a significant number of industrially produced chemicals, which may become pollutants, is non-specific. Furthermore, if inhibition is non-specific then a number of enzymes within the metabolic pathway may well be affected. The consequences of multiple inhibitions occurring within a metabolic pathway are that the inhibition of the first enzyme will then be amplified by the inhibition of the second enzyme etc. within that pathway, resulting in a cumulative effect on the whole pathway. A second issue to consider is that the level of inhibition of the differing enzymes in a metabolic pathway will not be the same i.e., different enzymes will be inactivated to different extents by the same pollutant. Utilising these concepts, it has been possible to demonstrate that the non-specific inhibition of an array of enzyme assays, suitable for biosensor construction, can be utilised in the detection of known pollutants. Artificial neural network software was used to interpret the pattern of inhibition from such an array of enzyme assays (Cowell et al., 1995). This approach has required a re-assessment of the design of conventional enzyme based assays. Consider the typical linked assay shown in Figure 7, where the product of the enzyme 1 reaction (product 1) is the substrate for enzyme 2 and is immediately removed to form product 2. The formation of the latter is used to monitor the assay. The construction of an analytical system to measure the concentration of the initial substrate would require that:

- the substrate itself was rate limiting i.e. below the K_m of enzyme 1;
- enzymes 1 and 2 were present in excess;
- enzyme 2 removed product 1 immediately on formation.

For the analysis of the activity of enzyme 1 in this system then:

- the substrate must be in excess;
- enzyme 1 must be rate limiting;
- enzyme 2 present in excess to ensure that product 1 is again removed immediately on formation.

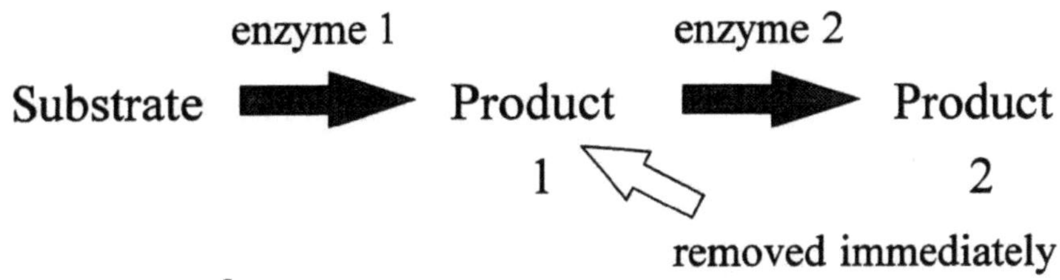

Assay type	substrate Xs	substrate Lim	enzyme 1 Xs	enzyme 1 Lim	enzyme 2 Xs	enzyme 2 Lim
Substrate		x	x		x	
Enzyme	x			x	x	
Amplified Inhibition		x		x		x

FIGURE 7. Comparison of the design of enzyme-based assays. (Xs = excess concentration or activity; Lim = limiting concentration or activity; "x" denotes the requirement.)

For amplified enzyme inhibition assays, where the type of inhibition may not be known for an unknown pollutant, it is important that the assay is constructed in a different format:

- the substrate must be rate limiting;
- enzyme 1 and enzyme 2 must be rate limiting

This ensures that irrespective of the type of inhibition of the enzyme or enzyme-substrate complex a decrease in signal due to non-specific inhibition will be measurable (Obst et al., 1988). To illustrate this approach, consider the linking of the two enzymes pyruvate kinase (PK) and lactate dehydrogenase (LDH) (Figure 8) and the construction of two sensors. The first sensor has only the enzyme LDH, the second has the two enzymes linked and both sensors are challenged by the two pollutants chromium (Cr) and the fungicide and insecticide, pentachlorophenol (PCP).

In sensor 1, LDH is inhibited by PCP and in the second sensor PK is inhibited by chromium. This example assumes complete conversion of each of the substrates so when no pollutant is present, 100% of the product lactate should be produced by both sensors. In the case of PCP inhibition of LDH, sensor 1 will have a reduced signal output and sensor 2 will have a similarly reduced signal. For chromium inhibition, sensor 2 will demonstrate inhibition but this will be larger. In the case of both pollutants being present, the signal output from sensor 2 will be considerably reduced (6 mM), due to the inhibition of PK being amplified by the inhibition of LDH. Sensor 1 will only demonstrate the effect of PCP being present.

		PK		**LDH**	
	PEP	→	pyruvate	→	lactate

Sensor 1	mM	% inhib	mM	% inhib	mM
No pollutant			100	0	100
Cr 50 mg/l			100	0	100
PCP 20 mg/l			100	35	65
Cr + PCP			100	35	65

Sensor 2	mM	% inhib	mM	% inhib	mM
No pollutant	100	0	100	0	100
Cr 50 mg/l	100	85	15	0	15
PCP 20 mg/l	100	0	100	35	65
Cr + PCP	100	85	15	35	6

FIGURE 8. Demonstration of a simplified, amplified enzyme inhibition assay based on pyruvate kinase (PK) and lactate dehydrogenase (LDH) incorporated in two sensors inhibited by chromium (Cr) and pentachlorophenol (PCP). (% inhib = percentage inhibition of the enzyme by the pollutant). In sensor 1 only LDH is inhibited; in sensor 2 both PK and LDH are inhibited.

Consequently it can be seen that the signal outputs provide different patterns depending on whether one or both of the pollutants are present. A study involving the pollutants cadmium, chromium, phenol, PCP and atrazine at levels three orders of magnitude above the maximum admissible concentrations (MAC) for potable waters, as defined by European Union legislation (Directive 80/778/EEC), demonstrated this concept of non-specific inhibition for an array of 12 enzymes as seen in Figure 9 (Cowell et al., 1995).

It is interesting to note that enzymes such as acetylcholinesterase, which are regarded as being specifically inhibited by the organophosphate insecticides, also demonstrate inhibition by chromium, cadmium and phenol. This is an issue that needs to be addressed when applying these systems for pesticide monitoring as the inhibition detected will be cumulative of all pollutants inhibiting the enzyme. Computer modelling and the use of artificial neural nets, utilising the data collected from six of the enzymes in Figure 9, demonstrated the possibilities of this approach with no pollutants being misclassified as false negatives. The false positive rate was minimal and mainly in the area of the organics rather than the heavy metals. The modelling also demonstrated the ability of the system to be semi-quantitative in indicating the level of pollutant. The outcome of this study identified that the system, as described, was not sensitive enough for the organic pollutants, particularly PCP.

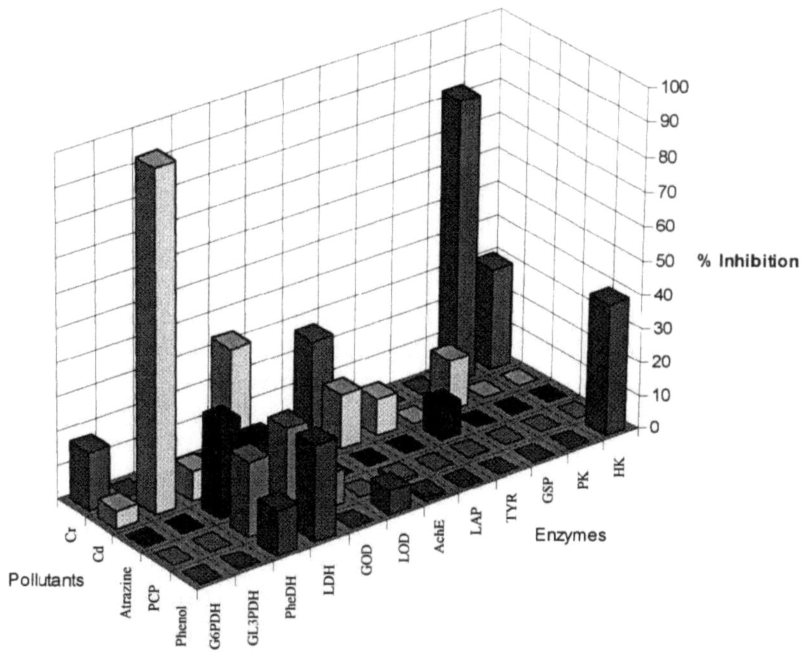

FIGURE 9. Inhibition of twelve enzymes by four pollutants demonstrating the non-specificity of enzyme inhibition. (reproduced from Cowell et al., 1995).

G6PDH = glucose-6-phosphate dehydrogenase; G3PDH = glycerol-3-phosphate dehydrogenase; PheDH = phenylalanine dehydrogenase; LDH = lactate dehydrogenase; GOD = glucose oxidase; LOD = lactate oxidase; AchE = acetylcholinesterase; LAP = leucine aminopeptidase; TYR = tyrosinase; GSP = glutathione peroxidase; PK = pyruvate kinase; HK = hexokinase.

Consequently, a preliminary study of a model system using the pollutant PCP and the enzyme LDH has been undertaken to further delineate ways in which greater sensitivity may be achieved.

2. MATERIALS AND METHODS

Analytical grade chemicals were used throughout (Sigma Chemical Company, Poole, UK) and the LDH enzyme (EC.1.1.1.27) from rabbit muscle (specific activity 459 U mg^{-1} protein) was supplied by Biozyme Laboratories, Blaenavon, UK. Assays were performed on a Cecil CE373, linear grating spectrometer (Cecil Instruments, Cambridge, UK). All reactants were prepared immediately prior to use and stored over ice; NADH was stored in the dark. Enzyme activity (specific activity) was determined from the rate of change ($\Delta A_{340} min^{-1}$) over the linear portion of the assay and expressed in terms of international units per mg of protein using the molar extinction coefficient at 340 nm for NADH and NADPH of $6.22 \times 10^3 cm^{-2} mol^{-1}$. All kinetic data from experimental results were analysed with the Enzpack for Windows computational software (Biosoft, Cambridge, UK).

2.1. Determination of the Michaelis-Menten Constants K_m, V_{max} and Inhibition Constant K_i

Enzyme assays were performed according to the method of Reeves and Fimognari (1963). This assay provided for the substrate, pyruvate (767 μM), and the co-factor, NADH (300 μM), to be in excess in the assay. Assays were carried out at 25 ± 0.1 °C in a 50 mM phosphate buffer pH 7.4. The assay was used for determining the specific activity of the enzyme and the Michaelis-Menten constants K_m and V_{max} for both substrate and co-factors. The type of enzyme inhibition exhibited by PCP was determined by repeating the K_m determination experiments in the presence of a fixed amount of PCP (70 μM). The resulting calculations of Michaelis-Menten constants K_m, V_{max} and the inhibition constant K_i were then used to determine the type of inhibition observed.

2.2. Inhibition Assays

For the inhibition assays, the pollutant PCP (>99%) was made up at 10 times the working strength in 50 mM phosphate buffer pH 7.4. Ethanol was required to solublise the PCP at 1 part to 24 parts phosphate buffer. This solution was diluted 1 in 10 for use in the assay. LDH was diluted in the phosphate buffer to an activity that produced a measurable reaction velocity without causing substrate depletion and was rate limiting under assay conditions. Assays were carried out at 25 ± 0.1 °C in a 50 mM phosphate buffer pH 7.4. To the buffer, PCP, NADH and pyruvate were added to achieve a final concentrations of 70 μM (200,000 times the MAC), 20 μM, 246 μM, respectively. The reactants were mixed and the baseline absorbance observed prior to initiating the reaction by the addition of LDH. To assess the reproducibility of the assay, ten assays without the addition of the pollutant PCP were run to ensure an adequate performance of the assay. These control assays were used to calculate the mean activity, standard deviation and coefficient of variation (CV) of the assay. The CV was required to be below 5% to satisfy reproducibility criteria. A minimum inhibition detectable (MID) parameter was calculated from this quality assurance data (Cowell et al., 1995). This corresponded to a 99% probability of inhibition being observed and was determined by the mean activity of the control assays minus 2.6 times the standard deviation. Inhibition was only recorded if the assay activity fell below this threshold value. Inhibition assays were expressed as a percentage of the mean control activity to normalise the data for comparative purposes.

2.3. Determination of the Detection Limit for PCP

The detection limit of the assay for PCP was determined by varying the concentration of the pollutant PCP under the fixed assay conditions of 246 μM pyruvate and 20 μM co-factor with the activity of the LDH being as low as possible commensurate with obtaining acceptable kinetic data. The detection limit was determined when the percentage inhibition was equal to the MID, as calculated from the quality assurance data.

3. RESULTS

3.1. Effect of Varying the Reactant Concentrations on Observed Enzyme Inhibition by PCP

Varying the substrate and co-factor concentrations for LDH, at a rate limiting activity, in the presence of 70 μM PCP had a significant effect on the percentage inhibition obtained (Figure 10). Although inhibition of the reaction was observed in the presence of excess reactants, reducing either the substrate or co-factor to a rate limiting concentration produced a significant increase in the level of inhibition. The most prominent effect was that of the co-factor NADH. When combining rate limiting concentrations of both the substrate, pyruvate, and the co-factor, NADH, little additional increase in the inhibition was observed over that shown for NADH alone.

3.2. Enzyme Kinetic Studies

The Michaelis-Menten constant K_m for pyruvate was 162 μM and for NADH 7.9 μM; these being close to the figures quoted in the literature of 164 μM and 10.7 μM respectively (Zewe and Fromm, 1962). Repetition of the K_m experiments in the presence of 70 μM PCP indicated that PCP was a competitive inhibitor of the substrate pyruvate with a K_i of 121 μM (Figure 11) whereas for the co-factor,

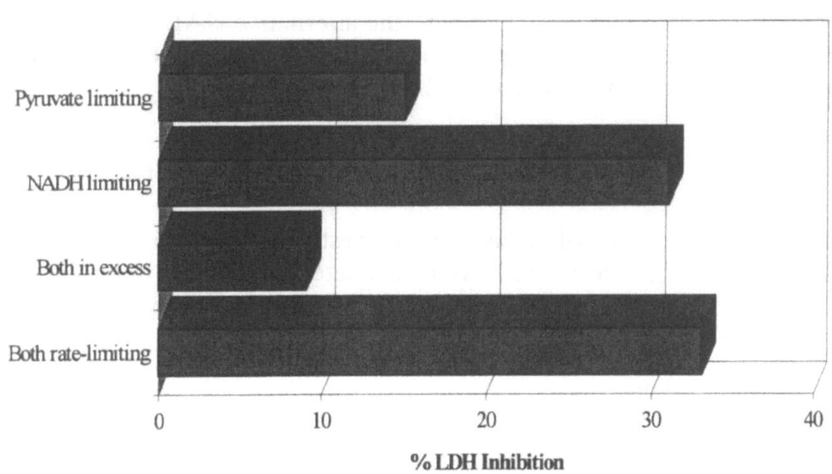

FIGURE 10. The percentage inhibition exhibited by varying the concentration of the substrate, pyruvate, and the co-factor, NADH, on the activity of LDH in the presence of 70 μM PCP. (Limiting and rate limiting refer to 1.5 times the K_m).

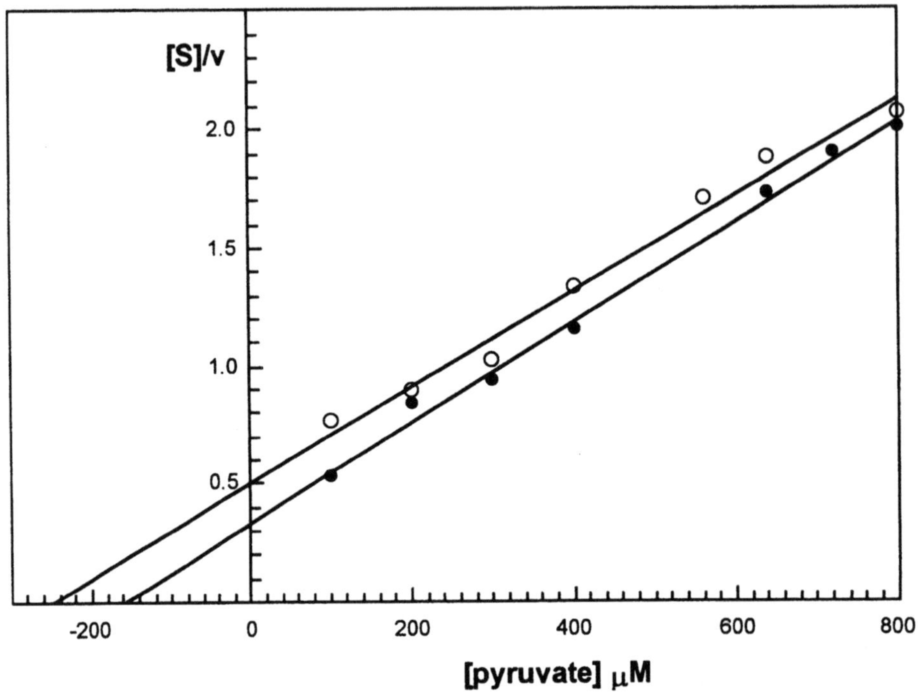

FIGURE 11. Determination of K_m, V_{max} for the substrate pyruvate (○-○) and K_i in the presence of the pollutant PCP at 70 μM (●-●) using Hanes plots.

NADH, a mixed competitive-noncompetitive inhibition was observed with a K_i of 75.8 μM (Figure 12).

Replacing the co-factor NADH with the alternative NADPH produced a K_m of 250 μM and in the presence of PCP, a mixed competitive-noncompetitive inhibition was again observed with a K_i of 58.1 μM (Figure 13).

3.3. Evaluation of the PCP Concentration on LDH Activity

The limit of detection of the assay was investigated over the range of 7–70 μM PCP (Figure 14). For the co-factor NADH, the detection limit, as determined by the point at which the curve crossed the MID, was 58 μM (15.5 ppm) whereas for NADPH the detection limited observed was 8 μM (2.1 ppm).

4. DISCUSSION

From Figure 10 it was apparent that varying the concentrations of the reactants, substrates or co-factors, could have a marked effect on the observed inhibition of LDH activity in the presence of the pollutant PCP. The validity of the

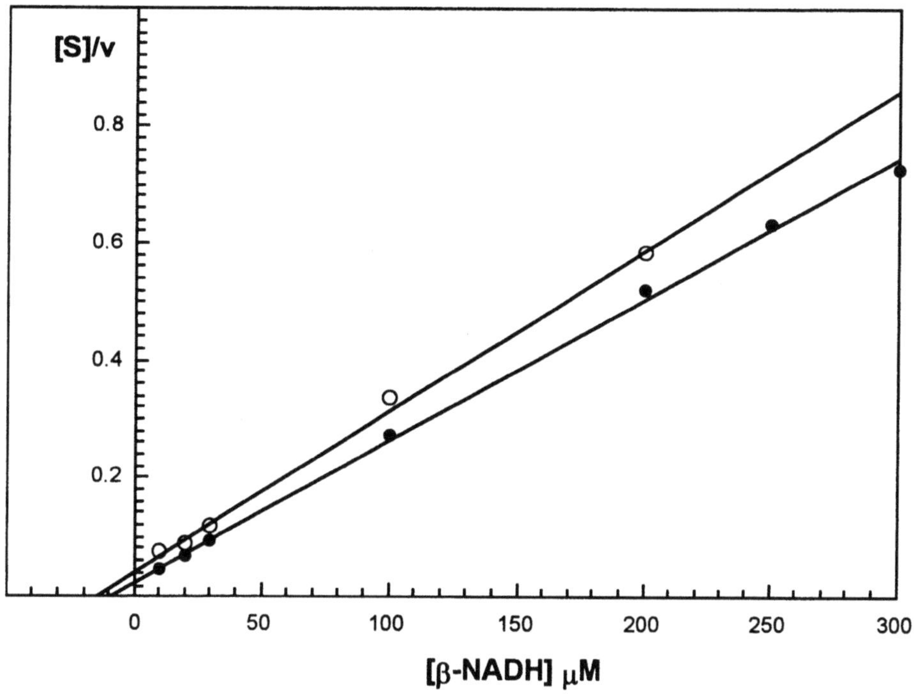

FIGURE 12. Determination of K_m, V_{max} for the co-factor NADH (○-○) and K_i in the presence of the pollutant PCP at 70μM (●-●) using Hanes plots.

observed data was confirmed by the close agreement of the Michaelis-Menten constants, K_m, with the literature.

4.1. The Effect of PCP on LDH Activity Related to Enzyme Kinetics

The differences observed in the detection limits between the two different co-factors can be explained in terms of the kinetic data obtained. The $K_i:K_m$ ratio for pyruvate was 1:1.4, indicating that the enzyme has a higher affinity for PCP than the substrate pyruvate and that PCP was a competitive inhibitor (Palmer, 1985). However, it should be noted that this ratio is close to 1 and consequently the effect of the pyruvate concentration would be minimal in any assay design. This is confirmed by the small increase in inhibition observed by rate limiting concentrations of pyruvate and NADH combined over rate-limiting NADH with pyruvate in excess (Figure 10). The $K_i:K_m$ ratio for NADH was 9.6:1, indicating that the enzyme had a higher affinity for the co-factor, NADH, than PCP. The mixed competitive-noncompetitive inhibition observed indicated that the PCP bound to the free enzyme and the enzyme-substrate (E-S) complex, in this case the substrate being the co-factor NADH. The inhibition observed when NADH was rate limiting

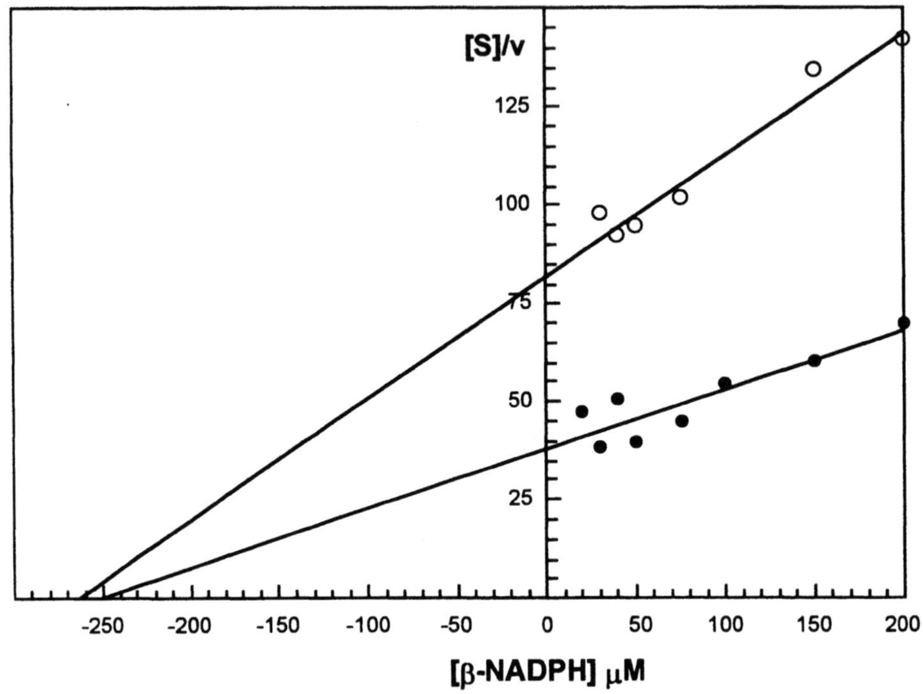

FIGURE 13. Determination of K_m, V_{max} for the co-factor NADPH (○-○) and K_i in the presence of the pollutant PCP at 70 μM (●-●) using Hanes plots.

(Figure 10) is difficult to interpret in relation to the $K_i:K_m$ ratio, but may be due to the PCP to NADH concentrations used (3.5:1) possibly favouring increased PCP binding to the free enzyme. For the alternative co-factor NADPH, the $K_i:K_m$ ratio is 1:4.3 indicating that the enzyme has a significantly higher affinity for PCP than NADPH. The mixed competitive-noncompetitive type of inhibition again indicates that PCP also binds to the E-S complex. As with NADH, this type of mixed inhibition causes the enzymatic reaction to be reduced, and the significantly higher affinity for the enzyme by PCP, as shown by the $K_i:K_m$ ratio, was further demonstrated by the much lower detection limit observed.

4.2. Future Developments to Lower the Detection Limit

This preliminary study has indicated that the use of a non-optimum co-factor, with a lower affinity for the enzyme has considerable potential to achieve further decreases in the detection limits when using inhibition assays. In terms of the LDH model, the substrate pyruvate could be replaced with compounds such as α-ketovalerate or α-ketobutyrate, both of which have significantly lower affinities for

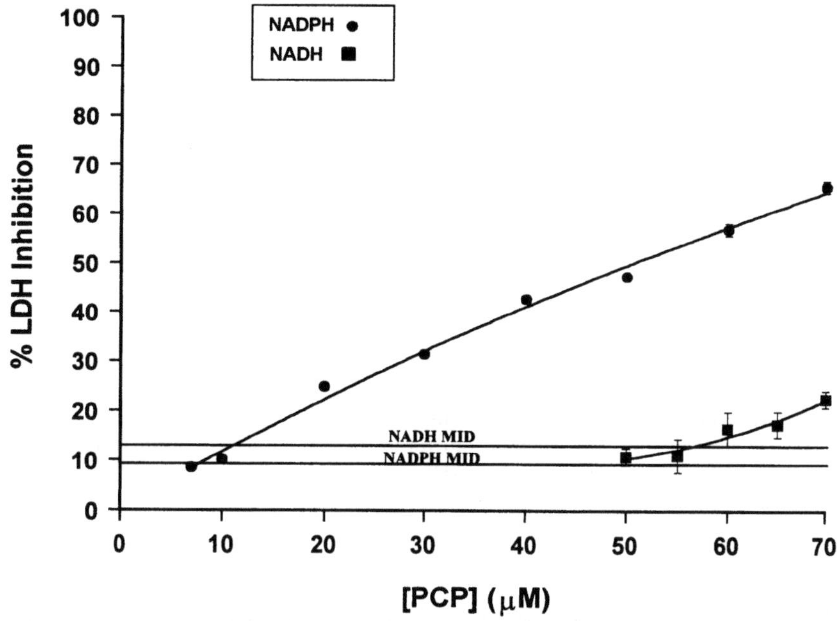

FIGURE 14. Percentage inhibition of LDH by the pollutant PCP using the co-factors NADH (■-■) and NADPH (●-●). The minimum inhibition detectable (MID) was calculated from the reproducibility study of assays in the absence of PCP.

the enzyme according to the literature (Meister, 1950). The current range of experiments were carried out with the co-factor NADPH at 20 μM. It is possible that reducing this concentration may reduce the detection limit further. Additional approaches to improving the detection limit may be to utilise electrochemical detection of the co-factor as the current procedures, based on spectrophotometry, are at their limit of resolution. This approach may enable the monitoring of the co-factor concentration at one order of magnitude lower. The possibilities also exist of using considerably lower levels of the co-factor by using enzyme amplification, ensuring regeneration of the co-factor, and allow monitoring of the signal via another enzymatic output as shown in Figure 15. In this scenario, LDH is still inhibited by PCP, but the oxidised co-factor, NAD^+, is regenerated via a second enzyme to ensure that is converted back to the reduced form, NADH. This would allow considerably lower levels of the co-factor to be utilised for the inhibition aspect of the assay as substrate and co-factor depletion effects would not be observed. Monitoring of the reaction could not take place via NADH as this would remain constant within the reaction due to the regeneration. By using a third enzyme, lactate oxidase (LO), to convert the product of the reaction, lactate, back to pyruvate with no reduction in the concentration of reactants for LDH activity, then it should be possible via a CoPC SPCE to monitor this reaction successfully. The only other substrate required for this reaction is oxygen.

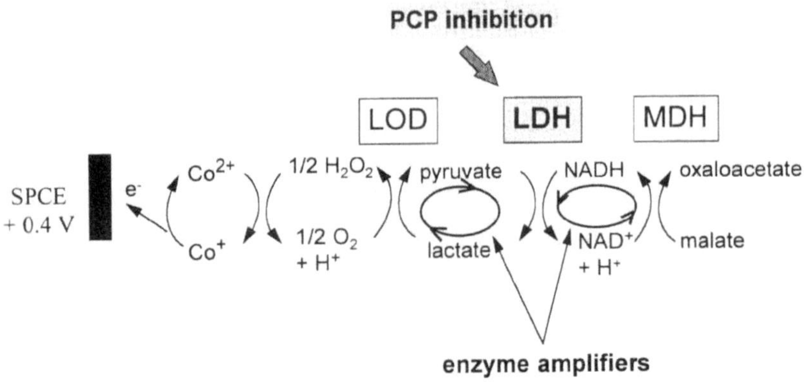

FIGURE 15. Schematic diagram of a proposed enzyme amplification system to lower the detection limit for PCP. Two enzyme amplifiers are proposed, one for the regeneration of pyruvate from lactate and the second for the regeneration of NADH from NAD^+.

4.3. Biosensors *vs.* Biomonitoring

The research to date has identified that biosensors can be used for monitoring environmental pollutants as single analytes. The possibilities also exist of developing biosensor arrays for the detection of a general range of pollutants. However, a serious debate still exists around the fact that biosensors do not monitor a whole biological system in comparison to for example, fish, daphnia, or shellfish monitors. There are a number of issues that should be considered for the future when comparing biosensors against these more classical forms of bioassay. Classical bioassays have a high cost of ownership, being complex monitoring systems requiring considerable investment in hardware and labour intensive maintenance. Perhaps more importantly they stress a living organism; an issue, which although dependent upon cultural and societal differences, is one that is in many cases unacceptable. As we move into a new millennium this issue is less likely to be so easily accepted by the general public. At the scientific level the major criticism of bioassay systems are their poor reproducibility. Reproducibility of bioassays is at best demonstrated by a CV of 10–15% (Kaiser and Palabrica, 1991; Brown et al., 1996). Translating the statistics into the realities of observing effects on these systems by pollutants, a CV of 15% becomes a minimum effect detectable of 39%. This figure is perilously close to the often quoted EC_{50} levels in many scientific papers (Brown et al., 1996). Thus, in reality, the detection of inhibition of the bioassay much below the EC_{50} is frequently not viable. Biosensors have significantly lower coefficient of variation, usually less than 5%. Additionally, the use of SPCEs equates with a low cost of ownership, the devices are disposable with calibration and reagent addition controlled at manufacture. Additionally, biosensors require only the use of semi or unskilled personnel, as has been demonstrated by diabetic patients monitoring their own blood sugar levels with a biosensor (Green and Hillditch, 1991).

This type of device consequently provides considerable possibilities for the environmental scientist to perform one off screening tests in the field with minimal skill and minimal equipment. The simple adaptation of this disposable technology using robotics for the almost continuous monitoring of river waters, effluent or other ecosystems is feasible using the same generic technology. The use of array SPCE biosensors in the area of continuous monitoring would allow not only the detection of single pollutants, but also the cumulative effect on these arrays of a multiplicity of pollutants. There is the admitted proviso with this technology that the artificial neural net must be "taught" to recognise the patterns of this inhibition and relate those to specific pollutants. However, the system also has the ability to recognise unknown pollutants, i.e., if a pattern of inhibition is observed which is not recognised, the neural net can record that inhibition has taken place, but the compound not identified. Consequently, we have a system that can be designed to look at a significant number of pollutants for monitoring purposes, but also to act as a "watchdog" for unknown pollutants being added to an ecosystem. Biomonitors act only as "watchdogs", giving no indication of the type of pollutant present.

Although with the research carried out to date, we have demonstrated the feasibility of SPCE biosensors being utilised successfully in the area of environmental monitoring of pollutants, it is recognised that we are only obtaining detection limits in the order of $10^{-6}\,\mu M$ (1 ppm) and that there is a considerable way to go to attain for example the levels required for PCP at $10^{-8}\,\mu M$ (0.1 ppb). However, the toolbox is not empty and we have a considerable number of possible avenues of research, utilising both electrochemistry and biological diversity of enzymes throughout the natural world to attain these levels.

5. REFERENCES

Brown, J.S., Rattray, E.A.S., Paton, G.I., Reid, G., Caffoor, I., and Kilham, K. 1996. Comparative assessment of the toxicity of a papermill effluent by respirometry and a luminescence-based bacterial assay. Chemosphere. **32**: 1553–1561.

Cowell, D.C., Dowman, A.A., and Ashcroft, T. 1995. The detection and identification of metal and organic pollutants in potable water using enzyme assays suitable for sensor development. Biosensors & Bioelectronics. **10**: 509–516.

Green, M.J., and Hillditch, P.I. 1991. Disposable single-use sensors. Analytical Proceedings. **28**: 374–376.

Gilmartin, M.A.T., Hart, J.P., and Birch, B.J. 1994. Development of amperometric sensors for uric acid based on chemically modified graphite-epoxy resin and screen-printed electrodes containing cobalt phthalocyanine. Analyst. **119**: 243–252.

Gilmartin, M.A.T., Hart, J.P., and Patton, D.T. 1995. Prototype, solid-phase, glucose biosensor. Analyst. **120**: 1973–1981.

Hartley, I.C., and Hart, J.P. 1994. Amperometric measurement of organophosphate pesticides using a screen-printed disposable sensor and biosensor based on cobalt phthalocyanine. Analytical Proceedings. **31**: 333–337.

Hart, J.P., and Wring, S.A. 1997. Recent developments in the design and application of screen-printed electrochemical sensors for biomedical, environmental and industrial analysis. Trends in Analytical Chemistry. **16**: 89–103.

Hart, J.P., Abass, A.K., Cowell, D.C., and Chappell, A. 1998. Development of an amperometric assay for NH_4^+ based on a chemically modified screen-printed NADH sensor. Analytica Chimica Acta. **373**: 1–8.

Kaiser, K.L.E., and Palabrica, V.S. 1991. *Photobacterium phosphoreum* toxicity data index. Water Pollution Research Journal of Canada. **26**: 361–431.

Lowe, C.R. 1985. An introduction to the concepts and technology of biosensors. Biosensors. **1**: 3–16.

Meister, A. 1950. Reduction of α-keto and α-keto acids catalysed by muscle preparations and by crystalline lactic dehydrogenase. Journal of Biological Chemistry. **184**: 117–129.

Obst, U., Holzapfel-Pschorn, A., and Weigand-Rosinus, M. 1998. Application of enzyme assays for toxicology water testing. Toxicity Assessment. **3**: 81–91.

Palmer, T. 1985. *Understaning enzymes.* 2nd Edition. Ellis Horwood Ltd., Chichester, England.

Pemberton, R.M., Hart, J.P., Foulkes, J.A. 1998. Development of a sensitive, selective electrochemical immunoassay for progesterone in cow's milk based on a disposable screen-printed amperometric biosensor. Electrochimica acta. **42**: 3567–3574.

Reeves, W.J., and Fimognari, G.M. 1963. Lactate dehydrogenase. Journal of Biological Chemistry. **238**: 3583.

Rippeth, J.J., Gibson, T.D., Hart, J.P., Hartley, I.C., Nelson, G. 1997. Flow-injection detector incorporating a screen-printed disposable amperometric biosensor for monitoring organophosphate pesticides. Analyst. **122**: 1425–1429.

Sprules, S.D., Hart, J.P., Wring, S.A., and Pittson, R. 1994. Development of a disposable amperometric sensor for reduced nicotinamide adenine dinucleotide based on a chemically modified screen-printed carbon electrode. Analyst. **119**: 253–257.

Zewe, V., and Fromm, H.J. 1962. Kinetic studies of rabbit muscle lactate dehydrogenase. Journal of Biological Chemistry. **237**: 1668–1675.

Zollner, H. 1993. *Handbook of enzyme inhibitors*. 2nd Edition. VCH, Weinhein; Basel; Cambridge; New York.

11

OPTICAL SENSORS AND BIOSENSORS FOR ENVIRONMENTAL MONITORING

PATRICIA SCULLY, REKHA CHANDY, ROBERT EDWARDS, DAVID MERCHANT, AND ROGER MORGAN

Division of Engineering and Science, Liverpool John Moores University, Byrom Street, Liverpool, L3 3AF UK

1. INTRODUCTION

The history and development of optical sensors for environmental monitoring can be traced back to the classical optical analytical techniques e.g. UV-Visible spectrophotometry, IR-spectrometry and spectrofluorimetry. Typically, the characteristic transmission, absorption or fluorescence spectrum of a chemical species is measured in order to determine its concentration or identity. Such instruments are big and bulky, cost in the region of $30K and require the sample to be taken to the instrument, which is operated by a highly skilled technician. If such instruments could be miniaturised, and used remotely, in situ or on line, then there would be advantages for uses such as on-line monitoring and control of rivers, water quality, water treatment works, biotechnological processes, etc.

To place biosensors in context with environmental monitoring, the development of monitoring techniques can be considered to take place at three levels (Marco and Barcelo, 1996). Firstly, the development of fast, highly sensitive and specific screening tools, easily adapted for use on site. Such techniques exploit the specificity and sensitivity of biological systems i.e bio-indicators and immunoassays. Secondly, the use of sophisticated and reliable analytical technologies that can unequivocally identify and/or quantify with a high accuracy, a broad range of pollutants at trace level. Finally, devices intended to combine sensitivity, reliability and flexibility of these two techniques are termed biosensors.

A biosensor can be defined as a miniaturised device which combines a biological sensing component with an appropriate transducer, in this case optical.

The biological or biochemical sensor can be bacterial cells, enzymes, antibodies and other bioreceptor proteins. The biological sensor generates or modulates a light signal, and this is measured and converted to an electrical signal, which can be datalogged, recorded, used to trigger an alarm. There are many excellent recent reviews of biosensors in general (Marco and Barcelo, 1996; Bassi et al., 1996) and optical biosensors in particular (Wolfbeis, 1991; Arnold and Wangsa, 1991; Rogers and Poziomek, 1996; Meadows, 1996). The optical transduction mechanisms include absorbance, fluorescence, total internal reflectance fluorescence (evanescence), chemiluminescence and bioluminescence.

Thus, constructing an optical biosensor involves adapting bio-indicators and immunoassays which exploit the specificity and sensitivity of biological systems, and immobilising them on an optical surface or device, to enable optical measurement and interrogation. Alternatively, the optical effects of cells and micro-organisms in solution, or growing on surfaces, can be measured by means of absorption, transmission and scattering of light. Finally, the chemical, physical and electrical effects of organisms can be measured optically, such as a chemical reaction which a biological organism causes or affects, or a physical change such as pressure, displacement, or changes in conductivity or other electrical properties.

Optical sensors have risen on the back of the burgeoning, multi-billion dollar optical fibre communications market, expanding throughout the Eighties, with associated mass production of optical and optoelectronic devices. The "opto-revolution" has lowered the cost of existing optoelectronic devices, and enhanced R&D interest in new optoelectronics, light sources, detectors and new optical materials. This has paved the way for a new generation of optical sensors and measurement techniques, which exploit the new, low cost and compact light sources, detectors and devices, arising from the optical communications revolution.

Existing optical and spectroscopic techniques can be adapted, using new optical components, to form new, miniaturised, low-cost and portable sensors and optical instruments for applications in medicine, process control and for environmental monitoring. New optical techniques, materials and devices are continually evolving due to the optical telecomms market-driven research, enabling the generation of new optical measurement and sensing systems.

The aim of this chapter is to explain the terminology of optical sensors and the rationale behind optical sensing. There are already numerous works in existence reviewing the mechanisms of optical biosensors for measurement of various analytes. This review aims to compliment these works, by enabling non-optical engineers to glean an idea of the components and techniques used, in order that biologists and environmentalists can work with optical engineers to optimise new devices. Finally, we will review some existing sensors including those developed in our own laboratories at Liverpool John Moores University.

2. OPTICAL SENSORS

Due to the diversity of techniques and technologies available both commercially and in research labs and reported in the literature, it is difficult to classify optical sensing and measurements. Different classification schemes have been used, ranging from those based on the physical or chemical quantities measured by the system, to those based either on the physics of the sensing mechanism, or the detection system, or the modulation of the light (Dakin and Culshaw, 1988).

Essentially, an optoelectronic system can be divided into its component parts, as shown in Figure 1; these comprise:

- The light source
- The light detector
- Coupling optics

The coupling optics can either be based on conventional lenses, feed and return optical fibres or waveguides. By using integrated optics, all three components can be placed on the same chip or substrate.

Optical fibre sensors form a large sub-set of the family of optical sensing and measurement techniques, and are particularly relevant to environmental monitoring, because of the ability to perform point and distributed measurements remotely from the interrogating electronics.

In an optical fibre sensor, the fibre forms the coupling optics, and transmits the light from the light source to the modulation zone as shown in Figure 1, where the properties of the light are modulated in response to a change in an external parameter, which can be physical, chemical or biological. The light is then transferred to the detector, where the perturbation in the light characteristics is converted into an electrical signal.

The advantages of optical fibre sensor systems over conventional electrical sensor systems have been well documented and are summarised as follows:

- Immunity to electromagnetic interference, since the signal is in the form of light travelling through a dielectric insulator rather than an electrical current or voltage, travelling in a metal conductor.
- Electrical isolation compatible with intrinsic safety requirements, rendering such sensors to be useful in wet, medical or flammable environments. The electrical circuitry and power supplies are remote and insulated from the optical sensor. The light signal cannot cause electrocution when the sensor is in contact with humans, nor generate sparks to ignite flammable atmospheres.

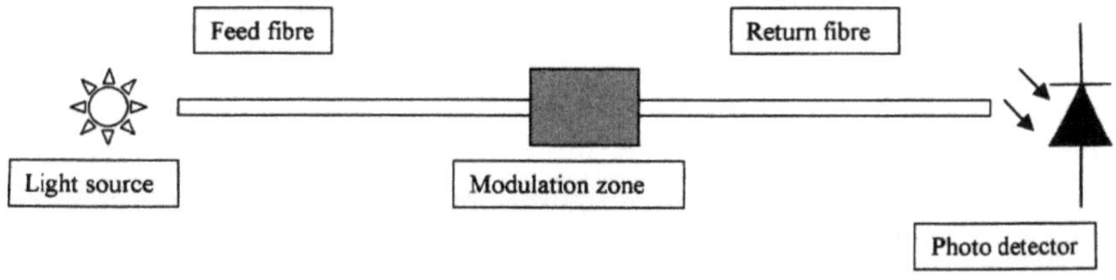

FIGURE 1. Schematic of the essential components of an optoelectronic system; an optical fibre sensor system with light source, feed fibre to the modulation zone, and return fibre to a photodetector.

- Passive operation, ensuring no power or electrical circuitry is required at the sensing point.
- Transmission of light over long distances up to 10s or 100s of km, enabling remote or distributed sensing due to the low losses achievable in optical fibres.
- Integration of sensing and telemetry in intrinsic optical fibre sensors.
- Glass and silicate materials used in optical sensors enables chemical immunity to corrosion and thus use in hostile environments.

Optical fibre sensors came into being because of the development of optical fibres in the 60s. Several types of optical fibres are currently available, based on silica glass, which can be manufactured in the form of single mode and multimode step index and graded index fibre. These fibres are now cheap due to large scale manufacture, resulting from the optical telecomms industry. Optical fibres made of materials other than silica are now available and can be used for sensing. Silica based fibres are well reviewed and described elsewhere (Senior, 1992; Wilson and Hawkes, 1997), but a number of new developments in fibres and optoelectronic devices, of particular interest to biosensors, are described as follows.

3. NEW DEVELOPMENTS IN OPTOELECTRONIC TECHNOLOGIES

3.1. Plastic Optical Fibre

Plastic Optical Fibres (POF) are thought to be the poor cousin of silica-based glass fibre in terms of their poor transmission properties (20 dB/km instead of 2 dB/km) making them only suitable for Local Area Networks (LANS) and crude sensor applications. However, recent advances in polymer materials in Japan (Koike, 1996; Berman, 1996; Polishuk, 1998), leading to low loss and high bandwidth, graded index POF, with improved attenuation characteristics and increased temperature tolerance, is posing a serious challenge to glass fibre. Such fibres are far easier and cheaper to terminate than silica glass based fibres, being easy to cut and polish. They usually have a 1 mm diameter core, which is easy to couple light into, and require low tolerance, injection moulded plastic couplers. Such fibres lend themselves to applications in the field and throw away optical sensors. Being polymer based, they are highly absorbing in the infra-red part of the spectrum (Figure 2A), where light sources and detectors for silica-based optical fibres are generally available, but rapid progress is being made in new visible LED and laser diode light sources (Section 3.3). Work at Liverpool John Moores University shows that POF can be imprinted with photo-induced gratings, tapered and incorporated into a multitude of sensors (Graydon, 1998; Merchant et al., 1998; Schmitt et al., 1996).

Fluorescence doped plastic optical fibres are a recent venture in POF, in which the core is doped with fluorescent dyes (Thevenin et al., 1986; Blumenfeld et al., 1987), originally for measurement of nuclear radiation by the French Atomic

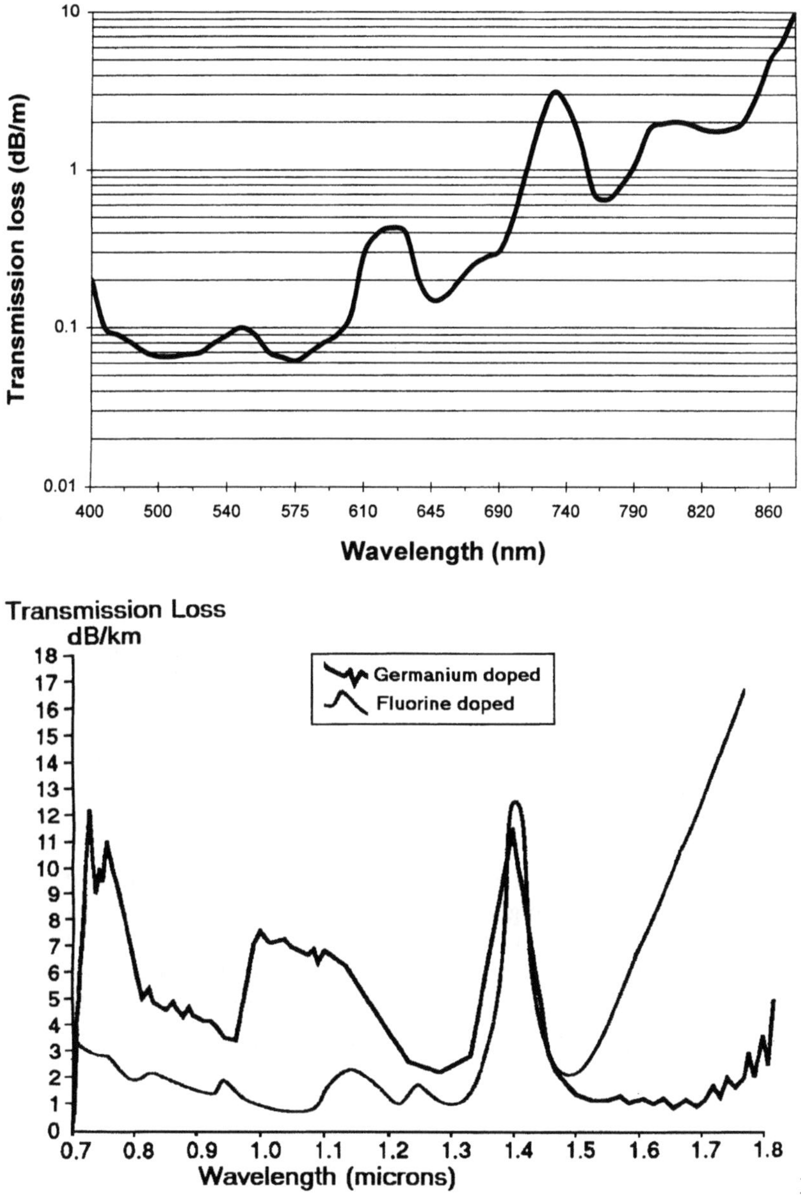

A

B

FIGURE 2. (A) Transmission spectrum of typical plastic optical fibre (POF). (B) Transmission spectrum of germanium-doped and fluorine-doped optical fibre.

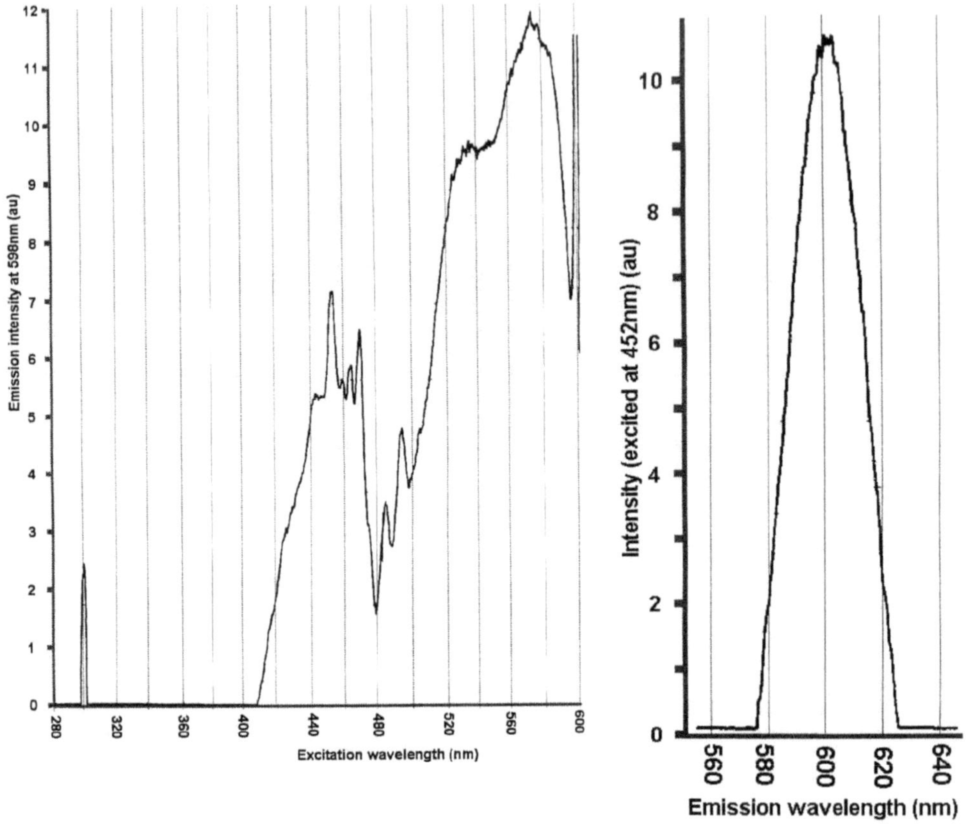

A B

FIGURE 3. (A) Absorption spectrum of red fluorescent plastic optical fibre, showing the excitation spectrum. (B) Emission spectrum of red fluorescent plastic optical fibre.

Commission. The application of FPOF has been extended into the area of physical sensors and decorative lighting and displays (Plastic Optical Fibres Conference Series). When such fibres are excited by a suitable wavelength, fluorescence occurs at a longer wavelength (Figure 3), and the fluorescent light is guided within the fibre core. Such fibres act as light detectors, amplifiers and wavelength shifters, and their applications in sensing are becoming apparent (Laguesse, 1993).

3.2. Other Specialist Fibres

Polymer clad silica fibres (PCS) are large diameter, step-index, multimode silica fibres, with a polymeric coating that can be chemically removed, exposing the cladding. Such fibres form the basis for multimode evanescent sensors (Deboux et al., 1995).

UV and IR transmitting optical fibres were developed for UV and IR optical fibre spectroscopy, since conventional silica fibres are damaged by, and are highly attenuating at UV wavelengths, and their use is restricted to the 300 to 1,800 nm wavelength region, which limits their use in sensors based on UV or IR spectroscopy. Fibres transmitting below 230 nm have been developed, with a transmission of 0.3 dB/m at 210 nm, enabling UV light sources such as a deuterium lamp to be used for fibre-based spectroscopic applications (MacCraith, 1996; Betlz et al., 1996). Fibres have also been developed for the mid-IR region, based on fluoride (0.5 to 5 μm), chalcogenide (1 to 6 μm), silver halide (3 to 15 μm) and germanium doped enhanced infra-red (Figure 2B). Hollow optical fibres have been developed for capillary electrophoresis (Baer, 1996).

3.3. New Developments in Light Sources

A major development, also driven by the Optical Telecomms industry, is the semiconductor laser diode or injection laser, producing monochromatic, coherent light at infra-red and red wavelengths. These were initially developed to match the transmission windows in optical fibres at 1.3 and 1.55 μm, but are now commercially available at near infrared (950, 850, 780 nm) and red wavelengths (660, 630 nm). Together with new developments in LEDs, such light sources have been mass produced at low price and are optimised for optical fibre coupling (pigtailed lasers etc.). They have the advantages of low power consumption, high stability and long lifetime, as well as being compact, robust and easily intensity modulated.

The availability of cheap laser diodes has fuelled interest in associated laser diode-driven light sources, such as fibre lasèrs, and diode-driven solid state lasers, giving an even greater range of wavelengths and intensities, and the possibility of using new optical materials and crystals for frequency doubling, and frequency mixing. Advances in semiconductor technology have led to green, blue and even white LEDs and laser diodes being commercially available (Figure 4), as well as the possibility of compact, cheap UV lasers (O'Reilly, 1996; Ball, 1997; Hill, 1997; Nakamura, 1998). Existing biomonitors and immunoassays require laboratory-based UV-visible and fluorescence spectrometers and colorimetric tests, with inherent bulky, heat-generating incandescent filament bulbs, interference filters and monochromators, or bulky, expensive, cooled, three-phase-driven visible and UV lasers (Argon-Ion, excimer). Small, compact light sources would enable on-line fibre-based spectroscopy systems to be developed. LEDs have a longer lifetime and are more efficient than incandescent bulbs, with a lifetime of up to 15 years, but a power consumption of just a tenth.

Tuneable lasers over the range 200 to 10,000 nm, can be constructed using optical parametric converters and difference frequency generation, and have important applications in spectroscopy. Diode-pumped solid-state lasers (DSP) and frequency-doubled DSPs are replacing helium neon and argon ion lasers (Marshall, 1994; Butcher, 1994).

Electroluminescent films (ELF), usually used for back lighting of displays, are available in a range of wavelengths, with high stability, and can be used to excite fluorescence in optical sensors. ELFs emit a diffuse pattern of light over

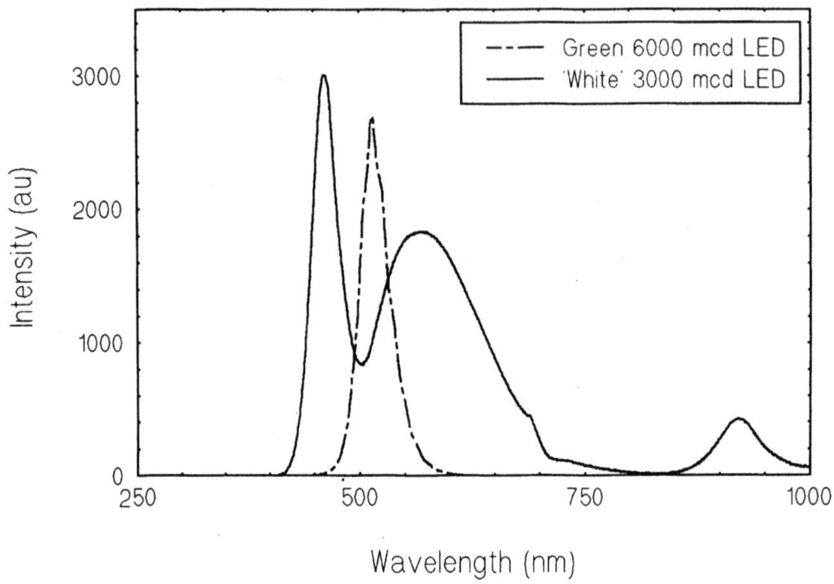

FIGURE 4. Emission Spectra of green and white Lucoled (luminescence conversion) LEDs. The white LED is based on a blue LED using gallium nitride to excite either organic dyes or inorganic phosphors in the yellow part of the spectrum.

their surface, and can be wrapped round glass vessels, to excite fluorescence of a measurand liquid. (Merchant et al., 1998).

3.4. Miniature Gratings and Spectrometers

For sensor applications based on absorption or fluorescence spectroscopy, dispersing elements such as a diffraction grating, or wavelength selective components such as optical filters are placed before the detector in order to select certain wavelengths or bandwidths which have been modulated by the measurand. The disadvantage of such optical devices is that they are lossy and do not utilise all the optical power available, but exploit only a narrow bandwidth. They require careful positioning and alignment, causing the optical sensor to be bulky and expensive. Compact scanning monochromators can be used, but an exciting recent development has been fibre-compatible CCD linear-array spectrometers, which are now commercially available from companies such as Zeiss, Ocean Optics and Microparts (Figure 5), with resolution of a few nm. Such devices provide spectral information in real time, have no moving parts and are compact in size. They cost in the order of \$1.5K and can be interfaced to a PC and customised for particular applications with choice of light source and software. They can be used as on-line spectrometers wherever a measurement of colour, absorption or fluorescence is required, making them highly versatile for a multitude of applications, in place of a non portable \$20K

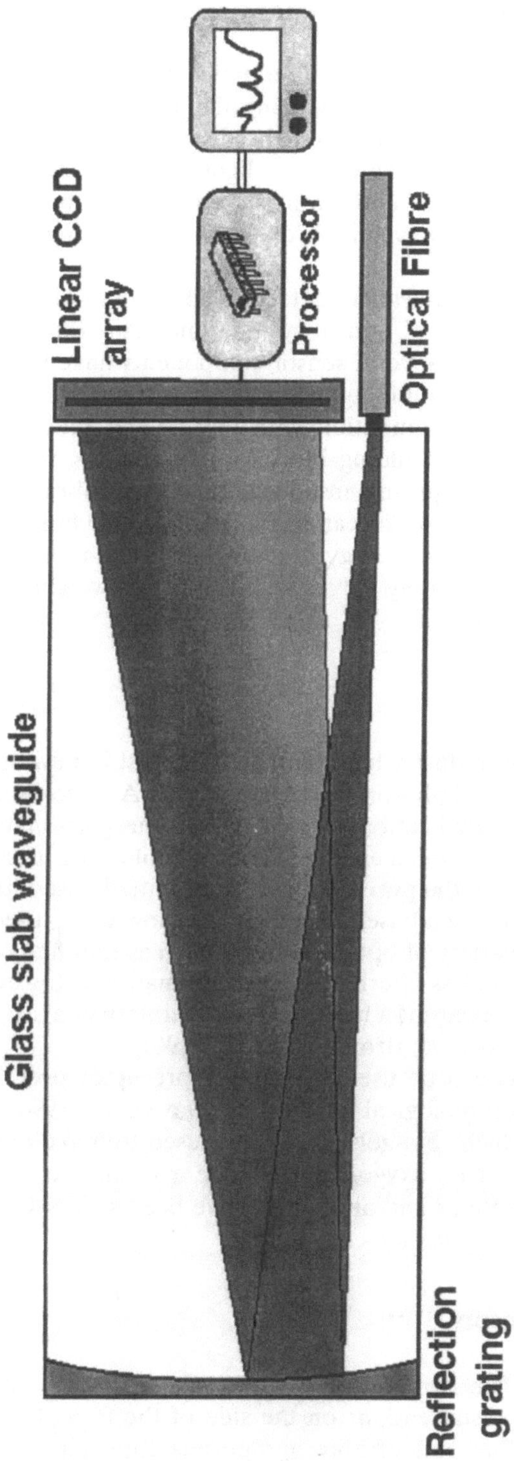

FIGURE 5. Miniature spectrometer showing light from an optical fibre being diffracted by a reflection grating, so that the wavelengths of the light are spatially separated, and collected by a linear CCD array. Processing electronics and output completes the compact spectrometer.

analytical spectrophotometer in a laboratory, operated by a trained technician carrying an array of samples in cuvettes.

Formation of a grating on the end of an optical fibre negates the requirement for a separate diffractive element, thus avoiding alignment and contamination problems, and such gratings can be formed on the distal ends of plastic optical fibres using a phase mask and 248 nm eximer laser light, and interrogated using a CCD linear array. Applications of fibre end gratings are numerous, including wavelength sensing and referencing and wavelength division multiplexing (WDM) (Schmitt et al., 1996).

Optical light sources, devices and detectors all on the same chip or substrate enable further miniaturisation of optical systems, including devices such as waveguides, gratings and couplers. Several sensor components have been integrated on a chip, by Texas Instruments, combining light sources, waveguides, detectors and coupling optics in a single device, compatible with a printed circuit board (Saini, 1996). Such devices can be used as building blocks for biosensors, replacing large and fragile optical components requiring an optical bench and vibration isolation, with a compact miniaturised device which can be manufactured in bulk and is assembled in a "ruggedised" form. This technology is analogous to the replacement of expensive, bulky, fragile, heat-generating valves with small, lightweight and rugged transistors and integrated circuits.

3.5. Sol-gels

The optoelectronics industry has stimulated interest in new optical materials, such as optical polymers and photorefractive materials. A material which is of particular interest for sensor applications, is sol-gel; a porous glass matrix produced at low temperatures (<100 °C), which enables chemical molecules, usually analyte sensitive dyes, to be trapped in the pores. Sol-gels were initially used to make a chemically porous glass tip for silica based fibres, but are now being used as chemically sensitive coatings for a variety of optical systems, such as thin film coatings on substrates and waveguides, and as alternative cladding materials for evanescent wave planar sensors and surface plasmon mode systems (Grattan et al., 1989; Klein, 1994; MacCraith et al., 1995 and 1996; Bromberg et al., 1996).

The sol-gel process can be used to entrap bioreceptor proteins, which have been shown to retain their biological functioning (Dave et al., 1994; Uttamchandani and McCulloch, 1996). Such "bio-gels" have been used to make optical biosensors for measurement of dissolved oxygen, nitric oxide and glucose. MacCraith (1996) reports that enzymes, proteins and antibodies have been successfully encapsulated in sol-gels for biosensor applications.

3.6. Optical Fibre Sensors

An optical fibre can be sensitised to a measurand, by immobilising the biological agent on the distal end, or on the side of the fibre. Configurations can be classified as end-of-fibre, side-of-fibre, and porous fibre (Figure 6).

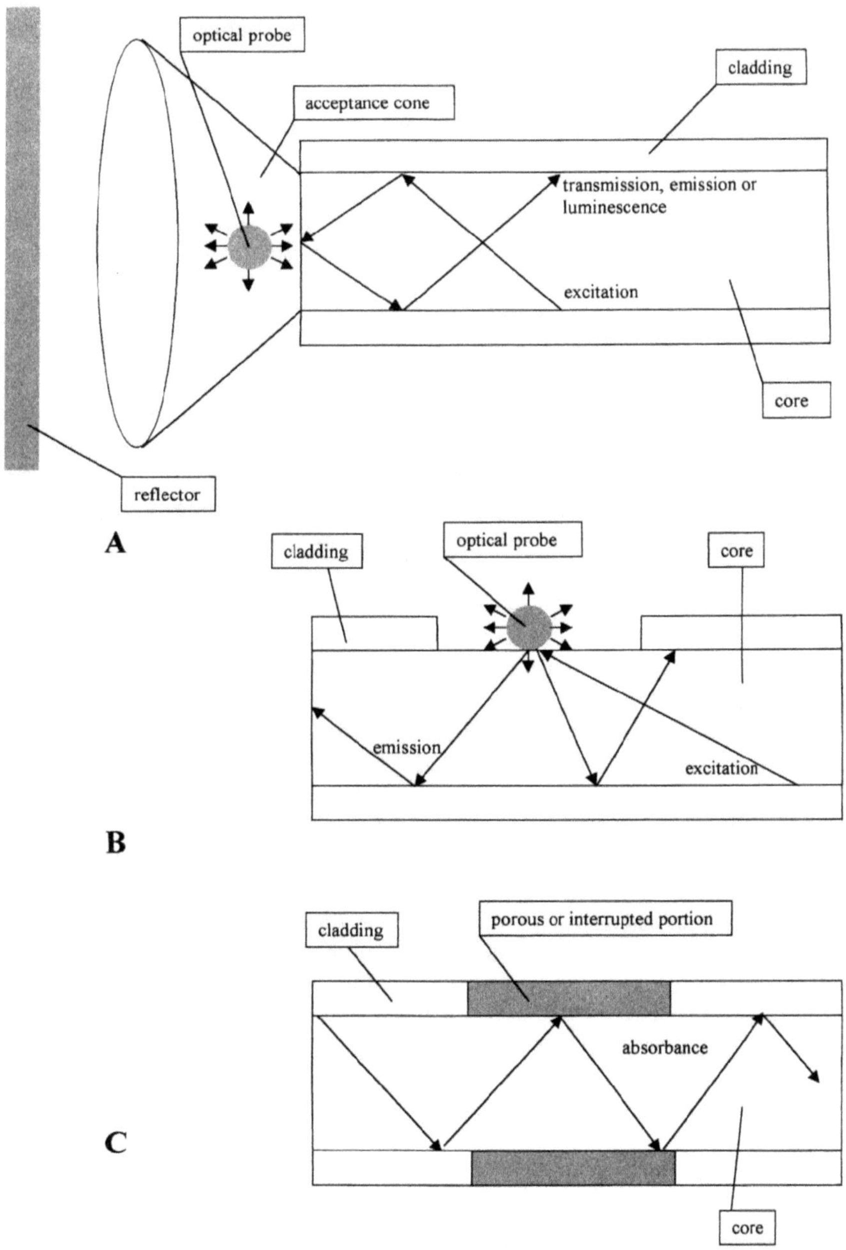

FIGURE 6. Fibre optic sensor configurations (A), end of fibre, (B) side of fibre, (C) porous fibre.

An end of fibre sensor uses the fibre to transfer the light to the sample. The light is modulated by absorbance or fluorescence of the analyte, indicator or analyte-indicator complex, which is trapped behind a membrane, in a polymer, or immobilised to the end of the fibre. In order to make absorbance measurements, the indicator can be placed between a reflector and the end of the fibre.

A side of fibre configuration usually uses the evanescent field, which is discussed in more detail in Section 4.1.1. Porous or interrupted fibre configurations incorporate the indicator directly into the structure of the fibre, and provide a large surface area for absorbance measurements. Fibres can be sensitised at a number of locations along their length for distributed measurement. Sol-gels can be used to form the porous coating, into which the indicator is introduced (MacCraith et al., 1995).

A fibre can be tapered in order to miniaturise the fibre tip (Uttamchandani and McCulloch, 1996) to form sub-micron optical fibre sensors; sensors that have spatial resolution below one micron, with faster response time and smaller analyte volumes of the order of pico-litres. Such a sensor can probe within individual biological cells, and within different parts of the cell such as the cell membrane or vacuoles. Silica optical fibres can be tapered by pulling the fibre, after melting in a fusion splicer, and have been shown to excite and detect fluorescence from fluorophore containing pore structures of the order of 10 microns. Sol-gels can be used to attach indicators to these probes.

Plastic optical fibres (POF) can be tapered by immersion in organic solvent (Merchant et al., 1998). Such tapered plastic fibres were used by the authors at Liverpool John Moores University to detect biofilm formation and scaling in closed loop water systems. A 1 mm diameter multimode plastic fibre with a poly-methyl-methacrylate (PMMA) core of 0.980 mm, surrounded by a thin cladding layer of fluorinated PMMA, had its cladding removed over a 5 cm length, using acetone. The fibre was tapered by immersion in a mixture of organic solvents, to increase the proportion of light present in the evanescent field. Light from a stabilised laser diode at 660 nm was transmitted through the fibre and its intensity was measured using a silicon photodiode. A non-tapered reference fibre was monitored to provide a common mode signal, using a 2:1 coupler. The tapered fibre was immersed in a solution of nutrients (1% yeast extract or 1% glucose dissolved in tap water) and left for 100 hours at room temperature for biofilm to form. After about 30 hrs, the optical signal change was 1.3 μW/hour, after which it levelled off after 70 hrs. The optical signal correlates well with measurement of biomass formation of biofilm with time.

3.7. Waveguide and Integrated Optics Sensors

Waveguide and integrated optics technology has given rise to a number of clever designs for sensors, including the deposition of sensing layers on waveguides, using a hollow capillary to act as waveguide and sample holder, and devices such as grating couplers adapted for sensing purposes.

Planar waveguides have advantages over optical fibres and are often used for biosensors in applications where the flexibility and remote capability of an optical fibre are not essential. These advantages include the following:

- Total internal reflection fluorescence (TIRF) can be used with lipid membrane structures stabilised onto optical surfaces in planar waveguides.
- Planar waveguides allow control over incident reflection angle, enabling control of penetration depth of evanescent field.
- Langmuir Blodgett deposition of membranes onto planar substrates is well characterised, but difficult onto cylindrical surfaces.
- Polarisation information is lost in multimode optical fibres, but retained in planar waveguides.
- The possibility of interferometric measurements.
- Single mode optical fibres would allow polarisation maintenance and interferometry but are not yet available at short wavelengths required for fluorescence in terms of fibre diameter or fibre material.
- Planar waveguide structures can be placed in flow cell modules.
- The ease of fabrication.

Robinson, 1995, describes how commercially-available optical biosensors are based on planar transducers.

4. TYPES OF OPTICAL BIOSENSORS

An optical biosensor incorporates a biologically active material, which alters its optical properties reversibly and selectively in response to the analyte, usually a chemical species. Biosensors form a new and rapidly expanding field of sensors, using isolated enzymes, intact bacterial cells, mammalian and plant tissue, antibodies and bioreceptor proteins.

Usually an optical biosensor uses the biologically active material as a catalyst which is immobilised at the surface of a single fibre, waveguide or fibre bundle, and the detected species is measured by absorbance, fluorescence or chemiluminescence. An analyte-consuming reaction is catalysed by the immobilised biocatalyst and a product of this reaction is monitored. The biocatalyst mediates between the analyte of interest and the transducer by converting the analyte to a detectable species. (Robinson, 1995).

Some configurations of optical biosensor are described in Sections 4.1 to 4.4.

4.1. Optical Immunosensors

These sensors (Vo-Dinh et al., 1991; Plowman, 1996) offer excellent selectivity via anti-body-antigen recognition, enabling measurement of important compounds at trace levels in complex biological samples. Optical immunosensors for detection of human and environmental exposure to toxic chemical and biological materials, are based on various spectrochemical mechanisms such as absorption, total internal reflection, fluorescence, evanescent field and surface plasmon resonance, using both optical fibres and waveguide configurations. Some examples are discussed as follows:

4.1.1. Evanescent Wave (EW) Immunosensors

EW Immunosensors comprise an optical fibre with cladding removed or a waveguide, upon whose surface the biological sensing agent is placed (Figure 7). A waveguide is an optical structure with a guiding layer of higher refractive index and a cladding layer of lower refractive index; an optical fibre is actually a waveguide with cylindrical geometry. The interrogative light is directed within the waveguide, and a portion of the light extends beyond the core to interact with the biological sensing layer, called the evanescent wave (EW) or field. Such immunosensors use a sensing layer made of fluorescently-labelled molecules which re-emit the absorbed evanescent field at a longer wavelength as fluorescence. The fluorescent light generates light in the core by evanescent field excitation, and the fluorescent light is thus transmitted to the photodetector.

The advantage of evanescent sensors is that interference from the bulk media is avoided since only substances directly and specifically adsorbed onto the core surface interfere with the evanescent field, which extends to a few hundred nanometres.

Immunosensors based on the evanescent effect have been developed to measure environmental pollutants such as atrazine to a detection limit of 2 nM or 0.1 $\mu g L^{-1}$ (Oroszlan et al., 1993), terbutryn to 15 nM or 0.1 $\mu g L^{-1}$ (Beir et al., 1992), imazephapyr to 0.3 $\mu g L^{-1}$ (Anis et al., 1993), polychlorinated phenols to 10 $\mu g L^{-1}$ (Zhao et al., 1995) and parathion to 0.3 $\mu g L^{-1}$ (Anis et al., 1992). Most of the devices use fluorescein to either label the antibody or the analyte, and competitive immunoassays.

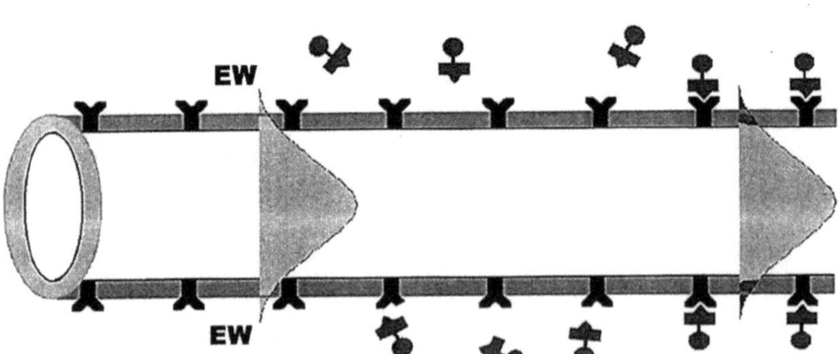

FIGURE 7. Waveguide evanescent wave (EW) immunosensor (after Marco and Barcelo, 1996). Light directed into the waveguide generates an electromagnetic field, with the evanescent region penetrating into the cladding of the fibre (labelled EW). The EW causes direct excitation of a fluorophore used as a label in the immnunoassay. The photons are re-emitted and part of the emission is coupled back into the waveguide, increasing the electromagnetic field within the fibre, and guided to the photodetector.

4.1.2. Grating Coupler Immunosensors

Optical devices such as grating couplers are used in integrated optics to couple light into and out of waveguides. The critical angle is the maximum angle of total internal reflection within the waveguide and determines the angle at which light couples into a waveguide. It is a function of the refractive index and thickness of the waveguide; if the surface of the waveguide is coated with a sensitive chemical and thus its refractive index changes in response to an immunoreaction, then a sensor can be constructed (Figure 8). Advantages include small size of sensing head and no moving parts. The commercial optical biosensor Bios-1™ from Biostar is an example of a grating coupler sensor using ligand-anti-ligand reactions (Robinson, 1995).

4.1.3. Surface Plasmon Resonance Immunosensors

These devices exploit an evanescent field generated at the surface of a metal conductor, usually silver or gold, when excited by impact of light at certain wavelength and angle (Figure 9A). Surface plasmons are generated by electrons at the metal surface, and produce a resonance at a different frequency to those in the bulk metal. The absorption of light energy by surface plasmons during resonance is observed as a sharp minimum in light reflectance (R) when the varying angle of incidence (θ) reaches a critical value, as shown in the graph of R versus θ. The critical value is affected by wavelength, polarisation state of incident light, and dielectric properties of the medium adjacent to the metal surface, so it is modulated by analytes binding to the surface. A number of commercial sensors are based on this technique including the Pharmacia BIAcore™, consisting of a gold-coated sensor chop mounted in a cassette that readily interfaces to the fluidics of the system used to detect atrazine on a direct immunoassay configuration to $0.05\,\mu g\,L^{-1}$ (Minunni and Mascini, 1993), and Fisons IAsys™, which is a resonant mirror sensor comprising a laminate of high and low refractive index materials coated with the biologically active layer as shown in Figure 9B.

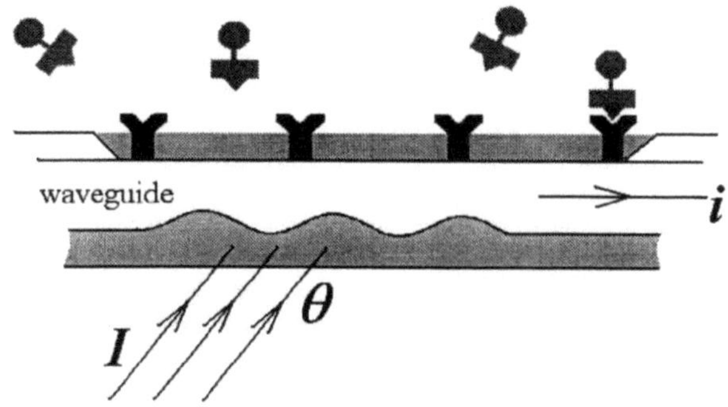

FIGURE 8. Grating coupler. The angle, (causing total internal reflection is strongly affected by molecules present within the evanescent field (shaded region). After Marco and Barcelo, 1996).

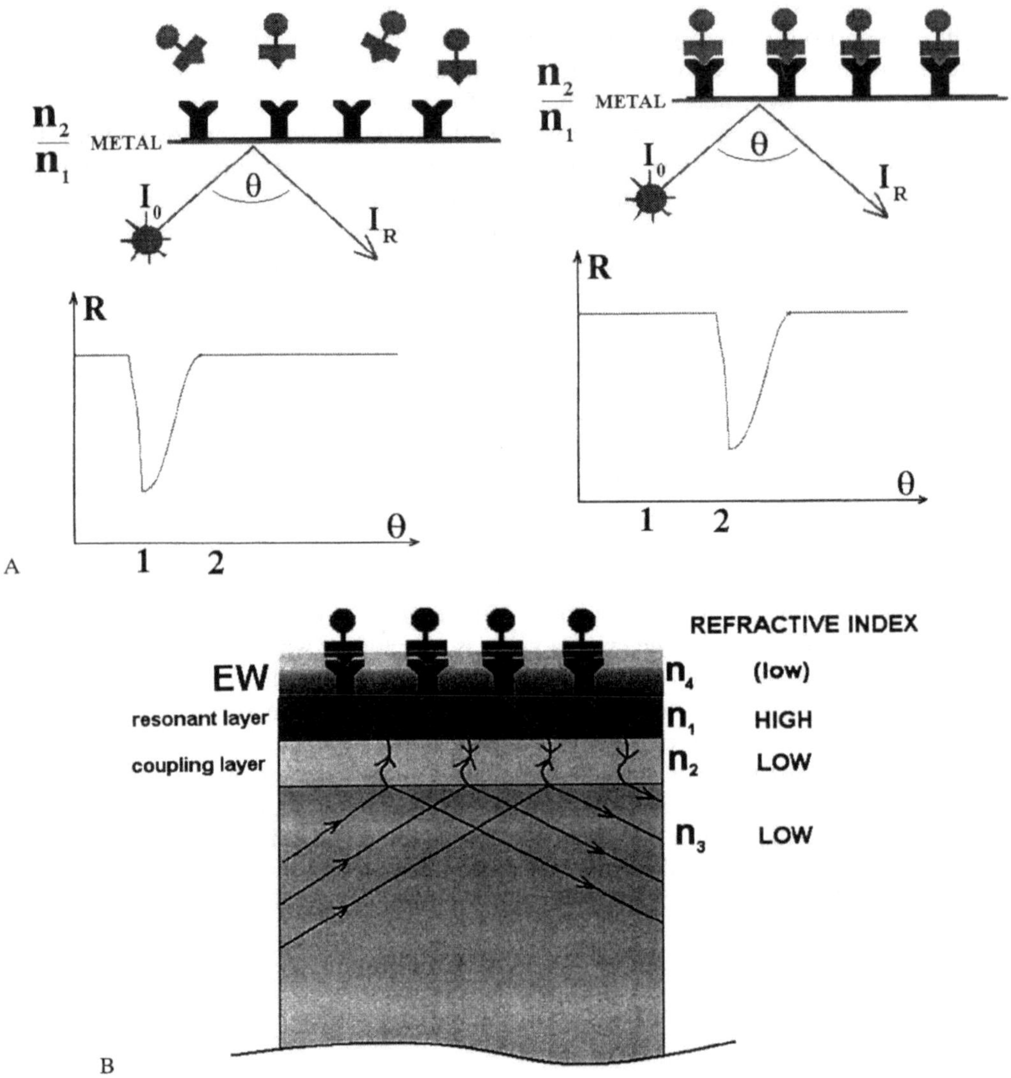

FIGURE 9. (A) Surface plasmon resonance (SPR) sensor. Surface plasmons of a metal conductor are excited by the light energy at a critical angle θ, causing oscillation and the excitation of an evanescent field as shown by the graph of R (reflectivity) versus θ, and thus a characteristic decrease of the reflected light intensity. The critical angle is strongly affected by molecules directly attached to the surface of the metal. It can be seen how θ shifts from position 1 to position 2 when molecules are attracted to the metal surface. After Marco and Barcelo, 1996. (B) Resonant mirror sensors (IAsys™). At a critical angle, light passes through a coupling layer (n_3) via the evanescent field and couples with a resonant layer (n_1, $n_1 > n_3$) generating a second evanescent field which propogates a certain distance along the surface before coupling back to the device. After Marco and Barcelo, 1996.

4.1.4. Reflectance Interference Spectroscopy (RIFS) Immunosensors

These devices exploit the partial reflection of light when a light beam passes through a thin film, or an interface formed by media of two different refractive indices. Thus, a single light beam incident on a thin film, will be reflected at the upper and lower interfaces, thus producing a number of light rays, each separated by a constant phase difference, which is related to the thickness of the film. The multiple beams interfere to form fringes, or spatial intensity modulation. The period of the fringes can be measured to give an indication of the film thickness and/or refractive index. Constructed such an immunosensor to detect atrizine using a hapten-derived immunosensor surface, without using any label.

4.2. Fourier Transform Infrared Spectroscopy

Fourier transform Infra Red spectroscopy (FTIR) has advantages over normal spectroscopy which requires dispersing elements. Advances in optical fibre, infrared detectors and new infrared laser diodes, as well as the application of digital fast Fourier analysis to the captured spectra, have led to new types of miniaturised sensors. FTIR can detect molecular bands due to their characteristic absorbance in the infrared (IR). Optical fibre sensors used with FTIR, are based on stripped cladding replaced with a medium which exhibits a concentration-dependent frequency shift in the IR on reaction with the analyte (Kerslake and Wilson, 1996).

4.3. Microbial sensors

These sensors use whole living organisms such as cells, bacteria, yeast and fungi, to exploit their metabolic functions, so that the organism is combined with an appropriate optical transducer. The advantage of using whole cells is that the extraction of single purified molecules as sensing elements, can be expensive. In addition, whole living cells can be easily isolated from river sediments, soil, activated sludge, etc. Other advantages include their lower sensitivity to inhibition from other compounds present on the matrix, and greater tolerance to variations in pH or temperature, and longer lifetime. Limitations include the fact that microbial sensors require longer response times than enzymes, and that selectivity is more difficult to establish than with single enzymes, due to the number of metabolic processes occurring in a living cell (Marco and Barcelo, 1996).

Optical microbial sensors have been developed using naturally occurring photobacteria or other genetically modified organisms. Luminescence is affected by changes in external conditions, leading to a variation in the concentrations of chemicals to produce light. Bioluminescent bacteria have been immobilised on sensors to produce BOD sensors (Hyun et al., 1993). A microbial sensor measures changes in the UV absorption of an immobilised bacteria, when placed on a flow cell and stressed by the presence of toxic compounds (Bains., 1994). The metabolic rate is reduced by the toxins, causing a change in the UV spectrum.

4.4. Toxicity Measurement with Eclox, Aquanox and Microtox

Both the Eclox (Aztec Environmental & Control Ltd) and Aquanox (Randox Laboratories) hand held monitors use an enhanced chemiluminescent reaction; free radical reaction of the oxidation of luminol in the presence of horse radish peroxidase enzyme, using p-iodophenol as an enhancer and to stabilise the reaction (Hayes and Smith, 1996). Any substance, such as an antioxidant pollutant, that inhibits the enzyme reaction, causes a reduction in light emission which is measured on a portable luminometer. The advantage of these units is that they are relatively cheap (Stg 1–4K), portable and can be used by inexperienced operators. Disadvantages include insensitivity to some toxins.

Microtox uses a freeze dried marine bacteria, Photobacterium phosphoreum, stored in a cooled, storage area within the instrument. In use, a standard amount is rehydrated and mixed with the water sample. The bacteria emit light under favourable conditions and reduce their emission when exposed to toxins. Microtox is widely used in the water industry but it is large, expensive (>$30K) and laboratory based, though an on-line monitor and a hand-held version are under development (Bartlett-Hooker, 1996).

5. OPTICAL TOXICITY AND FLUORESCENCE SENSOR

Research at Liverpool John Moores University has produced a novel fluorimeter based on optical fibre and waveguiding technology (Merchant et al., 1998). The sensor was initially designed for use with a toxicity assay, applied to live microorganisms, which produced fluorescein as the indicative measurand (Edwards et al., 1995; Grabowski et al., 1995). In this configuration, the sensor can be said to be a biosensor (Bogue, 1997).

The sensor uses new optoelectronic technologies such as fluorescent plastic optical fibres as the detectors, and an electroluminescent film as the light source. The sensor design is based on a flow-through optically transparent tube, which can be inserted within pipework removing the need for sampling or pumping systems. The sensor can be encased in a waterproof housing, enabling it to be immersed in water systems. Conventional devices for measuring fluorescence rely on costly and fragile light sources, detectors, optical filters, gratings and photomultiplier devices; for example the spectrofluorimeter.

5.1. Toxicity Assay

A colony of microorganisms such as bacteria, algae etc, are exposed to a solution containing fluorescein acetate, then an intracellular esterase enzymatic reaction hydrolyses the compound into fluorescein. The product of the reaction, fluorescein, emits over the wavelength range 515 to 550 nm when excited between 450 and 510 nm. Thus the intensity of fluorescence increases as the reaction proceeds, and the rate of increase of fluorescence indicates the reaction rate, which is affected by environmental parameters such as toxicity. Toxins affect the cytoplasmic pH and viscosity, metabolic rate and porosity, hence the light emitted by the reaction is a measure of the metabolic rate of the cells, and thus of the toxicity of the test solution.

Any organism can be used; algae, bacteria or animal cells which are suspended in the water under test for toxins. The time lag from adding fluorescein diacetate to measurement of metabolic rate is from 5 to 20 mins. The toxicity assay can be tailored to any ecosystem, by using marine or freshwater species, or robust or fragile organisms. An important application is for inhibition testing in activated sludge waste water treatment plants. The activated sludge can be used as the test colony, immobilised in a simple flow-through filter column, in order to predict the inhibitory effects of contaminated incoming water, which can be diverted to a storm tank. The bacteria mix in activated sludge is constantly varying, so standard toxicity tests are difficult to apply.

5.2. Sensor Design

The toxicity assay is intended for on-line monitoring of effluent and water quality and hence the sensor design has been optimised for use in continuous flow-through systems with low-quality water. It uses several novel techniques, described as follows.

The sensor (Figure 10) is based on a section of transparent tubing, glass or polymer. An inexpensive powder-deposition electroluminescent film provides the excitation light into the fluid. The sensor is believed to be the first and only analytical use of such light sources.

The separation of fluorescent light from the excitation signal is provided by selective waveguiding of the fluorescence within the tube walls as shown in Figure 11. This occurs due to the change in direction of the light when re-emitted from a fluorescent molecule and is independent of either wavelength. No optical filters, gratings or selective detectors are required making the sensor inexpensive, reliable and versatile.

The fluorescent light is detected by an array of polymer optical fibres (POF) embedded into the tube some distance from the light source (Figure 12). They are secured in grooves that do not compromise the internal surface of the tube, making it as resistant to physical and chemical damage from the fluid as the original tube material, and removing any obstructions to flow or sites for biofouling.

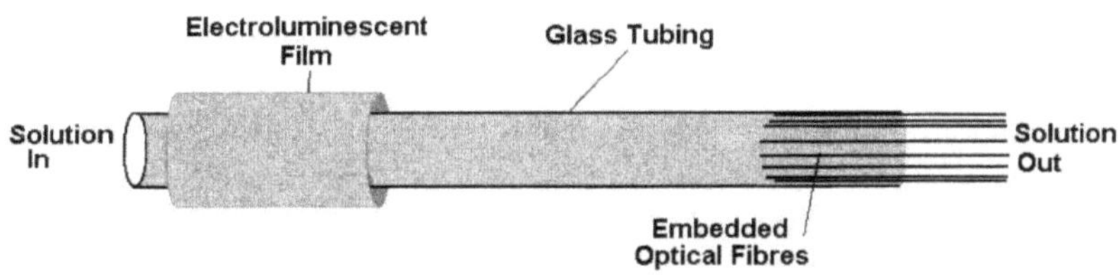

FIGURE 10. Schematic of the optical fibre toxicity and fluorescence sensor.

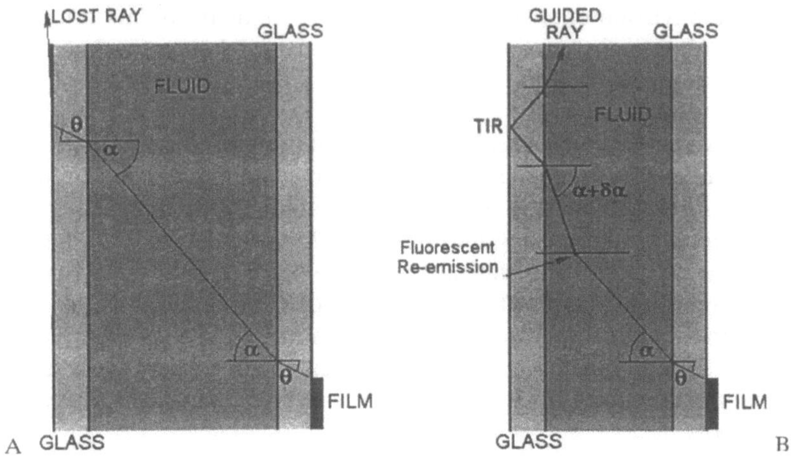

FIGURE 11. Wave guiding process in the sensor. (A) Ray from the electroluminescent film is lost, and guided outside the tube without interaction with a fluorophore. (B) Ray from the electroluminescent film interacts with a fluorophore, causing a fluorescent-emission event. The fluorescent light is emitted in all directions and some of the rays generated will be coupled into the walls of the glass tube, and wave-guided to the detecting fibres.

The POF array consists of a series of fluorescent injectors. A short length of polymer fibre with a doped fluorescent core material (polystyrene-based) is used in the embedded region, coupled to a clear PMMA-based fibre for conveyance of the light to a silicon photodiode detector unit (Figure 13).

As described in Section 3.1, fluorescent-doped polymer fibre (FPOF) is a recent development, originally for measurement of nuclear radiation by the French Atomic Commission, but the LJMU sensor is one of the first exploitations of the material for analytical measurement. The core of the FPOF acts to redirect light incident through the side walls of the fibre into guided light along the core at a far higher efficiency than conventional fibres. The combination of FPOF injector and POF link to a photodetector realises an extended cylindrical optodetector. The absorption spectrum of the dopant is selected to preferentially detect the fluorescent light wavelengths of the fluid under test and reject any stray light from the excitation source.

As with all conventional fluorescence detectors, samples with multiparameter variance (fluorescence, colour, turbidity) will not show responses to only one parameter. Using conventional instruments, extensive sample filtering, preparation and correction is required, to the point of human intervention. The JMU sensor has demonstrated the capability of measuring all three of these parameters and by use of several light source/ FPOF injector combinations, they can be separated by simple signal processing in real time. The sensor is unaffected by flow rate, direction, temperature, pressure or viscosity and can rely on an inherent flow rather than electrical pumping of the test fluid.

The JMU sensor has been characterised for all of these parameters to varying degrees:

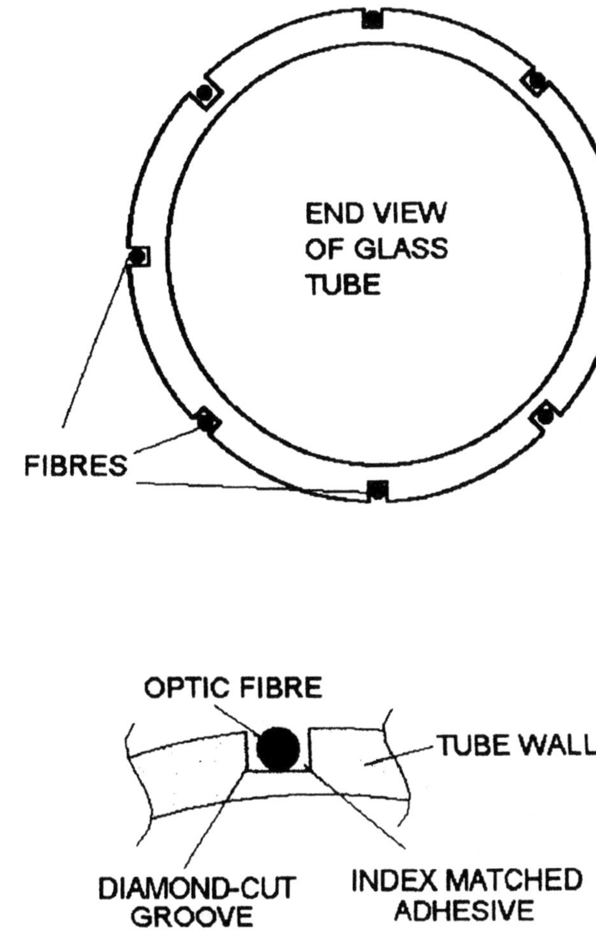

FIGURE 12. Details of the optical fibre mounting arrangement.

1. *Fluorescence*: Linear response for fluorescein sodium solutions, resolution of 0.2 ppb and full working range of 0.2 ppb to 100% (Figure 14). Will detect all visible-emitting fluorochromes.
2. *Turbidity*: Linear response over range 0.1 to 50 FTU, resolution 0.1 FTU for 0 to 1.0 FTU.
3. *Absorbance* (Color): Initial tests show the sensor will detect and differentiate between colours (absorption spectra) created by metal salt solutions in the ppm range.

The sensor offers a physical design ideally suited to on-line monitoring, as well as a low-cost and simple-to-produce, disposable sensor head unit. The detection and processing electronics are expected to be less costly than conventional instruments, as the sensor produces a large (several nW) optical signal and photomultiplier detectors are not required. The device has been shown to display equivalent sensitivity to laboratory equipment, but has several advantages. The fluid is only in

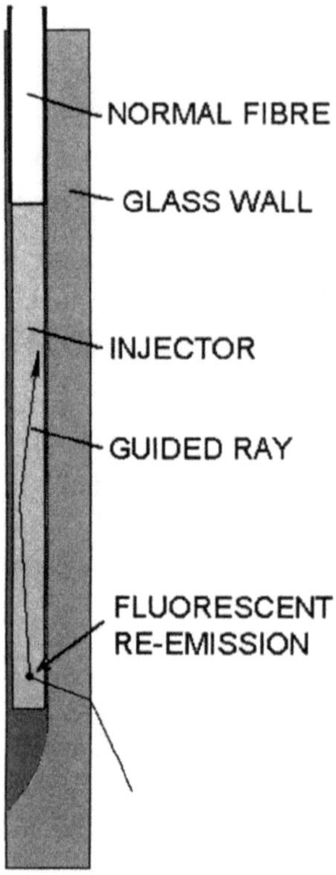

FIGURE 13. Cross-section of a groove showing fluorescent-core-doped-plastic optical fibre as an "injector", which absorbs the fluorescent light coupled into the walls of the glass tube, over its cylindrical surface. The light is absorbed and re-emitted as fluorescent light at a longer wavelength; light at the original excitation wavelength of the electroluminescent film is rejected. The fluorescent POF is butt-coupled to normal POF for transfer to the detector system.

contact with the tube, not any delicate optical surfaces. It is therefore very simple to clean, is rugged and can operate under pressure and vacuum. By using different electroluminescent films, it is possible to measure selected fluorochromes. The films can be multiplexed to measure several fluorochromes in real time with no moving parts. The entire device, including head, power unit and detectors, can be housed in a small waterproof casing, remaining resistant to impact, vibration and pressure. Tube wall thickness, tube size, bore and length can all be altered to suit the end application and the optimum concentration ranges. The design of the sensor has been filed for patent (96 10482.3GB/97 303 392.1EU).

6. CONCLUSION

Optical biosensors continue to expand, as new optoelectronic devices and materials become available, driven by the massive optical fibre communications and optoelectronics market, leading to novel sensors based on waveguides, integrated optics and new materials. New compact light sources in the visible and UV part of the spectrum, have become available due to advances in semiconductor and polymeric optoelectronic devices, replacing large, bulky and expensive lamps and lasers. Applications of biosensors to clinical and medical fields, and biotechnology and

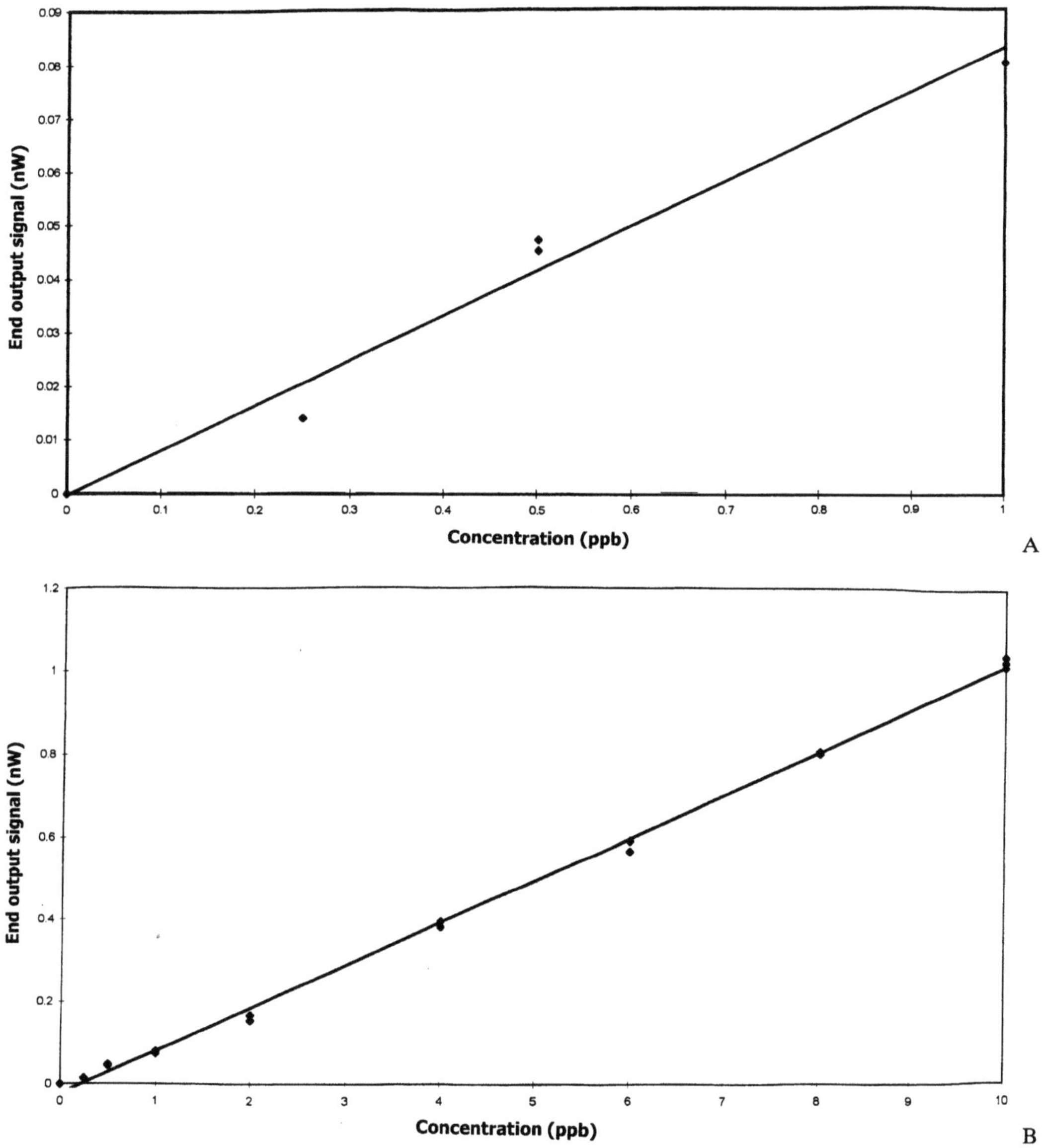

FIGURE 14. Plots of sensor optical output signal in nanowatts, versus concentration of fluorescein solution (ppb) in water. (A) Zero to 10 ppb concentration; (B) Zero to 1 ppb concentration.

process control industries are more advanced than their application to environmental monitoring. The former are more adventurous in exploiting new optical technology and many promising sensor principles and techniques are being encouraged, and will eventually filter through to more conservative industries. Optical techniques can offer advantages over existing diagnostic and monitoring techniques. Problems to be overcome by the optical designer include biofouling, calibration stability and selectivity.

7. REFERENCES

Anis N.A., Eldefrawi M.E., and Wong R.B., 1993. J. Agric. Food. Chem. 41, 848.

Anis N.A., Wright J., Rogers K.R., Thompson R.G., Valdes J.J., and Edelfrawi M.E., 1992. Anal. Lett., 25, 627.

Arnold M.A., and Wangsa J., 1991. Transducer based and intrinsic biosensors. Fiber Optic Sensors and Biosensors. Editor Wolfbeis O. Vol I and II. CRC Press, Boca Raton.

Baer T.M., 1996. Optics in biotechnology and health care. Optics & Photonics News. Jan 1996.

Badini et al., 1996. Characteristics of dye-impregnated tetrathylorthosilane (TEOS) derived sol-gel coatings. J. Sol-Gel Sci & Tech. 6, 262–272.

Bains W., 1994. Biosensors & Bioelectronics. 9, 111.

Ball P., 1997. Into the blue. New Scientist 2075, 28–31.

Bartlett-Hooker B., 1996. On-line toxicity monitoring. International Conference on Toxic Impacts of Waste on the Aquatic Environment. Apr 1996. Loughborough University, UK.

Bassi A.S., Tang D., Lee E., Zhu J.X., and Bergounou M.A., 1996. Biosensors in Environmental and Bioprocess Monitoring and Control. Food. Technol. biotechnol. 34(1), 9–22.

Beir F.F., Stoecklein W., Boecher M., Bilitewski U., and Schmid R.D., 1992. Sensors & Actuators, B7, 509.

Betlz M. et al., 1996. Water quality monitoring using fibre optics at wavelengths below 230 nm. Conference on Optical Science, Engineering & Instrumentation: Chemical, Biochemical & Environmental Fibre Sensors VIII, Denver, Aug 1996. Proc. SPIE 2836. 1996.

Berman E., 1996. Driving Bandwidth. POF'96. Fifth International Conference on Plastic Optical Fibres & Applications. Paris, France, Oct 22–24, 1996.

Blumenfeld H. et al., 1987. Plastic fibres in high energy physics. Nucl. Instrum. Methods. A287, p 603.

Bromberg A. et al., 1996. Optical fibre sensors for blood gases and pH based on porous glass tips. Sensors & Actuators B31, 181–191.

Bogue R.W., 1997. Novel fibre optic cell biosensor. Biosensors & Bioelectronics 12, XV.

Butcher S., 1994. Optical parametric oscillators open new doors to researcher. Photonics Spectra, May 1994 pp 133–138.

Dave B., Dunn B., Valentine J.S., and Zink J., 1994. Sol-gel encapsulation methods for biosensors. Anal. Chem 66, 1120–1127.

Deboux B. J.-C., Lewis E., Scully P.J., and Edwards R., 1995. A Novel Technique for Optical Fibre pH Sensing based on Methylene Blue Adsorption. J. Lightwave Technology. 13(7), 1407–1414.

Edwards R., El-Saadawy S., Gander M., Scully P., Young A., Baker P., and Grabowski J., 1995. The development of a biological toxicity based test for water quality using a fibre optic sensor. International Symposium on Analytical Chemistry, Hull, UK. July 1995.

Grabowski J., Baker P., and Scully P., 1995. Toxicity measurements using fluorogenic compounds. European symposium on Optics for Environmental and Public Safety. Munich. Germany. June 1995.

Grattan K.T.V. et al., 1989. Sol-gels for fibre optic sensors. Proc. 6th Int. Optical Fibre Sensors Conference, Paris. Arditty H.J., Dakin J.P., Kersten R.Th (Eds). Optical Fibre Sensors (Springer Verlag, Berlin) pp 436–442.

Graydon O., 1998. Mysterious chemistry tapers polymer fibres. Opto & Laser Europe, 52, July 1998.

Hayes E., and Smith M., 1996. Eclox: a rapid screening toxicity test. International Conference on Toxic Impacts of Waste on the Aquatic Environment. Apr 1996. Loughborough University, UK.

Hill P., 1997. White Light diodes are set to tumble in price. Opto-Laser-Europe Oct 1997, 17–20.

Hyun C.K., Tamiya E., Takeuchi T., and Karube I., 1993. Biotechnol. Bioeng, 41, 1107.

Kerslake E.D.S., and Wilson C.G., 1996. Pharmaceutical and biomedical applications of fiber optic biosensors based on infra-red technology. Advanced Drug Reviews, 21, 205–213.

Klein L.C. (Editor), 1994. Sol-gel Optics: Processing and Applications. Kluwer Academic Press.

Koike Y., 1996. Staus of POF in Japan. POF'96. Fifth International Conference on Plastic Optical Fibres & Applications. Paris, France, Oct 22–24, 1996.

Laguesse M., 1993. Sensor applications of fluorescent plastic optical fibres. POF'93 Second Annual Conference, The Hague, June 23–29.

Marco M.P., and Barcelo D., 1996. Environmental applications of analytical biosensors. Meas, Sci. Technol. 7, 1547–1562.

MacCraith B.D., McDonagh C.M., O'Keefe G., McEvoy A.K., Butler T., and Sheridan F.R., 1995. Sol-gel coatings for optical chemical sensors and biosensors. Sensors & Actuators B 29, 51–57.

MacCraith B.D., 1996. Water monitoring using advanced optoelectronic instrumentation. CIWEM Advanced Sensors for the Water Industry. 20 Nov 1996. Manchester, UK.

Marshall L., 1994. Biological Monitoring forseen with ultraviolet light source. Laser Focus World Apr 1994, p 83.

Meadows D., 1996. Recent developments with biosensing technology and applications in the pharmaceutical industry. Advanced Drug Delivery Reviews 21, 179–189.

Merchant D.F., Scully P.J., Edwards R., and Grabowski J., 1996. Optical fibre toxicity bio-sensor. Institute of Physics Applied Optics & Optoelectronics Conference, 16–19 Sept 1996. University of Reading.

Merchant D., Scully P.J., Edwards R., and Grabowski J., 1998. Optical fibre fluorescence and toxicity sensor. Sensors & Actuators B (Chemical) Vol B48, Nos 1–3., 476–484.

Merchant D.F., Scully P.J., and Schmitt N.F., 1998. Chemical Tapering of Polymer Optical Fibre. Eurosensors Conference. University of Southampton. Sept 1998.

Minunni M., and Mascini M., 1993. Anal. Lett. 26, 1441.

Nakamura S., 1998. Light Emission moves into the blue. Physics World, Feb 1998, 31–35.

O'Reilly E., 1996. Semiconductor lasers go into the blue. Physics World, 9 (4), p 19.

Oroszlan P., Thommen C., Wehrli M., Gert D., and Ehrat M., 1993. Anal. Methods. Instrum. I, 43.

Plastic Optical Fibres Conference Series; POF'93 to POF'98. POFIG Secretariat @ Information Gatekeepers Inc. 214 Harvard Ave, Boston MA 02134 USA.

Polishuk P., 1998. Price Attracts Industry to Plastic Optical Fibre. Fibre Systems Mar 1998, 17–21.

Robinson G., 1995. The commercial development of planar optical biosensors. Sensors & Actuators B 29, 31–36.

Rogers K.R., and Poziomek E.J., 1996. Fiber Optic Sensors for Environmental Monitoring. Chemosphere, 33, 6, 1151–1174.

Saini DPP, FCI Environmental Inc, Publicity Data, 1996.

Senior J., 1992. Optical Fiber Communications. Second Edition. Prentice Hall.

Schmitt N.F., Lewis E., and Scully P., 1996. UV photo-induced structures on plastic optical fibres. POF'96. Fifth International Conference on Plastic Optical Fibres & Applications. Paris, France. Oct 22–24.

Thevenin J.C. et al., 1986. Scintillating and fluorescent fibres for sensor applications. IEEE Trans. Nucl. Sci., 33, p 133.

Uttamchandani D., and McCulloch S., 1996. Optical nanosensors-towards the development of intracellular monitoring. Advanced Drug Delivery Reviews 21, 239–247.

Wilson J., and Hawkes J., 1997. Optoelectronics: An Introduction. Third Edition. Prentice Hall.

Wolfbeis O., 1991. Editor. Fibre Optic Chemical Sensors and Biosensors. Vol I and II. CRC Press, Boca Raton (1991).

Zhao C.Q., Anis N.A., Rogers K.R., Kline R.H., Wright J., Eldefrawi A.T., and Eldefrawi M.E., 1995. J. Agric. Food. Chem. 34, 2308.

12

RECOMBINATION AS INDICATOR FOR GENOTOXIC AND "NON-GENOTOXIC" ENVIRONMENTAL CARCINOGENS

RUDOLF FAHRIG

Department of Genetics, Fraunhofer-Institute for Toxicology and Aerosol Research, Hannover, Germany

1. INTRODUCTION

In contrast to genotoxic carcinogens, "non-genotoxic" carcinogens including tumor promoters or co-carcinogens, are assumed to operate only above a certain threshold dose and are, therefore, of reduced hazard to humans (Butterworth and Slaga, 1987). For this reason, risk assessment for carcinogens is an important issue. As the mechanisms by which "non-genotoxic" carcinogens induce cancer are not really understood, it may be helpful to detect those types of agents which act by non-mutagenic genetic toxicity. Under consideration are mechanisms like activation of oncogenes, DNA-methylation, receptor-mediated effects, cell-cell-communication, hormonal effects, peroxisome proliferation, stimulation of recombination, and others that may give selective advantages to initiated cells in the stage of tumor promotion. An additional mechanism is induction of mitogenesis/cellular proliferation as result of toxicity (Ames and Gold, 1990; Ames et al., 1993).

As the final product of genotoxic and "non-genotoxic" carcinogens is the same (a clone of genetically altered cells) it could be possible that "non-genotoxic" carcinogens may yield genotoxic events as a secondary result of cell toxicity having led to mitogenesis/cellular proliferation, or that genetic alterations are induced that are normally neglected in genotoxicity tests. In search for relevant genetic effects mainly stimulation of recombination, by "non-genotoxic" carcinogens, has been tested. Useful for the detection of "non-genotoxic" carcinogens being active as tumor promoters or co-carcinogens were the co- or anti-recombinogenic/mutagenic effects

Biomonitors and Biomarkers as Indicators of Environmental Change 2, Edited by Butterworth *et al.*
Kluwer Academic/Plenum Publishers, New York, 2000

of these agents in experiments using yeast, *Drosophila* or mice (Fahrig, 1979; 1984; 1987a,b; 1992a,b; 1993; 1996; De la Rosa et al., 1994). In contrast to the large number of chemicals which have been tested for their ability to enhance or reduce mutations, only few chemicals has tested for recombinogenicity, or co- and anti-recombinogenicity.

In general, mutagenic/carcinogenic substances are also able to induce recombination because the induction of both is stimulated by DNA damage. Whereas in most, but not in all cases, gene mutations may arise through a special DNA repair process that introduces errors as part of the repair mechanism, recombination itself is a DNA repair process. Two modes of exchange are involved in recombination, reciprocal exchange (crossing-over) and nonreciprocal exchange (gene conversion). Nonreciprocal recombination appears unidirectional, as the transfer of genetic information from one gene to its homologous allele leaves the donor gene unchanged.

The spot test (Russell and Major, 1957; Fahrig, 1975) is the best validated method for detection of genotoxic alterations, namely gene mutations and recombination, in somatic cells of mice in vivo. This is the particular importance of the spot test, because it is able to detect recombination as well as gene mutations (Fahrig, 1992a). Recombination is able to bring carcinogen induced mutations (Ames and Gold, 1990) to expression. In recent years the spot test as well as the yeast strain MP1 has gained importance in detecting co-recombinogenic effects, which are interpreted with regard to tumor promoting properties. (Fahrig, 1984; Fahrig, 1992b).

2. METHODOLOGY: MOUSE SPOT TEST

2.1. Guidelines for the Conduct of the Assay

This assay may be conducted in compliance with the OECD Guideline for testing of chemicals No. 484 (Genetic Toxicology: Mouse Spot Test, 1986), and with the Commission Directive 88/302/EEC, In vivo-Mouse Spot Test (somatic mutation), EC, L 133, 82–84, May 30, 1988.

2.2. Principle of the Test System

Embryos which are heterozygous for different recessive coat color mutations are treated in utero with a mutagen on the 9th day of embryonal development. This is performed by intraperitoneal injection of the mother animal with test substance. If this treatment leads in a pigment precursor cell to the alteration of the wild-type allele of one of the recessive coat color genes or to its loss, a color spot will develop after several cell divisions.

Embryos with the genetic constitution appropriate for the spot test can be obtained by crossing any *a/a*-strain (*a/a* produces a black coat and, thus, good perceptibility on this background) with the T-stock. Embryos of the genotype *a/a*; *b/B*;

p c^{ch}*/P C*; *d se/D SE*; *s/S* (dark gray coat, dark eyes) are obtained by the cross of a/a-strain C57BL/6J-females with T-strain males (*a/a*: "nonagouti"; *b/b* "brown"; *p* c^{ch}*/p* c^{ch}: "chinchilla" and "pink-eyed dilution"; *d se/d se*: "dilute" and "short ear"; *s/s*: "piebald spotting").

Spots of genetic relevance (SGR) can be induced by gene mutations, recombination, or chromosome aberrations. Of the numerical and structural chromosome aberrations only those that survive the filter of several mitoses cause a color spot. Chromosome aberrations leading to the death of pigment cells are expressed as genetically non-relevant white ventral spots (WMVS).

Using the cross C57BL/6J × T, it is possible in two cases to determine which of the different genetic alterations is responsible for the appearance of a color spot.

a) Detection of Gene Mutations

For the following reason gene mutations can only be detected as light brown hairs: c^{ch}/c^{ch} in combination with a/a results in a dark gray spot which doesn't contrast with the dark gray coat of the F_1-animals. Therefore, the only genetic alteration that can be detected is a mutation (in most cases a small deletion within the c-locus) to c, in combination with c^{ch} showing up as a light brown spot. It is not possible with the naked eye to decide if it is the expected spot. Only by microscopical analysis is it possible to achieve a clear picture.

b) Detection of Reciprocal Recombinations

It is possible to detect reciprocal recombination because the p and c loci are located on the same chromosome. The two reciprocal products are cells of the genotype *p* c^{ch}*/p* c^{ch} and *P C/P C*. A genetic alteration leading to *p* c^{ch}*/p* c^{ch} gives rise to near-white coat spots which can be distinguished from white and light gray ones. The most probable reason for appearance of near-white spots is reciprocal recombination due to mitotic crossing-over.

Cells of the genotype *P C/P C* show up as black spots and can easily be detected, at least on the abdomen. Under the influence of the heterozygous coat color mutations the level of pigmentation is lower in F_1-animals than in the mother animal. Appearance of both reciprocal products as a twin spot is a proof for appearance of reciprocal recombination. Experience with yeasts has shown that reciprocal recombination does not lead in every case to the appearance of both reciprocal products. Therefore, appearance of a twin spot is a relatively rare event.

2.3. Microscopical Pigment Analysis

Using microscopical analysis of hairs from a color spot it is possible to identify the gene loci affected by a genetic alteration. Microscopical hair analysis of spots of genetic relevance has been described in detail by Fahrig and Neuhäuser-Klaus (1985). Misdifferentiations can be distinguished from spots of genetic relevance by a bright yellow fluorescence, white spots by lack of pigment. The fine structure of mouse hairs is essentially similar for all hair types: a wide central medulla is surrounded by a narrow cortex that, in turn, is surrounded by a thin cuticle.

C57BL mice are homozygous for nonagouti (a/a), their medullary cells being almost totally filled with black pigment granules. Although (C57BL × T) F_1 mice are also homozygous for nonagouti, they have considerably less pigment granules than the mother animals. This fact, presumably caused by the heterozygous recessive coat-color alleles, allows detection of black spots on the dark gray coat. Such spots can be clearly recognized on the ventral side because it is less pigmented than the dorsal side. Black spots are, as already discussed, presumably the result of a reciprocal recombination (see Figure 1).

The single gene loci can be identified in the following way:

1. In most hairs of dilute black (*a/a*; *d/d*) mice, several clumps of melanin granules disrupt the otherwise normal pattern of pigment distribution. This leads through alterations in the refraction of light to a gray color.
2. Nonagouti pink-eyed dilution animals (*a/a*; *p/p*) show the same color of pigment but considerably reduced pigmentation. Moreover, the form of the pigment granules is changed to an irregular, shredded shape with flocculent clumping. The coat color resulting from these alterations is light gray.
3. Homozygotes *a/a*; c^{ch}/c^{ch} and heterozygotes *a/a*; c^{ch}/c are easily distinguished visually. While *a/a*; c^{ch}/c^{ch} results in a dull black or sepia color that is nearly indistinguishable from the (C57BL × T) F_1 black, *a/a*; c^{ch}/c gives a light brown color. Spots of such color can be induced only by gene mutations.
4. As *b/b* mutations show a clear brown and possess round granules instead of oval ones, light brown *a/a*; c^{ch}/c and brown *a/a*; *b/b* coat colors can easily be distinguished from one another.
5. Reduction in pigmentation can be observed in (*a/a*; *p* c^{ch}/*p* c^{ch})-mice. Some hairs contain no pigment at all, others only traces. Such near-white hairs are an indi-cation that reciprocal recombination has occurred.
6. A deletion covering the *c*-locus as well as the *p*-locus or loss of the whole chromosome would lead as well to black or near-white color spots. But it is questionable whether deletions of such size or chromosome loss would lead to viable cells. By contrast, recombination leads in every case to cells possessing complete vitality.

2.4. Determination of Toxicity

Criteria for determination of the highest dose of the test substance are embryotoxicity and toxicity to the mother animal. In a pretest, toxicity will be determined by treatment of maximum 30 non-pregnant C57BL/6J-females, 3–5 months old, with different concentrations of the test substance in ascending order. If for example, a single intraperitoneal dose of 100 mg/kg should be tolerable by mice, two C57BL/6J-females (P_1; P_2) will be treated by i.p. injection with 100 mg/kg test substance suspended in corn oil. Depending on the physical condition of the treated animals P_1 and P_2, two days later two other animals (P_3; P_4) will be treated with 150 mg/kg

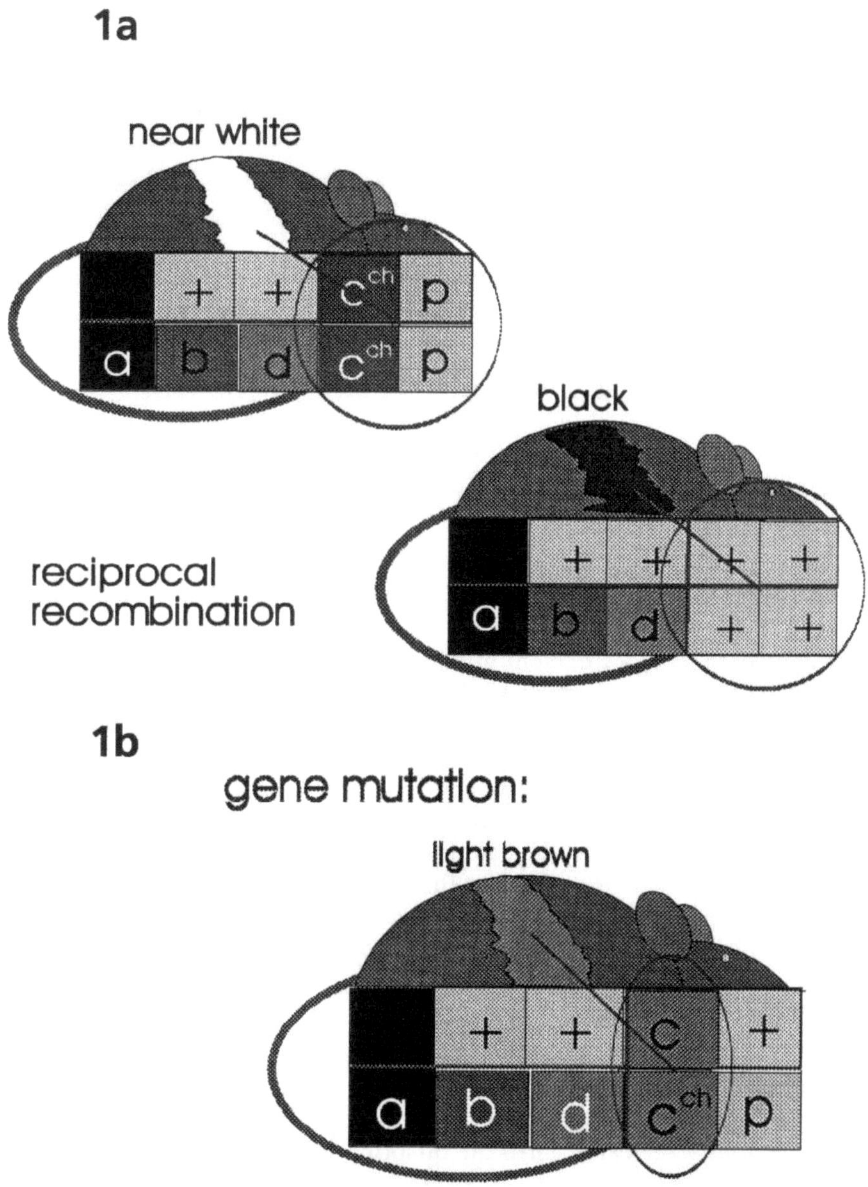

FIGURE 1. Diagrammatic representation of how to distinguish between induced gene mutation and induced reciprocal recombination using the cross C57BL × T.

Chromosome 7 of the mouse is heterozygous at the *c* and *p* loci. After replication the homologue chromosomes consist of two identical sister chromatids. Mitosis leads to two daughter cells with the parental heterozygous genotype, i.e., will lead to a coat with dark gray colour.

a: A crossing-over between centromere and the *p* and *c* loci leads to an exchange of the mutant and the wild-type alleles. Mitosis leads to homozygosity of the wild-type alleles in one of the daughter cells and to homozygosity of the mutant alleles in the other. Homozygous wild-type alleles can be seen as black spots on the dark gray coat, homozygous mutant alleles as near-white spots. If both reciprocal products survive, a twin spot may arise. No genetic alteration other than reciprocal recombination will result in twin spots.

b: A mutation of the wild-type allele at the *c*-locus is possible before replication. After mitosis this mutation leads to daughter cells with the two different mutant alleles *c* and c^{ch} (compound heterozygosity). This gives rise to a light brown spot. Light brown spots can be induced only by gene mutations, gray and brown spots also by the more frequent event of non-reciprocal recombination.

and, depending on the physical condition of the treated animals P_3 and P_4, two days later the dose will be enhanced to 200 mg/kg to animals P_5 and P_6. Thereafter or after administration of 400 mg/kg to animals P_7 and P_8 in further experiments, a dose will be selected just above the toxic dose. In case that 100 mg/kg dose is toxic, the doses will be reduced in descending order. The application of test substance is in corn oil or water.

Ten pregnant C57BL/6J-females will be treated with the dose just above the beginning of the toxic dose. In this experiment embryotoxicity will be determined.

2.5. Dose Selection of Test Substance

As stated in the guidelines, one of the doses chosen will be just above the beginning of the toxic dose. The second dose will depend on the litter size (toxicity) of the first dose.

2.6. Performance of the Main Test

For each dose level a sufficient number, i.e., a minimum of 300 offspring mice are required. Therefore, for each dose about 300 F_0 female mice are treated. About one third of the animals will be treated as control animals with corn oil alone, and about 20 animals with ethylnitrosourea (ENU) as positive control.

The test substance will be suspended in corn oil and given to the pregnant females orally or by intraperitoneal injection (0.25 ml per mouse $\cong$0.1 ml/kg body weight), that it can be assumed that the test substance or its metabolites will reach the embryonic target cells. Relevant for this estimation will be a reduction of litter size in the treatment groups, or appearance of white ventral spots (WMVS in the F_1-animals) being considered as result of pigment cell killing. Treatment will be performed on day 9 of pregnancy, whereby day 1 is the day of detection of the vaginal plug.

2.7. Evaluation of the Results

Three to four weeks after birth the offspring will be examined for color spots. Three categories of color spots can be distinguished:

a) White ventral spots which normally touch the ventral line. It is highly probable that such spots have been induced by pigment cell killing (WMVS, white midventral spots).
b) Yellow, agouti-like spots which are in all likelihood the result of misdifferentiations (MDS, misdifferentiated spots). It is likely that these spots are driven by toxicity.
c) Pigmented black, gray, brown or near-white spots randomly distributed over the whole coat, which are the result of genetic alterations (SGR, spots of genetic relevance). By microscopical hair analysis it is possible to assign a single spot to a single coat color mutation.

The test substance is classified as mutagenic if there is a statistically (Fisher's exact test, one-tailed) significant increase in the number of spots of genetic relevance (SGR). If it is apparent that the result is clearly negative or positive, no statistical analysis will be done.

3. METHODOLOGY: *SACCHAROMYCES CEREVISIAE*

3.1. Guidelines for the Conduct of the Assay

This assay may be conducted in compliance with the OECD Guideline for testing of chemicals No. 480 (Genetic Toxicology: Saccharomyces cerevisiae, 1986), and with the Commission Directive 67/548/EEC, Saccharomyces cerevisiae, EC, L 133, 55–60, May 30, 1988.

3.2. Principle of the Test System

Yeast strain. Cells of *Saccharomyces cerevisiae* strain MP1 (Fahrig, 1984) allow detection of reciprocal recombination between homologous chromosomes as well as nonreciprocal recombination between homologous chromosomes due to mitotic gene conversion and forward mutations.

MP1 is heterozygous for *ade2* and *his8*, heteroallelic for *trp5-12* and *trp5-21*. Induction of reciprocal recombination, i.e., of a mitotic crossing-over leading to homozygosis for the recessive markers *ade2* and *his8*, can be detected as a red sector in a normally white colony. Chromosome losses have the same effect but cannot be distinguished from reciprocal recombination. Red sectors or colonies that are only adenine-, but not histidine-dependent are based on mutation or nonreciprocal recombination. The system used to detect nonreciprocal recombination due to mitotic gene conversion is based on the induction of *trp*-prototrophic cells in a *trp*-auxotrophic population. Forward mutations can be determined by measuring the induction of cycloheximide-resistant cells in an cycloheximide-sensitive cell population. This system is the most sensitive for detecting mutations in yeast.

3.3. Test Procedure

About 10^3 yeast cells are inoculated into a 300 ml Erlenmeyer flask containing 100 ml YEP (2% Bacto peptone, 1% yeast extract and 2% glucose), set on a shaker and allowed to grow for 3 days at 25 °C into stationary phase. The cultures needed for one experiment are mixed together to obtain a similar spontaneous frequency of genetic alterations in both, experiment and control. The cultured cells are washed twice with distilled water and the cell titers adjusted to 5×10^8 cells per ml of 0.1 M phosphate buffer pH 7. These cell suspensions are incubated in a test tube on a shaker at 25 °C with different concentrations of the test substances +0.045

mmole TEM. After 24 h, treatments are stopped, and 0.1 ml aliquots of the suspensions (containing 5×10^7 cells) spread on four plates of solid media, selective for nonreciprocal recombinants and mutants, respectively. Similarly, 0.1 ml aliquots of 1×10^{-5} dilutions in distilled water (containing 5×10^2 cells) are plated out on ten plates of synthetic complete medium to ascertain the number of survivors (white colonies) and reciprocal recombinants (red colonies or red sectors). These cultures are incubated at 25 °C for 4 days.

In the experiments without metabolic activation, yeast cells are treated in phosphate buffer, pH 7. In the experiments with metabolic activation, they are treated in 85% S9-mix in phosphate buffer (Aroclor-induced rat liver S-9 plus cofactors).

4. RESULTS AND DISCUSSION

4.1. Validation of the Mouse Spot Test

Since 1975 experimental results with the spot test have been published for 80 different substances. Most published data deal with known mutagens; most experiments were aimed at checking the sensitivity and reliability of the spot test for detecting different classes of mutagens. The result to be upheld is that the spot test is not only suspectible to standard mutagens (like ethyl nitrosourea, methyl methanesulfonate) but also to mutagens which are difficult to detect *in vivo* (e.g. hydrazine and 4-chloro-o-toluidine), and specific point mutagens (such as 2-aminopurine and other base analogs, acridine orange, hycanthone and other frame-shift mutagens) (Fahrig, 1978; Russell et al., 1981; Fahrig et al., 1982; Lang, 1984; Styles and Penman, 1985; Neuhäuser-Klaus, 1991). Sometimes the reliability of data originating in the seventies is restricted by the fact that too few animals were used. In this way, even with strong mutagens, weak effects will result (Russell et al., 1981). Using high numbers of animals, even with different strains of mice, similar and reproducible results will be gained (Hart and Fahrig, 1985; Fahrig, 1988; Fahrig and Neuhäuser-Klaus, 1989).

In the course of an international collaborative study, pairs of substances have been tested which are similar with respect to their molecular structure, but only one of them being a carcinogen and the other a non-carcinogen (examples: benzo(a)pyrene as carcinogen, pyrene as non-carcinogen; 2-acetylaminofluorene as carcinogen, 4-acetylaminofluorene as non-carcinogen). The spot test was able to distinguish between carcinogen and non-carcinogen (Fahrig, 1988).

The published results on 60 chemicals and X-rays investigated until 1984 in the mouse spot test were compared with data on the same chemicals tested in the Ames test and lifetime rodent bioassays (Styles and Penman, 1985). The performance of the spot test as an in vivo complementary assay to the in vitro bacterial mutagenesis test revealed that of 60 agents, 38 were positive in both systems, 6 were positive only in the spot test, 10 were positive only in the bacterial test and 6 were negative in both assays.

The spot test was also considered as a predictor of carcinogenesis: 45 chemicals were carcinogenic of which 35 were detected as positive by the spot test and 3 out of 6 non-carcinogens were correctly identified as negative.

If the results are regarded in sequence, i.e. that a positive result in a bacterial mutagenicity test reveals potential that may or may not be realized in vivo, then 48 chemicals were mutagenic in the bacterial mutation assay of which 38 were active in the spot test and 31 were confirmed as carcinogens in bioassays. 12 chemicals were non-mutagenic to bacteria of which 6 gave positive responses in the spot test and 5 were confirmed as carcinogens.

These results provide strong evidence that the mouse spot test is an effective complementary test to the bacterial mutagenesis assay for the detection of genotoxic chemicals and as a confirmatory test for the identification of carcinogens. The main deficiency up to 1984 was the paucity of data from the testing of non-carinogens.

The data published since 1984 are summarized in Table 1. It is apparent that apart from known mutagens and genotoxic carcinogens, now non-carcinogens and nongenotoxic carcinogens/tumor promoters have been examined. A new aspect is the search for co- and anti-mutagens as well as for co- and anti-recombinogens.

4.2. Enhancement or Reduction of Mutations or Recombinations

The importance of recombination for the process of tumor development cannot be rated high enough. There is some indication that tumor promoters enhance the recombinogenic effects of carcinogens (Fahrig, 1979; 1984) or are recombinogenic per se (Huonma and Little, 1995), and there is some indication that loss of heterozygosity of tumor suppressor genes may perhaps be primarily the consequence of mitotic recombination (Gupta et al., 1997). Bertrand et al. (1997) observed an increase of spontaneous intrachromosomal homologous recombination from five- to 20-fold in mammalian cells expressing a mutant *p53* protein. Mekeel et al. (1997) determined that the rate of homologous recombination was suppressed by *p53*. Human tumor cell lines, mutant or null for *p53* had recombination rates 10,000-times greater than primary fibroblasts. They suggested that suppression of homologous recombination is the means by which *p53* maintains genetic stability.

In contrast to the large number of chemicals which have been tested for their ability to enhance mutations, only few chemicals have been tested for their ability to enhance recombination or show co-recombinogenic activity. Co-recombinogenic effects were observed in experiments using the yeast: tumor promoters were co-recombinogenic and anti-mutagenic, substances being tumor promoters as well as co-carcinogens were co-recombinogenic as well as co-mutagenic (Fahrig, 1979; 1984; 1987a, 1992a, 1996, 1997).

To confirm the experients with yeast, experiments with the spot test were performed (Fahrig, 1984). In these verification experiments, the co-recombinogen D-limonene and the anti-recombinogen catechol were used. Moreover, two substances which had not been tested or were ineffective in yeast were used: 12-O-tetrade-

TABLE 1. Published results with the mouse spot test since 1984.

Chemical	mouse strain	dose mg/kg	route	day	no. of F1 observed	F1-with SGR (%)	result	author(s)
Ethylnitrosourea (ENU)	C57BL	30	i.p.	9	325	54 (17%)	+	Fahrig, 1984
Catechol	× T	3 × 22	i.p.	9, 10, 11	216	2 (1%)	–	
D-limonene		3 × 215	i.p.	9, 10, 11	291	4 (1%)	–	
Catechol		3 × 22	i.p.	9, 10, 11	469	100 (21%)	comut	
+ ENU		+ 30		+ 9			agen	
D-limonene		3 × 215	i.p.	9, 10, 11	797	110 (14%)	antimu	
+ ENU		+ 30		+ 9			tagen	
control (buffer)		0	i.p.	9, 10, 11	288	3 (1%)		
Dimethylbenz-	C57BL	12.5	p.o.	10	106	2 (1.9%)	–	Shibuya and
anthrazene	× PW	25			108	4 (3.7%)	(±)	Murota, 1984
(DMBA)		50			132	9 (6.8%)	+	
control (oil)		0			103	1 (1%)		
DMBA		10–80					antimu	
+ phenobarbital		+ 80					tagen	
Dimethylsulfate	C57BL	50	i.p.	10	139	1 (0.72%)	–	Braun et al.
Ethylmethanesulfonate	× T	100			56	4 (7.1%)	+	(1984)
ENU		20			238	13 (5.5%)	+	
Methylnitrosourea		4			51	1 (2%)	–	
Trenimon		0.1			75	0	–	
Diethylsulfate		225			100	3 (3%)	(±)	
Methylmethanesulf.		125			125	2 (1.6%)	–	
Isoniazid		100			190	13 (6.8%)	+	
control (saline pooled)		0			1,710	9 (0.5%)		
ENU	NMRI	50	i.p.	8–11	465	99 (21%)	+	Hart, 1985
control (citric acid)	× PDB	0.1 ml	i.p.	8–11	356	2 (0.6%)		
Isoniazid		100	i.p.	10	116	3 (2.6%)	(±)	
		150			243	6 (2.5%)	+	
		200			93	1 (1.1%)	–	

Procarbazine		50	i.p.	10	72	7 (9.7%)	+	
cyclophosphamide		5	i.p.	10	151	4 (2.6%)	+	
4-NQO		15	i.p.	10	134	4 (3%)	+	
control (saline)		0	i.p.	10	504	4 (0.8%)		
Dimethylamino-	C57BL	50	i.p.	9	390	9 (2.3%)	–	Hart and
azobenzene, DAB	× T	100			407	13 (3.2%)	–	Fahrig,
	NMRI	150			307	4 (1.3%)	–	1985
	× PDB	200			272	3 (1.1%)	–	
4-Cyanodimethyl-	C57BL	100	i.p.	9	278	3 (1.1%)	–	
aniline, CDA	× T	200			317	5 (1.6%)	–	
	NMRI	150			310	9 (2.9%)	(±)	
	× PDB	200			293	4 (1.4%)	–	
control (oil)	C57BL × T	0	i.p.	9	224	3 (1.3%)		
	NMRI × PDB	0			479	3 (0.6%)		
Ethylnitrosourea	NMRI	10	i.p.	9	118	8 (6.8%)	+	Nielsen, 1986
(ENU)	× PDB	20			130	33 (25.4%)	+	
		40			128	41 (32%)	+	
control (saline)		0	i.p.	9	87	3 (3.5%)		
Dibromochloropropane	C57BL	106	i.p.	10	721	21 (2.9%)	+	Sasaki et al.,
ENU	× PW	50			189	45 (23.8%)	+	1986
control (oil)		0			643	4 (0.6%)		
TPA	C57BL	2 × 0.1	i.p	9, 10	478	8 (1.7%)	–	Fahrig, 1987b
	× T	1 × 1			72	2 (2.8%)	–	
control (buffer)		0			550	10 (1.8%)		
TPA		2×10^{-1}					comu	
+ ENU		+ 30					tagen	
Benzo(a)pyrene	T × HT	0	p.o.	9	559	12 (2.1%)		Neuhäuser-
(BP)		50			281	14 (5%)	+	Klaus, 1988
		100			272	11 (4%)	+	
		150			295	40 (13.6%)	+	

TABLE 1. *Continued*

Chemical	mouse strain	dose mg/kg	route	day	no. of F1 observed	F1-with SGR (%)	result	author(s)
Pyrene (PYR)	T × HT	0	p.o.	9	559	12 (2.1%)		
		200			220	4 (1.8%)	–	
		400			267	6 (2.2%)	–	
		600			197	4 (2%)	–	
2-Acetylamino-fluorene (2AAF)	T × HT	0	i.p.	9	734	12 (1.6%)		Neuhäuser-Klaus, 1988
		200			229	5 (2.2%)	+	
		400			268	15 (5.6%)	+	
		600			238	8 (3.4%)	+	
		800			178	9 (5.1%)	+	
	C57BL × T	0	i.p.	9	214	3 (1.4%)		Fahrig, 1988
		400			175	9 (5%)	+	
		0	p.o.	11	151	0		Hüttner et al., 1988
		446			195	1 (0.5%)	–	
4-Acetylamino-fluorene (4AAF)	T × HT	0	i.p.	9	734	12 (1.6%)		Neuhäuser-Klaus, 1988
		100			282	5 (1.8%)	–	
		200			310	7 (2.3%)	–	
		400			261	5 (1.9%)	–	
		600			157	3 (1.9%)	–	
	C57BL × T	0	i.p.	9	214	3 (1.4%)		Fahrig, 1988
		400			420	6 (1.4%)	–	
		0	p.o.	11	151	0		Hüttner et al., 1988
		446			191	0		
Caprolactam	T × HT	500	i.p.	9	1,018	58 (5.7%)	+	Neuhäuser-Klaus and Lehmacher, 1989
		700			122	6 (4.9%)	+	
control (saline)		0			637	22 (3.5%)		

Caprolactam	C57BL × T	400	i.p.	9	397	11 (2.8%)	(±)	Fahrig, 1989
		500			487	11 (2.3%)	(±)	
		500			490	17 (3.5%)	+	
control (buffer)		0	.		407	6 (1.5%)		
Acrylamide	T × HT	50	i.p.	12	213	14 (6.6%)	+	Neuhäuser-Klaus and Schmahl, 1989
		75		12	211	13 (6.1%)	+	
		3 × 50		10–12	196	26 (13.3%)	+	
		3 × 75		10–12	215	21 (9.8%)	+	
control (buffer)		0	i.p.	1–3×	437	11 (2.5%)		
Hypericum extract	NMRI × DBA/2J	1	p.o.	9	240	3 (1.3%)	–	Okpanyi et al., 1990
		5			236	3 (1.3%)	–	
		10			285	5 (1.8%)	–	
control		0			226	3 (1.1%)		
Tannic acid	C57BL × PW	500	p.o.	10	266	5 (2.2%)	–	Sasaki et al., 1990
control (aqua dest.)		0			566	5 (1%)		
Tannic acid + ENU		10–50 + 50					antimutagen	
Vanillin	C57BL × PW	3 × 500	p.o	10	220	1 (0.5%)	–	Imanishi et al., 1990
control (DMSO)		0			566	0		
Vanillin + ENU		3 × 50 + 10–50					antimutagen	
2-amino-N-hydroxyadenine	T × HT	16	i.p.	9	139	5 (3.6%)	–	Neuhäuser-Klaus, 1991
		20			133	11 (8.3%)	+	
		40			103	12 (11.7%)	+	
control (aqua dest.)		0			332	13 (3.9%)		
cyclophosphamide	C57BL × T	7.5	s.c.	9.5	206	6 (2.9%)	+	Braun and Hüttner, 1991
		7.5	s.c.	10.5	178	6 (3.4%)	+	
control (saline)		0	s.c.	9.5	222	1 (0.5%)		
		0	s.c.	10.5	238	1 (0.4%)		

TABLE 1. *Continued*

Chemical	mouse strain	dose mg/kg	route	day	no. of F1 observed	F1-with SGR (%)	result	author(s)
Dioxin mixture control (DMSO) TCDD + ENU	C57BL × T	0.128 0 3×10^{-3} + 30	i.p.	9	168 208	1 (0.6%) 3 (1.4%)	– co-recomb inogen	Fahrig, 1993
m-Phenylene-diamine control (aqua dest.)	C57BL × T	5 15 0	i.p.	9	488 116 117	3 (0.6%) 1 (0.6%) 0	– –	Umweltbundesamt/ Fahrig, 1994
1,2-dichlorethane control (oil)	C57BL × T	400 0	i.p.	9	579 298	4 (0.7%) 2 (0.7%)	–	Umweltbund./ Fahrig, 1994
(E)-5-(2-bromovinyl)-2′-deoxyuridine (BVDU)	C57BL × T	150 280 0	i.p.	9	498 513 472	2 (0.4%) 1 (0.2%) 4 (0.9%)	– –	Berlin-Chemie/ Fahrig, 1994
Diethylhexyl-phthalate (DEHP) + ENU	C57BL × T	5 ml + 30	i.p.	9	302 156 601 304	44 (14.5%) 23 (13.5%) 75 (12.5%) 32 (10.2%)	co-recomb inogen	Fahrig and Steinkamp-Zucht, 1996
Pentachloro-phenol (PCP) + ENU	C57BL × T	50 + 30	i.p.	9	182 565 128 284	21 (12%) 59 (10%) 14 (11%) 34 (12%)	– –	Fahrig and Steinkamp-Zucht, 1996

canoyl-phorbol-13-acetate (TPA) (Fahrig, 1987b) and 2,3,7,8-tetrachloro-dibenzo-p-dioxin (TCDD) (Fahrig, 1993). The direct alkylating agent ethyl-nitrosourea (ENU) served as mutagen. Both substances enhanced the recombinogenic activity of ENU.

In other in vivo-experiments using the spot test with mice (Fahrig and Steinkamp-Zucht, 1996), diethylhexylphthalate (DEHP) was co-recombinogenic and anti-mutagenic. In two independant experiments DEHP was able to increase in combination with ethylnitrosourea (ENU) the frequency of animals with spots of genetic relevance from about 12% (ENU alone) to about 15% (ENU + DEHP). This enhancement can be attributed to an enhancement of exclusively recombination: In historical as well as in current experiments, with astonishing accuracy about 8% of all spots induced with ENU alone are black, near-white or twin spots, i.e., presumed products of reciprocal recombination. Under the influence of DEHP about 20% of all color spots were black, near-white or twin spots. These co-recombinogenic effects are very strong if considering that the low frequency of twin spots with 0.3% (historical ENU-control) or 0.6% (current ENU-control) after ENU treatment alone has been enhanced up to a frequency of 3%. While ENU alone induces about 14% light brown color spots depending exclusively on gene mutations, under the influence of DEHP the frequency was only 7%. Therefore, DEHP was anti-mutagenic. In the experiments with yeast (Fahrig, 1997), with DEHP in combination with triethylene melamine (TEM) an increase in recombination but also a clear co-mutagenic effect could be observed.

In general, the in vitro results with the yeast could be confirmed by in vivo results with mice.

In genetic tests measuring only mutations, an anti-mutagenic effect cannot be distinguished from a simple desmutagenic effect (inactivation of mutagens by various mechanisms) if mutagen and anti-mutagen are given simultaneously. In the genetic tests described in the present work, however, the simultaneous increase in mutations and decrease in recombinations and vice versa suggests interference with DNA repair processes.

In an attempt to explain the effects described, a model was developed (Fahrig, 1979; 1992). This model proposes that nonreciprocal recombination, reciprocal recombination and gene mutation are correlated during mitosis, but that they are separate events:

(1) Essentially similar DNA lesions have different genetic consequences, mutation as well as recombination.

(2) The induction of mutation and recombination is correlated. Blockade of the pathway of mutation induction leads to the channeling of DNA lesions into the pathway of recombination induction, and vice versa (Hastings et al., 1976).

Anti-mutagenic/co-recombinogenic effects can be explained by the channeling of DNA lesions into pathways that lead to recombination by blocking those pathways which lead to mutations. Co-mutagenic/anti-recombinogenic effects can be explained by the channeling of DNA lesions into pathways that lead to mutations by blocking those pathways which lead to recombination. Substances like vanillin that are co-recombinogenic as well as co-mutagenic may act by enhancing both recombinogenic and mutagenic repair mechanisms. Since there exists a

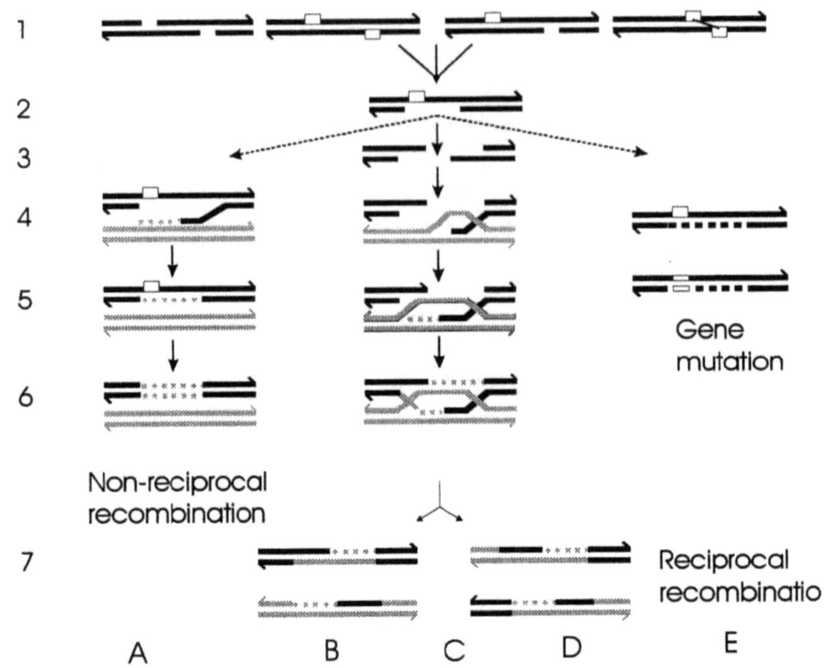

FIGURE 2. Co-recombinogenic and anti-mutagenic effects.

The basic assumption is that single-strand damages are not potentially mutagenic or recombinogenic because the undamaged strand can serve as template for restoring the original state. Damages affecting both strands at nearby positions, that is, double-strand breaks (1A), closely spaced lesions (1B, 1D), or cross-links (1E), are difficult or impossible to repair by error-free processes, and may therefore lead to genetic alterations (■■■■= repair synthesis; □= damaged base; □= wrong base).

Different systems of repair enzymes compete with each other for these double-strand damages leading to either nonreciprocal (A, B) or reciprocal (D) mitotic recombination, or to gene mutation (E). Excision repair enzymes remove damages from only one strand of DNA (2C). Due to the configuration of double-strand damage, the complementary strand cannot be used as template for resynthesis. If a homologue chromatid (4A) serves as template, the effect of this may be nonreciprocal recombination due to gene conversion (6A). In this case, induction of nonreciprocal recombination due to gene conversion is a result of correction of hybrid DNA (5A). As always the strand without damaged base is used as template for resynthesis, the frequency of nonreciprocal recombination is largely enhanced. If the complementary strand which includes damaged bases is used as template, wrong information will inevitably be inserted; this may be a gene mutation (5E). The same configuration can result in the formation of a gap. The two strand gap is repaired using information from an intact homologous duplex by a mechanism described by Szostak et al. (1983) in their double-strand break repair model, leading to nonreciprocal (7B) or to nonreciprocal plus reciprocal (7D) recombination.

According to the double-strand break repair model, (3C) a double-strand cut is made in one duplex, and a gap flanked by 3′ single strands is formed by the action of exonucleases. (4C) One 3′ end invades a homologous duplex, displacing a D-loop. (5C) The D-loop is enlarged by repair synthesis until the other 3′ end can anneal to complementary single-stranded sequences. (6C) Repair synthesis from the second 3′ end completes the process of gap repair, and branch migration results in the formation of two Holliday junctions. Resolution of the two junctions by cutting either inner or outer strands leads to two possible non-cross-over (7B) and two possible cross-over (7D) configurations. In the illustrated resolutions, the right-hand junction was resolved by cutting the inner, crossed strands.

By considering two different pathways to nonreciprocal recombination the mutation-recombination correlation model predicts that substances exist that enhance either nonreciprocal or reciprocal recombinations. The two kinds of recombination can be separated as shown in other studies (Roman and Fabre, 1983; Klein, 1984). The new conclusions drawn from the model are as follows.

(1) Essentially similar DNA lesions have different genetic consequences, i.e., mutation *and* recombination.
(2) The induction of mutation and recombination is correlated, that is, blockade of the pathway of mutation induction leads to channeling of DNA lesions into the pathway of recombination induction, and vice versa.

Co-recombinogenic effects can now be explained by channeling of DNA lesions into pathways that lead to recombinations by blockading those pathways which lead to mutations.

correlation between co-recombinogenic effects and activity as tumor promoter (Fahrig, 1984), it is possible that anti-mutagens with co-recombinogenic properties may act as tumor promoters.

5. REFERENCES

Ames, B., and L.S. Gold. 1990. Carcinogenesis debate, Science 250: 1498–1499.

Ames, B., M.K. Shigenaraand, and L.S. Gold. 1993. DNA lesions, inducible DNA repair, and cell division: Three factors in mutagenesis and carcinogenesis, Env. Health Perspect. 101: 35–44.

Bertrand P., D. Rouillard, A. Boulet, C. Levalois, T. Soussi, and B.S. Lopez. 1997. Increase of spontaneous intrachromosomal homologous recombination in mammalian cells expressing a mutant p53 protein. Oncogene 14: 1117–1122.

Berlin-Chemie /Fahrig. 1994. Spot test 210/211.

Bollag, R.J., and R.M. Liskay. 1988. Conservative intrachromosomal recombination between inverted repeats in mouse cells: association between reciprocal exchange and gene conversion, Genetics 119: 161–169.

Braun, R., E. Hüttner, and J. Schöneich. 1984. Transplacental genetic and cytogenetic effects of alkylating agents in the mouse. I: Induction of somatic coat color mutations. Teratogenesis, Carcinog. Mutagen. 4: 449–457.

Braun, R., and E. Hüttner. 1991. Influence of ethanol on genetic activity of cyclophosphamide in the mammalian spot test. Biol. Zbl. 110: 284–289.

Butterworth, E.B., and T.J. Slaga. 1987. Nongenotoxic Mechanisms in Carcinogenesis. 25 Banbury Report, Cold Spring Harbor Laboratory, Cold Spring Harbor, NY.

Commission Directive 67/548/EEC, Saccharomyces cerevisiae, Official Journal of EC No. L 133, 55-60, May 30, 1988.

Commission Directive 88/302/EEC, In vivo-Mouse Spot-Test (somatic mutation). Official Journal of EC No. L 133, 82-84, May 30, 1988.

De la Rosa, M.E., J. Magnusson, C. Ramel, and R. Nilsson. 1994. Modulation influence of inorganic arsenic on the recombinogenic and mutagenic action of ionizing radiation and alkylating agents in Drosophila melanogaster, Mutation Res. 318: 65–71.

Fahrig, R. 1975. A mammalian spot test: Induction of genetic alterations in pig-ment cells of mouse embryos with X-rays and chemical mutagens. Molec. Gen. Genet. 138: 309–314.

Fahrig, R. 1978. The mammalian spot test, a sensitive in vivo method for the detection of genetic alterations in somatic cells of mice. In: A. Hollaender, and F.J. de Serres (Eds.) Chemical Mutagens: Principles and Methods for their Detection, Vol. 5: 151–176, Plenum Press, New York.

Fahrig, R. 1979. Evidence that induction and suppression of mutations and recombinations by chemical mutagens in S. cerevisiae during mitosis are jointly correlated. Molec. Gen. Genet. 169: 125–139.

Fahrig, R. 1984. Genetic mode of action of cocarcinogens and tumor promoters in yeast and mice. Mol. Gen. Genet. 194: 7–14.

Fahrig, R. 1987a. Effects of bile acids an the mutagenicity and recombinogenicity of triethylene melamine in yeast strain MP1 and D61.M, Arch. Toxicol. 60: 192–197.

Fahrig, R. 1987b. Enhancement of carcinogen-induced mutations or recombinations by 12-O-tetradecanoyl-phorbol-13-acetate in the mammalian spot test, J. Cancer Res. Clin. Oncol. 113: 61–66.

Fahrig, R. 1988. Summary Report on the performance of the mammalian spot test. In: J. Ashby, F.J. de Serres, M.D. Shelby, B.H. Margolin, M. Ishidate, and G.C. Becking (Eds.) Evaluation of short-term in vivo tests for carcinogens. Report of the international programme on chemical safety's collaborative study on in vivo assays II, Cambridge University Press, Cambridge U.K.: 151–158.

Fahrig, R. 1989. Possible recombinogenic effect of Caprolactam in the mammalian spot test, Mutat. Res. 224: 373–375.

Fahrig, R. 1992a. Tests for recombinagens in mammals in vivo, Mutat. Res. 284: 177–183.

Fahrig, R. 1992b. Co-recombinogenic effects, Mutat. Res. 284: 185–193.

Fahrig, R. 1993. Genetic effects of dioxins in the spot test with mice, Env. Health Perspect. 101: 257–261.

Fahrig, R. 1996. Anti-mutagenic agents are also co-recombinogenic and can be converted into co-mutagens, Mutation Res. 350: 59–67.

Fahrig, R. 1997. Co-recombinogenic and co- or anti-mutagenic effects of non-genotoxic carcinogens in S. cerevisiae MP1, J. Env. Pathol., Toxicol. a. Oncol. (JEPTO) 16: 273–279.

Fahrig, R., G.W.P. Dawson, and L.B. Russell. 1982. Mutagenicity of selected chemicals in the mammalian spot test. In: F.J. de Serres, and M.D. Shelby (Eds.) Comparative chemical mutagenesis, Plenum New York: 713–732.

Fahrig, R., and A. Neuhäuser-Klaus. 1985. Similar pigmentation characteristics in the specific locus and the mammalian spot test: A way to distinguish between induced mutation and recombination. The Journal of Heredity 76: 421–426.

Fahrig, R., and A. Neuhäuser-Klaus. 1989. Positive response of Caprolactam in the mammalian spot test. Mutat. Res. 224: 377–378.

Fahrig, R., and A. Steinkamp-Zucht. 1996. Induction or suppression of SV40 amplification by genotoxic carcinogens, non-genotoxic carcinogens or tumor promoters, Mutation Res. 356: 217–224.

Gupta, P.K., A. Sahota, S.A. Boyadjiev, S. Bye, C. Shao, J.P. O'Neill, T.C. Hunter, R.J. Albertini, P.J. Stambrook, and J.A. Tischfield. 1997. High frequency in vivo loss of heterozygosity is primarily a consequence of mitotic recombination. Cancer Res. 57: 1188–1193.

Hastings, P.J., S.-K. Quah, and R.C. von Borstel. 1976. Spontaneous mutation by mutagenic repair of spontaneous lesions in DNA. Nature 264: 719–722.

Hart, J.W. 1985. The mouse spot test: Results with a new cross. Arch. Toxicol. 58: 1–4.

Hart, J.W., and R. Fahrig. 1985. Effects of 4-dimethylamino-azobenzene (DAB) and 4-cyanodimethylaniline (CDA) in the mammalian spot test. In: Parry, and Arlett (Eds.) Comparative Genetic Toxicology. The Macmillan Press LTD. London: 507–513.

Honma, M., and J.B Little. 1995. Recombinogenic activity of the phorbol ester 12-O-tetradecanoylphorbol-13-acetate in human lymphoblastoid cells. Carcinogenesis 16: 1717–1722.

Hüttner, E., R. Braun, and J. Schöneich. 1988. Mammalian spot test with the mouse for detection of transplacental genetic effects induced by 2-acetylaminofluorene and 4-acetylaminofluorene. In: J. Ashby, F.J. de Serres, M.D. Shelby, B.H. Margolin, M. Ishidate, and G.C. Becking (Eds.) Evaluation of short-term in vivo tests for carcinogens. Report of the international programme on chemical safety's collaborative study on in vivo assays II, Cambridge University Press, Cambridge U.K.: 164–167.

Imanishi, H., Y.F. Sasaki, K. Matsumoto, M. Watanabe, T. Ohta, Y. Shirasu, and K. Tutikawa. 1990. Suppression of 6-TG-resistant mutations in V79 cells and recessive spot formations in mice by vanillin. Mutat. Res. 243: 151–158.

Klein, H.L. 1984. Lack of assiciation between intrachromosomal gene conversion and reciprocal exchange, Nature 310: 748–753.

Lang, R. 1984. The mammalian spot test and its use for testing of mutagenic andcarcinogenic potential: experience with the pesticide chlordimeform, its principal metabolites and the drug lisuride hydrogen maleate. Mutat. Res. 135: 219–224.

Mekeel, K.L., W. Tang, L.A. Kachnic, C.M. Luo, J.S. DeFrank, and S.N. Powell. 1997. Inactivation of p53 results in high rates of homologous recombination. Oncogene 14: 1847–1857.

Neuhäuser-Klaus, A. 1988. Evaluation of somatic mutations in (T × HT) F1. In: J. Ashby, F.J. de Serres, M.D. Shelby, B.H. Margolin, M. Ishidate, and G.C. Becking (Eds.) Evaluation of short-term in vivo tests for carcinogens. Report of the international programme on chemical safety's collaborative study on in vivo assays II, Cambridge University Press, Cambridge U.K.: 168–172.

Neuhäuser-Klaus, A. 1991. Mutagenic activity of 2-amino-N6-hydroxyadenine in the mouse spot test. Mutat. Res. 253: 109–114.

Neuhäuser-Klaus, A., and W. Schmahl. 1989. Mutagenic and teratogenic effects of acrylamide in the mammalian spot test. Mutat. Res. 226: 157–162.

Neuhäuser-Klaus, A., and W. Lehmacher. 1989. The mutagenic effect of caprolactam in the spot test with (T × HT) F1 mouse embryos. Mutat. Res. 224: 369–371.

Nielsen, I.M. 1986. Induction of Coat Colour Spots in Non-agouti NMRI × PDB, F1 Dose Response Relationship with Ethylnitrosurea. Acta Pharmacol. et Toxicol. 48: 159–160.

OECD Guideline for testing of chemicals No. 484, Genetic toxicology: Mouse Spot-Test. Paris, Oct. 23, 1986.

OECD Guideline for testing of chemicals No. 480 (Genetic Toxicology: Saccharomyces cerevisiae, 1986), Paris, Oct. 23, 1986.

Okpanyi, S.N., H. Lidzba, B.C. Scholl, and H.G. Miltenburger. 1990. Genotoxizität eines standardisierten Hypericum-Extraktes. Arzneimittelforsch. 40: 851–855.

Roman, H., and F. Fabre. 1983. Gene conversion and associated reciprocal recombination are separable events in vegetative cells of Saccharomyces cerevisiae, Proc. Natl. Acad. Sci. USA 80: 6912–6916.

Russell, L.B., and M.H. Major. 1957. Radiation-induced presumed somatic mutations in the house mouse. Genetics 42: 161–175.

Russell, L.B., P.B. Selby, E. von Halle, W. Sheridan, and L. Valcovic. 1981. Use of the mouse spot test in chemical mutagenesis: Interpretation of past data and recommendations for future work. Mutat. Res. 86: 355–379.

Sasaki, Y.F., H. Imanishi, M. Watanabe, A. Sekiguchi, M. Moriya, Y. Shirasu, and K. Tutikawa. 1986. Mutagenicity of 1,2-dibromo-3-chloropropane (DBCP) in the mouse spot test. Mutat. Res. 174: 145–147.

Sasaki, Y.F., K. Matsumoto, H. Imanishi, M. Watanabe, T. Ohta, Y. Shirasu, and K. Tutikawa. 1990. In vivo anticlastogenic and antimutagenic effects of tannic acid in mice. Mutat. Res. 244: 43–47.

Shibuya, T., and T. Murota. 1984. Mouse spot tests with dimethylbenz[a]anthracene with and without phenobarbital pretreatment. Mutat. Res. 141: 105–108.

Styles, J.A., and M.G. Penman. 1985. The mouse spot test. Evaluation of its performance in identifying chemical mutagens and carcinogens. Mutat. Res. 154: 183–204.

Szostak, J.W., T.L. Orr-Weaver, R.J. Rothstein, and F.W. Stahl. 1983. The double-Strand-Break repair model for recombination, Cell 33: 25–35.

Umweltbundesamt/Fahrig. 1994. Spot test with m-phenylenediamine. UBA project 116 06 092/01.

Umweltbundesamt/Fahrig. 1994. Spot test with 1,2-dichlorethane. UBA project 116 06 087.

13

SOMATIC MUTATION AND RECOMBINATION TEST IN DROSOPHILA USED FOR BIOMONITORING OF ENVIRONMENTAL POLLUTANTS

JUDITH GUZMÁN-RINCÓN,[1] PATRICIA RAMÍREZ-VICTORIA,[1] AND LUIS BENITEZ[2]

[1]Department of Biology, ININ. Km.36.5 Carretera México-Toluca. CP. 05245. Salazar, Edo. de México, México. [2]Oncological Research Unit, Oncology Hospital, National Medical Center S-XXI, 06720. México, D.F., México

1. INTRODUCTION

The somatic mutation and recombination test (SMART) in *Drosophila melanogaster* has been designed to detect genetic damage in a rapid and inexpensive way. It is an in vivo system that uses an eukaryotic organism with metabolic machinery similar to that found in mammalian cells. Two different systems have been widely explored, i.e. the wing spot test and the eye spot test. Both of them are based on the fact that during the early embryonic development of the fly, groups of cells (imaginal discs) are set apart and proliferate mitotically during the whole larval period until differentiation into the different structures of an adult fly (Guzmán-Rincón and Graf, 1995).

A genetic alteration occurring in one of the imaginal disc cells during mitotic proliferation will be present in all the descendant cells, and these will form a clone of mutant cells. This phenomenon is called clonal expansion. If the genetic alterations lead to a visible change in phenotype, the mutant cell clone can be detected as a mosaic spot of mutant cells on the adult body, e.g. a spot in the eye or a spot on the wing. This phenomenon, permits the detection of a wide spectrum of genetic end points such as point mutation, deletions, certain types of chromosome aberra-

Biomonitors and Biomarkers as Indicators of Environmental Change 2, Edited by Butterworth *et al.*
Kluwer Academic/Plenum Publishers, New York, 2000

tions as well as mitotic recombination and gene conversion (Graf et al., 1984; Vogel and Zijlstra, 1987; Würgler and Vogel, 1986; Guzmán-Rincón et al., 1998).

The use of strains with increased cytochrome P450-dependent bioactivation capacity facilitates the detection of promutagens of different classes (Delgado-Rodríguez et al., 1995; Graf and van Schaik, 1992). At the present, more than 400 compounds have been tested in the eye and the wing SMART assays in *D. melanogaster* (Frei and Graf, unpubl.). These somatic assay systems have been shown to be sensitive to a broad spectrum of individual genotoxic agents as well as to dietary complex mixtures (Graf et al., 1994; Graf and Würgler, 1986; van Schaik et al., 1984; Guzmán-Rincón and Graf, 1995), or pollutants which are extracted from airborne particulate matter (Delgado-Rodríguez et al., 1995; Graf and Singer, 1989) and to different types of radiation (Guzmán-Rincón et al., 1994; Zambrano et al., in preparation).

Furthermore, special strains and crosses have been devised which posses an increased bioactivation capacity for indirectly-acting mutagens and carcinogens (Frölich and Würgler 1989, 1990a,b; Vogel et al., 1991). In addition, the use of some repair-deficient mutants in the SMARTs has also been investigated but without great success (Graf et al., 1990).

In order to clarify the applications of the SMART procedures for the analysis of the mutagenic activity of different compounds, we will present the results obtained in some of our experiments in which we tested the effect of different cigarette smoke condensates and some of it's components.

Cigarette smoke and cigarette-smoke condensate (CSC) contain genotoxic agents. More than 3,800 single components have been identified in the particulate and gas phase of tobacco smoke (IARC, 1986). The genotoxic effects of cigarette smoke condensate and its components have been reviewed by DeMarini (1998) and it has been demonstrated that they may cause DNA damage (Hess and Brandner, 1996), mutations (DeMarini et al., 1995; De Flora et al., 1995), increase of sister-chromatid exchange rates (Benedict et al., 1984), binucleated cell formation (Crebelli et al., 1991), mitotic disturbances, and chromosome abnormalities (DeMarini, 1998).

Most reports on cytogenetic findings, except for sister chromatid exchanges, were a bit sketchy because mitotic abnormalities, such as chromatin-bridge formation, increase of mitotic index, metaphase chromosome aggregation or scattering and other phenomena, were not classified into analyzable details (Hsu et al., 1991).

In animal models NNK and NNN cause tumors of the lung, liver, nasal mucous and pancreas (Hecht et al., 1986; Hecht and Hoffmann, 1988; Riverson et al., 1988). NNK is thought to mediate its carcinogenic effects by the formation of DNA adducts. The presence of DNA adducts leads to error in DNA replication and gene mutation. The carcinogenic process may also involve epigenetic changes such as modulating the expression of certain genes.

Previous studies of the comparative carcinogenicity of NNK and NNN have demonstrated higher activity for NNK than for NNN. In mice strain A, both compounds induced lung adenomas, but NNK induced more tumors per animal than NNN did. In F344 rats, subcutaneous injection of NNK resulted in tumors of the nasal cavity, lung, and liver, whereas NNN induced nasal cavity tumors only (Hecht et al., 1986).

The large variation in the spectrum and occurrence of cancer diseases between human populations demonstrates the importance of environmental rather than inherited factors for cancer induction. The identification of mutagenic and carcinogenic chemicals in the human environment is in itself an immense task and the introduction of interactions between agents evidently will complicate matters even more. The screening of such interactions between environmental agents will be facilitated by an understanding of the basic mechanisms involved, which in its turn requires suitable model systems. Most work along such lines has been performed on in vitro systems—microorganisms and cell cultures. One problem in that context is the fact that such in vitro systems do not reflect the complex, modifying synergistic and antagonistic system operating in vivo. The high costs and the time they take limit in vivo tests on mammals. The use of *Drosophila* and particularly the rapid and sensitive somatic mutation and recombination test (SMART) provides a suitable substitute or at least a complementary in vivo method to mammalian in vivo investigations.

2. METHODS

2.1. Markers for the Wing Spot Test

The marker *multiple wing hair* (*mwh*) is a completely, recessive, homozygous viable mutation, kept in a homozygous *mwh* strain. It is located in chromosome 3, near the tip of the left arm (3–0.03) and in homozygous condition produces multiple trichomes per cell instead of the normally unique trichome. The marker *flare*^{3} (*flr*^{3}) is also a recessive mutation, and it also affects the shape of the wing hairs (produces malformed wing hairs that have the shape of a flare) and is located on the left arm of chromosome 3 but in a proximal position (3–38.8). The three known alleles of *flr* that are recessive, zygotic lethals (zygotes homozygous for *flr*/*flr* are not capable to develop into adult flies), on the other hand, homozygous cells in the wing imaginal disc are viable and lead to mutant wing cells. Due to the zygotic lethality, *flr* alleles have to be kept in stocks over balancer chromosomes carrying multiple inversions *TM3* (*Third Multiple 3*), and a dominant marker which is a homozygous lethal *Bd*^{s} (*beaded-Serrate*). More detailed information is given by Lindsley and Zimm, 1992.

2.2. Crosses

Among the different crosses used so far with the best performance and the best status of validation are the standard one, in which virgin females are collected from the *flr*^{3}/*TM3*, *BD*^{s} strain and crossed with *mwh* males. The progeny will consist of trans-heterozygous *mwh flr*$^+$/*mwh*$^+$ *flr*^{3} and *mwh flr*$^+$/*TM3*, *Bd*^{s} larvae. They lead to two types of adults flies among which the inversion heterozygotes are marked with the *serrate* phenotype (clipped wing margins). Normally only the wings of the *non-serrate* flies are analyzed. This standard cross leads to a low spontaneous frequency of small mutant spots and a very low frequency of double-haired cells and is only slightly temperature sensitive (Graf, 1986).

To improve the performance of the wing spot test in the case of promutagens activated via cytochrome P450-dependent metabolic pathways, Frölich and Würgler (1989) constructed new tester strains. These strains carry chromosomes 1 and 2 from a DDT-resistant Oregon R(R) line exhibiting a constitutive high level of cytochrome P450-dependent enzymes of the xenobiotic metabolism (Hällström and Blanck, 1985). Since the third chromosomes are those of the standard tester strain, the scoring and data analysis is identical with the two crosses (Dapkus and Merrel, 1977). A number of promutagens show increased genotoxicity when the High Bioactivation (HB) cross (*ORR/ORR*; *flr*^{3}/*TM3*, *Bd*^{s} females and *mwh* males) is employed (Graf and Singer, 1992; Graf and van Schaik, 1992).

2.3. Treatments

Three tester strains are raised under standard conditions for *Drosophila* cultures at 25 °C. Virgin females are collected from the *flr*^{3}/*TM3*, *BD*^{s} or from the *ORR/ORR*; *flr*^{3}/*TM3*, *Bd*^{s} strain and males isolated from the *mwh* strain The parental flies of these two crosses are used to collect eggs during 8h in culture bottles containing a solid agar base (4% w/v) and a thick layer of live fermenting yeast supplemented with sucrose. At a pre-selected age of the cultures (for the standard procedure at a culture age of three days when the larvae are 72 ± 4h old), the larvae are washed from the bottles with tap water and collected in a strainer.

Larvae derived from the ST or the HB cross can be treated with the test compounds chronically (1 to 4d) or acutely (periods of 1–6h) by oral administration or inhalation (for volatile chemicals). In special cases, injection of larvae with test solutions may also be used (Graf et al., 1984). In the standard procedure for the wing spot test, 3-day-old larvae are fed for the rest of the larval development (approx. 48h) (Graf, 1994). For this chronic feeding, the larvae are put into vials with *Drosophila* Instant Medium (Carolina Biological Supply Co., Burlington NC) or any other type of medium (ordinary cornmeal medium, mashed potatoes, agar-yeast medium, etc.) containing the test compound(s) (Guzmán-Rincón et al., 1998). It is possible to administer single treatments or even co-treatments. It is also feasible to collect 2-day-old larvae and give them a pretreatment of 24h and then transfer them to fresh vials for the 48h chronic treatment. It is also possible to feed 48h larvae during 24h with a special treatment, and wash them to administrate a post-treatment during 24h (Figure 1). Or in special cases, when using physical agents like gamma rays or neutrons, which produce their effects immediately after their administrations, 4-day-old larvae can be collected and given the acute treatment in order to increase the sensitivity of the test, since at this moment, the cells of the imaginal disks are in the last cycle of division and they have reached the maximum number of cells (Zambrano et al., in preparation).

All the experiments are carried out at 25 °C and 65% rel. humidity. Strict control of temperature is essential as it affects the frequencies of spots (Graf, 1986; Katz and Foley, 1993).

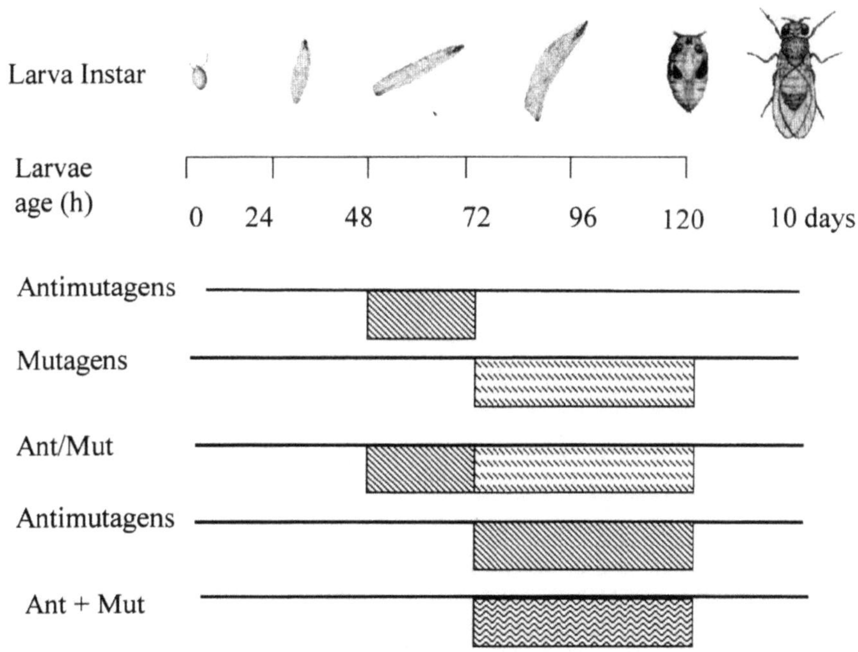

FIGURE 1. Protocols for the SMART test.

For feeding studies, solutions are used whenever possible. Water soluble compounds are dissolved in water and then tested by feeding. For non-water soluble chemicals, an appropriate solvent has to be used. Experience indicates that non-water soluble compounds may be dissolved in pure solvents or in mixtures of solvents (e.g. 5% ethanol plus 5% tween 80). In some cases, for compounds of very low solubility, sonification may help to obtain homogeneous suspensions. When solutions are mixed with the medium, compounds may precipitate, but usually a fairly uniform distribution within the medium is obtained. In spite of a possible increased variability in the results, dose related responses could be observed even under extreme conditions like this. To increase their stability, certain chemicals need to be dissolved in an appropriate buffer (e.g. phosphate buffer) (Guzmán-Rincón et al., 1998).

For volatile chemicals such as 1,2-dibromooethane or vinyl chloride, inhalation procedures may be used. Exposures may be chronic for days (Verburgt and Vogel, 1977) or acute for one to several hours (Graf et al., 1984; Kramers et al., 1985; Vogel and Zijlstra, 1987).

For the case of our experiments with CSC, five filtered commercial cigarettes (Viceroy) were smoked in a Filtrona Standard Smoking Machine, (calibrated for one smoke inhalation for two seconds, with a volume of 35 cc and a frequency of one smoke inhalation per minute at environmental conditions of 22 ± 2 °C and a relative humidity of 60 ± 5%). Trapping Cambridge filters are collected from the machine after each set of five smoked cigarettes. Extraction is performed using 2-propanol. Aliquots from this cigarette smoke condensate were used in our experiments. The NNN (CSL-90-289-72-20) and NNK (CSL-92-359-5-25) were obtained from Chemsyn Science Lab.

Chronic treatments for 48 h were administrated with the following concentrations: CSC 6%; NNK 0.6, 1.2 and 2.4 mM; NNN 7.2, 14.4 and 28.8 mM. Negative solvent control (H_2O) was included. We used 72 h-old trans-heterozygous larvae derived from ST and HB crosses. The larvae were placed in equal plastic vials containing 0.7 g of mashed potatoes (Maggi), this medium was hydrated with 5 ml of the mutagen solutions. Then, the SMART procedures were performed as it is mentioned above. At least three independent experiments were assayed in both crosses and they were treated under identical conditions 25 °C and 65% relative humidity.

2.4. Scoring of the Wings

The hatched flies are collected from the exposure vials and stored in 70% ethanol. Both the ST and HB cross produce two types of progeny which can be distinguished phenotypically based on the Bd^s marker: (1) marker-trans-heterozygous flies (*mwh flr*$^+$/*mwh*$^+$ *flr*^{3} phenotypically wild-type wings) and (2) balancer-heterozygous flies (*mwh flr*$^+$/*TM3*, *Bd*^{s}, phenotypically serrate wings). The wings are mounted in Faure's solution (gum arabic 30 g, glycerol 20 ml, chloral hydrate 50 g, and water 50 ml).

Both the dorsal and the ventral surfaces of the wings are analyzed under a compound microscope at 400X magnification for the occurrence of any of two categories of spots: single spots (*mwh* or *flr* phenotype) or twin spots (*mwh* clone adjacent to a *flr* clone. During the microscopic analysis of the wings, the position of the spots is noted according to the sector of the wing. Only the distal wing compartment is scored for clones. This is an area comprising a total of about 24,400 cells.

In each case the size of a spot is determined by counting the number of cells exhibiting the *mwh* or the *flr* phenotype. To arrive at uniform record of split clones and neighboring individual clones, it was decided to count two spots as separate spots as soon as they separated by three or more rows of wild-type cells. Because there is a correlation between the time of induction of a genetic change in the somatic cells and the size of the resulting spot (Graf, 1994), spot sizes are also recorded in addition to spot frequencies. Single spots can be caused by point mutation, deletion, specific types of translocation, mitotic recombination, and also monosomy; twin spots are most likely caused by recombination between *flr* and the centromere (Graf et al., 1984; see Figure 2).

Moreover, the analysis of the two different genotypes (*mwh*/*flr* and *mwh*/*TM3*) allows for a quantitative determination of recombinogenic activity of genotoxins. In individuals with structurally normal chromosomes both mutational and recombinational events lead to the formation of single and /or twin spots. In contrast, in the balancer-heterozygous individuals, all recombinational events are eliminated due to the multiple inversions, and only mutational events can give rise to single (*mwh*) spots (Frei et al., 1992a; Graf et al., 1992).

2.5. Statistical Analysis

For the evaluation of the genotoxic effect recorded, the frequencies of spots per individual fly rather than per wing of a treated series are compared to its concurrent negative (solvent) control series. A multiple-decision procedure is used to decide whether a result is positive, weakly positive, inconclusive, or negative (Frei and Würgler, 1995). The procedure is based on testing two hypotheses: (1) the mutation frequency (induced plus spontaneous) in the treated series is no higher than the mutation frequency in the appropriate control. (2) The induced mutation frequency in the treated series is no less than "m" times as high as the observed spontaneous mutation frequency in the control. The X^2-test for over-dispersion of the spots is applied to the different data sets in order to show that the spots are not normally distributed in most of the cases. For this reason, the non-parametric U-test is employed to test for statistically significant differences between control and treated series or between single and combined treatments.

The minimal number of analyzed wings must be 110, according with Frei and Würgler (1995), and for our experiments of CSC, NNK and NNN, a minimum of 120 wings per each experiment were scored for all the categories of spots. Besides, the same number of wings all recombinational events are eliminated due to the multiple inversions, and only was analyzed in all the experimental groups, in order to get an optimal test design.

3. RESULTS AND DISCUSSION

Normally, it is not possible to determine the dose of the compound reaching the target cells in the imaginal discs of the larvae. However, methods are available for a quantitative measurement of the amounts of the chemical compound taken up by individual larvae (Zordan et al., 1991). Moreover, it is also possible to determine quantitatively adducts in larval DNA as has been shown for benzo(a) pyrene treatments (Zordan et al., 1994).

Drosophila has a detoxification-activating system in many respects closely resembling the corresponding system in mammals (Hällström and Gräfstrom, 1981) which makes it possible to detect xenobiotics which require metabolic activation. Even more, with the newly developed tester strains that contain constitutively expressed high levels of cytochrome P450, the assays have been made much more sensitive for several classes of promutagens and procarcinogens, especially for the polycyclic aromatic hydrocarbons (PAH) (Delgado-Rodríguez et al., 1995; Frölich and Würgler, 1990a; Graf and Singer, 1992; Graf and van Schaik, 1992).

The use of two genes in the same chromosome makes it possible not only determine the frequency of recombination, but (using the adequate treatment

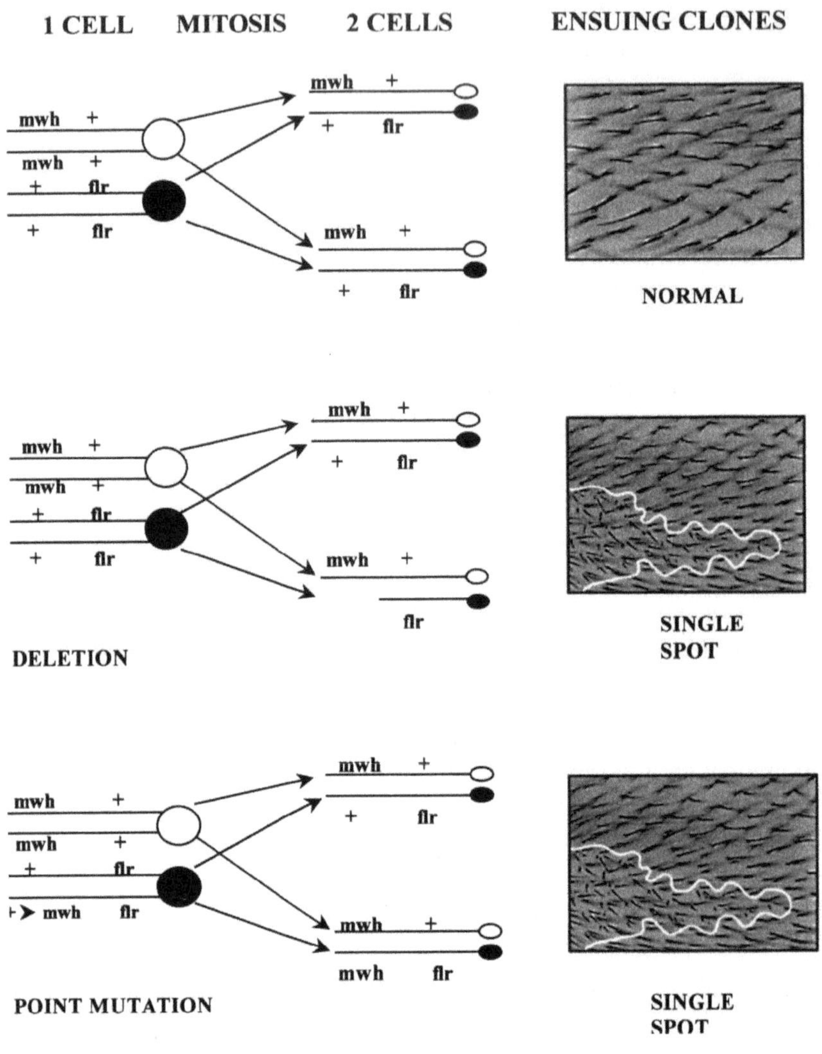

FIGURE 2. Genetic schemes showing different genotoxic events which lead to the formation of single and twin spots (modified from Graf et al., 1984).

patterns), to discern localization of recombination either to the (mostly heterochromatic) region between *flr* and the centromere or in the euchromatic interval between *flr* and *mwh*. Twin spots of *flr* and *mwh* cells occur as the result of a recombination in the interval close to the centromere. While spots with only *mwh* indicate recombination between *mwh* and *flr*. The frequency of chemically-induced twin and single spots varies between chemicals, indicating localized induction of somatic recombination.

Principally, three groups can be recognized: (1) induction only of single spots, suggesting an effect on the euchromatic interval between *flr* and *mwh*; (2) the occur-

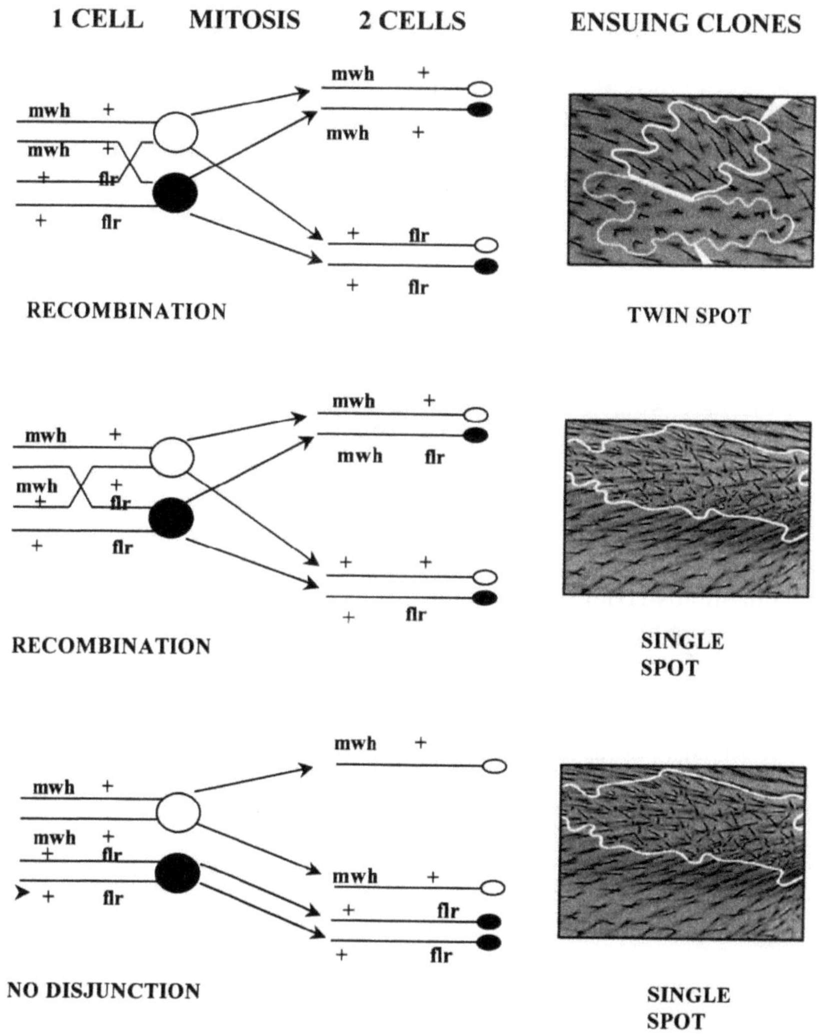

FIGURE 2. *Continued*

rence of both twin spots and single spots, with single spots the majority, suggesting a more or less random localization of the effect along the chromosome; (3) the induction of twin spots, suggesting a localized effect in the interval close to the centromere (Ramel and Magnusson, 1992).

Besides, in our laboratory, we have performed different protocols in order to study the genetic effect of different physical agents, and our results have demonstrated that the assay is an excellent tool to evaluate the mutagenic and recombinogenic effect of different types of radiation, and specially with gamma rays and neutrons, clear dose response relationship can be obtained (Zambrano et al., in preparation).

Moreover, this assay is very suitable in studies of antigenotoxicity. It is able to detect the modulation effect of different compounds called antimutagens, against different kinds of physical and chemical mutagens (Nakamura et al., 1997; Olvera et al., 1997). In our laboratory, we have demonstrated that third-instar larvae of *Drosophila* have in vivo nitrosation capacity, and it is possible to detect the mutagenic effect of the N-nitroso compounds formed by nitrosation, as well as the identification of inhibitors of endogenous nitrosation, which is relevant for primary prevention of gastric cancer (Guzmán-Rincón, et al., 1998). On the other hand, it has been possible to evaluate the inhibitory effect of catechin, ascorbic acid, as well as chlorophyllin against endogenous nitrosation (Guzmán-Rincón et al., unpubl.).

For our smoke experiments, no differences between triplicates were observed, therefore, the data were pooled. The results showed a clear dose-response relationship (Figure 3a and 3b, Tables 1 and 2). When the data were analyzed, we could observe that there was a reduction in the frequency of induced spots obtained in balancer-heterozygous individual wings with NNK, and NNN. In this genotype, all recombinational events are inhibited because of the multiple inversions present on the balancer chromosome. Only the frequencies and sizes of the *mwh* clones can be used for a quantitative comparison of the effects observed in the two genotypes and they are result from point mutations and deletions (Frei et al., 1992b; Graf and van Schaik, 1992; Graf and Würgler, 1996).

TABLE 1. Summary of results obtained in the *Drosophila* wing spot test after chronic treatments with NNK.

Compounds	wings	Small single spots m = 2	Large single spots m = 5	Twin spots m = 5	Total spots m = 2
Standard Cross					
H_2O	280	0.26 (73)	0.10 (29)	0.02 (6)	0.39 (108)
NNK (0.6 mM)	120	0.45+ (54)	0.41+ (49)	0.10+ (12)	0.96+ (115)
NNK (1.2 mM)	120	0.64+ (77)	0.47+ (57)	0.15+ (18)	1.27+ (152)
NNK (2.4 mM)	120	1.48+ (178)	0.74+ (89)	0.13+ (16)	2.36+ (283)
High Bioactivation Cross					
H_2O	318	0.31 (100)	0.13 (40)	0.02 (6)	0.46 (146)
NNK (0.6 mM)	120	0.98+ (118)	0.62+ (75)	0.11+ (13)	1.72+ (206)
NNK (1.2 mM)	120	1.65+ (198)	1.12+ (135)	0.12+ (15)	2.90+ (348)
NNK (2.4 mM)	120	2.32+ (278)	2.05+ (246)	0.14+ (17)	4.51+ (541)

Statistical diagnoses according to Frei and Würgler (1988) for comparisons with corresponding controls: + = positive; – = negative; i = inconclusive; m = multiplication factor.

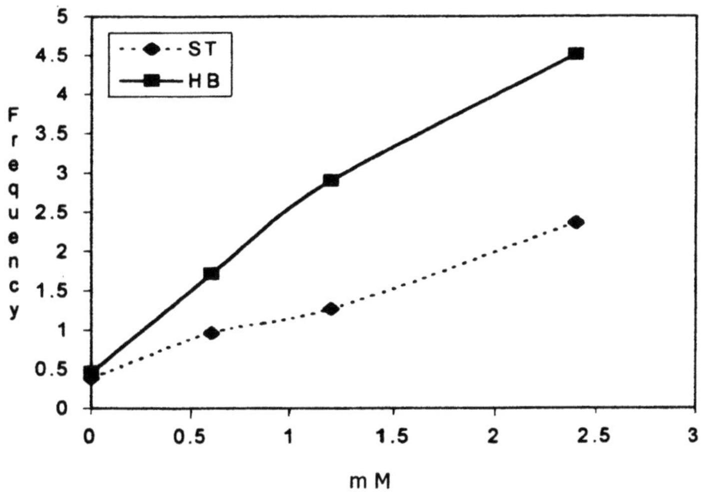

FIGURE 3a. Dose-response relationship of NNK in both crosses of *D. melanogaster*.

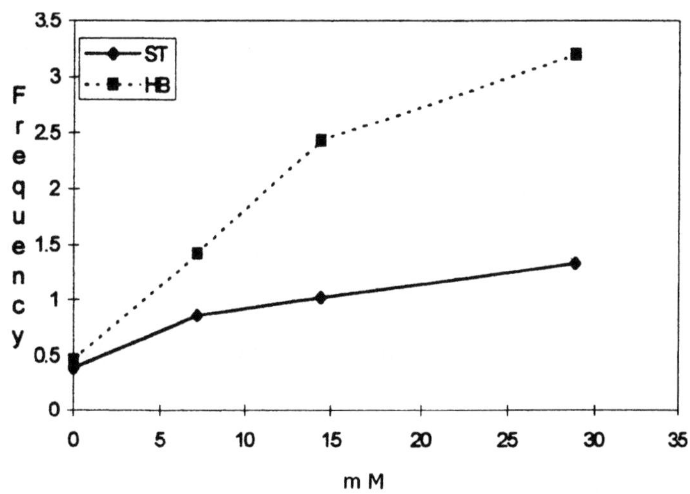

FIGURE 3b. Dose-response relationship of NNN in both crosses of *D. melanogaster*.

The proportion of mitotic recombination versus somatic mutation was calculated based on the total of *mwh* clone formation frequencies by linear extrapolation of the dose response relationships obtained for the two types of wings (Graf et al., 1992), and the comparison of the slopes of the curves gives an estimate of the proportion of recombination among the total genotoxicity induced by NNK and NNN. In the ST cross a value of 63% and 47% recombination, respectively, for NNK and NNN, and in the HB cross a value of 88% and 75% was obtained. These values

TABLE 2. Summary of results obtained in the *Drosophila* wing spot test after chronic treatments with NNN.

Compounds	wings	Small single spots m = 2	Large single spots m = 5	Twin spots m = 5	Total spots m = 2
Standard Cross					
H_2O	280	0.26 (73)	0.10 (29)	0.02 (6)	0.39 (108)
NNN (7.2 mM)	120	0.43+ (51)	0.34+ (41)	0.09+ (11)	0.86+ (103)
NNN (14.4 mM)	120	0.57+ (68)	0.37+ (44)	0.08+ (10)	1.02+ (122)
NNN (28.8 mM)	120	0.76+ (91)	0.47+ (57)	0.10+ (12)	1.33+ (160)
High Bioactivation Cross					
H_2O	318	0.31 (100)	0.13 (40)	0.02 (6)	0.46 (146)
NNN (7.2 mM)	120	0.82+ (104)	0.48+ (58)	0.07+ (8)	1.42+ (170)
NNN (14.4 mM)	120	1.50+ (180)	0.84+ (101)	0.09+ (11)	2.43+ (292)
NNN (28.8 mM)	120	1.98+ (238)	1.10+ (132)	0.12+ (14)	3.20+ (384)

Statistical diagnoses according to Frei and Würgler (1988) for comparisons with corresponding controls: + = positive; – = negative; i = inconclusive; m = multiplication factor.

show that the genotoxic effect of tobacco-specific nitrosamines is mainly the recombination (Figure 4a and 4b, Tables 3 and 4).

CSC showed a negative result for the small (0.32) and large single spots (0.15), inconclusive for twin spots (0.06) and weakly positive for total spots (0.53) with respect to the control group (0.31, 0.08, 0.02, and 0.40 respectively) for the ST cross. For the HB cross the values were negative for the small spots (0.37), weakly positive for the large spots (0.27), inconclusive for twin spots (0.04) and weakly positive for the total spots (0.68) in comparison with the control group with 0.26, 0.13, 0.01, and 0.40, respectively.

4. CONCLUSIONS

Our results and all the previous studies indicate that the SMART assays in somatic cells of *D. melanogaster* represent a rapid and inexpensive bioassay to evaluate the genotoxic activity of single compounds as well as of complex mixtures and physical agents like radiation. They are very sensitive to different classes of agents and are able to detect not only mutational events but also recombinogenic activity. For the applications of the test compounds, various protocols for single or combined as well as for sequential treatments of larvae by feeding or inhalation are available.

From our data we could demonstrate that NNK is a more potent agent than NNN. This is in agreement with previous studies in other assays, where the car-

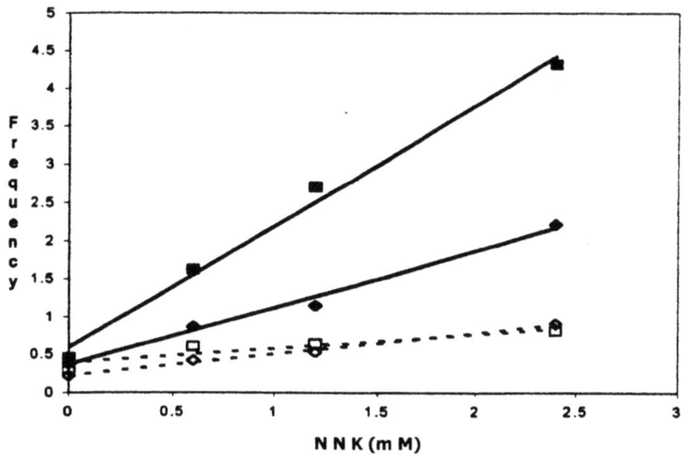

FIGURE 4a. Linear regression analysis of dose response relationship of NNK in both crosses of *D. melanogaster*.

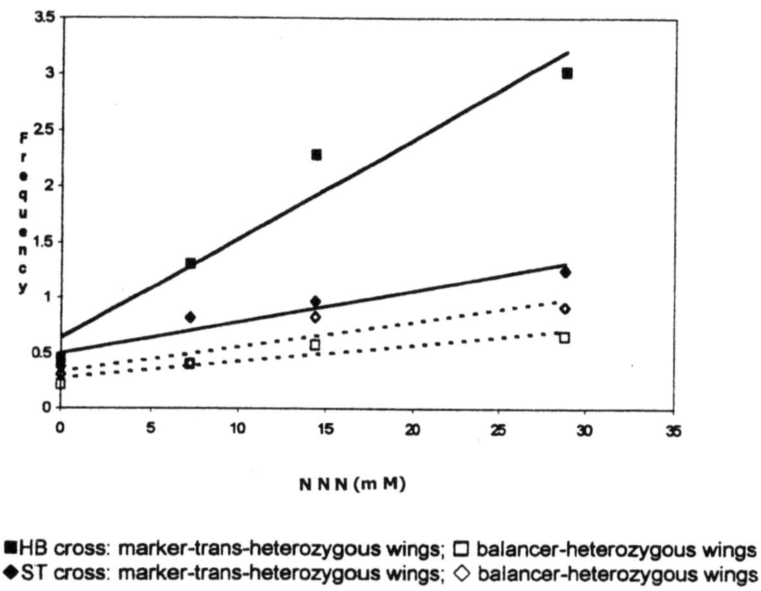

FIGURE 4b. Linear regression analysis of dose-response relationship of NNN in both crosses of *D. melanogaster*.

TABLE 3. Summary of data comparing the frequencies of *mwh* spots per wing observed in individuals with marker-trans-heterozygous wings and balancer heterozygous wings with NNK in standard and high bioactivation crosses.

TRANS-HETEROZYGOUS WINGS				BALANCER HETEROZYGOUS WINGS		
Compound (mM)	wings (N)	*mwh* clone (n/N)	main clone size (1)	wings (N)	*mwh* clone (n/N)	main clone size (1)
Standard cross						
H_2O	280	0.38	2.04	120	0.22	1.96
NNK (0.6)	120	0.87	2.73	120	0.42	2.16
NNK (1.2)	120	1.15	2.50	120	0.53	2.11
NNK (2.4)	120	2.21	2.27	120	0.91	1.67
High Bioactivation cross						
NNK (0.6)	120	1.62	2.38	120	0.60	2.25
NNK (1.2)	120	2.70	2.61	120	0.64	2.45
NNK (2.4)	120	4.31	2.56	120	0.81	2.40

cinogenic activities of NNK and NNN have been compared and where a higher activity for NNK has even shown (Hoffmann et al., 1981). There is no doubt that these tobacco-specific nitrosamines, in general require cytochrome P450-dependent metabolic activation to exhibit genotoxic activity. Our results have shown that they are more genotoxic in the HB cross than in the ST cross. However, we can say that the *in vivo* bioactivation capacity present in larvae of the ST cross is sufficient to produce at least a small amount of genotoxic metabolites of tobacco-specific nitrosamines.

TABLE 4. Summary of data comparing the frequencies of *mwh* spots per wing observed in individuals with marker-trans-heterozygous wings and balancer heterozygous wings with NNN in standar and high bioactivation crosses.

TRANS-HETEROZYGOUS WINGS				BALANCER HETEROZYGOUS WINGS		
Compound (mM)	wings (N)	*mwh* clone (n/N)	main clone size (1)	wings (N)	*mwh* clone (n/N)	main clone size (1)
Standard cross						
H_2O	318	0.45	2.01	120	0.31	2.05
NNN (7.2)	120	1.3	2.32	120	0.41	2.10
NNN (14.4)	120	2.28	2.38	120	0.83	2.3
NNN (28.8)	120	3.02	2.32	120	0.92	2.15
High Bioactivation cross						
NNN (7.2)	120	0.82	2.42	120	0.40	2.06
NNN (14.4)	120	0.97	2.41	120	0.58	2.16
NNN (28.8)	120	1.25	2.38	120	0.66	2.06

Moreover, our experimental results indicate that in the wing spot test the tobacco-specific nitrosamines exhibit their genotoxic activity mainly through the induction of recombinational events.

For all we have told before, we can support that the somatic genotoxicity assays in *Drosophila melanogaster* are excellent candidates to be used as biological monitors for genotoxic environmental contaminants.

5. ACKNOWLEDGMENTS

We are grateful to. Dr. Graf for his helpful discussions and the unconditional support he has given us. This work was supported by a grant from CONACYT.

6. REFERENCES

Benedict, W.F., A. Benerjee, K.K. Kangalingam, D.R. Dansie, R.E. Kouri, and C.J. Henry. 1984. Increased sister chromatid exchange in bone-marrow cells to exposed to whole cigarette smoke. Mutation Res. 136: 73–80.

Crebelli, R., L. Conti, S. Fuselli, P. Leopardi, A. Zijno, and A. Carere. 1991. Further studies on the comutagenic activity of cigarette smoke condensate. Mutation Res. 259: 29–36.

Dapkus, J., and D.J. Merrel. 1977. Chromosomal analysis of DDT resistance in a long-term selected population of *Drosophila melanogaster*. Genetics 87: 685–697.

De Flora, S., R. Balansky, L. Gasparini, and A. Camoriano. 1995. Bacterial mutagenicity of cigarette smoke and its interaction with ethanol. Mutagenesis 10: 47–52.

DeMarini, D.M. 1998. Mutation spectra of complex mixtures. Mutation Res. 411: 11–18.

DeMarini, D.M., M.L. Shelton, and J.G. Levine. 1995. Mutation spectrum of cigarette smoke condensate in Salmonella: comparison to mutations in smoking-associated tumors. Carcinogenesis 16: 2535–2542.

Delgado-Rodríguez, A., R. Ortiz-Martello, U. Graf, R. Villalobos-Pietrini, and S. Gómez-Arroyo. 1995. Genotoxic activity of environmentally important polycyclic aromatic hydrocarbons and their nitro derivates in the wing spot test o *Drosophila melanogaster*. Mutation Res. 341: 235–247.

Frei, H., and F.E. Würgler. 1988. Statistical methods to decide whether mutagenicity test data from Drosophila assays indicate a positibe, negative or inconclusive results. Mutation Res. 203: 297–308.

Frei, H., and F.E. Würgler. 1995. Optimal experimental design and sample size for the statistical evaluation of data from somatic mutation and recombination test (SMART) in Drosophila. Mutation Res. 334: 247–258.

Frei, H., H. Clements, D. Howe, and F.E. Würgler. 1992a. The genotoxicity of the anti-cancer drug mitoxantrone in somatic and germ cells of *Drosophila melanogaster*. Mutation Res. 279: 21–33.

Frei, H., J. Lüthy, J. Brauchli, U. Zweifel, F.E. Würgler, and C. Schlatter. 1992b. Structure/activity relationship of the genotoxic potencies of sixteen pyrrolizidine alkaloids assayed for the induction of somatic mutation and recombination in wing cells of *Drosophila melanogaster*. Chem. Biol. Interact. 83: 1–22.

Frölich, A., and F.E. Würgler. 1989. New tester strains with improved bioactivation capacity for the Drosophila wing spot test. Mutation Res. 216: 179–187.

Frölich, A., and F.E. Würgler. 1990a. Drosophila wing spot test: improved detectability of genotoxicity of polycyclic aromatic hydrocarbons. Mutation Res. 234: 71–80.

Frölich, A., and F.E. Würgler. 1990b. Genotoxicity of ethyl carbamate in the Drosophila wing spot test: dependence on genotype-controlled metabolic capacity. Mutation Res. 244: 201–208.

Graf, U. 1986. Temperature effect on *mwh* expression in the wing somatic mutation and recombination test in *Drosophila melanogaster*. Drosophila Inform. Serv. 63: 65.

Graf, U. 1994. Analysis of the relationship between age of larvae at mutagen treatment and frequency of size of spots in the wing somatic mutation and recombination test of *Drosophila melanogaster*. Experientia. 51: 168–174.

Graf, U., and D. Singer. 1989. Somatic mutation and recombination test in *Drosophila melanogaster* (wing spot test): Effects of extracts of airborne particulate matter from fire-exposed and non fire-exposed building ventilation filters. Chemosphere 19: 1094–1097.

Graf, U., and D. Singer. 1992. Genotoxicity testing of promutagens in the wing somatic mutation and recombination test in *Drosophila melanogaster*. Rev. Int. Contam. Ambient. 8: 15–27.

Graf U., and N. van Schaik. 1992. Improved high bioactivation cross for the wing somatic mutation and recombination test in *Drosophila melanogaster*. Mutation Res. 271: 59–67.

Graf, U., and F.E. Würgler. 1986. Investigation of coffee in Drosophila genotoxicity tests. Food Chem. Toxicol. 24: 835–842.

Graf, U., and F.E. Würgler. 1996. The somatic white-ivory eye spot test does not detect the same spectrum of genotoxic events as he wing somatic mutation and recombination test in *Drosophila melanogaster*. Environ. Mol. Mutagen. 27: 219–226.

Graf, U., C.B. Hall, and N. van Schaik. 1990. On the use of excision repair defective cells in the wing somatic mutation and recombination test in *Drosophila melanogaster*. Mutation Res. 271: 59–67.

Graf, U., S. Heo, and O. Olvera Ramírez. 1992. The genotoxicity of chromium (VI) oxide in the wing spot test of *Drosophila melanogaster* is over 90% due to mitotic recombination. Mutation Res. 266: 197–203.

Graf, U., A. Alonso-Moraga, R. Castro, and E. Díaz-Carrillo. 1994. Genotoxicity testing of different types of beverages in the Drosophila wing somatic mutation and recombination test. Food Chem. Toxicol. 32: 423–430.

Graf, U., F.E. Würgler, A.J. Katz, H. Frei, H. Juon, C.B. Hall, and P.G. Kale. 1984. Somatic mutation and recombination test in *Drosophila melanogaster*. Environ. Mutagen. 6: 153–188.

Guzmán-Rincón, J., and U. Graf. 1995. Drosophila melanogaster Somatic Mutation and Recombination Test as a biomonitor., in: F.M. Butterworth, L.D. Corkum, and J. Guzmán-Rincón (Eds). Biomonitors and Biomarkers as Indicators of Environmental Change: a Handbook, Plenum Press, New York pp. 169–181.

Guzmán-Rincón, J., J. Espinosa, and U. Graf. 1998. Analysis of the *in vivo* nitrosation capacity of the larvae used in the wing somatic mutation and recombination test of *Drosophila melanogaster*. Mutation Res. 412: 69–81.

Guzmán-Rincón, J., U. Graf, A. Varela, R. Policroniades, and A. Delfín. 1994. Somatic mutation and recombination induced by fast neutrons in the wing spot test of *Drosophila melanogaster*. Rev. Int. Contam. Ambient. 10. suppl. 1: 29–30.

Hällström, I., and A. Blanck. 1985. Genetic variation in cytochrome P450 system in *Drosophila melanogaster*. L. Chromosomal determination of some cytochrome P450-dependent reactions. Chem. Biol. Interactions 56: 157–171.

Hailstorm, I., and R. Gräfstrom. 1981. The metabolism of drugs and carcinogens in isolated subcellular fractions of *Drosophila melanogaster*. II. Enzyme induction and metabolism of benzo(a)pyrene. Chem. Biol. Interact. 34: 1455–1459.

Hess, R.D., and G. Brandner. 1996. DNA damage by filtered, tar-and aerosol-free cigarette smoke in rodent cells: a novel evaluation. Toxicol. Lett. 88: 9–13.

Hecht, S.S., and D. Hoffmann. 1988. Tobacco-specific nitrosamines, an important group of carcinogens in tobacco and tobacco smoke. Carcinogenesis. 9: 875–884.

Hetch, S.S., A. Riverson, J. Braley, J. Dibello, J.D. Adams, and D. Hoffmann. 1986. Induction of oral cavity tumors in F344 rats by tobacco. specific nitrosamines and snuff. Cancer Res. 46: 4162–4166.

Hoffmann, D., A. Castonguay, A. Riverson, and S.S. Hecht. 1981. Comparative carcinogenicity and metabolism of 4-(Methylnitrosamino)-1-(3-pyridyl)-1-butanone and N-nitrosonornicotine in Syrian golden hamsters. Cancer Res. 41: 2386–2393.

Hsu, T.C., L.M. Cherry, C. Bucana, L.R. Shirley, and C.G. Gairola. 1991. Mitosis-arresting effects of cigarette smoke condensate on human lymphoid cell lines. Mutation Res. 259: 67–78.

IARC (International Agency for Research on Cancer). Tobacco smoking. IARC. 1986. Monographs on the evaluation of the carcinogenic risk of chemicals to humans, Vol. 38, pp. 37–375. Lyon, France.

Katz, A.J., and T.A. Foley. 1993. Effect of temperature on frequencies of spots in Drosophila wing-spot assay. Environ. Mol. Mutagen. 22: 54–58.

Kramers, P.G.N., B. Bissumbhar, and H.C.A. Moout. 1985. Studies with gaseous mutagens in *Drosophila melanogaster*, in: M.D. Waters, S.S. Sandhu, J. Lewtas and L. Claxton (Eds.) Short-Term Bioassays in the analysis of Complex Environmental Mixtures, IV, Plenum, New York, pp. 65–73.

Lindsley, D.L., and G. Zimm. 1992. The genome of *Drosophila melanogaster*. Academic Press, San Diego. 1133pp.

Nakamura, K.Y., K. Kawai, H. Furukawa, T. Matsuo, K. Shimoi, I. Tomita, and Y. Nakamura. 1997. Suppressing effects of S-methyl methanethiosulfonate and diphenyl disulfide on mitomycin C-induced somatic mutation and recombination in *Drosophila melanogaster* and micronuclei in mice. Mutation Res. 385: 41–46.

Olvera, O., S. Zimmering, M.P. Cruces, E. Pimentel, C. Arceo, J. Guzmán, and M.E. de la Rosa. 1997. Antimutagenesis in somatic cells of Drosophila as monitored in the wing spot test. Food Factors for Cancer Prevention, Ohigashi, Osawa, Terao, Watanabe, Yoshikawa Eds. Springer-Verlag Tokyo pp. 567–571.

Ramel, C., and J. Magnusson. 1992. Modulation of genotoxicity in Drosophila. Mutation Res. 267: 221–227.

Riverson, A., D. Hoffmann, B. Prokopczyk, S. Amin, and S.S. Hecht. 1988. Induction of lung and exocrine pancreas tumours in F344 rats by tobacco-specific and areca-derived N'nitrosamines. Cancer Res. 48: 6912–6917.

Van Schaik, N., A. Grant, I. Rubenchik, and U. Graf. 1984. Use of Drosophila test systems for genotoxicity testing of herbal teas. *Inmunol.* Hematol. Res. 3: 199–202.

Vogel, E.W., M.J.M. Nivard, and J.A. Zijlstra. 1991. Variation of spontaneous and induced mitotic recombination in different Drosophila populations: a pilot study on the effects of poly-aromatic hydrocarbons in six newly constructed tester strains. Mutation Res. 250: 291–298.

Vogel, E.W., and J.A. Zijlstra. 1987. Mechanistic and methodological aspects of chemically-induced somatic mutation and recombination in *Drosophila melanogaster*. Mutation Res. 182: 243–264.

Würgler, F.E., and E.W. Vogel. 1986. *In vivo* mutagenicity testing using somatic cells of *Drosophila melanogaster*. In F.J. de Serres (Ed) Chemical Mutagens, Principles and Methods for Their Detection. Vol. 10 Plenum Press, New York, pp. 1–72.

Zambrano, F., J. Guzmán-Rincón, L. Paredes-Gutiérrez, and A. Delfin-Loya. 1998. Influencia de la edad larval en la inducción de la mutación y recombinación somática con neutrones de reactor en *Drosophila melanogaster*. In preparation.

Zordan M., M. Osti, S.S. Pavanello, R. Costa, and A.G. Levis. 1994. Relationship between benzo(a)pyrene-DNA adducts and somatic mutation and recombination in *Drosophila melanogaster*. Environ Mol. Mutagen. 23: 171–178.

Zordan M., U. Graf, D. Singer, C. Beltrame, L. Della Valle, M. Osti, R. Costa, and A.G. Levis. 1991. The genotoxicity of nitrolotriacetic acid (NTA) in a somatic mutation and recombination test in *Drosophila melanogaster*. Mutation Res. 262: 253–261.

14

A NEW WAY TO VIEW COMPLEX MIXTURES

Measurement of Genotoxic Effects of Mixtures of A Polychlorinated Biphenyl, A Polyaromatic Hydrocarbon, and Arsenic

RICHARD M. MCGOWEN,[1,2] D. CARL FREEMAN,[2] AND FRANK M. BUTTERWORTH[1,3]

[1]Oakland University, Rochester, MI 48309; [2]Wayne State University, Detroit MI 48204; and [3]current address: Institute for River Research International, Rochester, MI 48307;
[3]To whom correspondence should be sent

1. INTRODUCTION

Sediments in almost all natural waters are contaminated with anthropogenic chemicals (International Joint Commission, 1988), and many of these compounds are biomagnified on up the food web to humans. As a result, contaminated sediments are in large part the beginning of human exposure. Some of these compounds are highly toxic and stable such as pesticides, polynuclear aromatic hydrocarbons (PAHs), polychlorinated biphenyls (PCBs), furans, dioxins and toxic metals (Department of National Health and Welfare, 1992). The toxicity of many of these anthropogenic compounds tested alone is well known. But organisms in natural settings are rarely exposed to single compounds, emphasizing the need for biological systems to evaluate the overall toxicity of chemical mixtures (Butterworth, 1995). The biological effect of these pollutant mixtures has been measured in laboratory-exposure experiments and community analysis of in situ benthic macroinvertebrates (Hudson and Ciborowksi, 1995) and plants (Lovett Doust et al., 1994; Tracy et al., 1995; Sgro and Johansen, 1995), and in sentinel biomonitor studies (Rosenberg and Resh, 1993; Corkum et al., 1995). These studies indicate the enormity of the problem but they are not routinely or widely employed. Furthermore, there is no routine way to evaluate and compare the toxicity of complex mixtures.

The environment is in essence a mixture of chemical, physical, and biological components to which all living organisms are exposed. Knowing the influence of these components on an organism is a major goal of environmental regulatory agencies. However, as this paper will demonstrate, achieving this goal may be extremely difficult, and new monitoring paradigms will have to be developed. To attempt simplification in our experiments we have kept the physical and biological parameters constant, and in addition have restricted our scope to that of simulated sediments of natural waters. We used two natural strains of the same species to assay three pollutants in single, binary, and ternary mixtures over a five log concentration range. Results show: a. all three pollutants are genotoxic separately, and in combination, in all concentrations; b. positive dose responsiveness in most cases; c. strain-specific sensitivity to some compounds and mixtures; and d. unpredictability of some mixtures in one or both strains manifesting synergism, potentiation or antagonism. These three processes will be defined in terms of the experimental system in the discussion.

We chose to measure the effect of these compounds on chromosomal recombination. Because recombination is manifested mainly as *heterozygosity loss*, it is more important than other genotoxic events. Most organisms including humans are heterozygous for large numbers of recessive, and sometimes defective genes, but in this state these genes are not expressed. However, during development organisms exposed to recombinogens can have the defective genes expressed, through heterozygosity loss, when in development the normal alleles become active.

2. MATERIALS AND METHODS

2.1. Laboratory Procedures

Chemicals: Benzo[a]pyrene (BAP), Tween 80 and sodium arsenite (SA) were obtained from Sigma Chemical Co. (St. Louis, MO), 4,4′-dichlorobiphenyl (DCB) was obtained from Lancaster Synthesis Co. (NH, USA), and 4,4′-dichloro-3-biphenylol from Ultra Scientific (North Kingston, RI). The BAP and DCB were dissolved in alcohol and SA dissolved in distilled water prior to mixing with the ethanol/Tween 80 solvent mixture according to our published methods (Butterworth et al., 1995b; Pandey et al., 1995). The toxic chemicals used in the experiments were handled following strict institutional and laboratory protocols for chemical safety.

Strains: For most of the experiments two different wild-type strains were employed, Berlin K (BK) and Oregon K (OK). For further details about these strains consult Hällstrom et al., 1984; Zijlstra et al., 1984; and Lindsley and Zimm (1992). Marker genes were introduced into these strains (Vogel, 1980) to create the SMART assay system. The extent to which genotoxicity could be attributed to recombination was ascertained by using another wild-type strain Leiden Standard (LS) containing a large paracentric inversion $In\text{-}w^{+}/w$ ($In(1)sc^{S1L}\ sc^{8R} + dl\text{-}49$) on the X chromosome (Vogel, 1989; Vogel and Nivard, 1993). For a description of the symbols used see Lindsley and Zimm (1992).

SMART assay: The SMART assay or somatic mutation and recombination test for the eye (also known as the eye mosaic test) was developed by Vogel and Zijlstra (1987), Vogel (1989) and Vogel and Nivard (1993). It mainly detects changes in the rate of mitotic crossing over between homologous chromosomes occurring during the larval and pupal stages in the rapidly proliferating imaginal eye disk cells that undergo about five cycles of cell division. Hence, recombination will be used interchangeably with genotoxicity or mutational events. In addition to homologous recombination, the method detects unequal sister chromatid recombination, gene conversion, X-chromosome loss, deletion, or mutation in the *white* (w) locus. If one of the above genetic changes occurs to the normal chromosome in a developing eye, the reporter gene *white* (w) on the homologue is expressed resulting in a "white" spot in one or more facets of the adult eye. Modifications in the current experiments are described in Butterworth et al. (1995b) and Pandey et al. (1995).

Serial dilutions of the spiked solvent mixture were mixed with the medium to produce the desired concentration. No concentrations used had a negative effect on fertility, viability or development rate. Six virgin females and six males from the appropriate genotypes [virgin $y\ w^+/y\ w^+$ females (yellow body and wild-type eyes) and $y^+\ w$ males (wild-type body and white-eyes)] and strains were allowed to lay eggs on the spiked medium for 5–7 days. The parents were removed and the larvae heterozygous for w and w^+ developed, chronically exposed on the contaminated medium. Progeny flies were transferred to fresh, uncontaminated medium vials and their eyes were evaluated for spots according to protocols described by Butterworth et al. (1995b) and Pandey et al. (1995).

The effects of three anthropogenic compounds on the rates of recombination were evaluated singly and in mixtures, two and three at a time, using five sequentially increasing logarithmic concentrations on two genetic strains OK and BK (Tables 2–3). The concentrations of DCB and BAP ranged from 1.0 mM or 1 part per million (ppm) to 0.0001 mM or 100 parts per trillion (ppt). The concentration for SA ranged from 0.1 mM or 100 ppb to 10^{-5} mM or 10 ppt, except when the lowest response point was sought. Here the dilution was extended to 10^{-7} mM or 0.1 ppt for SA and in the case of binary mixtures with BAP (Table 2) and in ternary mixtures with BAP and DCB (Table 3). In all cases the mixed chemicals were in equal volumes. The number of eyes analyzed averaged around 200 at each concentration tested for the experimental animals and over 900 eyes for the animals tested on solvent alone. The recombination rate is given as an average number of spots observed per hundred eyes. First, the controls where the two, wild-type strains were fed medium mixed with solvent by itself give a recombination rate of 7% for the Oregon K strain and 7.1% for the Berlin K strain. These control values are an indication of the spontaneous recombination rates and are very similar to controls measured at other times in this laboratory.

2.2. Statistical Analysis

Data from each experiment were combined to yield a single mean, percent spot total. The percent recombination was calculated as a ratio of the total number

TABLE 1. Results of 4,4′-DCB, BAP, and SA administered alone to Oregon K and Berlin K strains.

Concentration; Chemical[a]	Oregon K		Berlin K	
	v[b]	%[c]	v	%
10^{0} mM DCB	4	21	3	9
10^{-1} mM DCB	5	19	3	8
10^{-2} mM DCB	5	17	5	7
10^{-3} mM DCB	3	12	4	7
10^{-4} mM DCB	3	11	4	8
10^{0} mM BAP	3	24	2	10
10^{-1} mM BAP	3	23	2	7
10^{-2} mM BAP	3	17	3	7
10^{-3} mM BAP	2	16	3	6
10^{-4} mM BAP	3	13	3	8
10^{-1} mM SA	4	33	2	27
10^{-2} mM SA	4	32	2	20
10^{-3} mM SA	3	25	3	18
10^{-4} mM SA	2	24	2	19
10^{-5} mM SA	2	25	2	18
10^{-6} mM SA		ND[d]	2	17
10^{-7} mM SA		ND	2	12
Control	9	7	8	7

[a] DCB = 4,4′dichlorobiphenyl; BAP = benzo[a]pyrene; SA = sodium arsenite.
[b] v = Number of vials counted.
[c] % = Number of spots per 100 eyes.
[d] ND = No data.

of spots counted over the total number of eyes observed (Tables 1–3). For each concentration studied an average of four vials of animals were counted. Each chemical was tested using at least 5 concentrations that were then transformed into common logarithms. Multiple regressions were performed on the multiple-vial data which also were transformed to common logarithms (Figures 1–3). The regression equations were compared using t tests. Regression equations were used to generate an expected percent of eye spots assuming that the concentration of the "pure" chemical would be increased 10 fold, and that the concentration of the combined mixtures would also be 10 fold. If the mixture influenced the frequency to the same extent as the "pure" chemical, then the ratio of mix/pure would be 1.0. If it were significantly greater than 1.0, it was called synergistic; if significantly less than 1.0, antagonistic.

3. RESULTS

The recombination rate increased significantly over the control rate, at every concentration, in each strain and each compound studied, with two major

TABLE 2. Results of binary mixtures administered to Oregon K and Berlin K strains.

Concentration; Chemical[a]	Oregon K v[b]	Oregon K %[c]	Berlin K v	Berlin K %
10^{0} mM DCB + 10^{0} mM BAP	3	35	2	7
10^{-1} mM DCB + 10^{-1} mM BAP	5	28	2	7
10^{-2} mM DCB + 10^{-2} mM BAP	5	26	2	8
10^{-3} mM DCB + 10^{-3} mM BAP	5	18	2	7
10^{-4} mM DCB + 10^{-4} mM BAP	4	15	3	8
10^{0} mM DCB + 10^{-1} mM SA	3	26	2	26
10^{-1} mM DCB + 10^{-2} mM SA	3	34	2	20
10^{-2} mM DCB + 10^{-3} mM SA	3	30	2	18
10^{-3} mM DCB + 10^{-4} mM SA	3	35	2	15
10^{-4} mM DCB + 10^{-5} mM SA	3	28	1	14
10^{0} mM BAP + 10^{-1} mM SA	2	29	2	28
10^{-1} mM BAP + 10^{-2} mM SA	3	48	2	22
10^{-2} mM BAP + 10^{-3} mM SA	2	29	2	20
10^{-3} mM BAP + 10^{-4} mM SA	2	27	2	15
10^{-4} mM BAP + 10^{-5} mM SA	3	26	2	15
10^{-5} mM BAP + 10^{-6} mM SA		ND[d]	2	15
10^{-6} mM BAP + 10^{-7} mM SA		ND	2	12
Control	9	7	8	7

[a] DCB = 4,4′-dichlorobiphenyl; BAP = benzo[a]pyrene; SA = sodium arsenite.
[b] v = Number of vials counted.
[c] % = Number of spots per 100 eyes.
[d] ND = No data.

TABLE 3. Results of ternary mixtures administered to Oregon K and Berlin K strains.

Concentration; Chemical[a]	Oregon K v[b]	Oregon K %[c]	Berlin K v	Berlin K %
10^{0} mM DCB, BAP, and 10^{-1} mM SA	3	37	2	27
10^{-1} mM DCB, BAP, and 10^{-2} mM SA	2	34	3	24
10^{-2} mM DCB, BAP, and 10^{-3} mM SA	3	33	2	20
10^{-3} mM DCB, BAP, and 10^{-4} mM SA	2	25	2	17
10^{-4} mM DCB, BAP, and 10^{-5} mM SA	3	26	2	15
10^{-5} mM DCB, BAP, and 10^{-6} mM SA	3	22	2	14
10^{-6} mM DCB, BAP, and 10^{-7} mM SA	3	18	3	12
Control	9	7	8	7

[a] DCB = 4,4′-dichlorobiphenyl; BAP = benzo[a]pyrene; SA = sodium arsenite.
[b] v = Number of vials counted.
[c] % = Number of spots per 100 eyes.

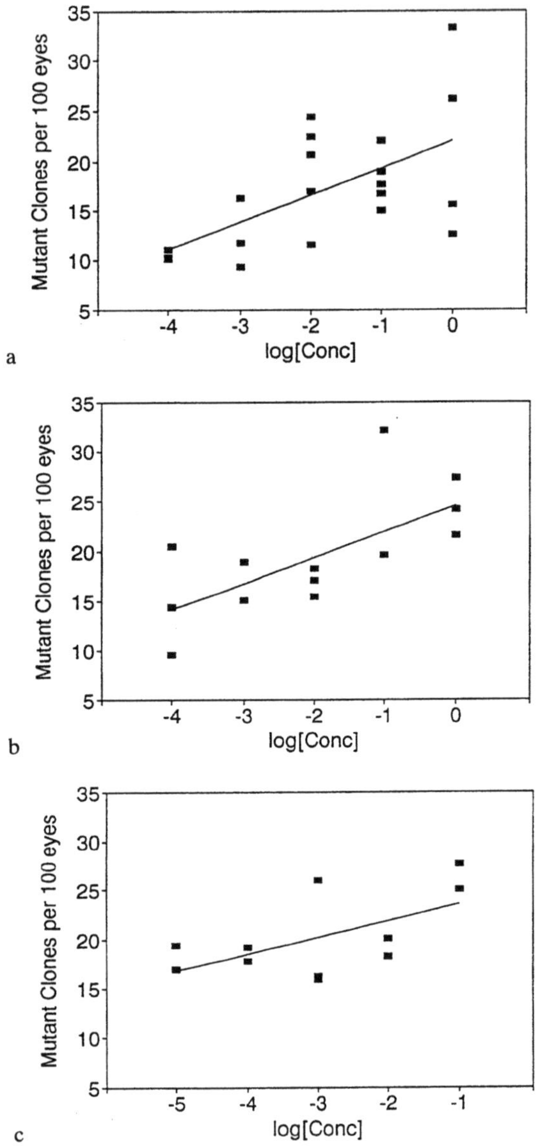

FIGURE 1a–c. Scatter plots of multiple vials with recombination rates as a percent of eye spots (mutant clones/100 eyes) regressed against log concentrations of test compound: A. dichlorobiphenyl (DCB) tested on the Oregon K strain, B. benzo[a]pyrene (BAP) tested on the Oregon K (OK) strain, and C. sodium arsenite (SA) on the Berlin K (BK) strain. Note that each compound displays a significant dose response over five log concentrations.

exceptions (Table 1, Figure 1). In the Oregon K (OK) strain both DCB and BAP in Figure 1a and b, respectively, exhibited a significant dose-response relationship ($R_{DCB} = 0.60$, $P < 0.01$; $R_{BAP} = 0.69$, $P < 0.01$). Both the slopes (2.76 for DCB and 2.59 for BAP) and intercepts (22.10 and 24.55 for DCB and BAP, respectively) were significantly greater than zero ($P < 0.01$). However, for DCB and BAP tested on the Berlin K strain, the recombination rates approximate those of the controls and had slopes, 0.48 and 0.49, respectively, that were not greater than zero. Earlier experiments demonstrated (Butterworth et al., 1995b) that both compounds require

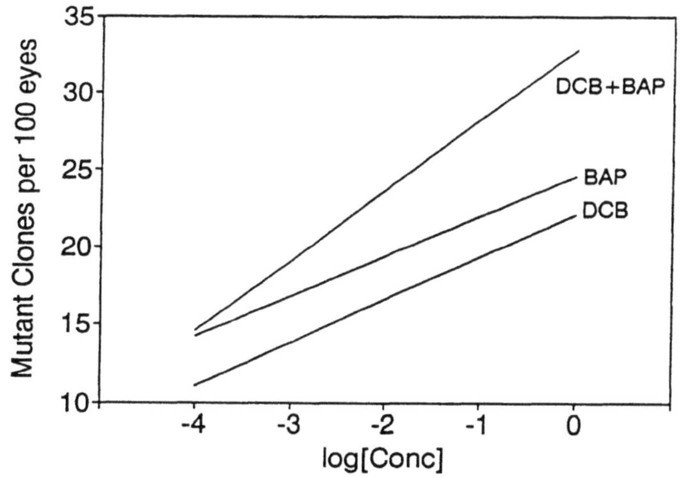

FIGURE 2. Regression curves showing the recombinogenic effects of increasing concentrations of selected mixtures of test compounds. Here DCB and BAP are compared to the mixture of the two compounds. Note that the mixture has a significantly greater effect than that of the compounds tested singly. The slopes of the compounds tested singly are the same, but the y intercepts (at **log[conc]0**) are not, indicating the OK strain is more sensitive than the BK strain for all three sets of experiments.

metabolic activation to become recombinogenic in the BK strain because only preoxidized compounds were recombinogenic when fed to BK animals. Thus, the lack of a dose-response relationship was to be expected, validating the earlier results. The OK strain is constitutive for CYP450 enzymes not requiring induction by feeding compounds such as phenobarbital or the PCB mixture Aroclor 1254 (Hällstrom et al., 1984; Zijlstra et al., 1984), and the enzymes in the BK strain are not induced by consuming certain congeners of PCBs (Butterworth et al., 1995b). The control medium containing solvent alone (as well as medium without additions) had no effect on recombination rates (Table 1).

Sodium arsenite (SA), on the other hand, increased recombination in both strains Figure 1c, Table 1, with slopes of 1.69 and 2.95, respectively, for BK and OK (Figures 3a and b, Table 1). Both slopes were significantly greater than zero ($P < 0.01$ for OK and $P = 0.057$ for BK). However, the effect appeared to be greater in OK (compare SA regression lines in Figures 3a and b). All three compounds produced effects at very low concentrations but most dramatically with SA as indicated in the BK strain (Table 1). SA significantly increased recombination rates at concentrations as low as 1×10^{-7} mM or 0.1 ppt in the BK strain. The other two compounds became effectively detectable over the control values at 10^{-4} mM, or 0.1 ppb in both strains. Thus, the concentrations chosen lay within the effective range of the three compounds tested for both strains.

When the test animals were exposed to the four possible mixtures (DCB + BAP; DCB + SA; BAP + SA; and DCB + BAP + SA) the results were unanticipated (Figures 2 and 3; Tables 2 and 3). For example, in Figure 2 the regressions for DCB and BAP were redrawn from Figure 1 along with that of the mixture, DCB + BAP.

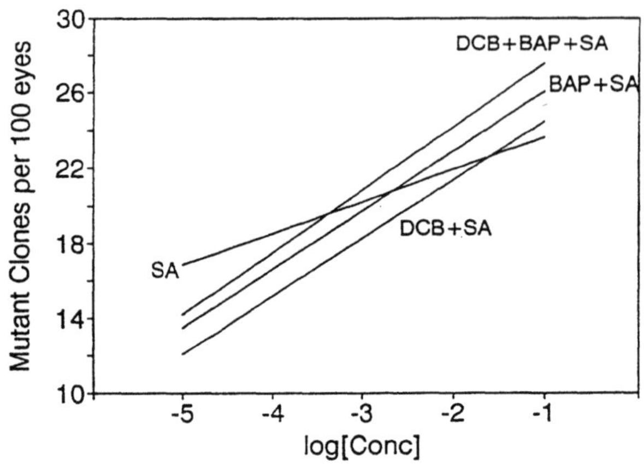

a

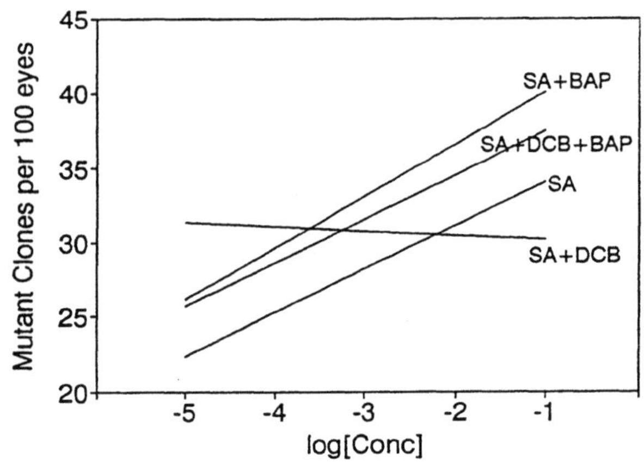

b

FIGURE 3a–b. Regression curves of arsenic and mixtures of DCB and BAP tested in two strains illustrate the differences and unpredictability. The y intercepts are read at **log[conc]-1**, the concentration of SA. However, recall that the concentrations of DCB and BAP are 10-fold higher than SA at each point. See text for details. A. Berlin K and B. Oregon K.

Comparing the mixture with DCB the slopes were not significantly greater ($t = 1.55$, $P < 0.1$) but the difference in the y intercepts was highly significant ($t = 1.74$, $P < 0.001$). Comparing the mixture with BAP the slopes and y intercepts were both significantly and highly significantly greater ($t = 52.7$, $P < 0.05$; and $t = 23.57$, $P < 0.001$, respectively).

In Figure 3a, when the BK strain was treated with the DCB + SA mixture, only the SA concentration entered into the regression equation ($R = 0.93$, $P < 0.0003$; i.e. of the other independent variables in the equation, SA accounted for the greatest proportion of the variance). The slope of the mixture (3.09) however, increased significantly compared with either DCB (slope = 0.48, $t = 4.05$, $P < 0.01$) or SA (slope = 1.69, $t = 2.29$, $P < 0.05$). Interestingly, the intercept of the DCB + SA mixture also increased, and differed significantly from that of the controls ($t = 11.25$, $P < 0.01$). Similarly when both BAP and SA were added as a mixture, only the concentration of arsenic entered into the equation at the 0.05 level of significance, but again, the slopes, and intercepts increased significantly compared to when either chemical was tested singly. BAP had a slope of 0.49 when tested singly, which did not differ significantly from zero, but did differ from the slope 3.15 obtained when BAP was combined with SA ($t = 4.22$, $P < 0.01$). Similarly, when tested singly, SA had a slope of 1.69, and this differed significantly from a slope of 3.15 when SA and BAP were combined ($t = 2.38$, $P < 0.05$). Thus, the effects of mixtures appeared to be considerably greater than the effect of chemicals tested singly. Furthermore, the effects were not additive, but are clearly synergistic (see Discussion). Not surprisingly the greatest slope, and highest intercept were obtained when all three chemicals were combined, but again, only the concentration of SA entered into the regression equation. In this case, the slope did not differ significantly from that obtained when only DCB and SA or BAP and SA were in the mixture.

One final aspect of our study on the BK strain bears mentioning. Every treatment that included SA had an intercept that was significantly elevated over the levels observed for controls (t values ranged from 6.9 to 11.7 and were highly significant: $P < 0.01$).

The OK strain appeared to be much more sensitive than the BK strain to both DCB and BAP. As mentioned above, significant dose-response relationships were obtained for both chemicals. Unlike what occurred in the BK strain the mixture of DCB and BAP significantly influenced the response ($R = 0.80$, $P < 0.01$), however only the BAP concentration entered into the regression equation. For the mixture the slope of 4.56 was significantly greater than when DCB (slope = 2.76, $t = 3.93$, $P < 0.01$) or BAP (slope = 2.59, $t = 5.66$, $P < 0.01$) were added singly. SA also exhibited a significant dose-response relationship with the percent of eyes bearing spots ($R = 0.66$, $P < 0.01$). Surprisingly, the combination of SA and DCB did not increase the response, indeed the dose response relationship degenerated ($R = 0.06$), and was not significant (Figure 3b). Similarly, the combination of BAP and SA exhibited a weaker dose-response relationship than either chemicals added singly ($R = 0.55$, $P < 0.065$, in combination versus $R = 0.69$ for BAP and $R = 0.66$ for SA singly). When all three chemicals are added in combination, a significant relationship was found ($R = 0.75$, $P < 0.01$), though only SA entered into the regression equation. While significant dose-response relationships were not always found, the intercepts for all seven experiments involving the OK strain were elevated compared to the controls (t values range from 3.93 to 11.24 and are highly significant).

Similarly, with the exceptions noted above, the combinations of two chemicals always produced steeper slopes than did the chemicals added singly (t values range from 15.84 to 187.4 and were highly significant. The three-way combination had a steeper slope than either DCB in combination with BAP, or DCB in combination with SA ($t = 218.82$, and 2.43, respectively) but not greater than SA and BAP in combination.

The test system using the inversion chromosome (which suppresses recombination) was used to ascertain the extent to which recombination caused increases in wing spot formation. As illustrated in Table 4 the inversion suppressed recombination for the three compounds tested: 71, 62, and 64% for DCB, BAP, and SA, respectively; indicating that recombination is the major genotoxic event caused by these compounds.

The metabolite of DCB, 4,4′-dichloro-3-biphenylol, that can be produced by the mammalian CYP450 system, was also tested in both strains. In OK strain the

TABLE 4. Results of DCB, BAP, and SA administered to the inversion strains.

Concentration; Chemical[a]	Inversion v^b	Normal Sequence %
10^0 mM 4,4′-DCB	4	14
10^0 mM BAP	11	29
10^{-1} mM SA	8	22

[a] DCB = 4,4′-dichlorobiphenyl; BAP = benzo[a]pyrene; SA = sodium arsenite; DCB and BAP data from Butterworth et al. (1995).
[b] % = Number of spots per 100 eyes.

metabolite produced 43% less recombinogenic activity than the parent compound (data not shown). This decrease suggests that if the OK strain produces 4,4′-dichloro-3-biphenylol, it is not the active, recombinogenic compound. The metabolite has no recombinogenic activity in the BK strain supporting the contention that the metabolite may not be produced by the OK strain, or it is not the active form.

4. DISCUSSION

4.1. Complex Mixtures

The standard assessment of the toxicity status of an environmental region (e.g. a river sediment) is to assay for and construct a list of known anthropogenic, polluting chemicals present whose toxicity has been determined earlier on a chemical-by-chemical basis. It has been assumed that the toxicity of these complex mixtures of chemicals is the sum of each of the individual chemicals present, when in fact the net toxicity is difficult, if not impossible, to predict. As a result, efforts have been made to understand the issue of complex mixtures and to test for their influence using biological monitoring systems (National Research Council, 1988; DeMarini, 1991; Butterworth et al., 1995a). The National Research Council realizing that complete characterization of every mixture is "not only impossible, but imprudent", suggests bioassay-directed fractionation, and strategies such as matrix testing and mathematical modeling to aid in prediction. But perhaps a totally new paradigm may have to be employed.

DeMarini et al. (1994) using bioassay-directed fractionation indicates, that although complex mixtures contain many components, only a small fraction are genotoxically important. In their studies of urban air, a base/neutral fraction that contained only 36% of the mass, accounted for 81% of the mutagenic activity in the modified Ames Salmonella assay. Similar results were obtained for diesel exhaust (Schuetzle and Lewtas, 1986), cigarette smoke condensates (Kier et al., 1974), and chlorinated drinking water (Meier et al., 1987; Kronberg and Vartiainan, 1988).

However, the conclusion that mixtures contain only a few toxic ingredients does not address the issue whether or not the behavior of these active ingredients changes in different mixtures, and that these effects differ in closely related strains. Our results show that such changes occur. For example, in the naturally-occurring BK strain the toxic effect of SA increased significantly in the presence of DCB or BAP even though neither of these chemicals had a significant effect by themselves. Results with the closely related, wild-type, OK strain were completely different. Thus, knowing the toxicity relationship of one strain did not enable us to accurately predict the toxicity results of the other strain, suggesting that bioassay-directed fractionation and single strain monitors are inadequate.

4.2. Chemical Interactions

Chemical interaction has been categorized and defined (National Research Council, 1988; McKee and Scala, 1994). Generally, *additivity*, occurs when two toxic chemicals produce an effect in a target organ or endpoint that does not exceed the sum of the separate effects of the two chemicals. *Synergism* results when the effects are greater than what would be expected based on the simple addition of doses. *Antagonism* occurs when the effects are less than what would be expected from the simple addition of doses. *Potentiation* results when a nontoxic compound is shown to increase the toxicity of another compound. These definitions are a good start, but use will necessitate modification.

Ample evidence exists for all four types of interactions of anthropogenic chemicals, including PCBs, PCB metabolites, BAP, SA and estrogen mimics (Balmer et al., 1976; Liang, T.T. and E.P. Lichtenstein, 1974; Kulkarni, 1976; Kluwe et al., 1978; Hertz-Picciotto et al., 1992; Bergeron et al., 1994; Arnold et al., 1996). In considering complex mixtures of chemicals, one has to also address the complexity of the dose and exposure, and the complexity of the biological system itself including closely related strains (Odkvist et al., 1980; Abddallah, 1963; Reynolds and Moslen; 1977; Graf and van Shaik, 1992; and Ramos-Morales and Rodriquez-Arnaiz, 1995). Terms such as "co-mutagenic" and "anti-mutagenic" better describe types of potentiation (Fahrig, 1995).

In our experimental system the terms additivity and synergism as defined above are inadequate. For example, when the concentration of a compound is doubled, the effect in the SMART assay cannot be distinguished from the original concentration (unpublished data). Only when the concentration of that compound is increased ten fold, can an increase be demonstrated statistically. This phenomenon was the same for all three compounds evaluated in both test strains. If one assumes that the separate compounds affect the same target, then by adding a second compound at the same concentration (i.e. in effect doubling the toxicant concentration), one would not expect it to be distinguished by the test system. Thus, the system in its present design would not be expected to demonstrate additivity. However, when two different compounds actually were mixed, the effects were in certain instances significantly greater than expected (i.e. greater than if one of the compound's concentration was increased ten fold) and hence, for this reason, we have called these effects "synergistic". We also have observed complexity in the chemical interaction. In one case, SA + DCB in the OK strain, the effect was synergistic at the lowest concentration and antagonistic at the higher concentrations. In another case, DCB + BAP in the OK strain, the synergism rate increased throughout the test range.

4.3. Mechanisms of Action

The simplest mechanism of action of recombinogenesis for all three compounds involves their direct association with DNA either by increasing the rate of strand breaks and/or interfering with DNA-repair synthesis. The clearest example is BAP which, as a result of CYP450 metabolism, becomes +anti-benzo[a]pyrene 7,8-dihydrodiol 9,10-epoxide (BPDE) that binds three different DNA sequences (Fountain and Krugh, 1995; Mao et al., 1995; Ponten et al., 1994). These DNA-BPDE adducts inhibit DNA-repair synthesis (Wei et al., 1995), are the site of G-T transversions (Yang et al., 1992) and are potential sites of strand breakage.

The mechanism of action of PCBs is not well understood. PCBs are known to bind to the Ah receptor which binds specific DNA sequences involved in the transcription of CYP 1A1, but it is unlikely that this process will increase recombination rates. A more likely mechanism involves arene oxides (PCB metabolites of CYP450 enzyme metabolism) which in turn form DNA adducts (Safe, 1994) that can interfere with DNA-repair synthesis and potentially increase strand breaks.

Arsenic also has a wide range of endpoints but it does interact directly with DNA by causing DNA strand breaks (Dong and Luo, 1993), inducing unscheduled DNA synthesis (Dong and Luo, 1994), and also inhibiting DNA-repair synthesis (Jha et al., 1992). Thus, all three compounds or their metabolites can influence recombination rates directly through their interaction with DNA.

Another requirement for recombination is mitosis. Curiously, all three agents are also carcinogens and mitogens (Léonard and Lauwerys, 1980; Montieth et al., 1990; Silberhorn et al., 1990; Soto et al., 1992) creating the conditions for the process, recombination, that they induce. In this regard here is a potential relationship with recombination and the mitogenic effects of steroid hormones. It is intriguing to realize that of some PCBs and their metabolites tested (Bergeron et al., 1994) two metabolites are endocrine mimics, all of a group of PCBs and metabolites tested were recombinogenic (this paper; Pandey, 1993; Pandey et al., 1995) and that one PCB (3,3′4,4′-tetrachlorobiphenyl) is both an endocrine mimic (Monosson et al., 1994) and a recombinogen (Pandey et al., 1995). Further, Fahrig (1996) found that estrogen, testosterone, and the estrogenic diethyl stilbesterol are potentiators of recombinogenesis (or what he calls co-recombinogens). He provided (Fahrig, 1992, 1995, and this volume) an explanation for this complex action of recombinogenic, co-recombinogenic and anti-recombinogenic activity by a "channeling" model where specific compounds can be both anti-mutagenic and co-recombinogenic *or* co-mutagenic and anti-recombinogenic. Thus, mutation and recombination are correlated during mitosis but are separate, mutually exclusive events. Clearly some of our results can be explained by co-recombinogenicity and anti-recombinogenicity. In one case, where the dose response result of SA + DCB is counterintuitive (Figure 3b), there are relatively high rates (synergism) of recombination at low concentrations and depressed rates (antagonism/anti-recombinogenesis) at high concentrations of the toxicant.

The DNA-binding model of recombinogenesis is consistent with synergism, if one assumes that there are DNA-binding sites specific for each of the agents and that, for example, the BAP binding sites are adjacent to the DCB sites. As the concentration of BAP increases, the BAP sites would be expected to fill up with BPDE increasing the recombination rate, until all sites are filled, and plateaus. Similarly, recombination would increase as DCB sites become filled with arene oxides. Synergism would occur if some of the two sets of sites were adjacent causing greater-than-expected instability, resulting in recombination greater than the additive rate.

Potentiation (or co-recombinogenesis) appears to occur in the BK strain. DCB and BAP alone or together are ineffective in the BK strain; but when SA is mixed with either compound or the mixture, there is a rate increase. Perhaps potentiation occurs by SA inducing metabolic CYP genes. This hypothesis is consistent with the knowledge that experiments with SA alone (and the SA + BAP and SA + BAP + DCB mixtures) were more active in the OK strain. Alternatively, biological systems methylate arsenic in order to more easily excrete the toxicant (Hopenhayn-Rich et al., 1996). Perhaps, the BK strain methylates SA more extensively than can the OK strain.

Finally, heterozygosity loss (Guzmán-Rincón and Graf, 1995) explains the potentially devastating effect of somatic recombination. Most organisms are heterozygous for large numbers of mutant genes, many of which, if homozygous could produce disease states, such as defective immune systems or cancer. Recombination in these tissues during normal development will create clones of cells homozygous for these mutant genes. Recombinogens will increase the rate of these phenomena, and steroid hormones and hormone mimics acting as co-recombinogens, would increase the rate of recombinogenesis (and heterozygosity loss) still further. Thus there would be the first-time expression of existing mutations. For example, it is known that both sets of compounds, PCBs and estrogen mimics, affect the immune system (Davis and Safe, 1990). Perhaps these agents by causing replication of specific immune cell types would result in somatic recombinations in clones of these cells, in turn resulting in the expression of existing deleterious genes through the loss of heterozygosity, that consequently impact negatively on cell function. Similarly cells that lose heterozygosity for mutant p53 or other tumor suppresser genes could become cancerous (Cavenee et al., 1983; see also Hussain and Curtis, this volume).

4.4. Concluding Remarks

Clearly this technique of complex mixture analysis and the data presented here are only a first step. More repetitions using this and other biomonitoring systems are needed. However, one cannot help asking whether this approach could be a new way to view complex mixtures. By determining the slope-intercept toxicities of a dilution series of a particular sediment, fish extract or air sample, one can form an idea about the quality of the mixture by comparing its toxicity to "slope-intercept catalogs" of dilutions of known mixtures tested with multistrain monitors such as the one used above.

5. ACKNOWLEDGMENTS

The authors are grateful for financial support from Oakland University, grants to OU by the US Environmental Protection Agency, the Research Excellence Fund of the State of Michigan, and for the helpful insights of Drs. Rudolf Fahrig, Ulrich Graf, and Patricia Ramos-Morales who read the manuscript.

6. REFERENCES

Abddallah, M.D. 1963. Interaction of some organophosphorus compounds in susceptible, resistant houseflies (*Musca domestica* L.) Mededele Landbouwhogesch. Wageningen, 63: 1–97.

Arnold, S.F., D.M. Klotz, B.M. Collins, P.M. Vonier, L.J. Guillette, Jr., and J.A. McLachlan. 1996. Synergistic activation of estrogen receptor with combinations of environmental chemicals. Science, 272: 1489–1492.

Balmer, M.F., F.A. Smith, L.J. Leach, and C.Y. Yuile. 1976. Effects in the liver of methylene chloride inhaled alone and with ethyl alcohol. Amer. Indust. Hygi. Assn. J., 37: 345–352.

Beckman, D.A., and R.I. Brent. 1984. Toxic nature of polychlorinated biphenyls. Ann. Rev. Pharmacol. Toxicol., 24: 483–509.

Bergeron, J.M., D. Crews, and J.A. McLachlan. 1994. PCBs as Environmental Estrogens: Turtle Sex Determination as a Biomarker of Environmental Contamination. Environ. Health Perspect., 102: 780–781.

Butterworth, F.M. 1995. Introduction to biomonitors and biomarkers as indicators of environmental change. In: *Biomonitors and Biomarkers as Indicators of Environmental Change.* Eds: Butterworth, Corkum, and Guzmán, Volume 50, Environmental Science Research, Series Editor: H.S. Rosenkranz, Plenum, New York.

Butterworth, F.M., L.A. Corkum, and J. Guzmán-Rincón (editors) 1995a. *Biomonitors and Biomarkers as Indicators of Environmental Change: A Handbook.* Volume 50, Environmental Science Research, Series Editor: H.S. Rosenkranz, Plenum Press, NY.

Butterworth, F.M., P. Pandey, R. McGowen, S. Ali-Sadat, and S.K. Walia. 1995b. Genotoxicity of polychlorinated biphenyls (PCBs): recombinogenesis by biotransformation products. Mutation Res., 342: 61–69.

Cavenee, W.K., T.P. Drya, R.A. Philips, W.F. Benedict, R. Godbout, B.L. Gallie, A.L. Murphree, L.C. Strong, and R.L. White. 1983. Expression of recessive alleles by chromosomal mechanism in retinoblastoma. Nature, 305: 779–784.

Corkum, L.D., J.J.H. Ciborowski, and Z.E. Kovats. 1995. Aquatic insects as biomonitors of ecosystem health in the Great Lakes areas. In: *Biomonitors and Biomarkers as Indicators of Environmental Change: A Handbook.* Eds: Butterworth, F.M., L.A. Corkum, and J. Guzmán-Rincón, Volume 50, Environmental Science Research, Series Editor: H.S. Rosenkranz, Plenum Press, NY.

Davis, D., and S. Safe. 1990. Immunosuppressive activities of polychlorinated biphenyls in C57BL/6N mice: structure activity relationships as Ah receptor agonists and partial antagonists. Toxicol. 63: 97–111.

DeMarini, D.M. 1991. Environmental mutagens/complex mixtures. Pp 285–302, In: Li, A.P., and R.H. Heflich (eds.) *Genetic Toxicology*, CRC Press, Boca Raton.

DeMarini, D.M., M.L. Shelton, and D.A. Bell. 1994. Mutation spectra in Salmonella of complex mixtures: Comparison of urban air to benzo[a]pyrene. Environ. Molec. Mutagen., 24: 262–275.

Department of National Health and Welfare. 1992. *Toxicity Profiles for Great Lakes Chemicals.* Technical Document 1992, Great Lakes Health Effects Division, Ottawa.

Dong, J-T., and X-M. Luo. 1993. Arsenic-induced DNA-strand breaks associated with DNA-protein crosslinks in human fetal lung fibroblasts. Mutation Res., 302: 97–102.

Dong, J-T., and X-M. Luo. 1994. Effects of arsenic on DNA damage and repair in human fetal lung fibroblasts. Mutation Res., 315: 11–15.

Fahrig, R. 1992. Co-recombinogenic effects. Mutation Res., 284: 185–193.

Fahrig, R. 1995. Anti-mutagenic agents are also co-recombinogenic and co-mutagenic agents are also anti-recombinogenic. Mutation Res., 326: 245–246.

Fahrig, R. 1996. Anti-mutagenic agents are also co-recombinogenic and can be converted into co-mutagens. Mutation Res., 350: 59–67.

Fountain, M.A., and T.R. Krugh. 1995. Structural characterization of (+)-trans-anti-benzo[a]pyrene-DNA adduct using NMR, restrained minimization, and molecular dynamics. Biochemistry, 34: 3152–3161.

Graf, U., and van Schaik. 1992. Improved high bioactivation cross for the wing somatic mutation and recombination test in *Drosophila melanogaster*. Mutation Res., 271: 59–67.

Graf, U., F.E. Würgler, A.J. Katz, H. Frei, H. Juon, C.B. Hall, and P.G. Kale. 1984. Somatic mutation and recombination test in Drosophila melanogaster. Environ. Mutagen., 6: 153–188.

Guzmán-Rincón, J., and U. Graf. 1995. *Drosophila melanogaster* somatic mutation and recombination test as a biomonitor. In: *Biomonitors and Biomarkers as Indicators of Environmental Change: A Handbook*. Eds: Butterworth, F.M., L.A. Corkum, and J. Guzmán-Rincón. Volume 50, Environmental Science Research, Series Editor: H.S. Rosenkranz, Plenum Press, NY.

Hällstrom, I., A. Blanck, and S. Atuma. 1984. Genetic variation in cytochrome P-450 and xenobiotic metabolism in *Drosophila melanogaster*. Biochem. Pharmacol., 33: 13–20.

Hertz-Picciotto, I., A.H. Smith, D. Holtzman, M. Lipsett, and G. Alexeeff. 1992. Synergism between occupational arsenic exposure and smoking in the induction of lung cancer. Epidemiology, 3: 23–31.

Hopenhayn-Rich, C., M.L. Briggs, D.A. Kalman, L.E. Moore, and A.H. Smith. 1996. Arsenic methylation patterns before and after changing from high to lower concentrations of arsenic in drinking water. Environ. Health Perspect., 104: 1200–1207.

Hudson, L.A., and J.J.H. Ciborowksi. 1995. Larvae of Chironomidae (Diptera) as indicators of sediment toxicity and genotoxicity. In: *Biomonitors and Biomarkers as Indicators of Environmental Change: A Handbook*. Eds: Butterworth, F.M., L.A. Corkum, and J. Guzmán-Rincón, Volume 50, Environmental Science Research, Series Editor: H.S. Rosenkranz, Plenum Press, NY.

International Joint Commission. 1988. *Upper Great Lakes Connecting Channel Study*. Vol. I: 1–50; Vol. II: 1–626, Environment Canada (Ottawa) and US Environmental Protection Agency (Washington), and IJC (Windsor), Ontario.

Jha, A.N., M. Noditi, R. Nilsson, and A.T. Natarajan. 1992. Genotoxic effects of sodium arsenite on human cells. Mutation Res., 284: 215–221.

Kier, K.D., E. Yamasaki, and B.N. Ames. 1974. Detection of mutagenic activity in cigarette smoke condensates. Proc. Natl. Acad. Sci. USA., 71: 4159–4163.

Kluwe, W.M., K.M. McCormack, and J.B. Hook. 1978. Potentiation of hepatic and renal toxicity of various compounds by prior exposure to polybrominated biphenyls. Environ. Health Perspect., 23: 241–246.

Kronberg, L., and T. Vartiainan. 1988. Ames mutagenicity and concentration of the strong mutagen 3-chloro-4-(dichloromethyl)-5-hydroxy-2(5H) furanone and of its genometric isomer E-2-chloro-3-(dichloromethyl)-4-oxo-butanoic acid in chlorine-treated tap waters. Mutation Res., 206: 177–182.

Kulkarni, A.P. 1976. Joint action of insecticides against houseflies. J. Toxicol. Environ. Health, 1: 521–530.

Léonard, A., and R.R. Lauwerys. 1980. Carcinogenicity, teratogenicity, and mutagenicity of arsenic. Mutation Res., 75: 49–62.

Liang, T.T., and E.P. Lichtenstein. 1974. Synergism of insecticides by herbicides: Effect of environmental factors. Science, 186: 1128–1130.

Lindsley, D., and G.G. Zimm. 1992. *The Genome of Drosophila melanogaster*. Academic Press, San Diego.

Lovett Doust, J., M. Schmidt, and L. Lovett Doust. 1994. Biological assessment of aquatic pollution: a review, with emphasis on plants as biomonitors. Biol. Rev., 69: 147–186.

Mao, B., J. Xu, B. Li, et al. 1995. Synthesis and characterization of covalent adducts derived from the binding of benzo[a]pyrene diol epoxide to a -GGG- sequence in a deoxyoligonucleotide. Carcinogenesis 16, 2: 357–365.

McKee, R.H., and R.A. Scala. 1994. Interactions: the effects of chemicals on each other. Toxic Subst. J., 13: 71–84.

Meier, J.R., R.B. Knoll, W.E. Colman, H.P. Ringhand, J.W. Munch, W.H. Kaylor, R.P. Streicher, and F.C. Kopfler. 1987. Studies on the potent bacterial mutagen 3-chloro-4-(dichloromethyl)-5-hydroxy-2(5H)-furanone: aqueous stability, XAD recovery and analytical determination in drinking water and in chlorinated humic acid solutions. Mutation Res., 189: 363–373.

Monosson, E., W.J. Fleming, and C.V. Sullivan. 1994. Effects of the planar PCB 3,3′,4,4′-tetrachlorobiphenyl (TCB) on ovarian development, plasma levels of sex steroid hormones and vitellogenin, and progeny survival in the white perch (*Morone americana*). Aquat. Toxicol., 29: 1–19.

Monteith, D.K., D. Ding, Y.T. Chen, G. Michalopoulos, and S.C. Strom. 1990. Induction of cytochrome P450 RNA and benzo[a]pyrene metabolism in primary human hepatocyte cultures with benzanthracene. Toxicol. Appl. Pharmacol., 105: 460–471.

National Research Council. 1988. *Complex Mixtures: Methods for In Vivo Toxicity Testing*. Eds: Doull, J., E. Bingham, et al. National Academy Press, Washington.

Odkvist, L.M., B. Larsby, J.M.F. Frederickson, S.R.C. Liedgren, and R. Tham. 1980. Vestibular and oculomotor disturbances caused by industrial solvents. J. Otolaryng., 9: 53–59.

Pandey, P. 1993. Genotoxicity of polychlorinated biphenyls (PCBs) and nitrobiphenyl and their biotransformed metabolites in a eukaryotic genome. Thesis. Oakland University, Rochester, MI.

Pandey, P., R.M. McGowen, E.W. Vogel, and F.M. Butterworth. 1995. Genotoxicity of polychlorinated biphenyl (PCB) and polynuclear aromatic hydrocarbon (PAH) mixtures in the *white/white*$^{+}$ eye-mosaic assay. In: *Biomonitors and Biomarkers as Indicators of Environmental Change*. Eds: Butterworth, Corkum, and Guzmán, Volume 50, Environmental Science Research, Series Editor: H.S. Rosenkranz, Plenum, New York.

Ponten, I., S.K. Kim, A. Graslund, et al. 1994. Synthesis and characterization of a (+)-anti-benzo[a]pyrene 7,8-dihydrodiol 9,10-epoxide-oligo-nucleotide adduct. IACR Scientific Publications, 125: 433–436.

Ramos-Morales, P., and R. Rodriquez-Arnaiz. 1995. Genotoxicity of two arsenic compounds in germ cells and somatic cells of *Drosophila melanogaster*. Environ. Molec. Mutagen., 25: 288–299.

Reynolds, E.S., and M.T. Moslen. 1977. Damage to hepatic cellular membranes by chlorinated olefins with emphasis on synergism and antagonism. Environ. Health Perspect., 21: 137–147.

Rosenberg, D.M., and V.H. Resh. 1993. Introduction to freshwater biomonitoring and benthic macroinvertebrates. Pp. 1–9. In: *Freshwater Biomonitoring and Benthic Macroinvertebrates*, Eds: D.M. Rosenberg, and V.H. Resh, Chapman and Hall, New York.

Safe, S.H. 1994. Polychlorinated biphenyls (PCBs): environmental impact, biochemical and toxic responses, and implications for risk assessment. Crit. Rev. Toxicol., 24, 2: 87–149.

Schuetzle, D., and J. Lewtas. 1986. Bioassay-directed chemical analysis in environmental research. Anal. Chem. 58: 1060A–1075A.

Sgro, G.V., and J.R. Johansen. 1995. Rapid bioassessment of algal periphyton in freshwater streams. In: *Biomonitors and Biomarkers as Indicators of Environmental Change: A Handbook*. Eds: Butterworth, F.M., L.A. Corkum, and J. Guzmán-Rincón, Volume 50, Environmental Science Research, Series Editor: H.S. Rosenkranz, Plenum Press, NY.

Silberhorn, E.M., H.P. Glauert, and L.W. Robertson. 1990. Carcinogenicity of polyhalogenated biphenyls: PCBs and PBBs. CRC Critical Reviews in Toxicol., 20, 6: 440–492.

Soto, A.M., T-M. Lin, H. Justica, R.M. Silva, and C. Sonnenschein. 1992. An "in culture" bioassay to assess the estrogenicity of xenobiotics (ESCREEN). Ch. 17. In: *Chemically Induced Alterations in Sexual and Functional Development: the Wildlife-Human Connection*. Eds: Colborn, T., and C. Clements. Princeton Scientific Publishing, Princeton.

Tracy, M., D.C. Freeman, J.M. Emlen, J.H. Graham, and A.H. Hough. 1995. Developmental instability as a biomonitor of environmental stress: an illustration using aquatic plants and macroalgae. In: *Biomonitors and Biomarkers as Indicators of Environmental Change: A Handbook*. Eds: Butterworth, F.M., L.A. Corkum, and J. Guzmán-Rincón, Volume 50, Environmental Science Research, Series Editor: H.S. Rosenkranz, Plenum Press, NY.

Vogel, E.W. 1980. Genetical relationship between resistance to insecticides and procarcinogens in two *Drosophila* populations. Arch. Toxicol., 43: 201–211.

Vogel, E.W. 1989. Somatic cell mutagenesis in *Drosophila*: recovery of genetic damage in relation to the types of DNA lesions induced in mutationally unstable and stable X-chromosomes. Mutation Res., 211: 153–170.

Vogel, E.W., and M.J.M. Nivard. 1993. Performance of 181 chemicals in a *Drosophila* assay predominantly monitoring interchromosomal mitotic recombination. Mutagen., 8: 57–81.

Vogel, E.W., and J.A. Zijlstra. 1987. Mechanistic and methodological aspects of chemically-induced somatic mutation and recombination in *Drosophila melanogaster*. Mutation Res., 182: 243–264.

Wei, D., V.M. Maher, and J.J. McCormick. 1995. Site specific rates of excision repair of benzo[a]pyrene diol epoxide adducts in the hypoxantheine phosphoribosyltransferase gene of human fibroblasts: Correlation with mutation spectra. Proc. Nat. Acad. Sci., 92: 2204–2208.

Yang, J.L., M.F. Chen, C.W. Wu, and T.C. Lee. 1992. Posttreatment with sodium arsenite alters the mutational spectrum induced by ultraviolet light irradiation in Chinese hamster ovary cells. Environ. Molec. Mutagen., 20: 156–164.

Zijlstra, J.A., E.W. Vogel, and D.D. Breimer. 1984. Strain-differences and inducibility of microsomaloxidative enzymes in *Drosophila melanogaster* flies. Chem. Biol. Interact., 48: 317–338.

15

DROSOPHILA IS A RELIABLE BIOMONITOR OF WATER POLLUTION

PATRICIA RAMOS-MORALES, MA GUADALUPE ORDAZ, ADRIANA DORANTES, HUGO RIVAS, PATRICIA CAMPOS, MOISES MARTINEZ, AND BLANCA HERNANDEZ

Lab. Genetica, Fac. Ciencias, Circuito Ext., Ciudad Universitaria, UNAM, Mexico D.F. 04510, Mex.

1. INTRODUCTION

The use of experimental systems, sensitive to environmental levels of pollutants provide an important alternative to biomonitoring studies, in particular, those involving whole, in vivo systems from which the fraction of organisms showing a particular change in relation with some causal factor, can be identified. Whether in some of the organisms this response is showed earlier than in the rest of the population exposed, it could be an alert that should to be attended to avoid more organisms being affected.

There are some disadvantages in the use of standardized protocols to find environmental mutagens: The apparent low sensitivity of plants and animals used to detect environmental levels of pollutants; the scarce development of approaches to distinguish the earliest response of the most sensitive organisms from the exposed population; and, the uncertainty to identify causal factors when complex mixtures are implied.

Attending the first point, a common practice is to concentrate environmental samples to increase the probability of pollutant detection by the chosen assay. An alternative practice is to extract, using the appropriate solvents, those pollutants contained in air (Suter, 1989a,b; Zwanenburg, 1989; Iwado et al., 1991; Lewtas et al., 1993), soil (Steinkellner et al., 1998) and water samples (Tabor and Loper, 1985; Wilcox et al., 1988; Pandey et al., 1995; Cernß et al., 1996) in order to identify the hazardous components and comparing their genotoxicity with those from the components assayed alone and from experimental mixtures planned in laboratory.

Kluwer Academic/Plenum Publishers, New York, 2000

However, this practice is limited because the multiple combinations of the components could be too numerous to reproduce the complexity of the environmental sample. Beside this, reports showing the use of unconcentrated, environmental samples to determine the effect of pollutants, and the use of in vivo systems providing information with a biological meaning are increasing (Jaylet et al., and 1986, 1987; Van Hummelen et al., 1989; Gomez-Arroyo et al., 1988 and 1997; Anwar and Gabal, 1991; Gimmler-Luz et al., 1992; Chroust et al., 1997).

Biomonitoring of environmental pollution sources has a high priority, mainly in zones in which the topography or the altitude make the control of emissions difficult.

1.1. *Drosophila*

Drosophila melanogaster is one of the most used organisms in mutagenicity studies to determine the effect of genotoxins on germinal and somatic cells (Zimmering et al., 1985; Vogel and Szakmary, 1990), with several advantages compared to bacterial systems due to its eukaryotic organization; its moderate cost, short life cycle and because in each generation numerous progeny are produced, as compared to rodents. Different genetic endpoints can be explored through particular methodologies as the induction of sex-linked, recessive-lethal mutations, chromosome translocations, spindle machinery disturbances, and chromosome integrity in germinal cells; but also mutation, deletion, chromosome exchange, and non-disjunction in somatic cells. The preparation of material under test is a crucial point and the use of solvents and procedures to concentrate samples, or to obtain specific extracts to be assayed have been documented and reviewed by Vogel (1988). The compounds can be administrated by contact-feeding (probably the most common), but also inhalation and injection are used; the exposure duration can be acute (<8 h), semichronic, chronic (all the life cycle), and also combined protocols can be used: an acute exposure to some factor followed by an acute or semichronic exposure to a different factor (Kilbey et al., 1981; Graf et al., 1984; Woodruff et al., 1985).

Using different experimental approaches with *Drosophila*, numerous types of chemicals, complex mixtures and other environmental factors have been studied to determine their mutagenicity potential, but also, this organism has been used to analyze the antigenotoxic potential of some of them. The concordance among conclusions from *Drosophila* studies and those provided from bacterial systems, rodents, plants, and cell cultures support that *Drosophila* is a reliable in vivo system for mutagenicity studies (Vogel, 1988).

The somatic mutation and recombination test (SMART) of *Drosophila* is a short and sensitive assay which detects mosaic spots in the eyes or in the wings of the flies, produced as cellular clones during larval development. The genetic mechanisms implied in spots production are: point mutation, deletion, mitotic recombination and chromosomal non-disjunction. Beside this, it is possible to estimate the relative contribution of mitotic recombination to the production of somatic spots. In the SMART wing version, progeny of two types are recovered: inversion-free flies and chromosome-balancer, heterozygous flies, both can be scored, and because of the suppression of mitotic crossing over in inversion heterozygous, the comparison of spot frequencies in the progeny of the two types allows for the estimation of the relative contribution of mitotic recombination to the total genotoxic activity (Graf et al., 1992). In the SMART eye version, natural differences based on *Drosophila*'s sex chromosomes can be used with additional advantages: both sexes have the same genetic background, and the investment of time and cost of setting up experiments, are substantially reduced. Furthermore, the genetic endpoints that the w^+/w assay can detect have an extensive literature (Vogel and Nivard, 1993). Occasionally, there is a heterogeneous response to genotoxins in organisms of the same strain, that have been explained through sex, individual proneness to spot formation, genetic polymorphism, developmental or environmental background, variable food intake, etc. (Frei and Würgler, 1995).

In spite of the plentiful reports about the SMART advantages and the relative low cost and time investment required to perform this assay, there are not many reports about the sensitivity of *Drosophila* to detect environmental levels of contaminants. Graf and Singer (1989) used the in vivo system of *Drosophila* to determine the effects of extracts of airborne particulate matter from ventilation filters on the frequency of somatic mutation and mitotic recombination, although only slight effects were found, they were in concordance with parallel studies using Salmonella (Suter, 1989b) and V79 cells from Chinese hamster (Zwanenburg, 1989).

1.2. Objective

Drosophila and the Somatic Mutation and Recombination Test (SMART) were analyzed for their sensitivity to detect mutagenicity in well water from the mining State of Hidalgo, Mex.

2. METHODOLOGY

2.1. Zone of Study

Hidalgo, a state with a centuries-old mining tradition, is located in the Sierra Madre Oriental and Mesa Central East, at 20°35′N Latitude; 99°25′W Longitude, and 1,800 to 3,000 m of altitude. The varied geographical features of the state provide for constantly changing climates, ranging from dry to moderately moist, with an average annual temperature of 19.2 °C, and 500.5 mm of annual precipitation (García and Falcon, 1993). Several years ago, Hidalgo was the most important mining state silver producer. Recently, this activity is renewed for gold, silver, copper, lead, zinc, and manganese (this manganese production being the most important in Mexico). The few sources that furnish drinking water in the Hidalgo range zone, been mainly through water wells and warm springs. There are numerous reports about the presence of arsenic in wells, in particular those from Armienta et al., 1993; Armienta et al., 1997; Arriaga and Daniel, 1996; C.N.A., 1992; Ramos Leal, 1996; SARH, 1977 who tried to determine the source of arsenic in each of the major wells, through a characteristic geochemical signature that would indicate whether the arsenic contamination is the result of dissolution of arsenic minerals in the massive sulfide skarn deposits or if it is due to the leaching process from the waste piles. Both mechanisms could contribute to the arsenic loading of the water of this region: probably the deep wells are polluted by dissolving minerals from the mining district, while the arsenic in the shallow, polluted wells comes from the mining waste piles (Ramos Leal, 1996). Due the geological features of this region, there is a variety of heavy metals that appear along with the arsenic in the well water.

2.2. Water Sample Collection

Ten wells with particular characteristics were chosen, all of them localized in three near zones of the mining state of Hidalgo (Table 1) (INEGI, 1990, 1994, 1995): 8 are wells more than 100 m deep; one (Pozo 3, Tierra Colorada) is a water well only 15 m deep, and the Manantial Benito Juárez is a superficial spring. On the other hand, 6 wells are from the Zimapan Region; three from the Tecozautla Region, at South of Zimapan; and one is from Benito Juarez, at North of Zimapan. The Muhi well was closed in 1992 (Armienta et al., 1997) due to a high level of arsenic (more than 1 mg/L) but the Muhi's outward pipeline is actually used to distribute the water from San Pedro well.

For each well, two water samples were collected in 1 liter, polypropylene bottles; 1 to 1.5 ml of 1% nitric acid was added as fixative in one of them, this sample was used for chemical determinations, the other without fixative was used to perform biological testing. The samples were kept refrigerated and care was taken to check that samples were chlorine-free using a commercial kit (Chlorine Test, Provedor Biológico, Mex.). This point is important because in preliminary studies using the SMART test of *Drosophila* we observed that chlorine in water samples increased the frequency of spots per imaginal disc. All the treatments were performed in the next 10 days after the sampling.

TABLE 1. Wells sampled for genotoxic determination.

Well's Name	Municipality	Deep (m)
Limoncito	Tecozautla	>120
Sabinas	Tecozautla	>100
Tecozautla	Tecozautla	>200
Muhi*	Zimapán	—
Pozo 1	Zimapán	>200
Pozo 3	Zimapán	15
Pozo 5	Zimapán	>100
Pozo Viejo	Zimapán	>100
San Pedro	Zimapán	300
Manantial Benito Juárez	Benito Juárez	Superficial spring

* Muhi well (more than 1 mg/L of Arsenic) was closed, but the Muhi's outward pipeline is actually used to distribute the water from San Pedro Well.

2.3. Chemical Analysis

Nitric acid-fixed water samples were used to determine environmental levels of the most frequent heavy metals related to mining: As, Ni, Zn, Pb, and Cr. Heavy metal concentrations were determined using flame atomic absorption spectrophotometry.

2.4. Treatment

Third instar larvae from standard crosses for eye and wing versions of the SMART, were used for a chronic treatment. Larvae were withdrawn by floating from culture bottles using a 20% sucrose solution (Nöthiger, 1970). Groups of 150 to 200 larvae were put into glass tubes containing 1 g of Carolina instant food (Carolina Biological Supply, Co.) and 4.5 ml of test water or distilled water. Well water samples were tested in four ways: unconcentrated, diluted with distilled water, filtered with a membrane (0.45 μ), and filtered-diluted. A strip of filter paper was put inside each tube to provide a drier surface and improve pupation. Larvae were kept in the glass tubes until the adults emerged.

2.5. Positive Controls

In the wing version of the SMART assay, the alkylating promutagen N-nitrosodimethylamine (DMN) and the alkylating mutagen N-nitrosodiethylamine (DEN) were used as positive controls. Both compounds were chosen due to their proven potential to induce somatic mutation and recombination in *Drosophila*. The concentrations were those used by Graf et al. (1984).

2.6. Nitric Acid

To determine whether the sample used for heavy metal analysis can be used to biological testing, too; the effect of nitric acid used as a fixative was determined. Third instar larvae from two versions of the SMART were exposed 48 h to instant food containing: 0.25, 0.5 and 1.0% nitric acid. Distilled water was used as dissolvent and negative control. In addition, a tap water sample was assayed to determine the effect of chlorination.

2.7. Strains

2.7.1. Eye Version

For the parental cross, Standard $y\ w^{+}/y\ w^{+}$ virgin females, with red eyes were mated with standard $y^{+}\ w/Y$ males, with white eyes (the strains were kindly provided by Dr. E. Vogel, Department of Radiation Genetics, University of Leiden, The

Netherlands). The marker *y* (*yellow*) is not used for the test, however it helps to distinguish progeny and parental genotypes.

2.7.2. Wing Version

For the standard cross (SC) *flr*^{3} / *In(3LR)TM3, ri p*^{p} *sep l(3)89Aa bx*^{34e} *e Bd*^{S} virgin females (briefly, *flr*^{3}/*TM3, Ser* flies) were mated with *mwh*/*mwh* males. The genetic markers *flr*^{3} (*flare*^{3}) and *mwh* (*multiple wing hairs*) are autosomic, recessive and located in the left arm of the third chromosome, at 38.8 and 0.03 m.u., respectively. In addition, *flr*^{3} is lethal when homozygous but viable in the heterozygous state and is conserved in combination with the *TM3* chromosome balancer, which includes the dominant, lethal mutation, *Serrate* (*Ser*), as chromosome marker. In homozygous, *Ser* is a lethal condition, but in heterozygous, it is observed as an indentation in the wing border. The two markers used in the SMART modified the phenotype of trichomes. Cells expressing the *mwh* marker produce multiple trichomes (>2 hairs), in contrast to only one in the wild type. On the other hand, cells expressing the *flr*^{3} marker produce irregular trichomes, like a flame. For detailed information on genetic markers see Lindsley and Zimm (1992).

Stocks were maintained in a constant temperature room at 25 ± 1 °C and 60% humidity. Food medium for the stocks was prepared with sugar (30.3%), agar (6.5%), corn meal (45.5%), yeast (14.3%), Nipagin in 10% ethyl alcohol solution (1.7%) and propionic acid (1.7%).

2.8. Score Criteria and Statistical Analysis

2.8.1. Eye Version

Following the scoring criteria of Vogel and Nivard (1993), F_1 flies from the Standard cross were used to score for *w* spots in the eyes, with one modification: the parental females are homozygous for the wild-type gene for red eyes (w^+/w^+) and the parental males are hemizygous for the mutant gene (*w*/Y) for *white* eyes. Due to the X-linked pattern of the *white* marker, F_1 females (w^+/w) and F_1 males (w^+/Y) which normally have red eyes can be scored for mosaic spots, indicating genotoxic events. Mutant spots in heterozygous females can be caused by point mutation, deletion, non-disjunction, and mitotic recombination; in contrast, males are hemizygous for the X chromosome and hence, except for illegitimate sister chromatid exchanges not informative for this assay, can only express somatic mutation produced from non-recombinant events (Pandey et al., 1995). Scoring of the eyes for the presence of spots was carried out on a solution consisting of Tween 80: ethanol: water, in a 1:9:90 ratio, under a dissecting microscope at 120 X magnification, and an optical fiber (Fiberlite M. 170D, Dolan-Jenner). Two spots were considered independent mutational events, if separated by at least four, wild-type ommatidial rows.

Mosaic spots were grouped into size classes according to the number of colorless ommatidia, i.e., 1–2, 3–4, 5–8, 9–16, 17–32, >32 ommatidia, respectively (the smaller spots often appear darker than the surrounding ommatidia). The comparison among frequencies of total spots in experimental and control series was done using the X^2 test for proportions (Frei and Würgler, 1995). To obtain an indirect estimation of genotoxic effectiveness in eye disc cells, the frequency of clones per 10^4 cells was estimated according to the formula $f = 2nm/NC$, where m is the mean clone size, n is the number of mosaic spots, 2 is a correction factor, C is the number of ommatidia per eye (»800) and N is the number of analyzed eyes.

2.8.2. Wing Version

During metamorphosis, each cell from the wing imaginal disc that will form the wing blade, produce a cuticular structure: one hair or trichome. The trichomes form two monocellular layers; and hence, spots located in dorsal and in ventral layers are independent events (Demerec, 1965). Spots in heterozygous progeny can be originated by point mutation, deletion, non-disjunction, and mitotic recombination.

Slides were prepared by putting a pair of wings of ten females and ten males using Faurè Solution, dried, covered with a slide and sealed with nail polish.

The spots were classified according to the phenotype of trichomes forming the spots in: singles, which can be *mwh*, or flr^3; and in twin spots, showing adjacent *mwh* and flr^3 areas. Single spots are produced either by point mutation, deletion, non-disjunction and mitotic recombination among *mwh* and flr^3 markers. Twin spots are produced exclusively by mitotic recombination between the proximal marker flr^3 and the centromere. The spots were also classified according to the number of trichomes in small (1–2 cells or trichomes) and large spots (>3 cells or trichomes) (Graf et al., 1984). The spots were grouped into size classes according to the number of trichomes, i.e. 1, 2, 3–4, 5–8, 9–16, 18–32, 33–64, 65–128, 129–256, 257–512 cells. As the size of the spots is a reversal function of larval age, then the smaller spots are produced at the end of larval development.

Only trichomes with 3 or more hairs were considered *mwh*, but trichomes with two hairs, that occasionally occur, were also counted when forming part of a spot. Two spots were considered independent mutational events, if separated by at least three, wild-type trichome rows (Graf et al., 1984).

For the analysis we used three different approaches. Commonly, in the wing version of the SMART statistical diagnosis can be made by contrasting two hypotheses: the null hypothesis (H_0: there are no differences among the frequency of spots in the experimental and control series) and an alternative hypothesis (H_1: the frequency of spots in experimental series is m times the frequency of spots in the control series) in which m magnitude is chosen according to the spontaneous occurrence of a particular spot, and means the minimal increase in the spot frequency to be considered as a positive diagnosis. So, $m = 2$ for small and total spots, and $m = 5$ for large and twin spots, which are rather rare events (Frei and Würgler, 1988, Frei et al., 1995).

In a second approach, we performed a qualitative analysis for the time of spot production. In several experiments in which very low concentrations of different mutagens were used, a modification in the time of spot production (observed as an increase in the size of spots) was detected before a quantitative increase in the spot's frequency (data not shown). That means that although the frequency of spots per wing can be the same as that of control flies, the occurrence of larger spots suggests genotoxic activity. This can be confirmed when higher concentrations, or longer times of exposure are used. This analysis can help to discard false negative responses, because we assume that in a true negative response no disturbance in the pattern of distribution of the size of spots should be observed when compared with concurrent negative controls. Commonly, in control flies, the highest frequency of spots corresponds to small ones (1–2 cells), and larger spots are rather rare events.

A third approach was done comparing the number of spots per fly. Commonly, in flies unexposed to genotoxins, there are no spots on their wings, but occasionally one spot can be present on one of the wings in some flies; the occurrence of two or more spots in the same organism are exceptional events. As quoted before, in experiments in which very low concentrations of mutagens were used, a rearrangement in the number of spots per fly occurs, too. Before a significant increase in the frequency of spots from flies exposed to a mutagen can be detected, the spots recovered are provided by a small fraction of treated flies, which probably represents the most susceptible organisms; when higher concentrations or longer exposures are used, the fraction of flies without spots decreases, but the number of flies with spots and the number of spots per fly, increases. For this approach, multiple comparisons were performed using the Kruskal-Wallis Test for the over dispersion in the number of spots per fly in control and experimental series.

The Kruskal-Wallis test is a non parametric test that compares three or more groups. The first step is to rank all the observations from low to high, i.e. How many flies showed 0, 1, 2, 3, . . . , n spots on their wings in control and experimental series? For data that have ties, the average of the two ranks for which they lie is taken for both. The smallest number gets a rank of 1, and the largest number, a rank of N, where N is the total number of values in all the series. Then, the sum of the ranks for each series is obtained and compared. The differences among the rank sums are used to obtain the Kruskal-Wallis statistic. If a large discrepancy among rank sums is found, a large value of the Kruskal-Wallis statistic is expected. After a positive diagnosis, the Dunn's post test was calculated to determine the chance that two or more series would have different medians (Steel et al., 1997; Rao, 1998).

3. RESULTS

3.1. Positive Controls

Positive controls N-nitrosodimethylamine (DMN) and N-nitrosodiethylamine (DEN) were assayed in the wing-spot test. Table 5 shows that at all concentrations tested, the alkylating promutagen DMN induces all types of spots: small and large spots, which can be singles, *mwh* or *flr*^{3}; or twin, *mwh-flr*^{3} ($P < 0.05$). In contrast, the alkylating mutagen DEN only induces large simple spots and twin spots in wing cells from 5 mM, although the sum of all types of spots increased in a significant manner, too. For two higher concentrations this compound exhibits a positive induction of all type of spots. In summary, DMN was more genotoxic than DEN (Figure 1). A qualitative comparison of the number of cells (trichomes) forming each spot recovered from both compounds is shown in Figure 2. The size of spots recovered in flies exposed to DMN or DEN were similar, but the frequency of total spots was one order of magnitude higher for DMN-treated flies. Pattern distribution of the number of spots per fly was different among treatments (Figure 3, Tables 6 and 7). For flies exposed to DMN, the number of spots per fly was concentration dependent ($P < 0.05$). Kruskal Wallis test indicates that dispersion induced from 12.5 to 50.0 mM were similar, although in the highest concentrations, flies showed more that 20 spots per fly. For flies exposed to DEN, the dispersion in the number of spots per fly was significant from 10 mM ($P < 0.05$), however in the highest concentration (20 mM) the range of spots (0 to 6) per fly was the maximum dispersion found.

3.2. Nitric Acid

The Figure 4 shows the corrected frequencies of spots per eye or wing obtained from larvae used in both versions of SMART. As it was expected, the treatment with nitric acid was toxic to *Drosophila*. The 1.0% nitric acid solution used as a fixative, and the dilutions assayed had a negative effect in the number of adult flies recovered. A genotoxic activity was detected for all concentrations in the wing-spot test and only for the two highest concentrations in the eye-spot test. However, the trends of curves from wing and eye imaginal disks, were similar. The tap water sample was checked to confirm the presence of chlorine in a qualitative manner, with a

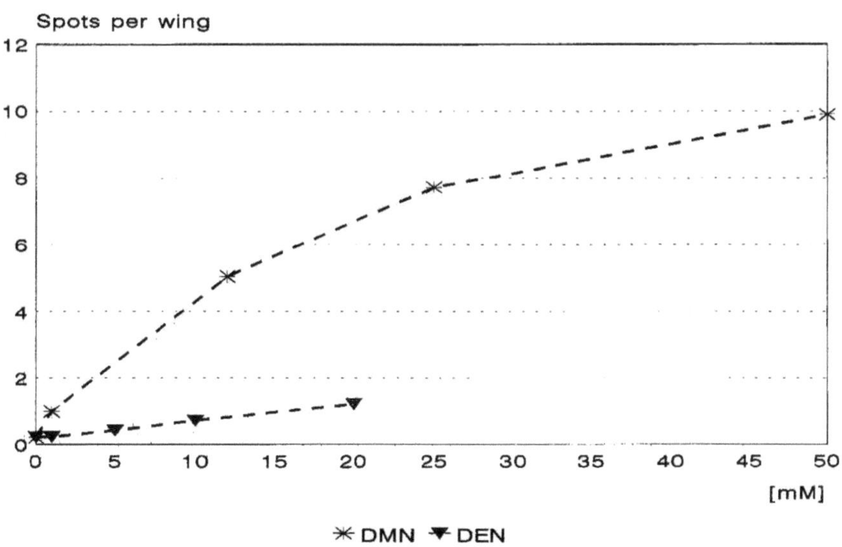

FIGURE 1. Frequency of total spots per wing induced by DMN and DEN. DMN induces small, large and twin spots at all concentrations assayed, but DEN induces large and twin spots from 5 mM, and all type of spots from 10 mM. DMN was an order of magnitude more genotoxic than DEN.

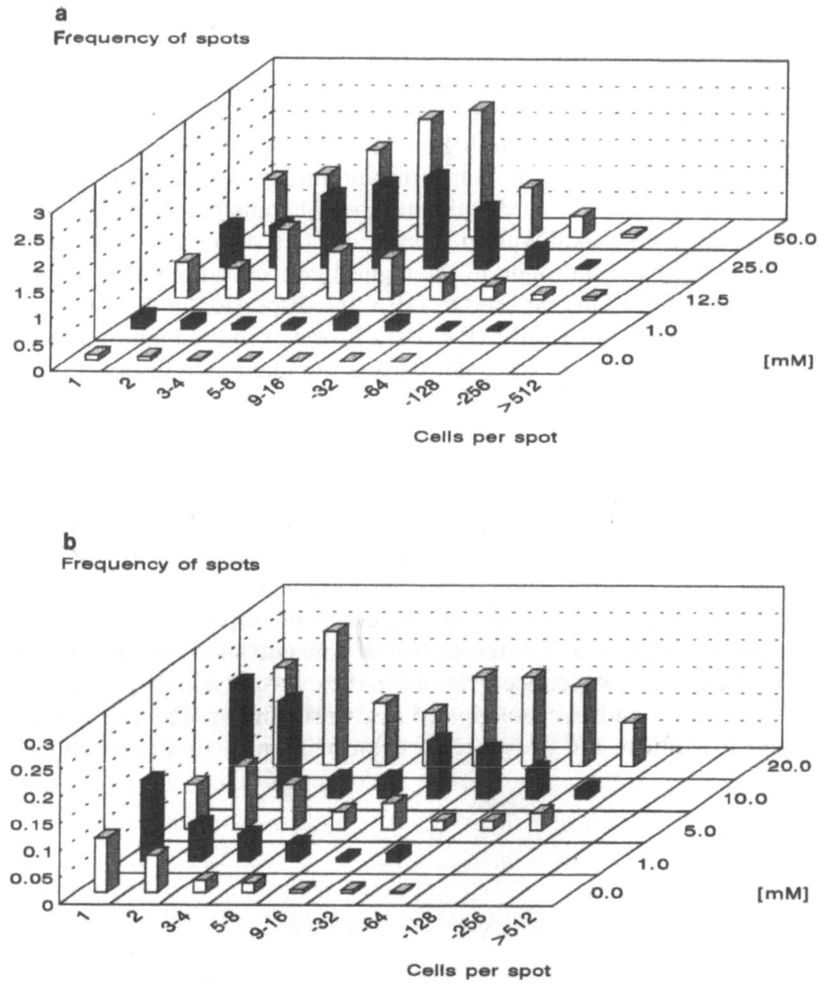

FIGURE 2. Frequency of cells per wing spot induced by a) DMN, and b) DEN. The spots were ranked by size (number of trichomes or cells forming each spot) which were similar for the two positive controls.

commercial kit. The frequency of spots in the wing-version flies was increased ($P < 0.05$) but no activity was detected in the eye-version flies. From these preliminary results, all samples assayed were first made nitric acid and chlorine free.

3.3. Physicochemical Parameters of Well Water Samples

Due to the geographical localization of the wells, all the samples were slightly alkaline (range between 7.22 to 8.50), with values of dissolved oxygen according to a range of 2 to 4 mg/L, expected for underground water (Meybeck et al., 1990). Only the Tecozautla well water was slightly turbid. Temperature varied between

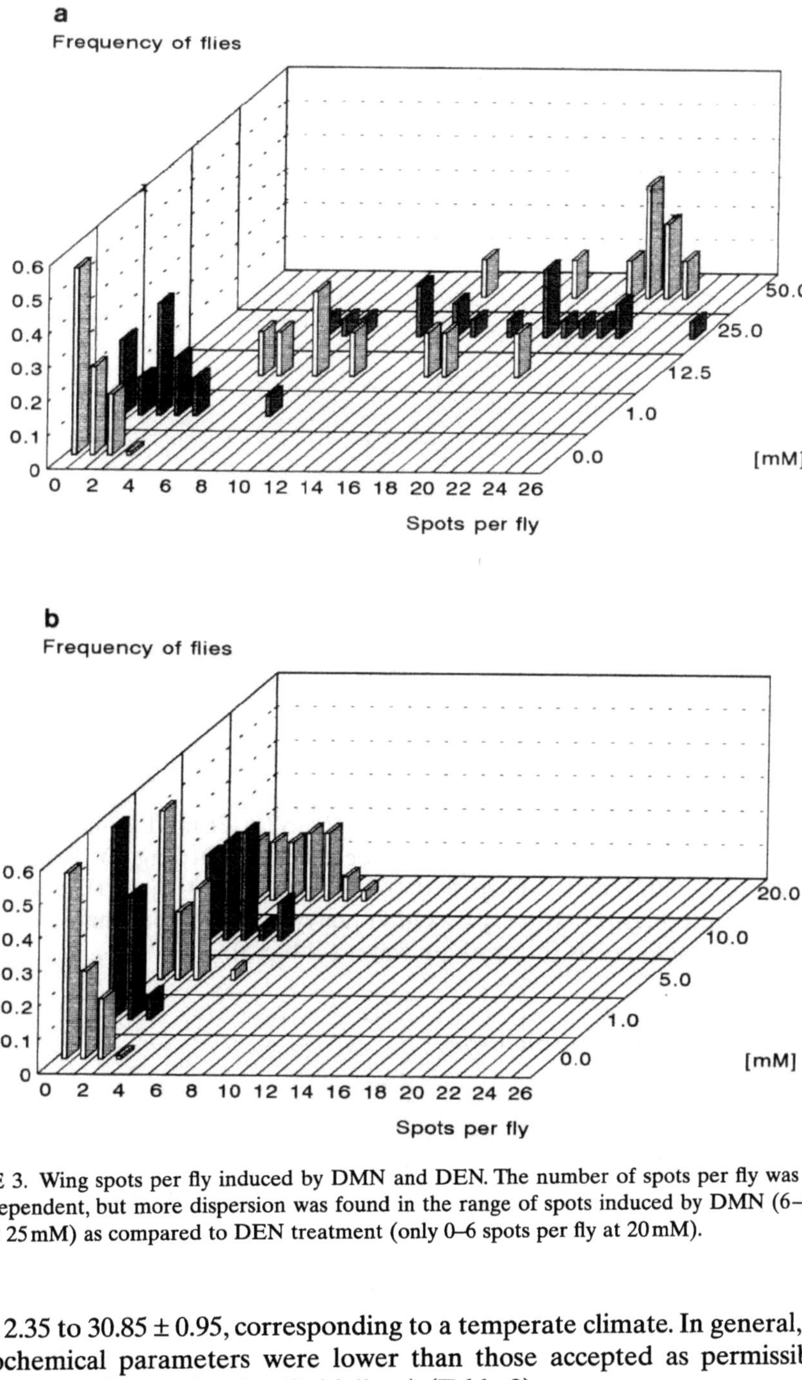

FIGURE 3. Wing spots per fly induced by DMN and DEN. The number of spots per fly was concentration dependent, but more dispersion was found in the range of spots induced by DMN (6–26 spots per fly at 25 mM) as compared to DEN treatment (only 0–6 spots per fly at 20 mM).

22.20 ± 2.35 to 30.85 ± 0.95, corresponding to a temperate climate. In general, all the physicochemical parameters were lower than those accepted as permissible for drinking water (Water Quality Guidelines) (Table 2).

The heavy metal levels in sampled wells were only above the permissible levels for the following metals and wells. Arsenic (0.05 mg/L) in: Tecozautla (0.118 mg/L),

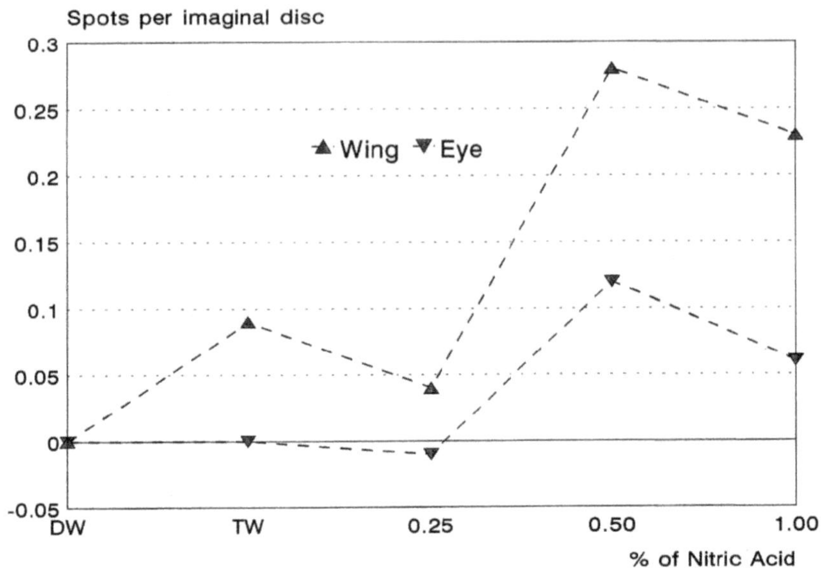

FIGURE 4. Corrected frequency of spots per eye or wing imaginal disc from water samples fixed with nitric acid. The 1% nitric acid solution reduced the survival of flies and the frequency of spots recovered in the two SMART versions. Genotoxic activity was detected in the wing test at all concentrations assayed, but only in the two higher in the eye test. The tap water (TW) increased the wing spots frequency. DW, distilled water.

Muhi (0.191 mg/L), Pozo 5 (0.129 mg/L), Pozo Viejo (0.138 mg/L); and for cadmium (0.005 mg/L) in: Limoncito (0.013 mg/L), Tecozautla (0.016 mg/L), Pozo 1 (0.019 mg/L), Pozo 3 (0.021 mg/L), Pozo 5 (0.021 mg/L), Pozo Viejo (0.025 mg/L), and Manantial Benito Juárez (0.019 mg/L). Other heavy metals determinated were lower than the permissible levels for drinking water (Table 3).

3.4. Genotoxicity Determinations

3.4.1. Eye Spot Version

Four well water samples were assayed in this version of the SMART (Table 4 and Figures 5 to 8). Data are shown separately by females and males.

Muhi. Both, females and males showed higher eye spot frequencies than those of respective controls ($P < 0.05$) (except for filtered-undiluted sample in males). Filtering reduced the activity of samples in females. Small spots (1–4 ommatidia) were mainly included in the distributions of the spots arranged by the number of ommatidia forming the spots from experimental males (except in the 50F treatment) (Figure 5c).

Pozo 1. All type of samples assayed failed to induce a significant increase in the spot frequency in both, females and males (Figure 6a). Size distribution of

TABLE 2. Physicochemical parameters of wells assayed.

Well's Name	Temp (°C) Mean ± s	pH	DO mg/L	C µS	TDS mg/L
Water quality criteria*			6.7–8.5		1,000
Limoncito	26.08 ± 1.85	8.40	3.80	612	303
Sabinas	30.85 ± 0.95	8.50	3.60	399	199
Tecozautla	30.25 ± 3.24	8.40	3.50	374	183
Muhi*	22.20 ± 2.35	8.21	2.75	399	209
Pozo 1	27.98 ± 0.75	8.40	2.05	488	257
Pozo 3	22.60 ± 0.93	7.49	2.65	623	313
Pozo 5	26.75 ± 1.06	7.22	2.20	513	254
Pozo Viejo	25.34 ± 0.96	7.35	2.10	486	238
San Pedro	23.78 ± 2.90	7.69	2.70	866	436
Manantial Benito Juárez	23.43 ± 1.06	7.37	2.20	475	239

* Gaceta Ecológica, 1990; DO, Dissolved Oxygen (mg/L); C, Conductance (µS); TDS, Total Dissolved Solids (mg/L).

recovered spots in males consisted mainly of small spots, except for flies exposed to the filtered-diluted sample (50F) (Figure 6b,c).

San Pedro. A significant effect was observed only in females exposed to unfiltered samples, being higher for the diluted sample [50] ($P < 0.05$); this treatment increased the frequency of spots in males, too ($P < 0.05$); however the size sample for this series was rather low (Table 4 and Figure 7a). The size of the spots from experimental females was smaller than in control flies (Figure 7b); this effect was stronger in male's distributions: in undiluted samples (100, 100F) only 1–2 ommatidia spots were recovered (Figure 7c).

Tecozautla. The samples assayed showed a significant activity in females and males ($P < 0.05$) (except for females from the 25 treatment), being higher in filtered

TABLE 3. Permissible levels of heavy metals (mg/L) in drinking water (Gaceta Ecológica, 1990, Meybeck et al., 1990) and heavy metals concentrations in wells assayed.

Well's Name	As	Cd	Cr	Cu	Ni	Pb	Zn
Permissible levels	0.050	0.005	0.050	1.00	0.100	0.5	5.00
Limoncito	0.034	**0.013**	0.003	0.008	0.014	0.07	0.013
Sabinas	0.017	0.005	0.009	0.014	0.042	0.11	0.018
Tecozautla	**0.118**	**0.016**	0.010	0.073	0.047	0.17	0.016
Muhi*	**0.191**	0.004	nd	0.107	0.012	nd	nd
Pozo 1	0.039	**0.019**	nd	nd	0.035	nd	nd
Pozo 3	0.009	**0.021**	0.007	0.013	0.079	0.14	0.017
Pozo 5	**0.129**	**0.021**	0.007	0.004	0.033	0.10	0.049
Pozo Viejo	**0.138**	**0.025**	0.014	0.017	0.056	0.18	0.039
San Pedro	0.031	0.004	nd	0.005	0.008	0.24	0.002
Manantial Benito Juárez	0.043	**0.019**	0.021	0.007	0.046	0.12	0.022

nd, not determinated; bold data, levels of heavy metals above the permissible level.

TABLE 4. Frequencies of mosaic clones in w^+/w females and w^+/y males treated with well water.

Sample Water (%)	No. of eyes Scored	No. of spots							Spots (%)			Corrected frequency	Clones Per 10^4 Cells
		Total	Size (ommatidia)						Total SPE	size (ommatidia)			
			1–2	3–4	5–8	9–16	17–32	>32		1–4	>4		
Females													
DW	596	68	45	11	5	1	5	1	0.11	82.24	17.64		13.12
Muhi													
50F	192	48	31	9	6		1	1	0.25+	83.33	16.17	0.14	30.75
100F	96	19	13	1	2	2	1		0.20+	73.68	26.32	0.09	22.41
50	108	40	27	6	4	2		1	0.37++	82.50	17.50	0.26	59.07
100	102	36	31	4		1			0.35+	97.22	2.78	0.24	19.59
Pozo 1													
25F	172	23	16	4	2		1		0.13–	86.96	13.04	0.02	11.63
50F	152	11	8	2			1		0.07–	90.90	9.09	–0.04	8.05
100F	182	29	24	2	1	2			0.16–	89.66	10.34	0.05	11.39
25	196	24	15	5	3	1			0.12–	83.33	16.67	0.01	9.70
50	160	18	13	2	1		2		0.11–	83.33	16.67	0.00	12.83
100	130	20	18	1	1				0.15–	95.00	5.00	0.04	8.85
San Pedro													
50F	90	12	10	1	1				0.13–	91.67	8.33	0.02	8.60
100F	166	21	16	1	3			1	0.13–	80.95	19.05	0.02	15.21
50	86	21	17	3	1				0.24+	95.24	4.76	0.13	14.53
100	110	24	20	3		1			0.22+	95.83	4.17	0.11	13.85
Tecozautla													
25F	184	32	19	7	6				0.17+	81.25	18.75	0.06	12.91
50F	356	82	63	12	3	2		2	0.23+	91.46	8.54	0.12	20.73
100F	258	72	53	13	3	2	1		0.28+	91.67	8.33	0.17	19.88
25	152	26	20	2	3	1			0.17–	84.62	15.38	0.06	11.85
50	284	64	46	12	3	3			0.23+	90.62	9.38	0.12	16.0
100	246	45	32	4	3	5	1		0.18+	80.00	20.00	0.07	17.29

Males													
DW	496	39	27	7	2	3			0.08	87.18	12.82		6.15
Muhi													
50F	144	30	24	4	2				0.21+	93.33	6.67	0.13	12.14
100F	66	9	7	2					0.14–	100.00		0.06	6.82
50	90	19	18	1					0.21+	100.00		0.13	10.82
100	74	11	9		2				0.15+	81.82	18.18	0.07	11.48
Pozo 1													
25F	168	14	12	2					0.08–	100.00		0.00	4.31
50F	168	19	15	2	1	1			0.11–	89.47	10.53	0.03	8.03
100F	198	16	15	1					0.08–	100.00		0.00	3.92
25	192	23	20	2				1	0.12–	95.65	4.35	0.04	11.08
50	218	16	13	3					0.07–	100.00		–0.01	4.02
100	162	18	15	3					0.11–	100.00		0.03	5.72
San Pedro													
50F	116	13	11	2					0.11–	100.00		0.03	6.47
100F	114	11	11						0.10–	100.00		0.02	4.39
50	78	29	13	12	3	1			0.37+	86.21	13.79	0.29	30.49
100	104	10	10						0.10–	100.00		0.02	4.33
Tecozautla													
25F	202	34	22	8	4				0.17+	88.24	11.76	0.09	11.61
50F	348	128	97	18	9	3	1		0.37+	89.84	10.16	0.29	25.10
100F	242	39	31	7	1				0.16+	97.44	2.56	0.08	9.19
25	146	28	19	3	3	2	1		0.19+	78.57	21.43	0.11	19.37
50	214	51	39	7	5				0.24+	90.20	9.80	0.16	14.84
100	232	36	28	3	5				0.16+	86.11	13.89	0.08	10.78

SPE, Spots per eye; DW, Distilled Water; Statistical analysis according to Frei and Würgler (1988). An indication at right of the frequency of spots per eye means the effect of treatment: ++, positive ($P < 0.01$); +, positive ($P < 0.05$); –, negative.

TABLE 5. Summary of type of spots obtained with samples water assayed.

% Sample Water	Type of spots								
	Number of wings	mwh		flr		twin		total	
		N	SPW	N	SPW	N	SPW	N	SPW
5% Sucrose									
	372	78	0.21	1	0.01	6	0.02	85	0.23
N-Nitrosodimethylamine (DMN)*									
1.0	44	33	0.75	5	0.11	6	0.14	44	1.00[a–d]
12.5	19	47	2.47	41	2.16	8	0.42	96	5.05[a–d]
25.0	42	221	5.26	82	1.95	21	0.50	324	7.71[a–d]
50.0	17	114	6.71	36	2.12	18	1.06	168	9.88[a–d]
N-Nitrosodiethylamine (DEN)*									
1.0	60	12	0.20	0	0.00	3	0.05	15	0.25
5.0	60	23	0.38	0	0.00	3	0.05	26	0.43[b,d]
10.0	56	34	0.61	1	0.02	6	0.11	41	0.73[a–d]
20.0	60	57	0.95	9	0.15	7	0.12	73	1.22[a–d]
Limoncito									
DW	120	41	0.34	4	0.03	1	0.01	46	0.38
50F	120	37	0.31	0	0.00	2	0.02	39	0.32
100F	120	42	0.35	0	0.00	0	0.00	42	0.35
50	120	45	0.37	1	0.01	1	0.00	46	0.38
100	120	40	0.33	1	0.01	2	0.02	43	0.36
Manantial Benito Juárez									
DW	80	33	0.41	0	0.00	0	0.00	33	0.41
25F	80	28	0.35	0	0.00	2	0.03	30	0.37
50F	62	43	0.69	0	0.00	0	0.00	43	0.69[a,d]
100F	80	30	0.37	0	0.00	2	0.03	32	0.40
25	80	21	0.26	0	0.00	0	0.00	21	0.26
50	56	21	0.37	1	0.02	2	0.04	24	0.43
100	80	41	0.51	0	0.00	0	0.00	41	0.51
Muhi									
DW	120	38	0.32	1	0.01	1	0.01	40	0.33
50F	80	34	0.43	0	0.00	0	0.00	34	0.43
100F	80	29	0.36	0	0.00	0	0.00	29	0.36
50	78	23	0.29	1	0.01	2	0.03	26	0.33
100	80	24	0.30	0	0.00	4	0.05	28	0.35
Pozo 1									
DW	80	26	0.32	0	0.00	0	0.00	26	0.32
25F	80	30	0.37	0	0.00	0	0.00	30	0.37
50F	80	21	0.26	0	0.00	2	0.03	23	0.29
100F	80	19	0.24	0	0.00	0	0.00	19	0.24
25	80	29	0.36	0	0.00	0	0.00	29	0.36
50	80	29	0.36	0	0.00	0	0.00	29	0.36
100	80	16	0.20	0	0.00	0	0.00	16	0.20
Pozo 3									
DW	80	23	0.29	0	0.00	3	0.04	26	0.32
25F	80	25	0.31	0	0.00	4	0.05	29	0.36
50F	80	34	0.43	0	0.00	1	0.01	35	0.44

TABLE 5. *Continued*

% Sample Water	Type of spots: Number of wings	mwh N	mwh SPW	flr N	flr SPW	twin N	twin SPW	total N	total SPW
100F	80	36	0.45	0	0.00	1	0.01	37	0.46
25	80	31	0.39	0	0.00	2	0.03	33	0.41
50	80	33	0.41	1	0.01	2	0.03	36	0.45
100	80	40	0.5	0	0.00	0	0.00	40	0.50[a]
Pozo 5									
DW	80	31	0.39	1	0.01	0	0.00	32	0.40
50F	80	62	0.77	0	0.00	0	0.00	62	0.77[a,d]
100F	80	38	0.47	8	0.10	0	0.00	46	0.57
50	76	35	0.46	0	0.00	0	0.00	35	0.46
100	80	31	0.39	0	0.00	0	0.00	31	0.39
Pozo Viejo									
DW	80	27	0.34	2	0.03	0	0.00	29	0.36
50F	80	20	0.25	0	0.00	0	0.00	20	0.25
100F	80	38	0.47	0	0.00	0	0.00	38	0.47
50	80	41	0.51	0	0.00	4	0.05	45	0.56[d]
100	80	28	0.35	0	0.00	0	0.00	28	0.35
Sabinas									
DW	80	33	0.41	0	0.00	0	0.00	33	0.41
25F	80	23	0.29	0	0.00	1	0.01	24	0.30
50F	80	47	0.59	1	0.01	5	0.06	53	0.66[c,d]
100F	80	20	0.25	0	0.00	2	0.03	22	0.28
25	80	33	0.41	0	0.00	0	0.00	33	0.41
50	80	35	0.44	0	0.00	1	0.01	36	0.45
100	80	33	0.41	0	0.00	1	0.01	34	0.43
San Pedro									
DW	120	39	0.32	3	0.03	2	0.02	44	0.37
25F	80	18	0.22	1	0.01	2	0.03	21	0.26
50F	80	36	0.45	1	0.01	1	0.01	38	0.47
100F	80	27	0.34	1	0.01	1	0.01	29	0.36
25	80	43	0.54	0	0.00	0	0.00	43	0.54[a,d]
50	80	33	0.41	1	0.01	0	0.00	34	0.43
100	80	32	0.40	0	0.00	2	0.03	34	0.43
Tecozautla									
DW	120	35	0.29	3	0.03	1	0.01	39	0.32
25F	120	32	0.27	0	0.00	4	0.03	36	0.30
50F	120	42	0.35	0	0.00	0	0.00	42	0.35
100F	120	62	0.52	0	0.00	2	0.02	64	0.53[a,d]
25	120	53	0.44	0	0.00	0	0.00	53	0.44
50	120	46	0.38	1	0.01	0	0.00	47	0.39
100	120	36	0.30	1	0.01	1	0.01	38	0.32

N, Number of spots; SPW, Spots per Wing; *, Positive Controls: N-nitrosodimethylamine (DMN), N-nitrosodiethylamine (DEN), [mM]; DW, distilled water; Statistical analysis according to Frei and Würgler (1988). A superscript at right of the frequency of spot per wing means that significant differences ($P < 0.05$) were find for: [a], small single spots (1–2 cells); [b], large single spots (≥3 cells); [c], twin spots; [d], total spots. The single spots can shown one of the two trichome phenotypes: *mwh* or *flr*^{3}, but in the twin spots , *mwh* and *flr*^{3} trichomes forming part of the spot.

TABLE 6. Summary of data from flies treated with DMN.

Group [mM]	Number of points	Median	Minimum	Maximum
5% Sucrose	89	0.00	0.0	3.0
1.0	18	2.00	0.0	8.0
12.5	8	9.00	5.0	19.0
25.0	18	18.00	6.0	26.0
50.0	9	21.00	12.0	23.0

KW = 91.907, P < 0.0001.

Dunn's Multiple Comparisons Test for DMN results.

	0.00	1.0	12.5	25.0
1.0	–29.059*			
12.5	–66.775*	–37.715**		
25.0	–75.726***	–46.667**	–8.951	
50.0	–83.504***	–54.444**	–16.729	–7.778

*, P < 0.05; **, P < 0.01; ***, P < 0.0001.

samples (Figure 8a). The highest effect was found in males from the filtered-diluted sample (50F). The dispersion in the distributions of size of spots was similar to respective controls for both, females and males (Figures 8b and c).

Corrected frequencies for the complete samples (100%) are shown in Figure 9. In females, the order of spot induction was: Muhi > San Pedro > Tecozautla > Pozo 1; and in males: Tecozautla = Muhi > Pozo 1 = San Pedro.

TABLE 7. Summary of data from flies treated with DEN.

Group [mM]	Number of points	Median	Minimum	Maximum
5% Sucrose	89	0.00	0.0	3.0
1.0	30	0.00	0.0	2.0
5.0	30	0.50	0.0	4.0
10.0	28	1.00	0.0	4.0
20.0	30	2.50	0.0	6.0

KW = 40.056, P < 0.0001.

Dunn's Multiple Comparisons Test for DEN results.

	0.00	1.0	12.5	25.0
1.0	7.227			
5.0	–9.589	–16.817		
10.0	–38.771*	–45.999*	–29.182	
20.0	–65.223***	–72.450***	–55.633**	–26.451

*, P < 0.05; **, P < 0.01; ***, P < 0.0001.

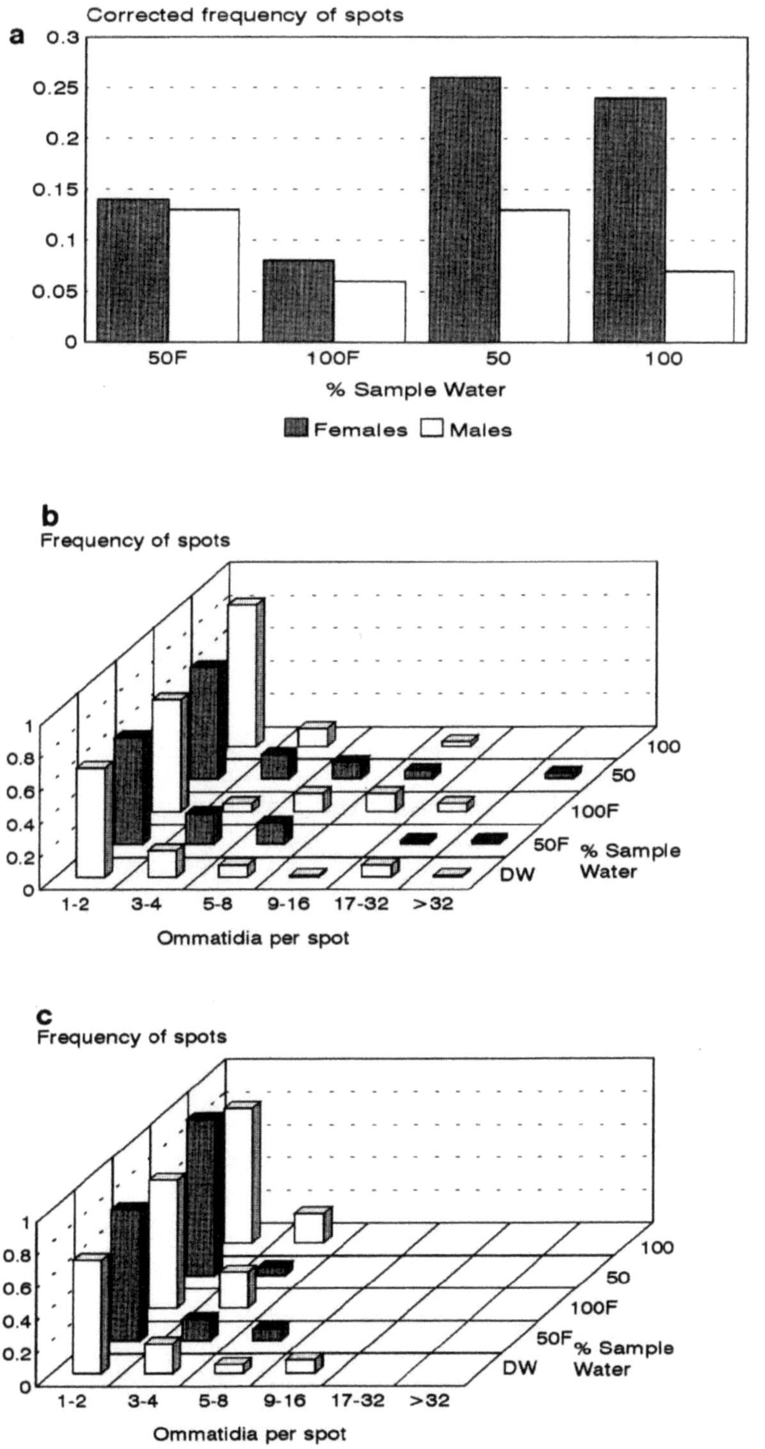

FIGURE 5. Spots per eye in flies exposed to the Muhi well water. a) Corrected frequency of spots in females and males; Distribution of ommatidia per spot (size of spots) in b) females, and c) males. The size of spots in males was smaller (1–4 ommatidia) as compared to the control size spots (1–16 ommatidia). DW, distilled water. 100F and 50F, filtered and filtered-diluted samples, respectively.

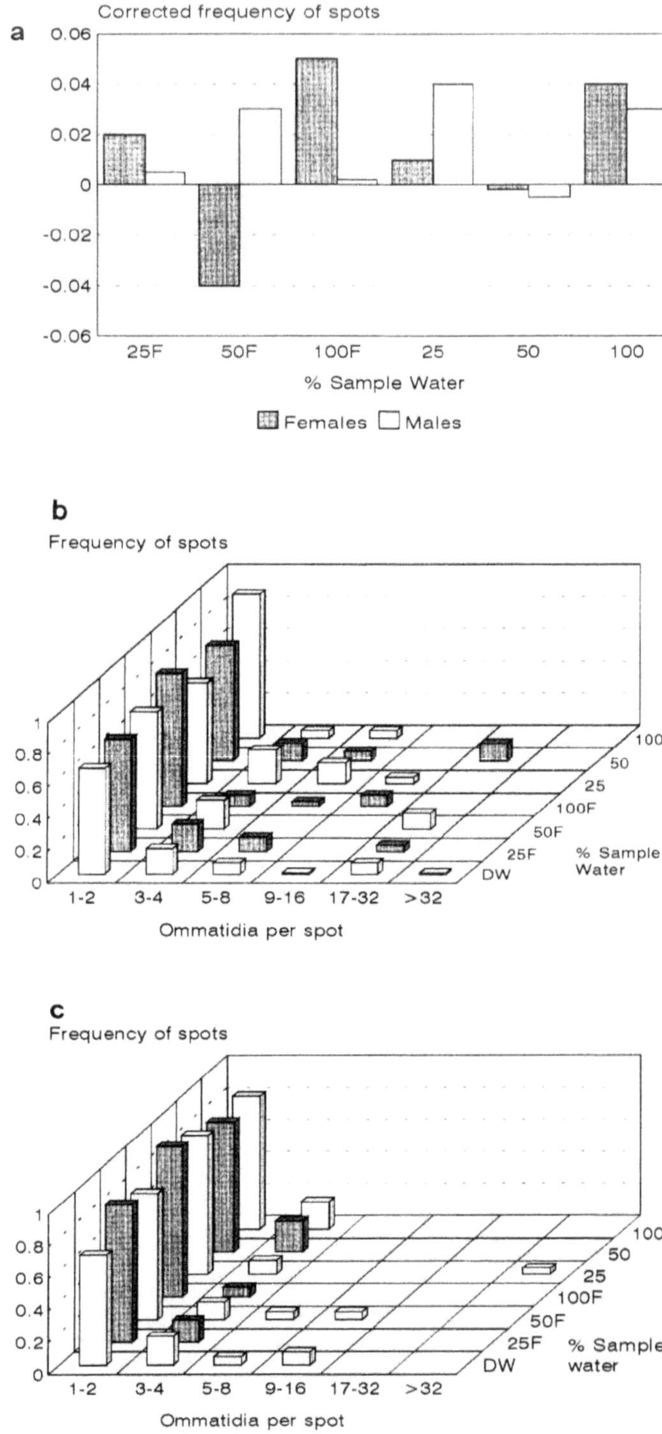

FIGURE 6. Spots per eye in flies exposed to the Pozo 1 well water. a) Corrected frequency of spots in females and males. Distribution of ommatidia per spot (size of spots) in b) females, and c) males. The size of spot tended to reduce in both sexes exposed to the complete sample (100). DW, distilled water. 100F, filtered sample; 50F and 25 F, filtered-diluted samples.

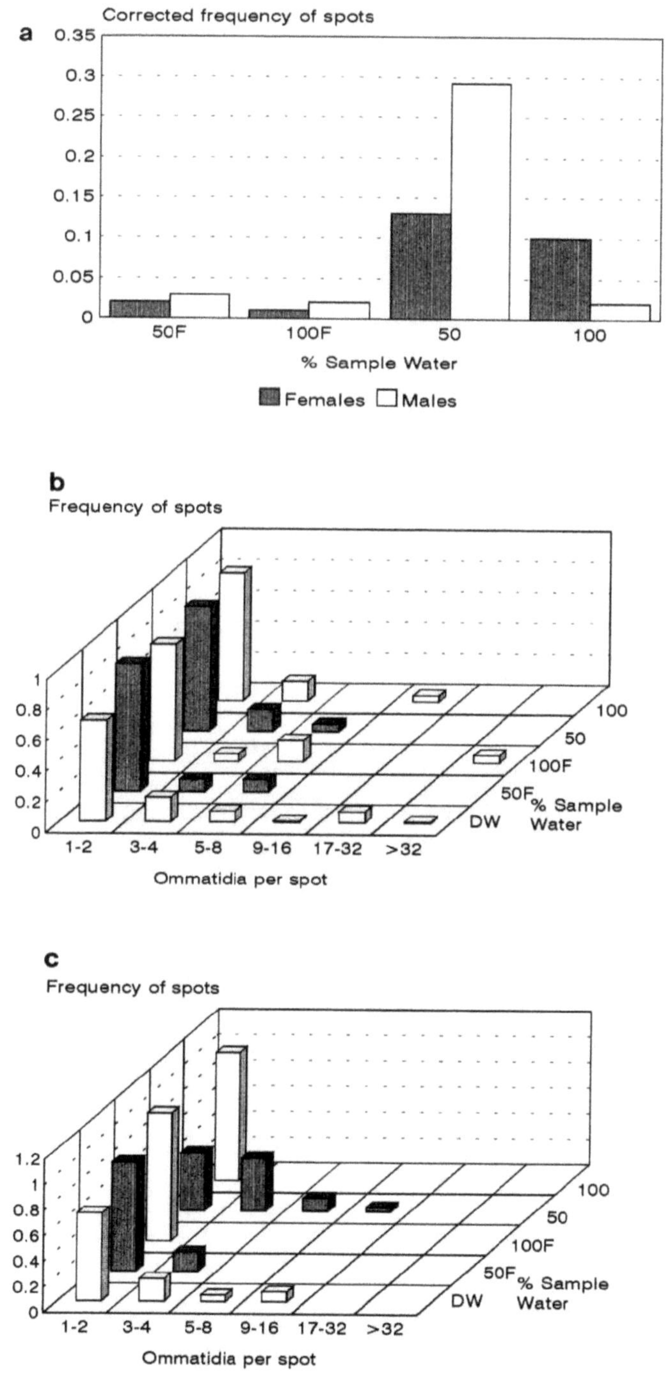

FIGURE 7. Spots per eye in flies exposed to the San Pedro well water. a) A significant increase in the corrected frequency of spots was detected in flies exposed to 50% sample, b) The size of spots (ommatidia per spot) in experimental females was smaller than that from control females, c) Small spots were only recovered in experimental treatments (except in 50% sample water).

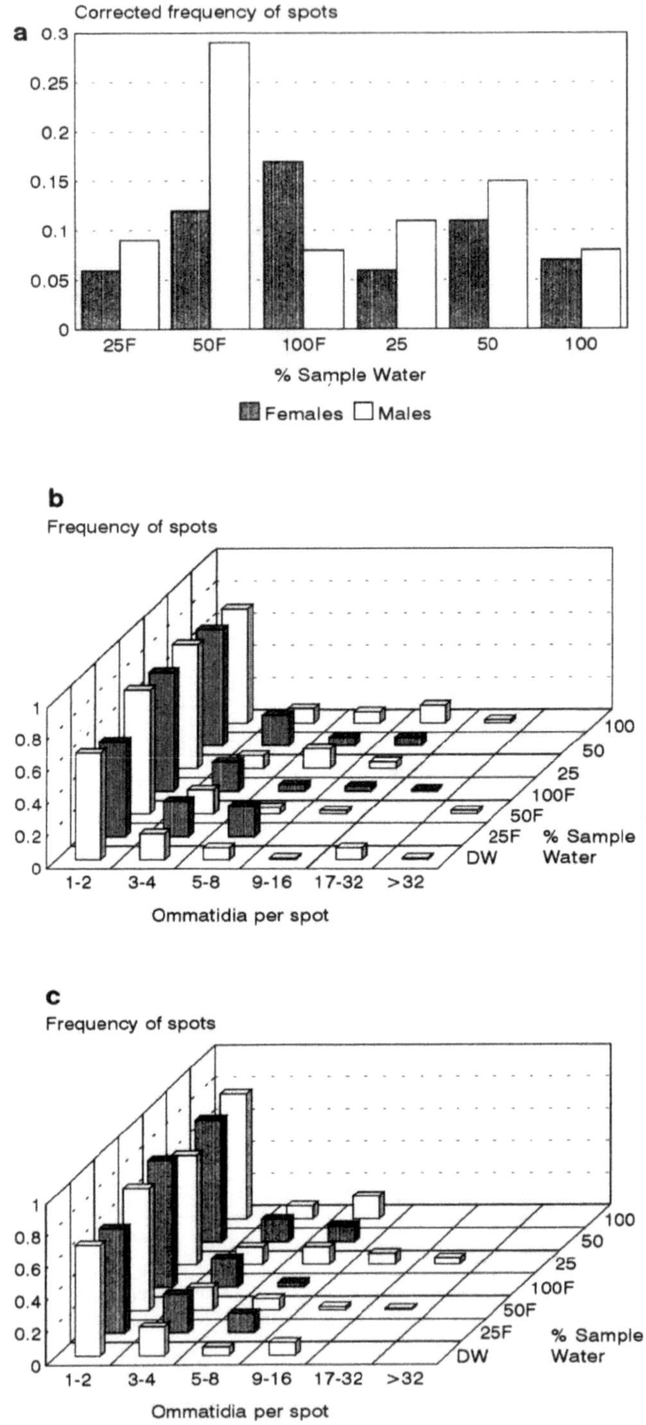

FIGURE 8. Spots per eye in flies exposed to the Tecozautla well water. a) The frequency of spots was increased, being higher in females exposed to filtered samples, and in males exposed to diluted samples, being the filtered-diluted (50F) sample the most important. The size of spot in females (a) and males (b) was similar to respective controls. DW, distilled water.

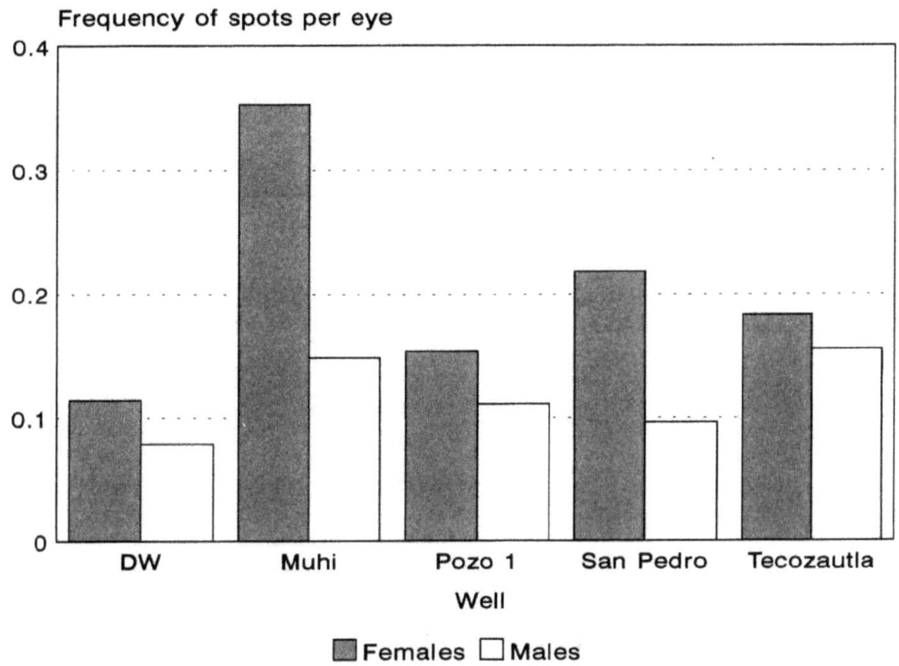

FIGURE 9. Spots per eye in flies exposed to well water. Comparison of the effect of complete samples (100%) in females and males. DW, distilled water.

3.4.2. Wing Spot Version

Data obtained in this version of the SMART analysis were analyzed for three approaches: The increase in the frequency of spots per wing, the size of spots scored, and the number of spots per fly (Table 5 and Figures 10 to 19). In addition, the results were grouped by geographic region: Tecozautla, Zimapan, and Benito Juárez Municipalities. Limoncito, Las Sabinas and Tecozautla are located in Tecozautla Municipality.

Limoncito. No increase in the frequency of spots was recovered in the treatments ($P > 0.05$) (Figure 10a). Distribution of the number of cells per spot and the number of spots per fly from experimental flies was similar to control flies in every case (Figures 10b and 10c).

Las Sabinas. The unfiltered samples assayed failed to cause an increase in the frequency of spots (Figure 11a). Filtering affected the activity of samples: the frequency of spots in flies from the filtered samples 25F and 100F were lower than the control frequency, but an increase in the frequency of twin and total spots ($P < 0.05$) was obtained from flies exposed to the filtered-diluted sample (50F). The distribution of the number of cells per spot showed that more small spots (1–2 cells) were recovered in 50F, 25, 50 and 100% samples (Figure 11b). All the experimental series showed flies with 3 spots and in 50F series, flies with 5 and 17 spots were recovered (Figure 11c).

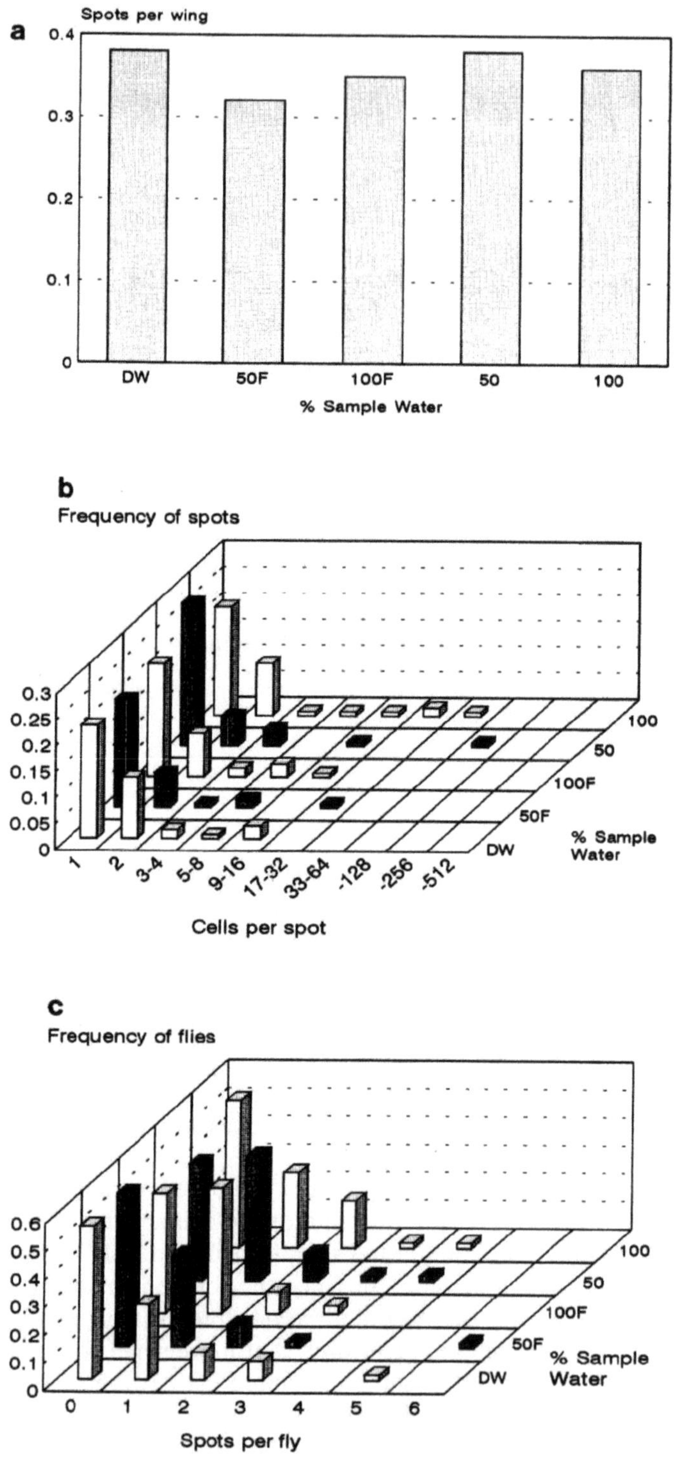

FIGURE 10. Spots per wing in flies exposed to the Limoncito well water. No differences were found in: a) the frequency of spots per wing, b) the size of spots (cells per spot) distribution, or c) the number of spots per fly in experimental and control series. DW, distilled water.

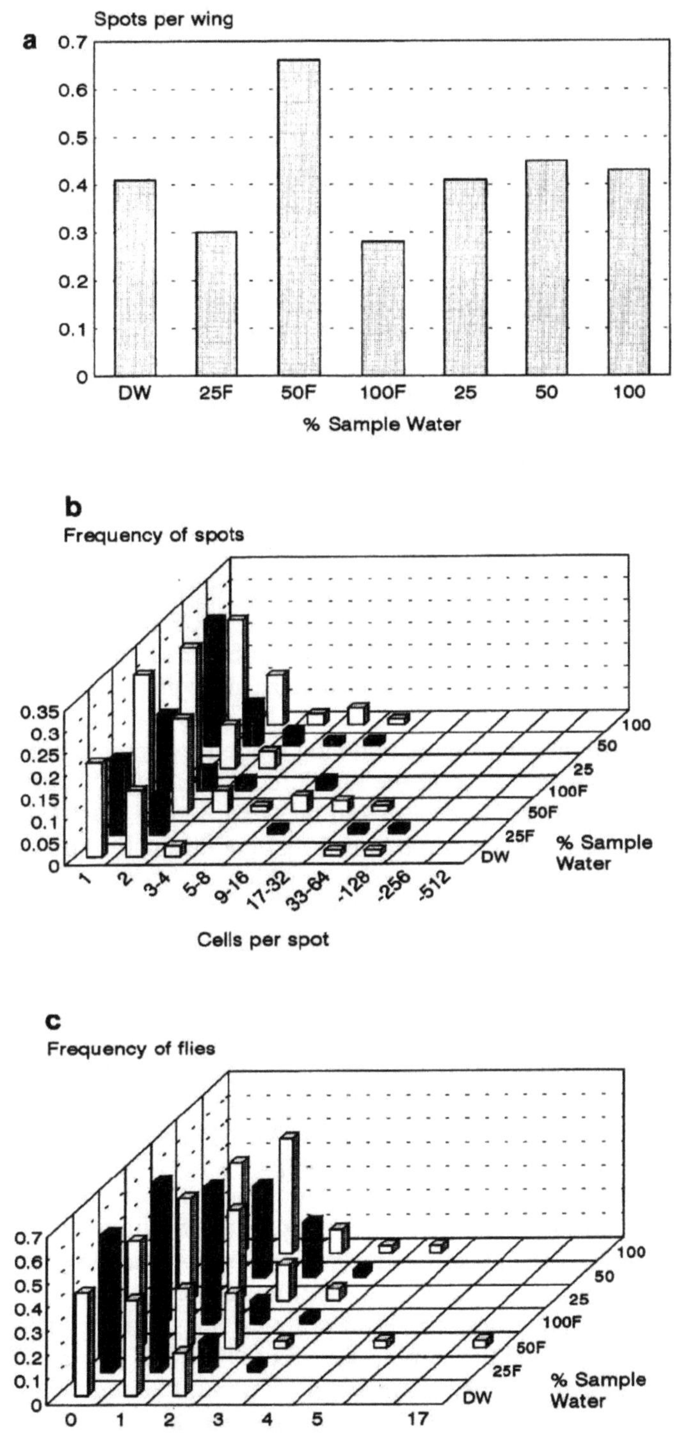

FIGURE 11. Spots per wing in flies exposed to the Sabinas well water. a) Filtering affected the activity of samples, the highest activity was found in 50F, and the lowest in 25F and 100F series; b) More small spots (1–2 cells) were recovered (except in 25F and 100F series), c) In the experimental series 3 spots per fly were found, and in the 50F treatment, flies with 5 and 17 spots per fly were observed.

Tecozautla. The dilution of samples increased slightly their activity ($P > 0.05$) (Figure 12a). Filtering modified the activity of samples, being significantly higher in the undiluted sample (100F) ($P < 0.05$). The frequencies of one-cell spots in flies exposed to 50F, 100F and in all unfiltered samples, were higher as compared to that from control (Figure 12b). On the other hand, significant dispersion of experimental distributions ($P < 0.05$) and tends to increase the number of spots per fly were detected. However, the Dunn's Multiple Comparison test failed to distinguish particular differences among series (Figure 12c).

With exception to Manantial Benito Juárez, the remaining wells are located in Zimapan Municipality.

Muhi. No induction of spots was observed in experimental series (Figure 13a). In flies exposed to filtered samples, more small spots were recovered, but not in a significant number (Figure 13b). The fraction of flies in the experimental series showing two or more spots each one were higher, but not significant as compared to control distribution (Figure 13c).

Pozo 1. There was no induction of spots in experimental flies (Figure 14a). However, the frequencies of spots from undiluted samples (100F and 100) were lower than the control frequency, and those from diluted samples were similar. Figure 14b shows that compared with control distribution, the experimental distributions showed irregular shapes, mainly those from 100F and 100 samples. A similar and significant response was observed for the number of spots per fly ($P < 0.05$), however the Dunn's test failed to distinguish particular differences among this series (Figure 14c).

Pozo 3. The frequency of spots per wing tended to increase in the experimental series, being higher in flies from the undiluted samples (100F and 100), but only significant for small spots (1–2 cells) from flies exposed to the 100 treatment ($P < 0.05$) (Figure 15a), as it is shown in Figure 15b. In addition, the size of spots tended to increase in all the experimental series in which an important fraction of spots with two and more than 32 cells per spot were found. Compared to the control distribution, the fraction of flies without spots was lower in the experimental series. The fraction with more than two spots was higher but not significant (Figure 15c).

Pozo 5. The frequency of spots tended to increase when the well water sample was filtered and also when diluted. Only a significant increase was observed for small spots and for the sum of all spots recovered in flies exposed to the filtered-diluted sample (50F) (Figure 16a). It is important to note that the experimental distributions from the number of cells per spot showed less dispersion compared to the control distribution, with two exceptions: a large spot (≈256 cells) recovered in the 100F treatment, and the important fraction of one cell spots in the 50F treatment (Figure 16b). Significant differences were found in the number of spots per fly ($P < 0.05$); in addition, significant differences were confirmed by the Dunn's test between the 50F and 100F series ($P < 0.05$) (Figure 16c).

Pozo Viejo. In flies from unfiltered samples, the frequency of spots was significantly higher only when the sample was diluted (50) ($P < 0.05$). Filtering modified the activity of samples, being higher in filtered-undiluted sample (100F) and lower in filtered-diluted sample (50F), as compared to the control frequency of spots (Figure 17a). The number of cells per spot increased in flies from the unfiltered

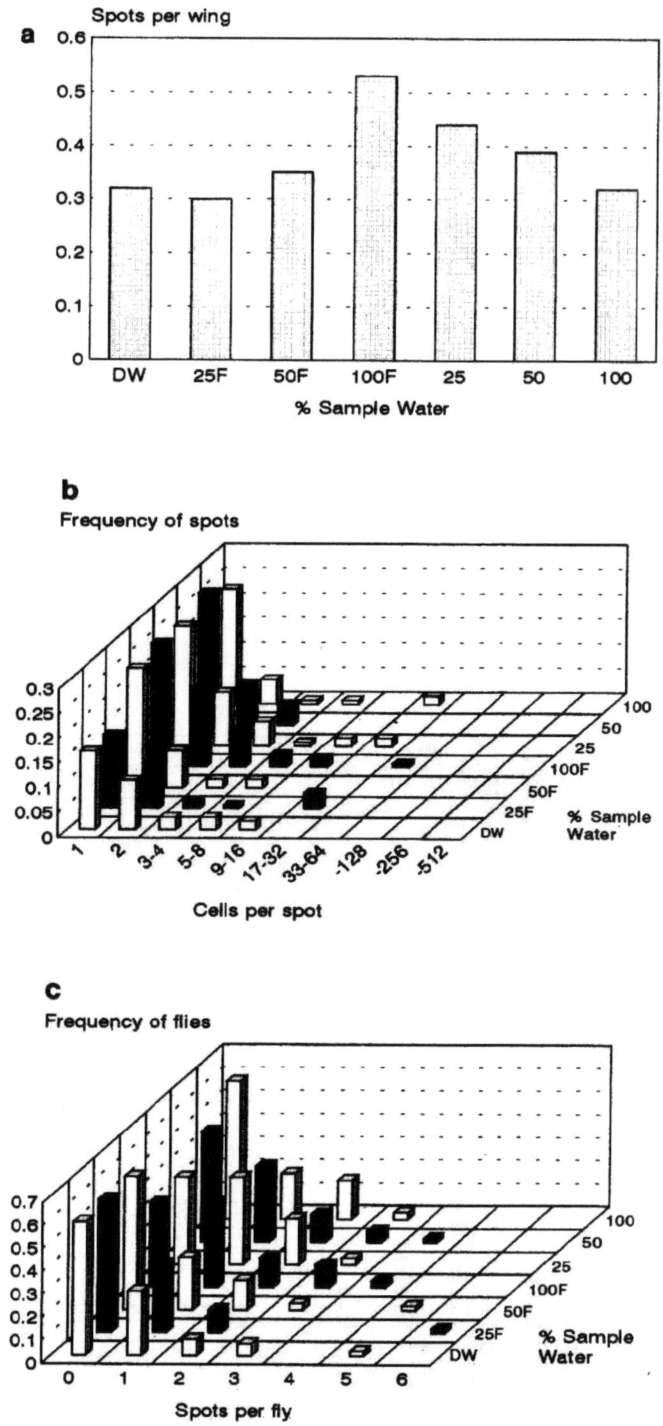

FIGURE 12. Spots per wing in flies exposed to the Tecozautla well water. The frequency of spots tended to increase with dilutions, but in the 100F, filtered sample, the highest increase was found; b) Mainly small spots were recovered in flies from 50F, 100F, and unfiltered samples; c) The number of spots per fly tends to increase, too. DW, distilled water.

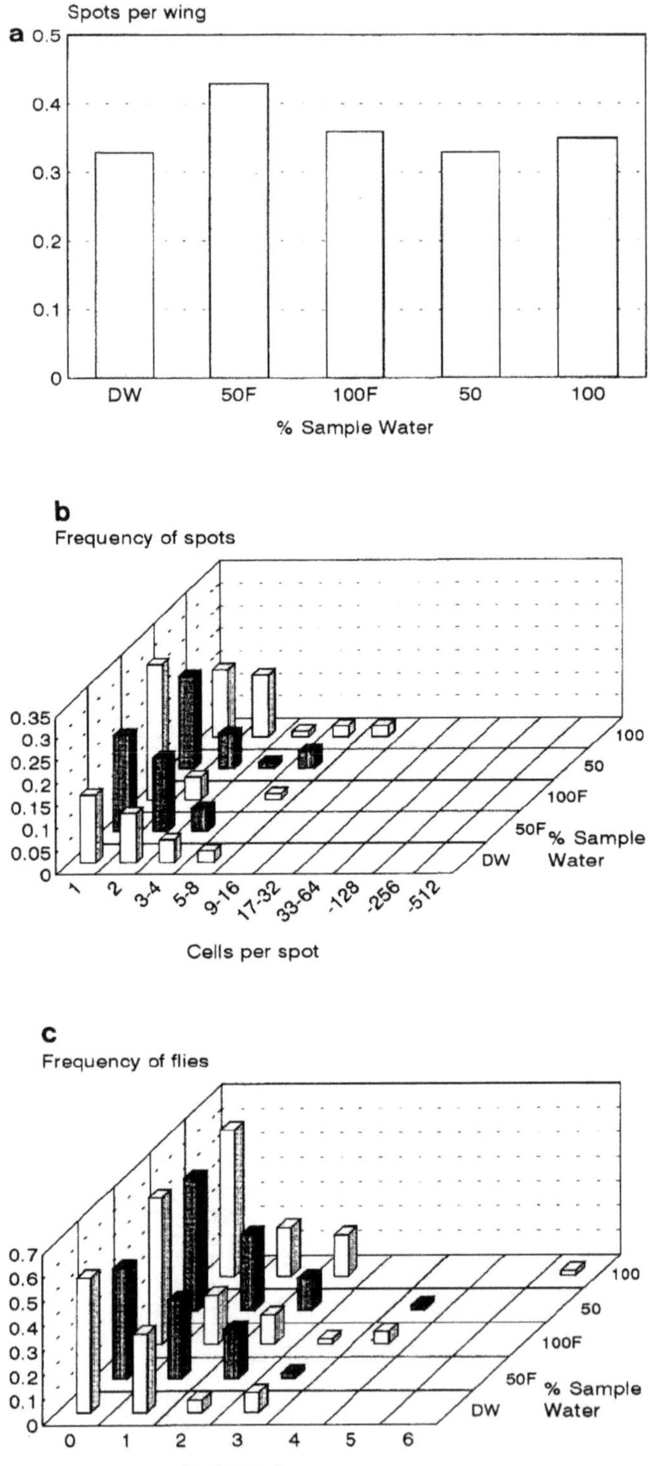

FIGURE 13. Spots per wing in flies exposed to the Muhi well water. a) The frequency of spots per wing was not increased, b) In undiluted samples (100 and 100F) slight differences in the proportion of small spots (1–2 cells) were observed, c) A slight increase in the frequency of flies with two or more spots each fly was observed in the experimental series.

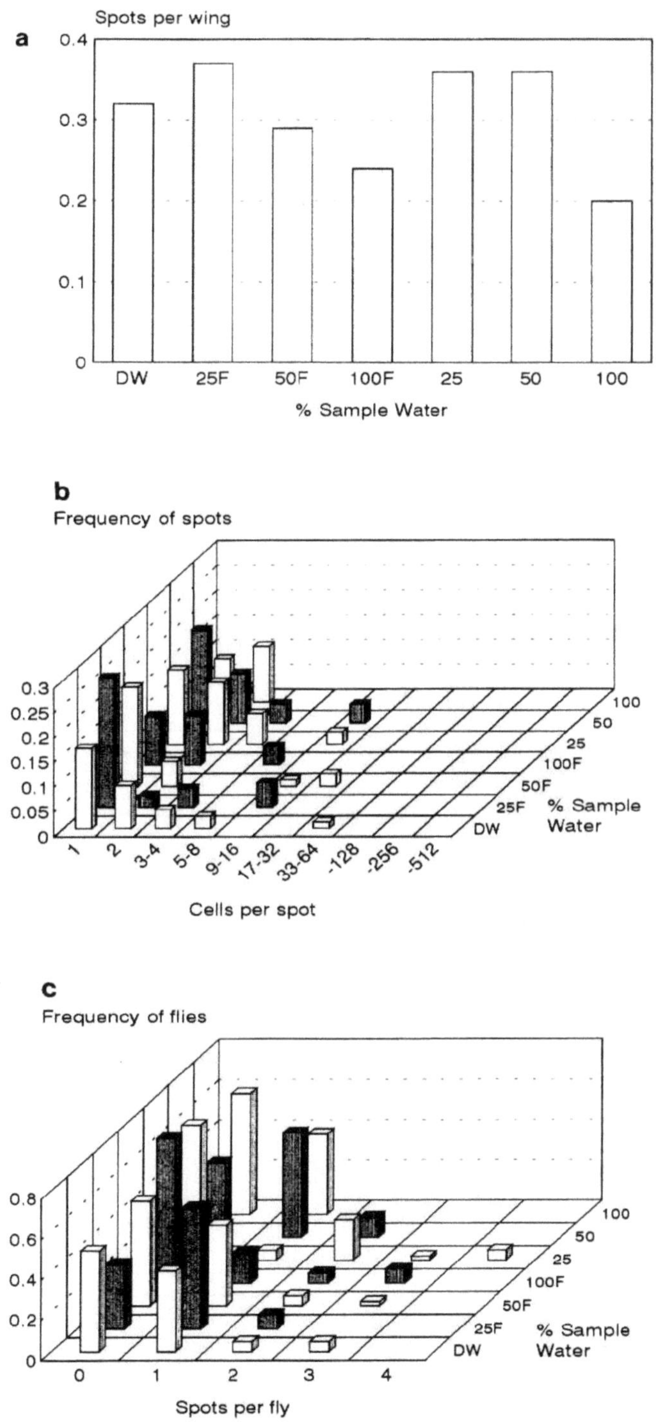

FIGURE 14. Spots per wing in flies exposed to the Pozo 1 well water. a) Frequencies of spots per wing lower than the control frequency were found in flies from undiluted samples (100 and 100F); b) A different proportion in the size of spots recovered was observed, too; c) A similar and significant response was confirmed for the number of spots per fly.

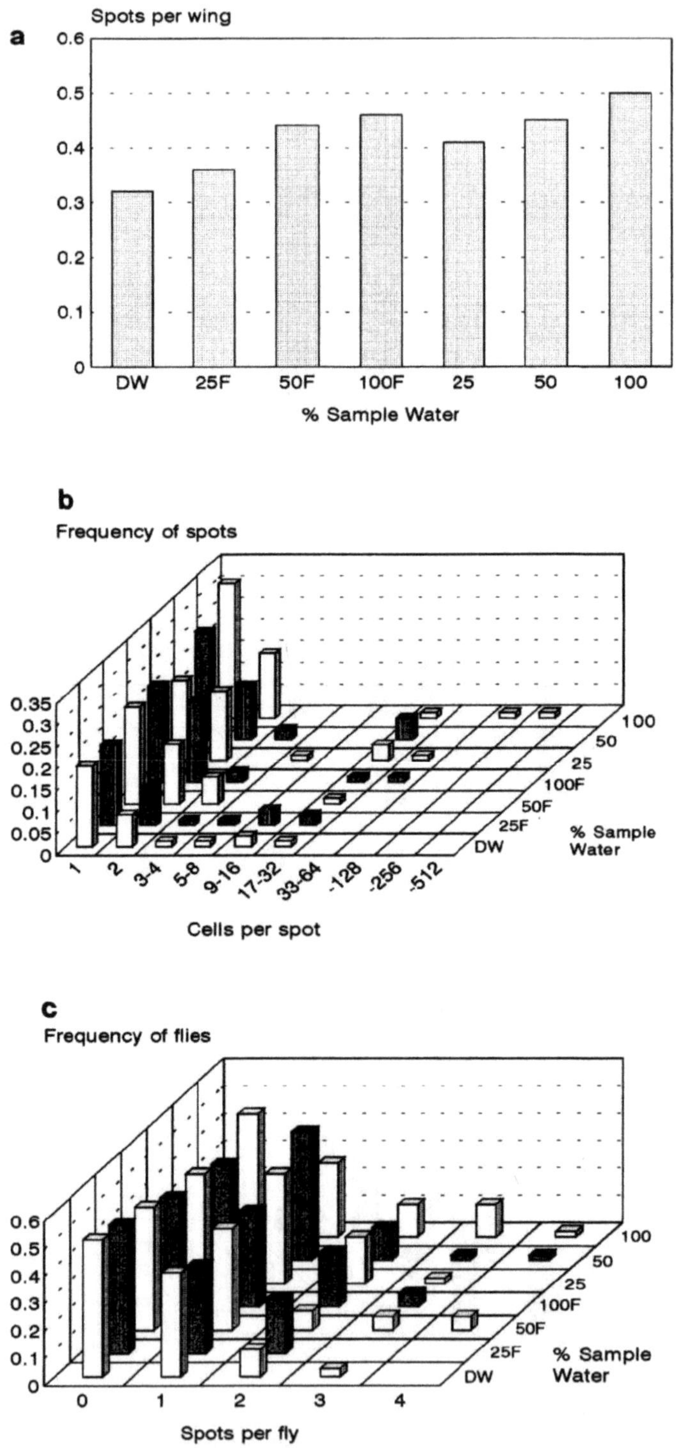

FIGURE 15. Spots per wing in flies exposed to the Pozo 3 well water. The frequencies of spots per wing tended to decrease with dilution, only in 100 series the effect was significant; b) More large spots (>17 cells per spot) were observed from the experimental series; c) The fraction of flies with more than two spots each fly was higher, but not significant as compared to control series.

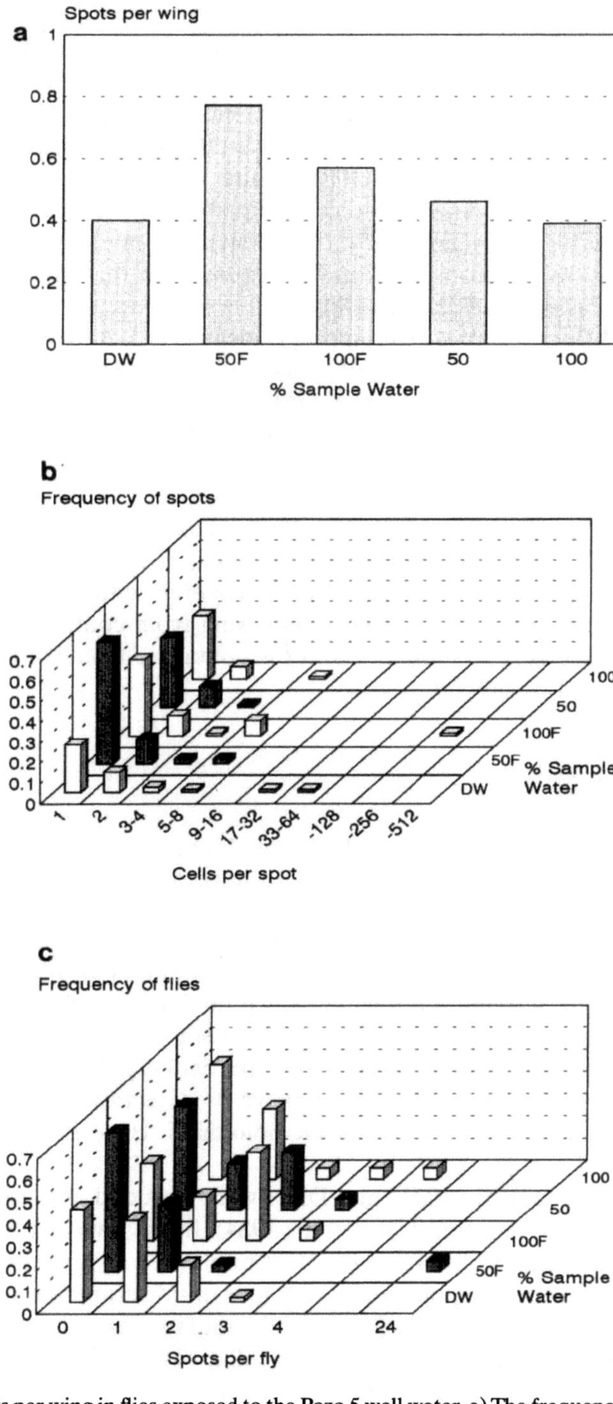

FIGURE 16. Spots per wing in flies exposed to the Pozo 5 well water. a) The frequency of spots increased with dilution and filtering, been significant in the 50F treatment; b) The size of spots from experimental flies was smaller than in control flies, except by one spot with >256 cells in 100F and an excess of small spots (one cell) in 50F; c) Significant differences in the number of spots per fly were observed, and 24 spots per fly were recovered in the 50F treatment.

samples and decreased when filtering (Figure 17b). The comparison among the number of spots per fly showed a significant dispersion in the experimental distributions ($P < 0.05$), and the Dunn's Multiple Comparison test confirmed differences between distributions from 50 and 50F treatments ($P < 0.05$) (Figure 17c).

San Pedro. The frequency of small and total spots was increased only in the diluted-unfiltered sample (25) ($P < 0.05$) (Figure 18a). The analysis of the size spot distribution showed that compared to control pattern, more small spots occurred in the experimental series (except in 25F treatment). Although more dispersion and a trend to show spots were observed in flies exposed to this sample, nonsignificant differences were confirmed (Figure 18c).

Manantial Benito Juárez. This spring is located in Benito Juarez Municipality. In unfiltered samples, the frequency of spots recovered decreased from complete to diluted samples and the frequency from flies exposed to the diluted sample (50) was similar to that of the control flies (Figure 19a). In filtered samples, the frequencies of small and total spots increased only in flies from the 50F treatment; the frequencies of spots from the remaining treatments were like the control. The analysis of size spot distribution confirmed these data (Figure 19b). Nonsignificant differences were detected in the number of spots per fly. However, it is important to note that in the 50F treatment flies were recovered with 24 spots! (Figure 19c).

The corrected frequencies of spots per wing from complete samples assayed are shown in Figure 20. The summary of results obtained from the two SMART versions are presented in Table 8.

4. DISCUSSION

The use of laboratory methodologies for biomonitoring requires one to evaluate their advantages and disadvantages to have a sensitive, reliable and reproducible system. Conditions should be evaluated to determine acceptable standards concerning cost, size of sample, type of exposure, age of the organisms at treatment, statistical procedures, time needed for the assay, number of repetitions, etc. Also, it is important to determine the form in which the biomonitoring will be done, i.e.: Is it convenient to bring the organisms for the assay to the emission source? Is the selected organism susceptible to detect some particular pollutant? Is it necessary to concentrate the pollutant to allow or improve its detection? Is it necessary to use special solvents to extract the potential pollutant? How should the sample be conveyed?

Two positive controls DMN and DEN were included and assayed using concentrations based on literature (Graf et al., 1984). In the SMART the activity of the two alkylating agents was: DMN > DEN, in one order of magnitude. The qualitative analysis of the size of spots recovered does not discriminate between the time of spots production by the promutagen and the mutagen, as it is expected when very high concentrations are used, and also when cell division delay or cell death to occur (Gonzalez and Ramos-Morales, 1997). However, in flies from the DMN treatment more spots per fly were recovered. This effect cannot be explained through the highest potency of the DMN: the number of spots per fly from the lowest and highest

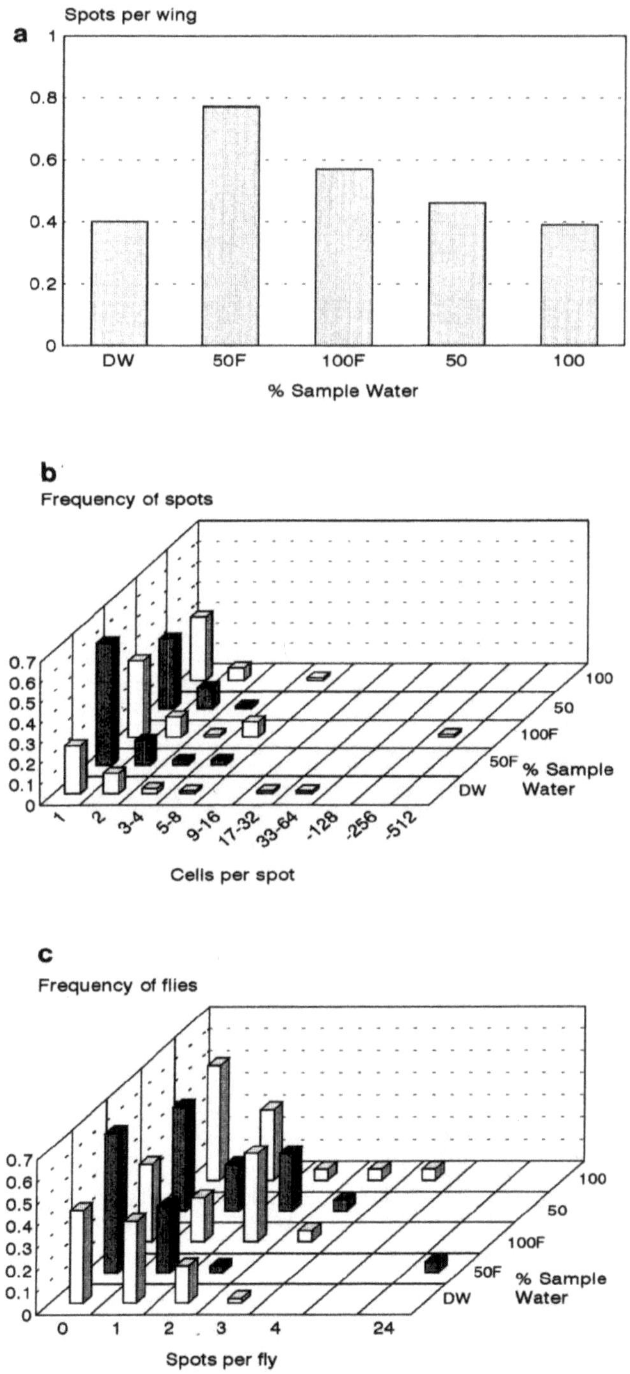

FIGURE 17. Spots per wing in flies exposed to the Pozo Viejo well water. a) The frequency of spots per wing increased in the diluted (50) and filtered (100F) samples, but was lower than that from control in the 50F treatment; b) More large spots were found in unfiltered samples (50, 100); c) Significant dispersion in the distribution of the number of spots per fly was found in the 100F, 50 and 100 treatments ($P < 0.05$, Kruskal-Wallis test).

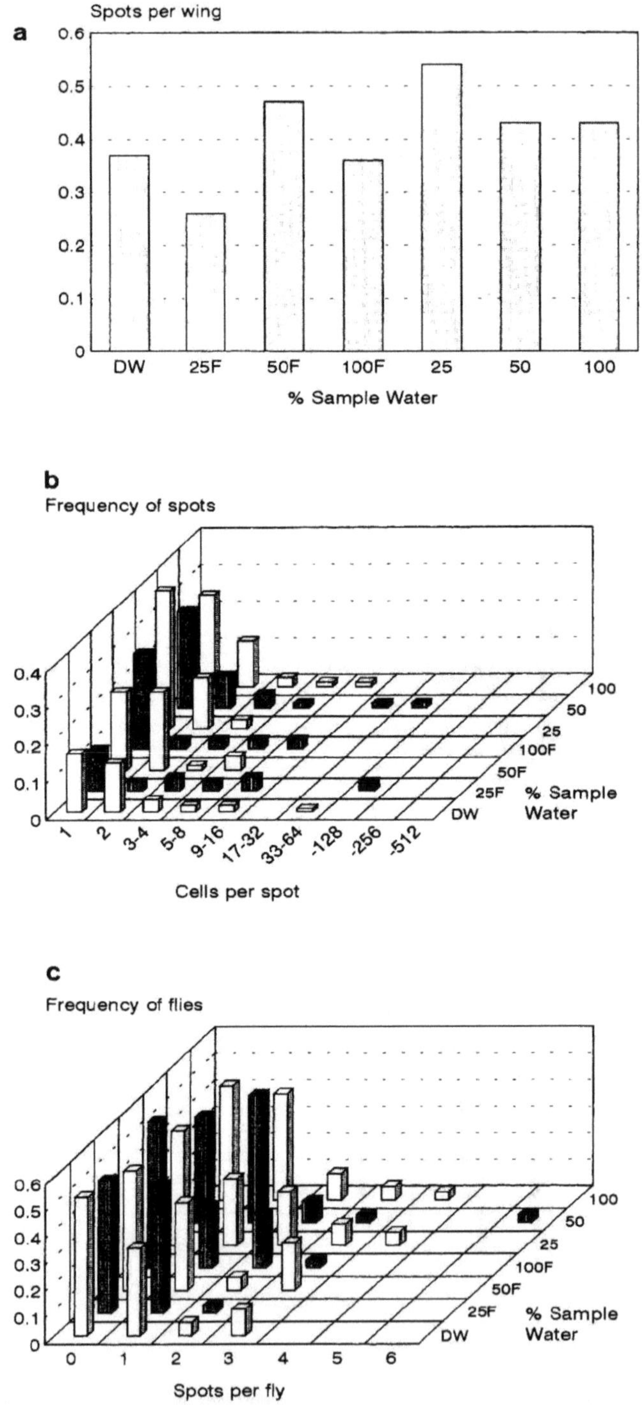

FIGURE 18. Spots per wing in flies exposed to the San Pedro well water. a) The frequency of spots increased in the diluted-unfiltered sample (25); b) More small spots were found from experimental series (except in 25F); and, c) more flies with spots were recovered, too.

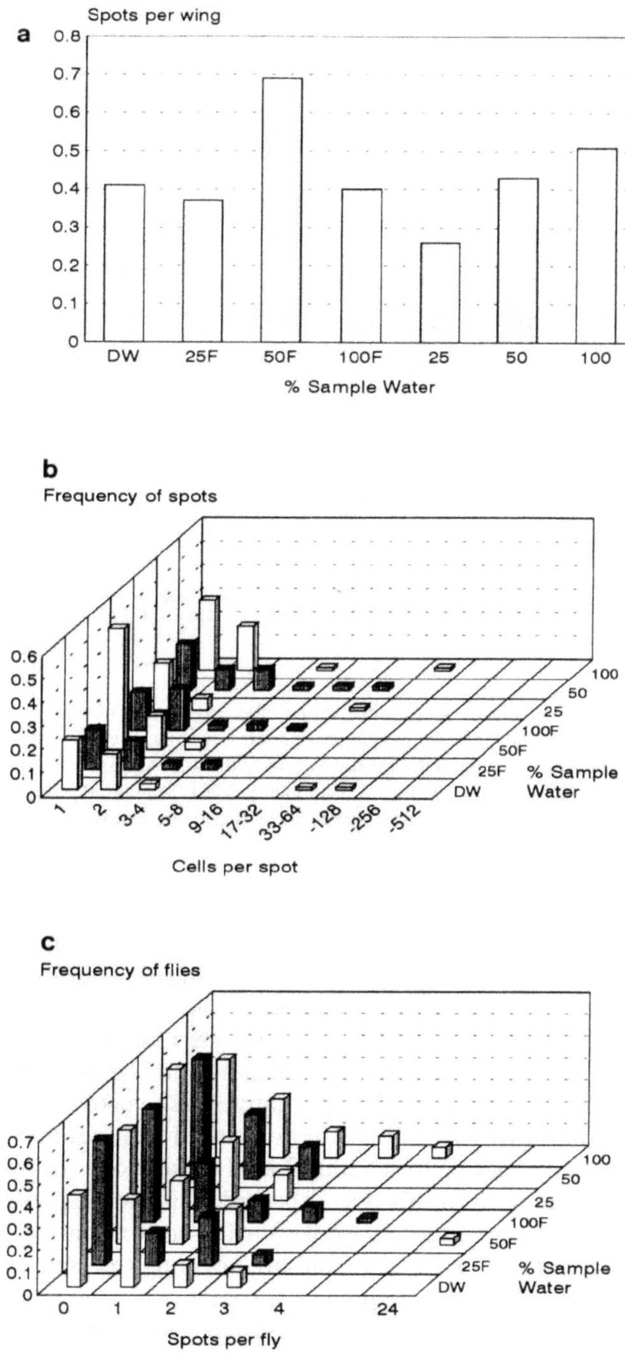

FIGURE 19. Spots per wing in flies exposed to the Manantial Benito Juarez water. a) The frequency of spots increased in the filtered and diluted treatment (50F), b) Mainly small spots (one cell) was induced by the 50F treatment; c) The distribution of the number of spots per fly was similar in experimental and control series, but one fly with 24 spots was found in the same treatment (50F).

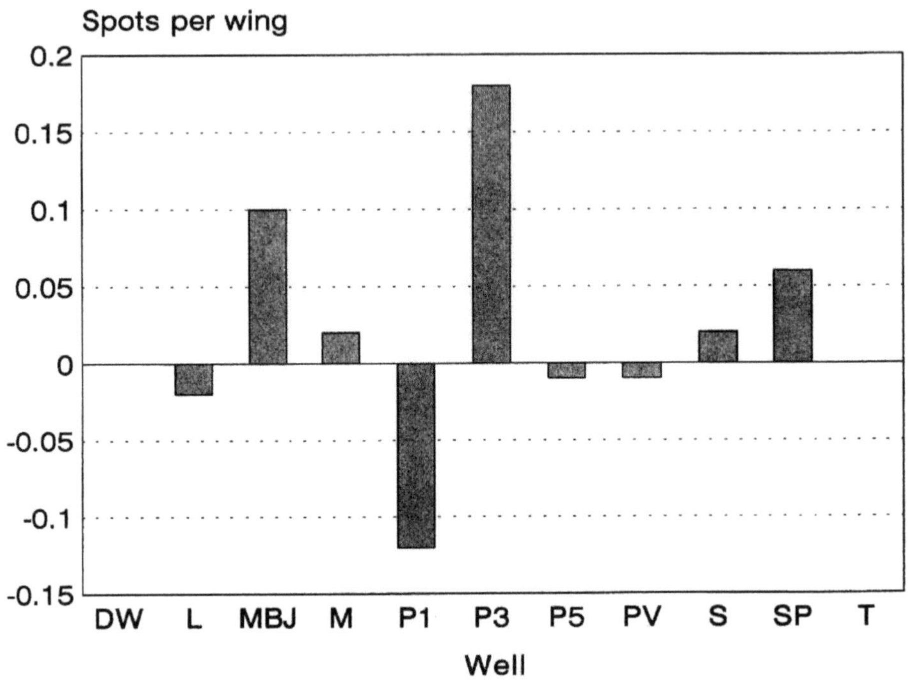

FIGURE 20. Corrected frequencies of spots per wing in flies exposed to well water. Comparison of the effect of complete samples (100%) in the frequency of spots per wing. The P3 treatment was significant, in the P1 treatment the frequency of spots was lower, and in the T treatment was the same, as compared to control's frequency. DW, distilled water; L, Limoncito; MBJ, Manantial Benito Juarez; M, Muhi; P1, Pozo 1; P3, Pozo 3; P5, Pozo 5; PV, Pozo Viejo; S, Sabinas; SP, San Pedro; and T, Tecozautla wells.

concentrations assayed ranged in 0–8 and 12–23 spots per fly, respectively; in contrast, in the DEN highest concentration, only 0–6 spots per fly were found.

In preliminary experiments we discarded the use of the SMART of *Drosophila* for extra laboratory practices due that temperature variations and the transportation of organisms modified the frequency of spots and the life cycle length, which interfered of course with this assay. An alternative was to take samples of water and convey them to laboratory for testing. The toxicity of fixatives makes necessary to take independent samples, one for chemical analysis and one for testing. Although the two type of larvae used in this assay showed a similar response to nitric acid, the wing spot test was more informative. This acid was toxic in the highest concentration assayed (commonly used as a fixative), affecting the survival of flies, but it was also cytotoxic, reducing the frequency of somatic damage recovered. This was confirmed by the irregular size of spots observed in either the eye or the wing assay. The other point that needs attention, is to make the samples chlorine free, as this compound increases the frequency of somatic spots.

To determine the way in which the water samples should be assayed, we compared the effect of filtering and dilution of samples. It is important to say that based in the higher contents of different salts, we expected higher (but not significant) frequencies of spots than those from distilled water, even in flies exposed to water samples with permissible levels of pollutants. The general assumption of this study was, that in spite of little variations in the basal frequency of spots, flies exposed to nongenotoxic samples of water should maintain a pattern of frequency, size of spots, and spots per fly similar to control flies.

From four wells assayed in the eye version of the SMART, only the sample from Pozo 1 does not induce any genotoxic response in females and males, while the other three: Muhi, San Pedro, and Tecozautla were detected as positive in both sexes of *Drosophila*.

Based on data from complete samples, genotoxic effects were in this order: Muhi > San Pedro > Tecozautla, in females; and Muhi = Tecozautla, in males. However, the qualitative analysis of the size of the spots recovered suggested that part of the spots could be missing by either, cell division delay or cellular death. Commonly in control flies from the eye version of the SMART assay, the size of spots recovered is rather small, 1–2 ommatidia per spot, followed by less than half of spots with 3–4 ommatidia per spot, but larger spots occur in a very low frequency. The size of spots is larger in females than in males, however the same relation is true.

To confirm the genotoxicity of samples, they were assayed diluted, filtered, or filtered first and then diluted (filtered-diluted). In general, diluted samples showed more genotoxicity (but less toxicity) in females and males (except in flies from the Pozo 1 treatment). This could mean that dilution of samples allows to recover more spots per eye, probably due to the lower genotoxicity of diluted samples, which also seems to interfere less with cell proliferation. This was confirmed by the presence of larger spots recovered (>5 ommatidia per spot). In males exposed to the Muhi water, spots with more than one ommatidia affected were observed, only in diluted samples. The hemizygous condition of males, makes them particularly sensitive to damage because the lack of homology of sexual pair prevents recombinant activity, which seems to interfere with DNA repair too. This has been observed for clastogenic compounds which induce somatic spots produced by recombinant activity in inversion-free flies, but failed to induce spots in inverted-chromosome carrier flies (which interferes with recombinant activity) (Gonzales and Ramos-Morales, 1997).

Sample filtering reduced the frequency of spots per eye in flies from the Muhi and San Pedro treatments, and increased those from Tecozautla treatment, in females; but failed to modify the observed activity in males. In aquatic systems, part of the chemicals are dissolved but and important fraction of them forming colloids are associated to organic matter. It is possible that the flux of water through the membrane contributes to increase the availability of some of the pollutants (Koren, 1991). It is probable that sample filtering using a membrane (0.45 μ), as it is recommended in standard methodologies for chemical analysis (Meybeck et al., 1990), may contribute to increase the concentration of pollutants in the water assayed. At present we are testing alternative ways to filter samples for biomonitoring. The lack of effect on males seems to confirm the toxicity of samples.

The combination of filtering and dilution of samples, increased the frequency of spots in females from the Muhi treatment, and males from the Tecozautla treatment. In both treatments, the size of spots recovered tended to increase. For Tecozautla water an additional dilution was carried out. The frequency of spots per eye in females and males was only slightly higher than that of concurrent controls, and the size of spots recovered was similar to their respectives controls.

From results obtained using the eye version of the SMART we conclude that the comparison of frequencies in experimental and control series must be accompanied with the analysis of size of spots. In consequence, a frequency of spots similar to that of controls should have a control similar pattern of spot size. In addition, the separate scoring of females and males is recommended due to the heterozygous condition of females that allows them to recover spots induced by breakage's that are missing in males which are hemizygous, and hence, more sensitive to this type of damage. We also suggest to determine the survival index of females and males from experimental and control series to confirm whether or not the treatment affects the viability.

From wells assayed in the wing's version of the SMART, three were detected as non genotoxic. The Limoncito well's water did not modify the frequency of spots, and also control similar distributions of size of spots were recovered in experimental series. The fraction of flies with at least one spot was slightly increased, but the pattern of distribution did not change in a significant manner, and although in this sample the concentration of cadmium was over the permissible level, the results obtained were according with our expectations about a nongenotoxic sample.

In flies exposed to the Muhi's water, the frequencies of spots for unfiltered samples were similar to the control's frequency, but the proportion of spots with two or more cells tended to increase in flies exposed to the complete sample, which agrees with the results obtained in the eye version of the SMART. In flies exposed to the complete sample of Pozo 1, the frequency of spots per wing was lower than on the control's frequency (which indicates toxicity), but again, the size of spots recovered tends to decrease. The dilution of the sample increased slightly the frequency of spots recovered and their size. Data from Muhi and the Pozo 1 seem to indicate that they are really polluted wells as it was detected by the eye version of the SMART for the Muhi treatments. In addition, although no apparent genotoxicity was detected in the Pozo 1 sample through the eye version, in experimental series we recovered fewer and smaller spots that in control, maybe as a consequence of cell division delay or death cell. The results from flies exposed to the San Pedro and Tecozautla water wells in the wing version of the SMART confirmed those obtained in the eye version. In flies exposed to this samples, there was a significant increase in the frequency of small spots; and the dispersion on the number of spots per fly differs from that of control distribution.

The remaining wells samples were assayed only in the wing version of the SMART and a positive diagnosis was obtained for all of them. In the flies exposed to complete samples from pozo 3, the highest frequency of spots per wing was observed, and dilution reduced genotoxicity of the samples. The size of spots recovered in experimental series was higher than those of control flies, and the fraction of flies with spots in the wings tended to increase. It is probably that in this well, the type, concentration or combination among pollutants was less toxic that in the others wells.

The toxic effect of water from Pozo 5 was evident not only in the frequency of spots per wing but in the size of spots recovered, which were smaller than in control flies, only a larger spot was recovered in the filtered-undiluted sample. One of the reasons that support our suggestion to analyze the number of spots per fly was observed in this treatment. The number of spots per fly tended to increase in experimental flies, but among those exposed to the filtered-diluted sample (50F), two organisms with 24 spots each one were recovered. This data require special mention because the spontaneous frequency of the SMART reported in numerous studies varied from 0.25 to 0.45 spots per wing, that means an average of 35 spots per one hundred of wings. In data from Pozo 5 treatment, only two organisms contributed to double the spontaneous frequency! It is important to detect an early response in the most susceptible organisms from a population exposed to some particular genotoxic or hazardous conditions. However, assuming that a particular condition is affecting a whole population may not be true when it is based on a response from only two organisms. The same was found in flies exposed to water from Las Sabinas well and Manantial Benito Juarez in which flies with 17 and 24 spots per fly were recovered, respectively.

In the last well, Pozo Viejo, the frequency of spots per wing in flies exposed to the complete sample was the same than that from control flies, and again, the dilution and the filtering of samples showed that this initial appreciation was wrong. The analysis of the number of spots per fly showed that there were significant differences among distributions from flies exposed to diluted, and filtered-diluted samples.

With these results we showed the convenience to include the three different, but complementary analysis to make any decision about genotoxicity. Comparison of corrected frequencies of spots per imaginal disc are shown in Figures 9 and 20. Based in the spots induced by complete samples only, positive diagnosis were possible for the Muhi, San Pedro and Tecozautla wells for females; and the Muhi and Tecozautla wells for males in the eye SMART version; and for the Pozo 3 in the wing SMART version. In Table 8 we summarized the results obtained from complete, filtered and diluted samples, and from the three types of analysis used in the wing version. Based on this, we suggest that only the water from the Limoncito well can be considered as not polluted; from the results of the Pozo 1 water we cannot make any decision about its genotoxicity; finally, we considered that the contents of the remaining wells were genotoxic to *Drosophila*. A relationship between chemical determinations and genotoxic responses cannot be established, i.e. in the Limoncito well the concentration of cadmium was detected above permissible levels, however no correspondence with genotoxicity was found. In contrast, in the Sabinas and San Pedro wells none of the pollutants levels determined were above permissible concentrations. We determined only the most common in that mining place, but also it is possible that other pollutants can be implied and the synergism among particular pollutants may be modifying genotoxicity. In summary, data support the conclusion that the *in vivo* system of *Drosophila* is sensitive to environmental levels of pollutants.

As for the selected methodology to assay the water samples, we conclude that is important to determine not only the effect on the complete samples because in

TABLE 8. Summary of results obtained from the two SMART's versions and diagnosis.

Well	Pollutants detected		Eye Spot Version				Wing Spor Version		Biol. Diag.
			Females		Males		Unfiltered	Filtered	
	As	Cd	Unfiltered	Filtered	Unfiltered	Filtered			
L	–	+	nd	nd	nd	nd	–	–	–
MBJ	–	+	nd	nd	nd	nd	–	+	+
M	+	–	+	+	+	+	–	–	+
P1	–	+	–	–	–	–	–	–	i
P3	–	+	nd	nd	nd	nd	+	–	+
P5	+	+	nd	nd	nd	nd	–	+	+
PV	+	+	nd	nd	nd	nd	+	–	+
S	–	–	nd	nd	nd	nd	–	+	+
SP	–	–	+	–	+	–	+	–	+
T	+	+	+	+	+	+	–	+	+

Statistical diagnosis according to Frei and Würgler (1988) for the increase in the frequency of spots per eye or per wing, as is mention in methodology. The symbol means the effect of samples (complete, filtered, diluted, and filtered-diluted) assayed: +, positive; –, negative; nd, not determinated. Biol. Diag., Biological diagnosis suggested by authors: +, positive; –, negative; i, inconclusive.

the case of toxic samples, this factor can mask the real hazard of the suspected sample. Regarding the use of the micropore membrane, we are not sure of its convenience. It is necessary to compare different methods of filtering to determine their advantages and disadvantages in order to recommend or discard the use of the micropore membrane.

One alternative to increase the sensitivity of this assay is to reduce the age of larvae at the treatment. For a chronic exposure the parental flies may put to lay directly on instant food enriched with the test sample, one disadvantage for this method is that a low palatability of the food may reduce drastically the number of eggs produced, other inconveniences are the manipulation needed for the egg counting to obtain the male/females survival index which may cause alterations in the organisms development, and make the assay tedious. At this moment, we prefer to use third instar larvae because we can provide a similar amount of larvae for each sample with minimal manipulation. However, the use of chronic exposures should be recommended based on the objective of the research.

By comparing the two versions of the SMART, we found that the same conclusion can be obtained from both. Some disadvantages of the eye version are: the material cannot be stored for posterior analysis; the time required to scoring prevents that numerous samples can be assayed simultaneously; although one part of the statistical analysis is quantitative, it is difficult to compare the occurrence of multiple spots in the eyes. Some important advantages are that males are really sensitive to genotoxins, and the comparison with their sibs facilitates the interpretation of the potential hazard of a suspected sample.

On the other hand, the wing version disadvantages are: the material must be mounted to scoring the spots; for continuous monitoring the size of sample recommended in mutagenicity determinations seems to high (110 wings per treatment) (Frei and Würgler, 1995), an alternative is to use only 80 wings, however this reduces the possibility to detect slight differences. The advantages are that biological material can be stored to further analysis, this is important specially in this type of studies; multiple samples can be assayed simultaneously and scored later; the statistical analysis is easy and allows to determine the earliest response from the most susceptible organisms from a population which is an useful tool for biomonitoring.

It is important to say that both assays require to keep acceptable laboratory practices and that training is necessary to their performance, however, the information obtained from the *in vivo* system of *Drosophila* justifies them.

5. ACKNOWLEDGEMENTS

To Dr. Felipe Vazquez from Instituto de Ciencias del Mar, UNAM, to give the support to performance the chemical determinations for this study. To Alfredo Delgado, for technical assistance.

6. REFERENCES

Anwar W.A., and Gabal M.S. (1991) Cytogenetic study in workers occupationally exposed to mercury fulminate, Mutagenesis, 6: 3, 189–192.

Armienta M.A., Rodríguez R., and Villaseñor (1993) Estudio de reconocimiento de la contaminación por arsénico en la zona de Zimapán, Hidalgo. Report to the Municipality of Zimapán: Geophysics Institute, Universidad Nacional Autónoma de México, 85 pp.

Armienta M.A., Rodríguez R., and Cruz O. (1997) Arsenic content in hair of people exposed to natural arsenic polluted groundwater at Zimapan, Mexico, Bull. Environ. Contam. Toxicol., 59, pp. 583–589.

Arriaga N., and y Daniel R. (1996) Bases geológicas del origen y movilidad del arsénico en el Valle de Zimapán, Hidalgo, Tesis Ing. Geol., Escuela Superior de Ingeniería y Arquitectura, IPN, México.

C.N.A. (1992) Estudio de prospección geohidrológica y exploración geofísica en la zona de Tecozautla-Vizarron, Estados de Hidalgo y Querétaro. IGAMSA, Informe Técnico.

Cernß M., Pastorkovß A., SmÝd J., Bavorovß H., OcadlÝkovß D., Rossner P., and Zavadil J. (1996) Genotoxicity of industrial effluents, river waters, and their fractions using the Ames test and *in vitro* cytogenetic assay, J. Toxicol. Lett., 88: 1–3, 191–197.

Chroust K., KuglÝk P., Relichovß J., Holoubek I., Cβsasy J., Veselskß R., Ryskovß M., and BenedÝk J. (1997) *Drosophila melanogaster*, Vicia faba and Arabidopsis thaliana short-term bioassays in genotoxicity evaluation of air and soil samples from sites surrounding two industrial factories in the Czech Republic, Folia. Biol. (Praha) 43: 2, 71–78.

Demerec M. Ed. (1965) Biology of *Drosophila*, Hafner Publishing Co., 632 p.

Frei H., and Würgler F.E. (1988) Statistical methods to decide whether mutagenicity test data from *Drosophila* assay indicate a positive, negative, or inconclusive result, Mutation Res., 203: 297–308.

Frei H., and Würgler F. E. (1995) Optimal experimental design and sample size for the statistical evaluation of data from somatic mutation and recombination tests (SMART) in *Drosophila*, Mutation Res., 334: 247–258. [33]

García E., and Falcon Z (1993) Atlas, Porrua, Mexico, 50–51 p.

Gimmler-Luz M., Erdtmann B., and Balbueno R. (1992) Analysis of clastogenic effect of Porto Alegre drinking water supplies on mouse bone marrow cells, Mutat. Res., 279: 4, 227–231.

Gomez-Arroyo S., Armienta M.A., Cortes-Eslava J., and Villalobos-Pietrini R. (1997) Sister chromatid exchanges in Vicia faba induced by arsenic-contaminated drinking water from Zimapan, Hidalgo, Mexico. Mutat. Res., 394: 1–3, 1–7.

Gomez-Arroyo S., Hernandez-Garcia A., and Villalobos-Pietrini R. (1988) Induction of sister-chromatid exchanges in Vicia faba by arsenic-contaminated drinking water, Mutat. Res., 208: 3–4, 219–224.

Gonzalez-César E., and Ramos-Morales P. (1997) Sodium azide induces mitotic recombination in *Drosophila melanogaster* larvae, Mutation Res., 389: 157–165.

Graf U., Würgler F., Katz A., Frei H., Juon H., Hall C., and Kale P. (1984) Somatic mutation and recombination test in *Drosophila melanogaster*, Environmental Mutagenesis 6: 153–188.

Graf U., and Singer D. (1989) Somatic mutation and recombination test in *Drosophila melanogaster* (Wing spot test): effects of extracts of airborne particulate matter from fire-exposed and non fire-exposed building ventilation filters. In: Suter K.E., Gruntz U., and Schlatter Ch. (Eds.) (1989) Analytical and toxicological investigations of respiratory filters and building ventilation filters exposed to combustion gases of the chemical warehouse fire in Schweirzehalle, Chemosphere, 19(7): 1019–1109.

Graf U., Heo O.S., and Olvera O. (1992) The genotoxicity of chromium (VI) oxide in the wing spot test of *Drosophila melanogaster* is over 90% due to mitotic recombination, Mutation Res., 266: 197–203.

Hebert P.D., and Luiker M.M. (1996) Genetic effects of contaminant exposure–towards an assessment of impacts on animal populations. Sci. Total Environ., 191: 1–2, 23–58.

I.N.E.G.I. (1990) Zimapán de Reyes, Estado de Hidalgo, Cuaderno Estadístico Municipal. H. Ayuntamiento de Zimapán de Reyes, 86 pp.

I.N.E.G.I. (1994) Zimapán de Reyes, Estado de Hidalgo, Cuaderno Estadístico Municipal. H. Ayuntamiento de Zimapán de Reyes, 92 pp.

I.N.E.G.I. (1995) Tecozautla, Estado de Hidalgo, Cuaderno Estadístico Municipal. H. Ayuntamiento de Tecozautla, 96 pp.

Iwado H., Naito M., and Hayatsu H. (1991) Mutagenicity and antimutagenicity of air-borne particulates, Mutat. Res., 246(1): 93–102.

Jaylet A., Deparis P., and Gaschignard D. (1986) Induction of micronuclei in peripheral erythrocytes of axolotl larvae following *in vivo* exposure to mutagenic agents, Mutagenesis, 1: 3, 211–215.

Jaylet A., Gauthier L., and Fernandez M. (1987) Detection of mutagenicity in drinking water using a micronucleus test in newt larvae (Pleurodeles waltl), Mutagenesis 2: 3, 211–214.

Kilbey B.J., MacDonald D.J., Auerbach C., Sobels F.H., and Vogel E.W. (1981) The use of D. *melanogaster* in tests for environmental mutagens, Mutation Res., 85: 141–146.

Koren H. (1991) Handbook of Environmental Health and Safety. Principles and Practices, Lewis Publishers, 2 a. ed. Vol. I, pp. 109, 147–148, 243–244, 252.

Lewtas J., Claxton L., Mumford J., and Lofroth G. (1993) Bioassay of complex mixtures of indoor air pollutants. IARC Sci. Publ., 109: 85–95.

Lindsley D.L., and Zimm G.G. (1992) The genome of *Drosophila melanogaster*, Academic Press, New York, 1133 p.

Meybeck M., Chapman D., and Helmer R. Eds. (1990) GEMS: Global Environment Monitoring System. Global Freshwater Quality. A first Assessment, WHO, UNEP, 306 pp.

Nöthiger E. (1970) Sucrose density separation: A method for collecting large numerous of *Drosophila* larvae. DIS 45: 177.

Pandey P., McGowen R.M., Vogel E.W., and Butterworth F.M. (1995) Genotoxicity of polychlorinated biphenyl (PCB) and polynuclear aromatic hydrocarbon (PAH) mixtures in the white/white+ eye-mosaic assay, In: Biomonitors and biomarkers as indicators of environmental change. Eds: F.M. Butterworth, L.D. Corkum, and J. Guzmán-Rincón, Plenum, New York, 1995, 183–191.

Ramos Leal J.A. (1996) Parámetros estructurales que controlan la hidrodinámica de las aguas subterráneas en el área de Zimapán, Hgo., Tesis Maestría en Ciencias de la Tierra, UNAM, México.

Rao P.V. (1998) Statistical Research Methods in the Life Sciences, Duxbury, New York, 889 p.

S.A.R.H. (1977) Estudio geohidrológico, zona Ixmiquilpan—Zimapán, Estado de Hidalgo. Geohidrológica Mexicana, S.A. Reporte Técnico.

Steel G.D., Torrie J.H., and Dickey D.A. (1997) Principles and Procedures of Statistics. A Biometrical Approach, McGraw-Hill, New York, 666 p.

Steinkellner H., Mun-Sik K., Helma C., Ecker S., Ma TH., Horak O., Kundi M., and Knasmuller S. (1998) Genotoxic effects of heavy metals: comparative investigation with plant bioassays. Environ. Mol. Mutagen, 31: 2, 183–191.

Suter W. (1989a) Ames test: mutagenicity determination of material extracted from respiratory masks worn at the site of the fire. In: Suter K.E., Gruntz U., and Schlatter Ch. (Eds.) (1989) Analytical and toxicological investigations of respiratory filters and building ventilation filters exposed to combustion gases of the chemical warehouse fire in Schweirzehalle, Chemosphere, 19(7): 1019–1109.

Suter W. (1989b) Ames test: mutagenic activity of airborne particulate matter from fire-exposed and non fire-exposed building ventilation filters. In: Suter K.E., Gruntz U., and Schlatter Ch. (Eds.) (1989) Analytical and toxicological investigations of respiratory filters and building ventilation filters exposed to combustion gases of the chemical warehouse fire in Schweirzehalle, Chemosphere, 19(7): 1019–1109.

Tabor M.W., and Loper J.C. (1985) Analytical isolation, separation and identification of mutagens from nonvolatile organics of drinking water, Int. J. Environ. Anal. Chem., 19: 4 281–318.

Van Hummelen P., Zoll C., Paulussen J., Kirsch-Volders M., and Jaylet A. (1989) The micronucleus test in Xenopus: a new and simple '*in vivo*' technique for detection of mutagens in fresh water, Mutagenesis 4: 1, 12–16.

Vogel E. (1988) Summary report on the performance of the *Drosophila* assays. En: Ashby J., de Serres F., Shelby M., Margolin B., Ishidate M., and Becking G. (1988) Evaluation of short-term tests for carcinogens. WHO, Cambridge University Press, 2.227–2.285.

Vogel E., and Szakmary A. (1990) Basic principles and evaluation of results of assays measuring genotoxic damage in somatic cells of *Drosophila*, Mutation and the Environment, Part b: 149–158.

Vogel E.W., and Nivard M.J.M. (1993) Performance of 181 chemicals in a *Drosophila* assay predominantly monitoring interchromosomal mitotic recombination, Mutagenesis 8(1): 57–81.

Wilcox P., Williamson S., Lodge D.C., and Bootman J. (1988) Concentrated drinking water extracts, which cause bacterial mutation and chromosome damage in CHO cells, do not induce sex-linked recessive lethal mutations in *Drosophila*, Mutagenesis, 3: 5, 381–387.

Woodruff C.R., Mason J.M., Valencia R., and Zimmering S. (1985) Chemical mutagenesis testing in *Drosophila* V. Results of 53 coded compounds tested for the National Toxicology Program, Environmental Mutagenesis 7: 677–702.

Würgler F.E., and Vogel E.W. (1986) *In vivo* mutagenicity testing using somatic cells of *Drosophila melanogaster*, In: Chemical Mutagens, Vol. 10, F.J. de Serres (ed.), Plenum, New York, pp. 1–59.

Zimmering S., Mason J.M., Valencia R., and Woodruff R.C. (1985) Chemical mutagenesis testing in *Drosophila*. II. Results of 20 coded compounds tested for the National Toxicology Program. Environ. Mutagen., 7: 1, 87–100.

Zwanenburg T.S.B. (1989) Chromosomal aberrations induced by extracts of airborne particulate matter from fire-exposed and non fire-exposed building ventilation filters. In: Suter K.E., Gruntz U., and Schlatter Ch. (Eds.) (1989) Analytical and toxicological investigations of respiratory filters and building ventilation filters exposed to combustion gases of the chemical warehouse fire in Schweirzehalle, Chemosphere, 19(7): 1019–1109.

16

A NEW MULTISPECIES FRESHWATER BIOMONITOR FOR ECOLOGICALLY RELEVANT SUPERVISION OF SURFACE WATERS

ALMUT GERHARDT

LimCo International, An der Aa 5, D-49477 Ibbenbüren, Germany

1. INTRODUCTION

Biomonitoring can be seen as an applied field of ecotoxicology, where passive biomonitoring includes measurements of toxicant levels in selected organisms at the site of investigation, and active biomonitoring comprises transplantation of test organisms from a reference site or a culture to the site of investigation and measurement of toxic substances in their body (Arndt, 1992). On-line biomonitors, continuous (dynamic) biotests, biotest automates or biological early warning systems use the reactions of selected organisms, especially physiological and behavioral reactions to chemical or physical stress as a sensitive alarm system to detect exposure to toxicants within short exposure times and provide data about possible permit violations and need for secondary treatment. Biomonitors based on subcellular biological responses, e.g. protoplasten biotest, are better called biosensors (Hansen et al., 1994). Continuous biotests are based on the permanent renewal of test water in a flow-through system, which is realized in *Daphnia*-tests, fish tests, mussel tests and some bacteria tests. In semicontinuous biotests the water flow is stopped during a defined contact time (±30 min.) where the biological parameters are measured (Schmitz et al., 1994). Detailed overviews over different biotests are given elsewhere (e.g. VDI, 1987; Gruber and Diamond, 1988; Pluta et al., 1994; Gerhardt, 1999). Biological systems measure a summation parameter, "biological effect" of the water to be tested in contrast to chemical monitoring, which measures concentration levels of selected chemical compounds. Biological systems are a good complement to chemical monitoring as they integrate the potential toxic effects of

Biomonitors and Biomarkers as Indicators of Environmental Change 2, Edited by Butterworth *et al.*
Kluwer Academic/Plenum Publishers, New York, 2000

different chemical compounds and their degradation products and toxic effects on the organism over time and they indicate the overall effects on the biocoenosis of the aquatic ecosystem (Gerhardt, 1995). On-line biomonitors offer a fast and cheap method to evaluate potential negative effects of pollutants on organisms, especially when chemical monitoring is impossible (unknown degradation products of organic toxicants) or unrealistic (syngergistic or antagonistic effects of toxicants in a toxic cocktail). The biological reactions of the organisms to a toxic cocktail can be responses to or effects of the pollution. A behavioral reaction can be 1) a trial to maintain homeostasis in the body by compensatory responses or 2) an overt effect at toxic concentrations above the maintenance capacity (Depledge, 1989; Wilson et al., 1994).

The development of on-line biomonitors follows the demands of the Agenda 21, where protection of the restricted water resources is highlighted. The basic idea of the use of automated biological sensor systems for water quality management was first proposed by Cairns et al. (1970). In Germany and the Netherlands the development and installation of on-line biomonitors has mainly started after the Sandoz accident at the River Rhine in 1986. The following biomonitors are used in Germany since 1990: Dynamic *Daphnia*-test (at 27 sites), Fish-Rheotaxis-test (at 14 sites), Koblenzer Verhaltensfischtest (at 1 site), Dreissena-Monitor (at 11 sites), Mossel-Monitor (at 1 site) at 14 rivers. Moreover, several algae- and bacteria-tests are in use since 1995 at ca. 8 sites (LAWA, 1996). Fish tests (Fish-rheotaxis-test, WRc-fishmonitor and Koblenzer Verhaltensfischtest) are no longer recommended, due to lack of sensitivity, high costs and ethical aspects (LAWA, 1996). The Deutsche Kommission zur Rinhaltung des Rheins (DK) recommends to replace fishtests with other, more sensitive biotests. One of the existing *Daphnia* tests and one algae-test should be used in a biotest-battery including different trophic levels. However, these systems still have shortcommings and problems/failures, e.g. *Daphnia* tests: reproduction of *Daphnia* sp. in the test cuvette falsifies signals, filtration of the test water is needed to inhibit clogging of the sieves. For the different algae- and bacteria- tests, there is still a lack of information about on-line operation in the field (LAWA, 1996).

2. METHODS

2.1. Principle of the Multispecies Freshwater Biomonitor (MFB)

The baisc idea of using different species in an automated biomonitor has already been mentioned by Morgan et al. (1988). However, no multispecies system has been used in the field by now. The difficulty is to develop a technique by which all kinds of aquatic organisms, independently of their habitat preferences can be applied. Measurements of changes in the electrical field seem to be robust and widely applicable: Electrode chambers based on bipolar impedance conversion technique and connected to a Wheatstone Bridge circuit have earlier been built (Spoor et al., 1971; Swain et al., 1977) and been used for the measurement of behavioral patterns of marine fish (Wingard and Swanson, 1992), chironomids and daphnids (Heinis and Swain, 1986) and copepods (Gill and Poulet, 1986). The WRc fish monitor is based on the measurement of small voltage oscillations from the musculature of the fish by electrodes immersed in the water (Stein et al., 1994). The MFB, however, is based on quadropole impedance conversion. In this technique a high frequency alternating current is generated over a test chamber filled with medium (e.g. water, sediment) by a pair of electrodes attached on opposite chamber walls. A second pair of non-current-carrying electrodes measures the changes in the impedance due to the organisms' movements in the chamber. The organism generates signals, which mirror its movement patterns (Gerhardt et al., 1994). The construction and position of the two electrode pairs in the chambers is variable as long as a homogenous electrical field is achieved in the whole chamber: e.g. platinum thread windings arranged interwoven in each other or stainless steel plates arranged on opposite chamber walls (Figure 1). The chambers can be equipped with more than one electrode configuration, e.g. in order to measure the vertical migration of plankton organisms or swimming distances of fish. Moreover, the chambers can be placed in the water column, on the sediment or submerged in the sediment. For the control and collection of emerging adults, the chambers can also be equipped with emergence cages. The size, form and construction of the test chamber is variable which makes it possible to adapt to the ecological and behavioral demands of all kinds of aquatic organisms (Gerhardt et al., 1994; Gerhardt et al., 1998).

2.2. Description of Hard- and Software of the MFB

The MFB consists of a PC, a measuring instrument with up to 96 channels and the test chambers (Figure 2). The measuring instrument contains a power supply, a signal-generation card and signal-processing cards. The latter generate the sine signal over the 96 chambers with adjustable amplitude, and collect data from the test chambers with a sampling frequency of 20 Hz.

The data analysis algorithm is based on the spectrogram for impedance variations which is calculated by splitting the signals sucessively in intervals of 64 samples each and calculation of the discrete fast Fourier transform (FFT) with the Hamming function (Gerhardt et al., 1998). After application of the criteria for distinguishing between noise/inactivity and activity of an organism (Gerhardt et al., 1998), the result of the organisms' movements is a frequency histogram, which includes the relative amounts of slow movements at low frequencies (e.g. crawling) and fast movements at high frequencies (e.g. gill ventilation) for frequency bands of 0.5 Hz in width between 0–10 Hz (Figure 3).

The operation of the MFB-program written for Windows-NT 4.0 is easy and userfriendly. Up to 96 channels can be defined and named individually. For each channel or a sum of x (≤12) channels the following alarm criteria can be defined: 1) x % deviation in the last measurement of a fequency band from the running mean, which can be defined out of the last y values, 2) a distinction between day- and night-time activity, 3) mortality after z number of measurements with no activity and 4) a weighting of each frequency band, e.g. the higher frequencies, typical for

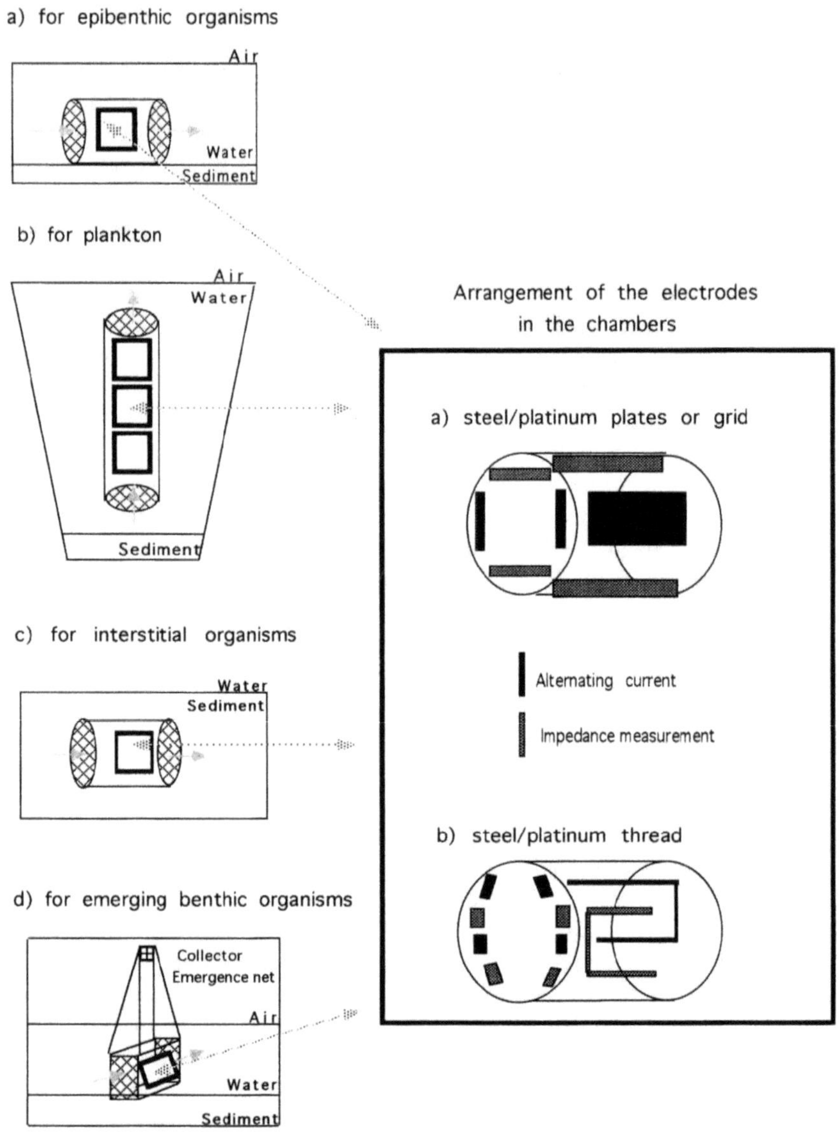

FIGURE 1. Different types of flow-through chambers for the MFB with different possibilities of arranging the electrodes: a) a small tube chamber with screwed lids with nylon netting at the in- and outflow placed on the stream bottom, b) a long vertical tube chamber with electrodes in different levels, c) a small tube chamber filled with sandy sediment, which can be placed in the stream sediment and d) a rectangular chamber with in- and outflow as described for the tube chambers, however being open on top, so that emerging insects can leave the water and are caught in the collector of the emergence net. The arangement of the elcetrodes can be done either with stainless steel or platinum plates or grids arranged on the opposite chamber walls or with thin steel or platinum threads interwoven in each other.

FIGURE 2. Setup of the MFB including test chambers, measuring instrument and PC.

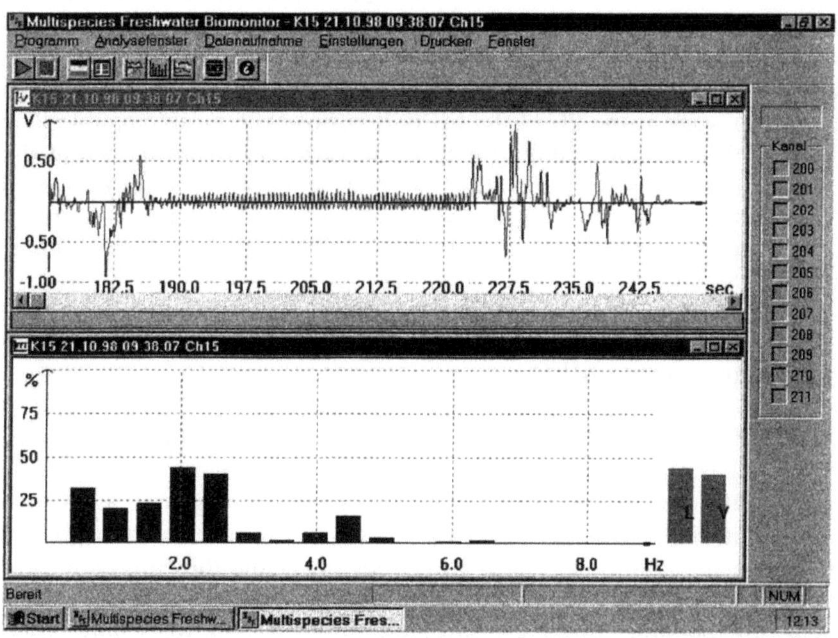

a

b

FIGURE 3. Normal activity pattern of *Gammarus pulex*. a) Behavioral signals of locomotion and ventilation in *Gammarus pulex*. Locomotion is characterized by high signal amplitudes and low frequencies (below 2.5 Hz), ventilation is characterized by low signal amplitudes and high frequencies (above 2.5 Hz). b) Frequency histogram of a recorded signal: Relative magnitudes of the occurrence of the frequencies 0.5 to 10 Hz, summarized in intervals of 0.5 Hz width, in the record of *Gammarus pulex*.

ventilation could be more important as an alarm criterium than the lower frequencies, typical for locomotion. A small change in ventilation could thus be defined as of higher alarm value than a big change in locomotion. Moreover, the background noise level, which depends mainly on the conductivity of the water, the flow-through conditions etc., can be adjusted. It is also possible to adjust the amplification of the signals (1, 2 and 4x). Also the calculation of the FFT can be adjusted according to different noise and threshold values. The threshold value determines the limit of how often and how long a fequency band has to occur in the signal in order to be taken into account for the FFT calculation. Also the mathematical method of the FFT can be chosen (Normal or Hamming). After these adjustments, the program starts to measure all selected channels on line for up to 4 mins. In case 96 channels are active, the data transfer and the calculations need ca. 5 min. after each measurement. In case fewer channels are active, the data transfer is faster and thus the on-line registration time can be longer than 4 min. The actual measurements of selected test chambers, the FFT-histograms and selected longterm diagrams, i.e. selected frequency bands over time are actualised after each measurement. An alarm bar gives the status of the "water quality alarm" according to the previously defined alarm criteria in 3 steps (o.k., warning, alarm). Such alarms that are generated for each species and can be compared and evaluated for an ecologically relevant "ecosystem alarm", e.g. in case three different locally important species have been chosen such as a fish, a crustacean and an insect and only the benthic insect shows a "warning" in one of two different behaviors no real "ecosystem alarm" would be given. However, if all three species would show alarms in one or two different behaviors a serious "ecosystem alarm" would be given. For the moment, the user has to work at the PC of the MFB. It is, however, easily possible to extend the system with a standard remote data administration software, so that the generated data, the actual screen and all the graphs can be watched on another PC via a GSM net. If the user wishes that the MFB announces an actual alarm on the remote PC, the PC has to be equipped with specific software, which still has to be developed.

2.3. Advantages of the MFB

For the first time in Europe, the use of freshwater benthic insects and crustaceans as well as marine invertebrates in an online biomonitor is possible. Benthic macroinvertebrates have been attractive targets for pollutant risk assessment and trend biomonitoring, because they are a diverse group reacting strongly and predictably to aquatic pollution (Cairns and Pratt, 1993) and they are important links in the aquatic foodweb (Garmendia-Tolosa and Axelsson, 1993). With the use of the MFB it is now possible to relate trend biomonitoring data to on-line monitoring data, which strenghtens the conclusions and evaluations for risk and hazard assessment, thus closing the gap between these two different approaches. Each test species has different thresholds of toxic effects and reaction times. For example, a reaction of a plankton organism to a water soluble toxicant may occur before the response of a benthic organism. On the other hand, a sediment-bound toxicant may result in a fast response of benthic organisms compared to swimming organisms. As all the test species are measured with the same method (quadropole impedance conversion), these different reactions can be compared and analysed in an intelligent, stepwise and ecologically-relevant "ecosystem alarm".

Moreover, the movement pattern of a test organism is composed of different types of behaviors, either performed sequencially or simultaneously, such as swimming, ventilation, cleaning, net spinning, looping etc. (Gerhardt et al., 1994). This allows for a multiparameter monitoring in contrast to the mono-parameter measurement in different biotests (e.g. Dreissena-Monitor, Fish-Rheotaxix-test). Different behaviors of different species that have different toxicant sensitivity, which allows for stepwise definition of an alarm-gradient according to the sensitivity of the species and the sensitivity of different behaviors. For example, the benthic freshwater macrocrustacean, *Gammarus pulex* (Amphipoda, Gammaridae) reacts to dissolved metal ions, salinity or acidity first by increased swimming behavior ("avoidance", trial to "escape" from the pollution source as first threshold of stress). As escape from the closed test chamber is not possible, *G. pulex* reacts now with increased gill ventilation, both in time and in ventilation frequency as a second threshold of stress (Figures 4 and 5). This series of stress responses can be described in a "stepwise stress threshold model" for acute stressors based on stress

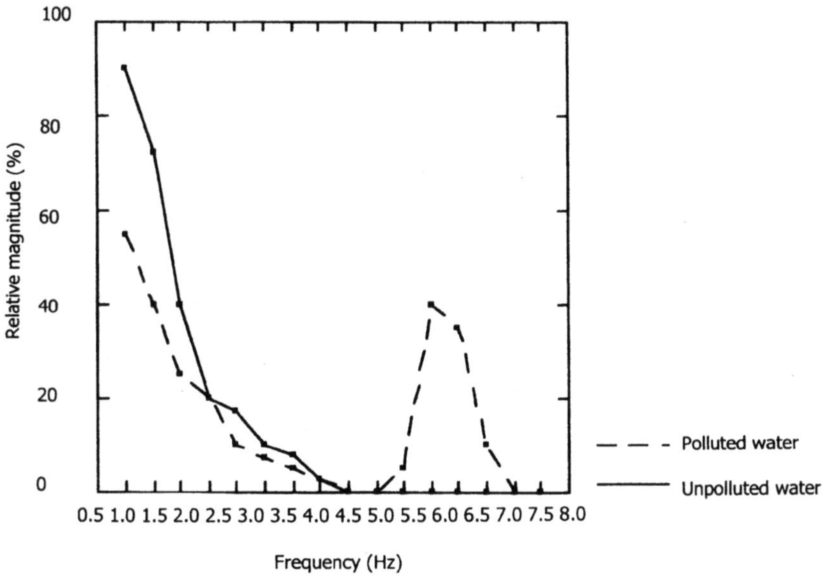

FIGURE 4. Comparison of the behavior of *Gammarus pulex* in unpolluted and in polluted water after 15 min. of exposure. Pollution, which was simulated in three different experiments with increased acidity, salinity and metal ions, always led to the observed typical changes in the frequency histogram of the behavioral signals, such as decreased low frequency behavior (locomotion), increased high frequency behavior (ventilation) and increased ventilation frequency. These changes are the basis for the alarm criteria for Gammaridae in the MFB.

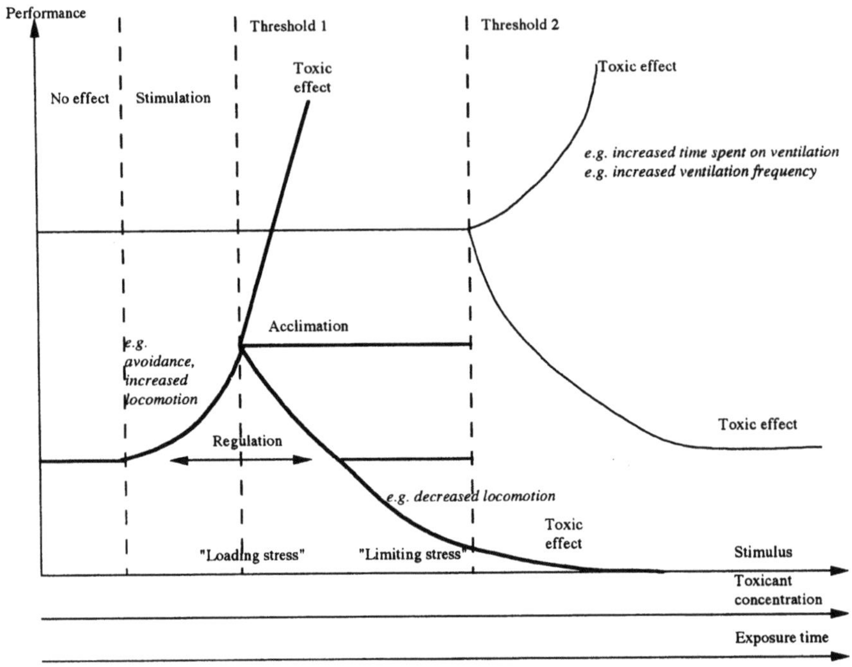

FIGURE 5. Stepwise stress model for behavioral responses to chemical stressors. The theoretical reaction of organisms in a certain performance (e.g. a certain behavior) in relation to either increasing stress stimuli or toxicant concentration in acute intoxication scenarios or to increasing exposure time of toxicants in chronic intoxication scenarios. An increasing stress stimulus provokes regulative responses ("loading stress"). Above a certain stimulus level, however, several reactions are possible: 1) the homeostasis cannot be maintained and a toxic effect in the organism occurs ("limiting stress"), 2) the organism can acclimate to the increased stress level and 3) the organism decreases the performance of the response and increases the performance of another response to the stimulus. If the first stress response decreases more than to the original level, a toxic effect occurs ("limiting stress").

descriptions by Selye (1936, 1973) Schweizer and Arndt (1996). In the future, the MFB has to be calibrated in field tests with as many aquatic species as possible in order to get background information for the evaluation of a software based on the above mentioned stress model for multiparameter and multispecies stress responses. As all kinds of aquatic organisms can be used in the MFB, it is also possible to take locally occurring key-species at the study site for use in the MFB, e.g. below a chemical factory the population density of *G. pulex* is drastically diminished, whereas it is the most abundant species above the factory. It is also possible to adapt the choice of test species to the seasonal abundance variations of the macroinvertebrates in the stream. For example, during the summer with decreased surface water flow, toxicant concentrations can be elevated. This should be monitored with the species being abundant at that time in the stream in order to achieve ecologically relevant water quality alarms. An "alarm" generated by a certain clone of the waterflea *Daphnia* sp., which are all genetically equal and have the same health condition, gives no relevant information to the potential damage of a pollution pulse in the ecosystem, which it is supposed to protect. In the ecosystem, however, individuals of a certain species are genetically different and some are parasitised, some are in good, some in bad health. Of course, this enlarges the variation in the measured responses, however, the measured responses are real and ecologically relevant. It might be that the MFB using locally occurring organisms reacts less sensitive to detect a toxic wave than using a standard laboratory cultured organism, however, the main application of on-line biomonitoring in surface waters is to protect the ecosystem from effects of toxic loads and not to indicate the lowest toxic concentration with the most sensitive laboratory organism. The latter case could however be senseful in drinking water preparation plants.

As all kinds of aquatic organisms can be used, the MFB system might be applied for monitoring of surface waters, drinking water as well as in purification plants for waste water inflow and purified water outflow and at oil platforms.

3. RESULTS AND DISCUSSION

3.1. Determination of Test-specific Sensitivity

The MFB based on quadropole impedance conversion technique has been built and tested with different aquatic organisms (Gerhardt et al., 1994). Changes of the behavioral pattern due to toxicants have been proven in acute toxicity tests (LC_{50}) and as early warning responses in static toxicity bioassays (Tables 1 and 2). In most cases, changes in the two behaviors locomotion and ventilation were found at metal concentration levels of ca. 1/100 of the respective LC_{50} values. Early warning responses were found within 30 to 90 min. depending on the types of toxicants, their site and mode of action and the sensitivity and reaction time of the test organism. For example, *G. pulex* exposed to a single dose of $60\,\mu g\ Cu^{2+} L^{-1}$ provoked changes in the number of active organisms, time spent on locomotion and ventilation (Gerhardt et al., 1998). Sensitivity also depends on the choice of population and origin of the test species. A population of *G. pulex* from a clean mountainous stream showed lower activity levels in the test stream than the local population, adapted to the water quality parameters of that stream (Gerhardt et al., 1998).

3.2. Long-term in Situ Operation of the MFB

The MFB has been used twice for three-week-long field tests, once along a small stream polluted by copper and agricultural pesticides (Gerhardt et al., 1998) and once along the River Rhine in winter 1997 (Gerhardt unpubl.). The MFB has been used as single species monitor with *G. pulex* (Gerhardt et al., 1998) in the first field test and as multispecies monitor with eight individuals of local *Chaetogammarus sp.* (Crustacea), four individuals of the predatorous insects *Dinocras cephalotes* (Plecoptera) and the waterbug *Aphelocheirus sp.* (Heteroptera) from a nearby tributary at the River Rhine in 1997 (Gerhardt

TABLE 1. Laboratory toxicity tests with unfiltered stream water under static renewal (#), recirculating (∧) conditions and with flow-through of tap water (″).

Metal	Species	LC_{50} (120h)	Effect (120h)	Ref
Cd (#,∧)	*L. marginata* (Ephemeroptera)	5 mgL^{-1} (pH 5 and 7)	≥0.05 mgL^{-1}: decreased locomotion	*,1
	P. flavomaculatus (Trichoptera)	>2.5 gL^{-1} (pH 5)	≥100 mgL^{-1}: increased ventilation	*
Pb (#)	*B. niger* (Ephemeroptera)	1.0 mgL^{-1}	≥0.01 mgL^{-1} (72h): decr. locomotion	*
	L. marginata (Ephemeroptera)	1.3 mgL^{-1} (pH 5) 7.1 mgL^{-1} (pH 7)	≥0.5 mgL^{-1}: decr. locomotion no effect	2
	G. pulex (Crustacea)	>0.5 mgL^{-1} (pH 7)	≥0.01 mgL^{-1} (1h): incr. ventilation ≥0.05 mgL^{-1} (24h): decr. locomotion	3
Fe (#)	*L. marginata* (Ephemeroptera)	63.5 mgL^{-1} (pH 4.5) 98.0 mgL^{-1} (pH 7.0)	40.2 mgL^{-1} (EC50): decr. locomotion 70.0 mgL^{-1} (EC50): decr. locomotion	2
Cu (″)	*A. auriculata* (Ephemeroptera)	0.18 mgL^{-1} (96h)	≥0.046 mgL^{-1}: incr. abdomen undulations ≥0.277 mgL^{-1}: incr. ventilation ≥0.231 mgL^{-1}: incr. locomotion	4
	B. stenochorias (Gastropoda)	0.07 mgL^{-1} (96h)	≥0.06 mgL^{-1}: snails leave the water	4

*: Gerhardt (unpubl.), 1: Gerhardt (1992), 2: Gerhardt (1994), 3: Gerhardt (1995), 4: Gerhardt & Palmer (1998).

TABLE 2. Early warning responses of freshwater organisms to polluted surface water measured in the MFB under static (#) and flow-through (∧) conditions.

Pollution type	Species	Reaction time	Type of response	Ref.
1) Laboratory studies with unfiltered stream water				
Cu (3–50 μgL^{-1}) (#) (Ståstorpsbäcken)	*G. pulex* (Crustacea)	15 min. 15 min	decreased locomotion increased ventilation	1
Organics, metals (#) (Ybbarpsån)	*G. pulex* (Crustacea)	60 min.	decreased locomotion	2
(∧) (#)	*H. angustip.* (Trichoptera)	60 min. (night) 15 min. 15 min.	increased activity decreased ventilation lower ventilation frequency	2
PAH, metals (#) (Molengracht)	*Ch. thummi* (Chironomidae)	60 min.	increased variance in locomotion	3
Cd, Zn, Pb (#) (Dommel)	*Ch. thummi* (Chironomidae)	60 min.	decreased ventilation	3
Mining effluent (#) (metals, salts)	*O. mykiss* (Pisces)	90 min.	≥20% effluent conc.: decreased locomotion increased ventilation	4
2) *in situ* experiment in Ståstorpsbäcken				
Cu (70 μgL^{-1}) (∧) simulated peak	*G. pulex* (Crustacea)	60 min.	increased activity increased ventilation	5

The names of the streams are given in brackets.
1: Gerhardt (1995), 2: Gerhardt (1996), 3: Gerhardt (1995a:PHD), 4: Gerhardt (1998), 5: Gerhardt *et al.* (1998).

unpubl.). In both studies, unfiltered stream water was pumped in a bypass through aquaria in which the test chambers, each containing one test organism, were attached. The whole construction occupied a $2\,m^2$ place in a field laboratory next to the stream. The organisms survived in the chambers (3.2 cm in diameter, 9 cm long, 144.7 mL volume) until the end of the experiment after three weeks, survival being checked weekly and the chambers being cleaned twice a week from clogging with fine sediment and detritus with a water pistol. Local *G. pulex* reacted within 30 minutes to a simulated pollution peak of $60\,\mu g\ Cu^{2+}\,L^{-1}$ with increased number of active organisms, increased time spent on locomotion and ventilation, whereas no reaction in the drift of *G. pulex* was found in the stream when exposed to the same pollution peak (Gerhardt et al., 1998). In conclusion, the organisms in the biomonitor were more sensitive than those in the stream, thus being a good "alarm-indicator" for the protection of the natural stream population. During the long-term test at the River Rhine in 1997, a "water quality alarm" neither occurred nor was simulated. Consequently neither aberrant behavior nor increased mortality was found (Gerhardt unpubl.). At present, the MFB is being tested at several universities and water authorities in Germany, the Netherlands and China. The following topics are under investigation: 1) The use of *Daphnia* sp. in the above mentioned test chamber for plankton organisms in the MFB in the field; 2) Behavioral responses to varying abiotic factors in *Gammarus pulex* in comparison with the measurement of biomarkers in order to relate behavioral changes to biochemical causes; 3) Long-term field tests of the MFB with *Gammarus pulex* and other invertebrates in order to study seasonal activity changes.

3.3. Selection of Test Species

The MFB principially allows for the monitoring of all kinds of aquatic (in)vertebrate species.

However, several species-specific problems restrict the choice. Circadian activity rhythms are known for several aquatic macroinvertebrate species, which have often been measured as drift in the field (Elliott, 1968; Brittain and Eikeland, 1988; Cowan and Peckarsky, 1994). Increased nocturnal activity in the test chambers of the MFB as well as increased drift of organisms in the stream, where the biomonitor was exposed, was found for two different populations of *G. pulex* (Gerhardt et al., 1998). Increased nocturnal locomotory activity in the biomonitor could also be measured for the trichopteran *Hydropsyche* sp. (Figure 6). Increased locomotion as response to polluted water was found for larvae of *Hydropsyche angustipennis* in the test chambers of the biomonitor only during nighttime measurements (Gerhardt, 1996). The behavior of a single organism in its test chamber can change during prolonged exposure due to lack of competition and predation pressures, lack of food, lack of appropriate habitat structures, flow conditions, size of chamber etc. Examples for such changes have been found for *Hydropsyche siltalai*, exposed in a flow-through chamber for 72 hours, which showed decreased ventilation frequency and longer phases of ventilation towards the end of the recording period probably due to a lack of oxygen supply (Gerhardt et al., 1994). Rainbow trout *Oncorhynchus mykiss*

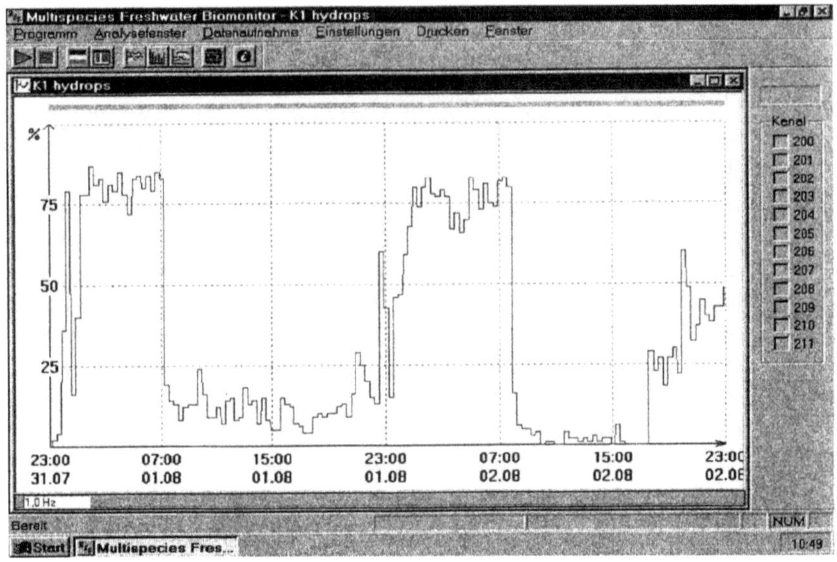

FIGURE 6. Diurnal locomotory activity rhythms of *Hydropsyche* sp. The relative magnitude of the frequency range 0.5–1 Hz, typical for locomotion signals is monitored every 10 min. over a period of 3 days.

decreased locomotion during 10 hours of exposure in a test chamber, probably due to a lack of swimming space (Gerhardt, 1998). Decreased activity of *Dionocras cephalotes* exposed in flow-through chambers with water from the Rhine showed less activity after 3 weeks than in the first week, probably due to a lack of prey stimuli and food supply (Figure 7). These changes can be independent of changes in water quality and have to be eliminated as much as possible in a reliable early warning system. Seasonal activity rhythms, activity decreases just before moulting, different inherent activities, habitat and food choices of different instars (e.g. Hydropsychidae) and availability of organisms throughout the year are further biological variables to be considered for the evaluation in an early warning system.

Many of these problems can be avoided by adapting the early warning system to the site and the season by choosing the respective dominant test species in its respective stage of life cycle, by taking a high number of replicates and by exposing the system directly in the stream, where unfiltered streamwater together with organismic drift can flow through the test chambers thus delivering food for the organisms. Moreover, optical and olfactorical presence of predators and competitiors can be simulated and the flow through conditions and light regime are realistic by in-stream exposure. The more artificial the conditions in a biomonitor, the more aberrant and variable the behavior of the test organisms, the more stressed the organisms and thus the more unrealistic a "water quality alarm" (more false alarms, alarms not interpretable). Some of the inherent biological behavioral variables could be taken into account by the use of expert systems and intelligent, self-learning software, e.g. neuronal nets and Fuzzy logics as being developed for the MFB.

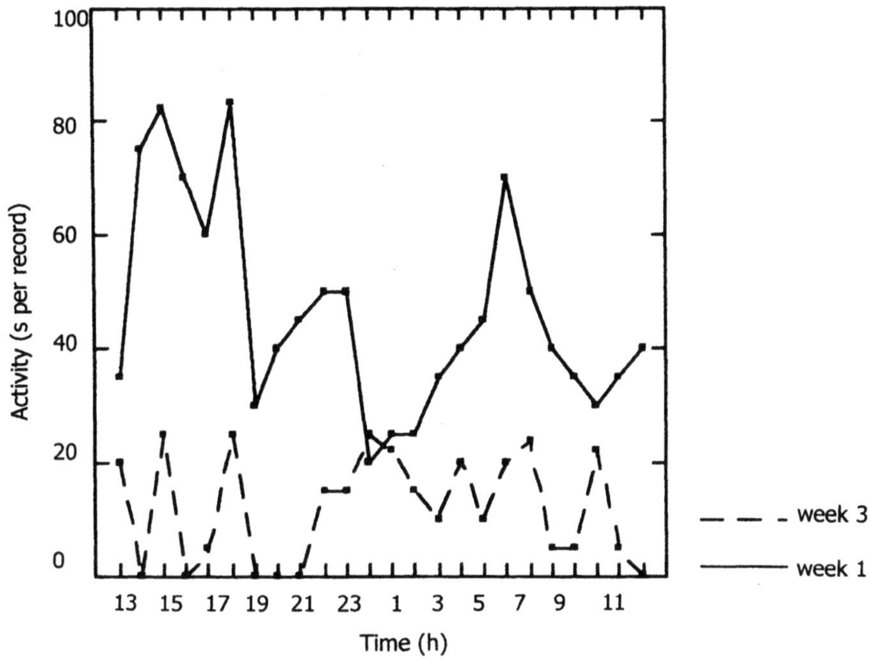

FIGURE 7. Diurnal activity of *Dinocras cephalotes* in week one and week three of *in situ* exposure to Rhine water. The curves represent means of eight organisms measured every hour for a recording time of 120s.

3.4. Comparison with Other Biotests

The MFB is a continuous biotest for aquatic (in)vertebrates and can therefore only be compared to other continuous biomonitors, such as *Daphnia*-tests, mussel and fish tests (Gerhardt, 1999). Following the criteria of the "Wirkungstest Rhein" (Umweltbundesamt, 1994) and the LAWA (1996), the MFB has a reaction time of ≥10min depending on the test species and type of pollution. The uninterrupted operation time (stand alone time) is ≥ one week, thus comparable to the *Daphnia*-tests, mussel and fish tests. The operation is easy and one-day training is sufficient for successful operation of the system. However, experienced and trained biologists are demanded for the evaluation of the data, especially the interpretation of the alarms. The maintenance requirements including cleaning of the system, collecting and replacing test organisms are estimated to 2–5h per week depending on the survival of the test species in the MFB, its abundance and ease of collection in the field. Similar times of weekly work efforts are demanded for the other biotests. The refractory time, i.e. the time needed to reinstall the whole system after damage of the organisms is ≤1 day. The alarm analysis of the MFB relies on dynamic thresholds as the other above mentioned biotests. The whole data analysis is well transparent and alarms can be followed up backwards stepwise through the calculation procedure until the original data, thus allowing for high plausibility control. Background data for stress responses to pollution are available, longterm drift data mostly for amphipods. The costs for a MFB-system with 96 channels are comparable to those of the *Daphnia*-tests and fish tests, which are only able to use much fewer specimens of only one species. The MFB offers multispecies and multiparameter biomonitoring with a simple, robust technique. The MFB can be used in marine and freshwater biomonitoring. The different adjustments in hard- and software make the system flexible and allow for definition of ecologically relevant alarm criteria based on different sensitivities of different species and their different behaviors. However, the system is more complicated than a single species and single parameter biomonitor and needs more time to be calibrated for each new species to be tested. As the MFB is able to monitor the stress behavior of fish, *Daphnia* sp. and other invertebrates, it is an attractive choice to replace the simpler and less sophisticated tests in future.

4. CONCLUSIONS

The MFB is a robust and flexible multispecies biomonitor, thus allowing for different applications for monitoring surface water, drinking water and waste water with respectively selected test organisms. The MFB monitors all kinds of aquatic organisms which allows for locally-adapted, optimal-alarm evaluation. The use of local species in the MFB allows for a more ecological relevant water quality alarm than the alarm given by a biotest with a species, which doesn't occur in situ but originates from a standard laboratory culture. Drinking water might be monitored by groundwater amphipods, which might be more sensitive than standard biotest organisms and thus a better protection for human health. Finally, waste water flows in purification plants might be controlled with Tubificidae or Chironomidae instead of standard biotest organisms, which would not survive these conditions.

5. ACKNOWLEDGMENTS

Herewith I would like to thank Beckmann & Farwerk Ingenieurgesellschaft for cooperation in the recent development of the MFB soft- and hardware. I am grateful to "Landesanstalt für Umweltschutz Karlsruhe" to allocate space and assistance for the field test at the River Rhine in 1997.

6. REFERENCES

Arndt, U. 1992. Einführung in die Bioindikation. In: Bioindikatoren für Umweltbelastungen. Neue Aspekte und Entwicklungen. Hrsg. Kohler, A. and U. Arndt. Hohenheimer Umwelttagung, 24: 13–19.

Brittain, J.E., and T.J. Eikeland, 1988. Invertebrate drift—a review. Hydrobiologia 166: 77–93.

Cairns, J. Jr., K.L. Dickson, R.E. Sparks, and W.T. Waller, 1970. A preliminary report on rapid biological information systems for water pollution control. J. Water Pollut. Control. fed. 42(5): 685–703.

Cairns, J. Jr., and V.H. Pratt, 1993. A history of biological monitoring using benthic macroinvertebrates. In: Freshwater biomonitoring and benthic macroinvertebrates. Eds.: Rosenberg, D.M. and W.A. Resh, Chapman & Hall, New York, 488 pp.

Cowan, C.A., and B.L. Peckarsky, 1994. Diel feeding and positioning periodicity of a grazing mayfly in a trout stream and a fishless stream. Can. J. Fish. Aquat. Sci. 51: 450–459.

Depledge, M.-H. 1989. The rational basis for detection of early effects of marine pollutants using physiological indicators. Ambio 18(5): 301–302.

Elliott, J.M. 1968. The daily activity patterns of mayfly nymphs (Ephemeroptera). J. Zool. London 155: 201–221.

Garmendia-Tolosa, A.J., and B. Axelsson, 1993. Gammarus, their biology, sensitivity and significance as test organisms. Swedish Environmental Research Institute, IVL-report B 1095 88 pp, Stockholm.

Gerhardt, A. 1992. Acute toxicity of Cd in stream invertebrates in relation to pH and test design. *Hydrobiologia* 239: 93–100.

Gerhardt, A. 1994. Short term toxicity of iron (Fe) and lead (Pb) to the mayfly *Leptophlebia marginata* (L.) (Insecta) in relation to freshwater acidification. *Hydrobiologia* 284: 157–168.

Gerhardt, A., M. Clostermann, B. Fridlund, and E. Svensson, 1994. Monitoring of behavioural pattern of aquatic organisms with an impedance conversion technique. Environment International 20(2): 209–219.

Gerhardt, A. 1995. Effects of metals on stream invertebrates. phD thesis at Lund University, dept. of Ecology, Ecotoxicology. 188 pp.

Gerhardt, A. 1995. Monitoring behavioural responses to metals in Gammarus pulex (L.) (Crustacea) with impedance conversion. Environ. Sci. & Pollut. Res. 2(1): 15–23.

Gerhardt, A. 1996. Behavioural early warning responses to polluted water.—Performance of Gammarus pulex L. (Crustacea) and Hydropsyche angustipennis (Curtis) (Insecta) to a complex industrial effluent. Environ. Sci. & Pollut. Res. 3(2): 63–70.

Gerhardt, A., A. Carlsson, C. Ressemann, and K.P. Stich, 1998. New online biomonitoring system for Gammarus pulex (L.) (Crustacea): In situ test below a copper effluent in South Sweden. Environm. Sci. & Technol. 32(1): 150–156.

Gerhardt, A. 1998. Whole effluent toxicity testing with Oncorhynchus mykiss (Walbaum 1792): Survival and behavioural responses to a dilution series of a mining effluent in South Africa. Arch. Environm. Contam. Toxicol. 35: 309–316.

Gerhardt, A., and C. Palmer, 1998. Copper tolerances of *Adenophllebia auriculata* (Insecta) and *Burnupia stenochorias* (Mollusca) in indoor artifical streams. *The Science of the Total Environment* 215: 217–229.

Gerhardt, A. 1999. Recent Trends in Online Biomonitoring for Water Quality Control. In: Biomonitoring of Polluted Water—Reviews on Actual Topics. Ed.: A. Gerhardt, Environmental Research Forum 9 (2000), 95-119, Trans Tech Publications LtD, Switzerland.

Gill, C.W., and S.A. Pulet, 1986. Utilization of a computerized micro-impedance system for studying the activity of copepod appendages. J. Exp. Mar. Biol. Ecol. 101: 193–198.

Gruber, D., and J. Diamond (eds.), 1988. Automated Biomonitoring: Living Sensors as Environmental Monitors. Ellis Horwood Limited Publishers, Chichester, 206 pp.

Hansen, P.D., Stein, P., and H.J. Löbbel, 1994. Entwicklung, Erprobung und Implementation von Biotestverfahren zur Überwachung des Rheins. Teil 3. Institut für Wasser, Boden und Lufthygiene des Bundesgesundheitsamtes Berlin.

Heinis, F.W., and R. Swain, 1986. Impedance conversion as a method of research for assessing behavioural responses of aquatic invertebrates. Hydrobiol. Bull. 19: 183–192.

LAWA 1996. Empfehlungen zum Einsatz von kontinuierlichen Biotestverfahren für die Gewässerüberwachung. Länderarbeitsgemeinschaft Wasser (LAWA) (Hrsg.), Kulturbuchverlag, Berlin, 37 pp.

Morgan, E.L., R.C. Young, and J.R. Wright, Jr., 1988. Developing portable computer-aided biomonitoring for a regional water quality surveillance network. In: Automated Biomonitoring: Living Sensors as Environmental Monitors. Eds.: Gruber, D., and J. Diamond, John Wiley & Sons, New York, Chapt. 9: 127–145.

Pluta, H.J., J. Knie, and R. Leschber (eds.), 1994. Biomonitore in der Gewässerüberwachung. Schriftenreihe des Vereins für Wasser-, Boden- und Lufthygiene 93, Gustav Fischer Verlag, Stuttgart.

Schmitz, P., U. Irmer, and F. Krebs, 1994. Automatische Biotestverfahren in der Gewässerüberwachung. In: Biomonitore in der Gewässerüberwachung. Eds. Pluta, H.J., J. Knie, and R. Leschber. Schriftenreihe des Vereins für Wasser-, Boden-, und Lufthygiene 93, Gustav Fischer Verlag, Stuttgart.

Schweizer, B., and U. Arndt, 1996. Auswertung der Waldschadensforschungsergebnisse (1982–1992) zur Aufklärung komplexer Ursache-Wirkungsbeziehungen mit Hilfe systematischer Methoden. Hier: Bioindikation mit synökologischen Aspekten. Umweltbundesamt Bericht 104 01 110: 293–382.

Selye, H. 1936. A syndrome produced by diverse nocuous agents. Nature 138: 32.

Selye, H. 1973. The evolution of the stress concept. Am. Sci. 61: 693–699.

Spoor, W.A., I.W. Neiheisel, and R.A. Drummond, 1971. An electrode chamber for recording respiratory and other movements of free-swimming animals. Trans. Am. Fish. Soc. 100: 22–28.

Stein, P., P.D. Hansen, and H.J. Löbbel, 1994. Erprobung des WRC-Fischmonitors zur Störfallüberwachung. In: Biomonitore in der Gewässerüberwachung. Eds. Pluta, H.J., J. Knie, and R. Leschber, Gustav Fischer Verlag, Stuttgart.

Swain, W.R., R.M. Wilson, R.P. Neri, and G.S. Porter, 1977. A new technique for remote monitoring of activity of freshwater invertebrates with special reference to oxygen consumption by naids of Anax sp. and Somatochlora sp. (Odonata). Can. Entomol. 109: 1–8.

Umweltbundesamt 1994. Continuous Biotests for Water Monitoring of the River Rhine. Summary, Recommendations, Description of Test Methods. Umweltbundesamt Texte 58, Berlin.

VDI (Verein Deutscher Ingenieure) 1987. Grundlagen zur Kennzeichnung vollständiger Meßverfahren.—Begriffsbestimmungen VDI-Richtlinien 2449. VDI Handbuch Reinhaltung der Luft, Band 5.

Wilson, R.W., H.L. Bergman, and C.M. Wood, 1994. Metabolic costs and physiological consequences of acclimation to aluminium in juvenile rainbow trout (Oncorhynchus mykiss). 2. Gill morphology, swimming performance and aerobic scope. Can. J. Fush. Aquat. Sci. 51: 536–544.

Wingard, C.J., and C.J. Swanson, 1992. Ventilatory responses of four marine teleosts to acute rotenone exposure. J. Appl. Ichthyol. 8: 132–142.

17

CYTOGENETIC AND CYTOTOXIC DAMAGE IN EXFOLIATED CELLS AS INDICATORS OF EFFECTS IN HUMANS

MARIA E. GONSEBATT, PATRICIA GUZMAN, AND JULIANA BLAS

Dept de Genética y Toxicología Ambiental, Instituto de Investigaciones Biomédicas, AP 70228, Ciudad Universitaria, México D.F. 04510

The advantages of using exfoliated epithelial cells as indicators of effects in individuals at risk is discussed. Due to the normal desquamation of epithelia, cells can be collected with little or no discomfort for the donor and by using cytogenetic and cytopathological criteria, useful information with regards to the geno- and cytotoxic consequences of exposure at a given target tissue, can be obtained. Thus, dose-response models could be generated if tissue levels of toxics are determined at the same time in collected cells. Although the implementation of the assay requires some training, laboratory equipment demands are very basic and samples can be easily transported and stored making the assay an ideal one for field studies.

1. INTRODUCTION

Epithelial tissues cover the external surface of the human body and the inner surface of the hollow organs such as nose, mouth, stomach, bladder, renal tubules, vagina, etc. Cells are directly attached to the basement membrane (gastrointestinal tract, nasopharynx, etc.), in simple or single-layered epithelia, or they can be stratified and only the basal cells are in direct contact with the basement membrane and the remaining ones overlap each other in a variable number of layers (oral cavity, esophagus, vagina, bladder, etc.). Cells shed after a period of time which varies from organ to organ. For example, the average life span in the nasopharynx is 15 days while in the bladder it could take 60 days. Replacement takes place due to the

Biomonitors and Biomarkers as Indicators of Environmental Change 2, Edited by Butterworth *et al.*
Kluwer Academic/Plenum Publishers, New York, 2000

continue mitotic divisions of basal undifferentiated cells. After division, epithelial cells mature and differentiate. These normal changes as well as pathologic alterations can be observed in desquamated or exfoliated cells, that is why they have been widely used in cytopathologic exams to detect abnormal, premalignant changes and cancer (Naib, 1996). Epithelia are in constant contact with air and water pollutants, and with ingested toxics as well as with their metabolites in body fluids such as urine, suffering the most direct and damaging encounters with the external world. According to Cairns (1975), most human cancers are epithelial in origin, 92% being derived from the external and internal epithelia, i.e. the skin, bronchial epithelium and the epithelia lining the alimentary canal.

Among the manifestations of a premalignant or a malignant condition are the chromosome aberrations and the association between specific cytogenetic alterations and tumorigenesis is strong (Mitelman, 1994). It is this relationship that is used as one justification for including cytogenetic endpoints in toxicological evaluation of industrial chemicals, and in the development of new pharmaceutical and therapeutic compounds (Tucker and Preston, 1996). Cytogenetic toxicity data, such as chromosomal aberration or micronuclei (MN) frequencies, are used also for ecological and environmental monitoring assessment and cleanup, workplace hazard evaluation and for cancer risk assessment (Petras et al., 1995; Backer et al., 1995; Hagmar et al., 1998). MN, these small round-shaped structures that stain like the cell nucleus (Müller and Streffer, 1994), are formed by the loss of whole chromosomes or portions of chromosomes from daughter nuclei at mitosis and exist separately from the main nucleus of a cell. The presence of MN is easier to score than chromosome aberrations and does not require metaphase cells as chromosome analysis does, so they have been observed and induced in several types of cells such as bone marrow, lymphocytes, erythrocytes, plant cells, etc. Müller and Streffer (1994) have published an excellent review of the existing micronucleus assays. In theory then, any proliferative tissue may generate MN due to abnormal chromosome segregation or to unrepaired chromosome breaks.

Another important event when evaluating exposure risk, is the induction of cell death because it can produce compensatory, regenerative cell proliferation that may provide a growth stimulus to spontaneously occurring preneoplastic cells, allowing them to overcome normal growth regulation. Increased cell division may also induce genetic damage by increasing the number of genetics errors due to rapid cell division and reduced time for DNA repair. (Cunningham, 1996). As mentioned above, structural changes in exfoliated cells due to normal natural aging or differentiation as well as those due to inflammation, viral or bacterial infections, premalignancy or malignancy and also changes due to exposure to radiation and other types of therapy, can be evaluated following cytopathologic criteria (Naib, 1996). These alterations not only include the generation of MN, but also the presence of binucleated cells or of nuclear anomalies such as "broken-eggs" or budding nuclei, karyorrhesis, pycnosis, karyolysis and nuclei with condensed chromatin. Different from MN, these nuclear atypia do not require cell division to occur and in some cell types the nuclear fragmentation observed in karyorrhesis (Figure 1, D and E) is associated with apoptosis (Muller and Streffer, 1994). Nuclear atypia appear at elevated levels in response to cellular injury (Livingstone, 1990; Naib, 1996; Tolbert, 1992), for example, cellular manifestations of cell death, such as karyorrhesis and pyknosis were frequently seen in postirradiation smears (Murad and August, 1985). Thus, it is possible to evaluate both cytogenetic as well as cytotoxic effects in exfoliated cells. Also important is the fact that because the potency of carcinogens for different tissues is not only dependent on differential exposure patterns, but also on the tissue-specific genotype or phenotype of target cells (Williams, et al., 1996), this assay offers the possibility of providing information of biological effects in tissues in direct contact with the environment and to calculate effective doses for safe limits determination in the assessment of risk.

Cytogenetic field studies in human, and in other species as well, can take advantage of the fact that these tissues have life-time mitotic activity. Since cells samples can be obtained with little or no discomfort, repetitive sampling is possible. DNA isolation and/or cell preservation is also a possibility if future studies are envisaged.

2. HUMAN MONITORING USING EXFOLIATED EPITHELIAL CELLS

Stich et al. (1982a, 1982b, 1983) and Stich and Rosin (1983) were the first to adapt the MN test to exfoliated human cells. According to them, their frequency appears to increase in carcinogen-exposed tissues long before any clinical symptoms are evident, although an increase in the frequencies of MN does not necessarily indicate the formation of preneoplastic lesions or carcinomas (Stich and Rosin, 1984). Also, in patients with Bloom syndrome or Ataxia Telangiectasia, genetic diseases characterized by unstable chromosomes and an increased susceptibility to cancer, high levels of micronucleated cells in buccal and urothelial cells have been observed (Rosin and German, 1985; Rosin and Ochs, 1986; Rosin et al., 1989). Moreover, agents such as radiation, known to induce chromosome damage in peripheral blood lymphocytes of exposed individuals, also induce micronuclei in irradiated epithelial tissues (Tolbert, 1992). The assay has been performed in exfoliated cells from the esophagus (Muñoz et al., 1987), the cervix (Fontham et al., 1986; Stich F.M., 1987), the nose (Sarto et al., 1990b; Ballarin et al., 1992) and the urinary bladder (Reali et al., 1987; Ribeiro et al., 1990; Anwar and Rosin, 1993; Gonsebatt et al., 1997). Elevated frequencies of MN have been demonstrated in these population groups with exposures associated with an elevated risk for the development of cancer in the studied tissues, validating the assay as a tissue-specific dosimeter of carcinogen exposure in humans (Rosin, 1992). For example, increased frequencies of MN were found in oral mucosa cells from smokers and tobacco chewers (Stich and Rosin, 1983; Mandar et al., 1987; Sarto et al., 1987; Livingston et al., 1990) and betel nut chewers (Stich et al., 1982b). Ethanol drinking and smoking seemed to be acting synergistically in buccal cells, while heavy smoking and coffee drinking elevated the frequency of MN in urinary bladder cells (Stich and Rosin, 1984). Workers exposed ethylene oxide (Sarto et al., 1990b), formaldehyde (Ballarin et al., 1992) or individuals drinking water with high levels of arsenic showed higher prevalences of cells with MN in urinary bladder and oral mucosa cells (Moore et al., 1996; Gonsebatt et al., 1997). Due to the continued renewal of the epithelium, it is possible to observed a reduction in the damage after an acute exposure or by using protective agents. In intervention studies designed by Stich and Rosin (1984) and Sarto et al. (1990a), it was possible to demonstrate a reduction of MN frequencies when individuals ceased receiving radio and chemotherapy, or were given chemopreventive agents such as β-carotene and retinyl palmitate (Stich and Rosin, 1984), or the antiparasitic agent praziquantel to treat bladder infestations by *S. haematobium* (Anwar and Rosin, 1993). A similar observation was made by Moore et al. (1997) in urinary bladder cells, when the ingestion of arsenic in the drinking water was reduced. It is also possible to measure numerical chromosome changes in epithelial tissues. For that purpose, Moore et al. (1993, 1996) developed the use of fluorescence in situ hybridization with centromeric chromosome probes.

Background MN frequencies show a great variability among reports, as can be noticed from Table 1. Since the degenerative processes produce anomalies difficult to distinguish from MN (Müller and Streffer, 1994), Tolbert et al. (1992) developed important protocol modifications to ensure reproducible MN frequencies among observers, which reduce the scoring of false-positive MN and will be discussed in the Methodology section. She also described pycnotic, condensed chromatin, karyorrhetic and karyolytic nuclei as nuclear anomalies related to necrosis and mentioned that pycnosis, condensed chromatin and karyorrhesis (but not karyolysis) appear as early stages of apoptosis and may also be an expression in the latter case, of genotoxic damage, and in the earlier stage of tissue injury or cytotoxic damage. Such changes are presumed to be reversible, if the source of irritation is eliminated. However, chronic irritation of the epithelial surface may eventually override cellular defense mechanisms leading to further changes such as metaplasia, dysplasia, and neoplasia (Livingstone, 1990).

Both cytogenetic and cytotoxic damage can thus give valuable information when evaluating risk exposure. Using cytopathological criteria for nuclei atypia classification we evaluated samples of epithelial cells from individuals at risk by exposure to arsenic, to observe the impact of air pollution in urban areas and to determine genotoxic effects in painters. The prevalence of nuclear anomalies in buccal cells in smokers was also investigated.

2.1. Arsenic Exposure by Drinking Water

Arsenic is a human carcinogen. Several thousands of individuals around the world, are exposed to arsenic (As) mostly by drinking well water with high arsenate/arsenite levels. They have an increased incidence of skin cancer and cancer of internal organs such as bladder and liver (Chen et al., 1985). To investigate the possibility of using the MN assay as an indicator of arsenic exposure, the prevalence of MN and of nuclear anomalies was investigated in exfoliated cells from the oral mucosa and from bladder in two groups of individuals: one living in Mexico (Gonsebatt et al., 1997) and a smaller one living in Chile (Aposhian et al., 1997). They have been chronically (life-time) exposed to high levels of arsenic in drinking water and in Mexico they have skin lesions associated with arsenic poisoning (Cebrián et al., 1984). Cells were collected in the field and immediately sent by air to the lab, where they were processed. A 4-fold increase in the frequency of

TABLE 1. Mean prevalence of micronucleated cells in exfoliated epithelial tissues of individuals at risk.

Exposure to:	Mean MN/1000 cells			Reference
	Cell type	Controls[1]	Exposed	
Bettel nut chewers	Buccal mucosa	4.1	46.8	Stich et al., 1985
Bettel quid chewers	"		72.5	"
Smoking and ethanol drinking	"		22.9	"
Smoking and coffee drinking	Urinary bladder	5.6	47	"
Smoking	"	0.66	1.16	Sarto et al., 1987
	Urinary bladder	0.5	2.9	Fontham et al., 1986
	Buccal mucosa	0.7	2.4	"
	Bronchus	1.2	2.8–3.6	"
	Uterine cervix	0.7	1.2–3.7	"
Tobacco chewers	Buccal mucosa	2.7	22.2	Livingston et al., 1990
Bloom's syndrome, heterozygotes	Buccal mucosa	2.8	4.4	Rosin and German, 1985
Bloom's syndrome, homozygotes	"		41.5	
Bloom's syndrome, heterozygotes,	Urinary bladder	4.2	3.0	
Bloom's syndrome, homozygotes	"		44.4	
Ataxia Telangiectaxia homozygotes	Buccal mucosa	2.9	10.2	Rosin et al., 1989
Ataxia Telangiectaxia heterozyg.	"		15.1	
Ataxia Telangiectaxia homozyg.	"	0.71	0.65[2]	Tomanin et al., 1990
Upper digestive tract cancers	"	1.3	2.6	Mandard et al., 1987
Schistosoma haematobium	Urinary bladder	5.3	71.7	Rafaat et al., 1984
Schistosoma haematobium	Urinary bladder	1.2	8.4	Anwar and Rosin, 1993
Smokers		0.9	9.5	"
Non-smokers		1.3	7.9	"
Radiotherapy patients	Buccal mucosa	0.3	5.2	Tolbert et al., 1992
Aromatic amines				
Smokers	Urinary bladder	3.0	4.2	Ribeiro et al., 1990
Non-smokers	"	2.6	3.2	
Leather tanning (chromate)	"			
Smokers	"	9.7	10.5	González Cid et al., 1991
Non-smokers	"	4.0	9.6	
Formaldehyde	Nasal	0.25	0.90	Ballarin et al., 1992
Ethylene oxide	Nasal	0.44	0.77	Sarto et al., 1990b
	Buccal mucosa	0.51	0.48[2]	"
Chromic acid	"	0.51	0.23	"
	Nasal	0.44	0.50[2]	"

[1] Control values reported by authors in Reference.
[2] No significant differences observed.

TABLE 2. Genotoxic and Cytotoxic endpoints in exfoliated epithelial cells in individuals exposed to arsenic in drinking water (Mean ± S.E.).

	As in water (μg/L)	As in urines (μg/L)	Buccal cells				Urothelial cells			
			MN	Karyorrhesis	Pycnosis	Karyolisis	MN	Karyorrhesis	Pycnosis	Karyolisis
Mexico										
Controls	29.88	34.0 ± 35.3	**0.56 ± 0.13**	22.93 ± 5.85	ND[1]	14.78 ± 2.45	**0.48 ± 0.10**	16.06 ± 2.94	0.03 ± 0.02	18.68 ± 2.61
Males			0.55 ± 0.15				0.58 ± 0.17			
Females			0.57 ± 0.15				0.43 ± 0.12			
Exposed	408.17	739.8 ± 361.4	**2.21 ± 0.47**[2]	23.72 ± 5.56	ND	14.15 ± 2.14	**2.22 ± 0.99**[2]	10.04 ± 2.94	2.00 ± 1.00	21.55 ± 3.17
Males			3.08 ± 0.80[2]				4.18 ± 2.35[2]			
Females			1.28 ± 0.37[2]				1.24 ± 0.72[2]			
Chile										
Controls	21.0	91.0 ± 17.0	**0.52 ± 0.31**	22.10 ± 10.75	0.74 ± 0.36	115.40 ± 40.32				
Exposed	593.0	605.0 ± 81.0	**1.96 ± 0.42**[2]	42.52 ± 10.80[3]	4.98 ± 0.77[3]	158.90 ± 31.60				

[1] Not determined;
[2] t-Test, $p < 0.05$;
[3] Mann-Whitney U-test, $P < 0.05$.

micronuclei in both epithelia was observed. Males were more affected than females, probably because they are more exposed to arsenic as field questionnaires revealed that men work in the fields and drink more water than women. A similar consideration was made by Hsueh et al. (1995) in Taiwan, who found a higher prevalence of skin cancer among men which they attributed to the fact that males drink more water or to a higher susceptibility to skin cancer than women. A significant prevalence of nuclear anomalies associated to an increased cell toxicity such as karyolysis, pycnosis and karyorrhesis was found, specially in the group living in Chile, where the levels of arsenic in water were higher (Table 2). This could result in a toxic insult to the oral mucosa and a more keratinized epithelium (Naib, 1996) which could explain, together with the smaller sample size, the lower frequency of MN observed in Chile as compared with the group studied in Mexico.

2.2. Air Pollution

Air pollution in the Valley of Mexico where Mexico City is located, represents a risk to human health due to the high levels of ozone, acidic sulfates, nitrogen oxides, heavy metals and polycyclic aromatic hydrocarbons present in the monitored air. Epidemiological studies have found an increase incidence of adverse respiratory symptoms among the residents of Mexico City (Castillejos et al., 1992, Fortoul et al., 1995). Histopathologic changes and DNA strand breaks have been observed in the nasal cells of residents of an area with high ozone levels (Calderón-Garcidueñas et al., 1996). Since the nasal respiratory epithelium is directly exposed to air pollution we decided to investigate the frequency of MN in this tissue and in buccal cells, in a group of 60 young adults (20–25 years old) that live in Mexico City and to compare it with the frequency of MN observed in a similar group of 60 individuals living in Mérida, an urban area of the Yucatan peninsula with lower air pollution. Volunteers completed questionnaires to give information about their health status, smoking habits and previous exposure to confounding factors such as radiation. Preliminary unpublished results from the analysis of twenty nasal cell samples from Mexico City residents and a similar number from Mérida residents, showed a significant prevalence of condensed, karyorrhetic and karyolytic nuclei among Mexico City residents but the MN frequencies were not significantly higher. The higher prevalence of nuclear anomalies observed in the samples obtained from Mexico City residents could represent a cellular injury induced by agents present in the complex mixture of polluted air. Sarto et al. (1990b), suggested that a cytostatic effect in basal dividing cells could prevent chromosomal damage to be expressed as MN.

2.3. Nuclear Anomalies in Smokers

If nuclear anomalies represented toxic insults in epithelial cells (Sarto et al., 1990b; Naib, 1996), smokers should have a higher prevalence of nuclear anomalies in buccal cells. We compared the frequency of broken-eggs, karyorrhetic, condensed

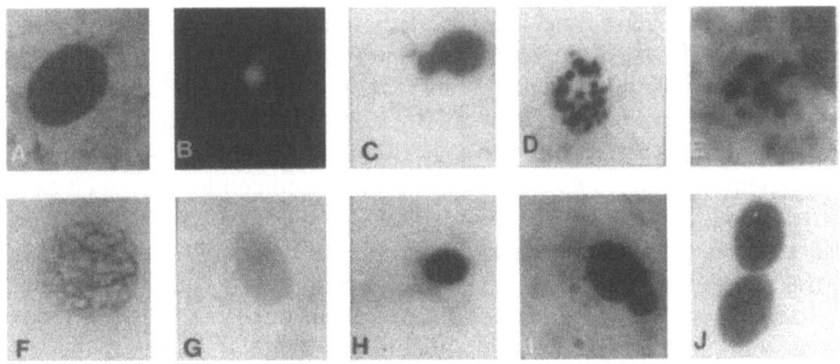

FIGURE 1. A: normal nucleus, B: main nuclei and a small MN (Hoescht 33258 stained); C and I: broken-eggs or budding nuclei; D and E: karyorrhetic nuclei; F: condensed chromatin; G: karyolisis; H: pycnotic nucleus; J: binucleation. Micronuclei, binucleation and broken eggs can be considered genotoxic events while picnosis, condensed chromatin, karyorrhesis and karyolysis could be associated with cytotoxity (Tolbert et al., 1992; Naib, 1996).

and karyolytic nuclei in a group of 10 individuals who smoked more than 20 cigarettes per day, with the prevalence observed in 36 nonsmokers. A significant elevation of some of the nuclear atypias as broken-eggs or budding nuclei, binucleated cells and nuclei with condensed chromatin or karyorrhetic (Figure 2) was observed. Livingston (1990) reported a similiar increased frequency of karyorrhetic nuclei in individuals exposed to smokeless oral tobacco (snuff). As mentioned above, nuclei with clumped or condensed chromatin or karyorrhetic (Figure 1) are considered cytological manifestations of cell injury (Livingstone, 1990; Tolbert, 1992), tissue changes which in this case were induced by cigarette smoking.

2.4. Painters

To investigate health risk in a group of outdoor painters that used paints with high lead content we conducted a monitoring study. The application of paints was performed with brushes or by aerosol spraying while the removal was mostly done by sand-blasting and burning the surfaces coated with lead paint without protection (masks or gloves). Samples of blood and of buccal cells were obtained from 25 painters and from 25 age- and gender-matched controls after group selection where individuals exposed to confounding factors such as heavy smoking and drinking were excluded. While organic solvents and/or their metabolites were not detected in blood samples, lead in blood was significantly elevated among painters. In 13 out of the 24 blood samples obtained from painters, lead levels were above the recommended limit of 10 μg/dL, while in the control group only 2 individuals showed levels higher than 10 μg/dL. Buccal cells showed higher frequencies of MN ($p < 0.01$) which were positively associated with exposure time (years working as an outdoor painter) and age ($r^2 = 0.453$; $p < 0.001$) (Table 4; Pinto et al., 1999 submitted).

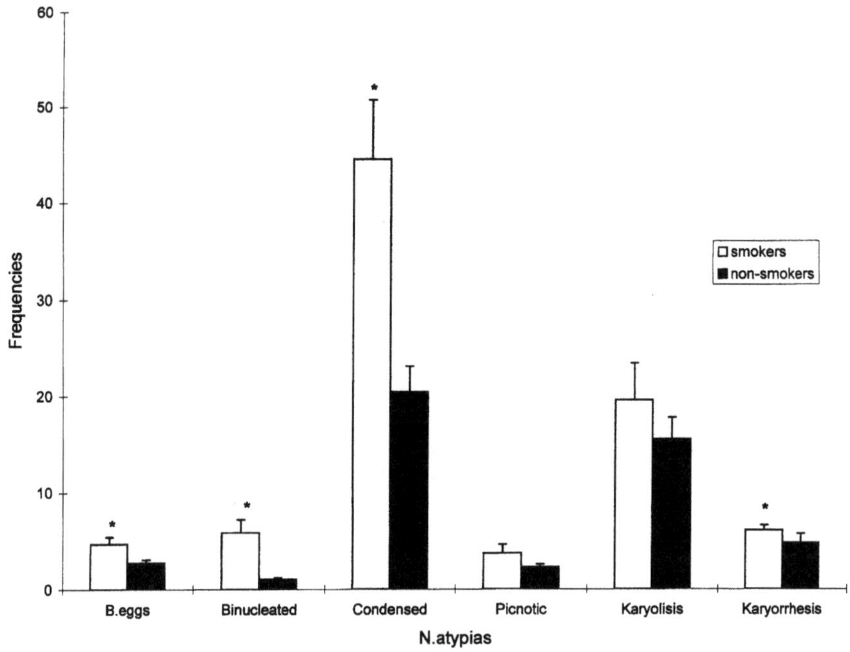

FIGURE 2. Frequency of nuclear atypia or anomalies in smokers and non-smokers. *Mann-Whitney U-test, $P < 0.005$.

3. EXPERIMENTAL PROTOCOL

To monitor a group at risk, it is important to identify a similar group that could serve as the non-exposed population to obtain the background prevalence of micronucleated cells. Since MN and nuclear anomalies occur following exposure to genotoxic and/or cytotoxic agents, a questionnaire should be completed by the volunteers where an estimation of the type and length of exposure can be made, specially if the presence of the agent cannot be determined in body fluids or tissues.

TABLE 3. MN and nuclear anomalies in young adults living in urban areas with different levels of air pollution.

Group	MN	Karyorrhesis	C. Chromatin	Karyolisis	Total % anomalies
Mexico City (n = 20)					
Nasal cells	0.20 ± 0.06	18.69 ± 3.51[1]	46.94 ± 15.70[1]	64.77 ± 24.72[1]	139.98 ± 36.77[1]
Buccal cells	0.26 ± 0.06	8.97 ± 1.51[1]	33.10 ± 3.16[1]	5.60 ± 0.84[1]	53.60 ± 4.11[1]
Merida (n = 20)					
Nasal cells	0.14 ± 0.02	3.33 ± 0.36	10.13 ± 0.89	1.57 ± 0.31	25.12 ± 2.05
Buccal cells	0.16 ± 0.03	3.50 ± 0.45	16.47 ± 1.12	3.79 ± 0.48	32.02 ± 2.35

[1] Mann-Whitney U-Test, $P < 0.05$.

TABLE 4. Lead blood levels, and MN frequencies in painters (Average ± Standard Error).

	Buccal cells (MN/1000)	Blood Lead (μg/dL)
Controls (n = 25)	0.32 ± 0.23	7.10 ± 2.79
Painters (n = 25)	1.19 ± 0.29[2]	10.48 ± 3.13[1]

[1] t-Test, $P < 0.05$,
[2] $P < 0.01$.

Previous exposure to potentially confounding factors such as X-rays and drinking and smoking habits must also be included. Since epithelial cells have different life spans, the renewal time of the tissues to be studied has to be known. For chronic exposures this should not be a problem. Most exposures are by ingestion or inhalation so buccal, nasal and urothelial cells would be in tissues in direct contact with toxics or their metabolites.

3.1. Exfoliated Cells Collection

3.1.1. Exfoliated Uroepithelial Cells

Cells are collected from fresh urine samples by centrifugation (1000–1200 rpm during 10 min.) in 50 ml conical tubes. Urine samples larger than 50 ml are advised to collect enough number of scorable cells. After centrifugation, the cell pellet is washed twice in 0.9% NaCl.

a) The cell pellet can be kept in this solution at low temperatures (1–4 °C), and be sent overnight to the lab (if sampling is done in the field) where the final analysis is going to the be performed or,
b) The cell pellet can be then fixed in the following way:
   - 2 or 3 consecutive washes with NaCl solution to remove mucus and debris,
   - remove the NaCl solution, add absolute methanol and gently resuspend the cells with a Pasteur pipette to allow a better fixation.

3.1.2. Exfoliated Buccal Cells

Cells are collected using the following procedure:

1) The donor is asked to vigorously rinse his/her mouth twice, with tap water
2) The inside of the mouth is rubbed two or three times with a premoistened wooden applicator. Sampling both cheeks should yield enough cells. You can practice by collecting your own buccal cells.
3a) The applicator is dipped in a 5 ml-tube, containing a buffer made of 0.1 M EDTA, 0.01 M Tris HCl, 0.02 M NaCl, pH 7.0 (Moore et al., 1993), and gently tapped against the bottom of the tube to allow cells to fall into

the solution. Gentle pipetting of cells in this solution reduces clumping. Samples are kept cool while sent to the lab.

3b) Another procedure is to dip the rod in a 5 ml-tube containing absolute methanol, tap against the bottom of the tube and pipette to reduce clumping. Fixation is done for 30 minutes and fixative is changed before storing cell pellets until analysis.

3.2. Slide Preparation

An adequate number of cells in the slides will facilitate scoring. Moore et al. (1993) recommended counting cells and placing a density of 1–2 million cell/ml per slide. Cell pellet can also be resuspended in a volume of fixative to obtain a cloudy suspension and then monitor the slide preparation under a microscope. A good preparation has no clumps and each cell can be scored separately from others. For immunohistochemical staining, slides should be treated with l-lysine to give a better adhesion to cells to avoid losing them during the procedure.

3.3. Nasal Cells

These cells are collected with a sterile cytological brush (frequently used for cervical smears), introduced in the upper turbinate. The procedure is a little uncomfortable but with a single sampling a large number of cells from the pseudostratified epithelia can be obtained. Cells are deposited in clean glass slides by gently pressing the brush on the slide. Fixation of cells is performed by placing the slides in methanol.

3.4. Materials and Solutions Needed

The amount would depend on the number of samples obtained:

50 ml plastic or glass conical tubes for centrifugation of urine samples, 5 ml plastic or glass tubes with cap, wooden applicators, masking tape (to label samples), pen, 0.09% NaCl solution, absolute methanol (Baker), solution made of: 0.1 M EDTA, 0.01 M Tris HCl, 0.02 NaCl, pH 7.

3.5. Staining and Scoring

Feulgen staining is performed as described by Stich et al., (1982a) with minor modifications. Slides are placed in 1 M HCl for 5 min and then 1 M HCl at 60 °C for another 5 min, they are rinsed in distilled water and put into Schiff's reagent for 10–20 min, rinsed again in tap water and air dried. Counterstaining with 1% fast green is optional since cytoplasm limit visualization can be performed using contrast phase illumination in combination with brightfield illumination. Giemsa staining can also be employed but since it is less specific for DNA containing structures, Feulgen is preferred. Flourescent DNA dyes such as Hoescht 33258 (20 minutes in a 0.05% Hoechst 33248 solution, then rinsed with tap water and covered with a few drops of Sorensen's phosphate buffer (pH = 6.8) and a coverslip) can also be employed, but then a fluorescent microscope with adequate filters is needed.

Developing a reliable scoring criteria is very important. We decided to use that recommended by Tolbert (1992): the cell has to have an intact cytoplasm with little or no overlapping with adjacent cells. A micronucleus is required to be rounded, with a smooth perimeter suggestive of membrane, smaller than a third the diameter of the main nucleus but large enough to discern shape and color, positive to Feulgen stain or to fluorescent DNA staining and staining similarly to the main nucleus, with the same intensity and texture, in the same focal plane as the nucleus and having no overlapping or bridge to the nucleus. To describe the nuclear anomalies the following criteria are used: binucleated cells present two nuclei in the same cytoplasm, both nuclei are very similar in shape, texture and staining proprieties. Broken eggs or budding nuclei are a MN-like portion of the main nucleus that is bridged to it (Montero et al., 1997). Pycnotic nuclei appear as shrunken nuclei heavily stained. Condensed chromatin appears aggregated in discrete bodies inside the nucleus. Karyorrhesis occurs when the nucleus disintegrates and the nuclear membrane disappears. Karyolysis is a ghost-like structure remaining of the nucleus as a Feulgen-negative structure in the cytoplasm (Figure 1).

3.5. Statistical Analysis

In most cases, the frequency of micronucleated cells can be described by a binomial or a Poissonian distribution, whereas the frequency of MN per cell shows overdispersion, pointing to a nonrandom distribution of MN among the daughter cells during mitosis (Müller and Streffer, 1994). Norman et al. (1985) reported that the natural logarithm of micronucleus frequency was more or less normally distributed, so that after transformation, tests based on normal distribution can be used. Due to the fact that the occurrence of MN is a rather rare event, Tolbert et al. (1992) suggested that if less than 5 micronucleated cells are observed after counting 1000 cells, an additional 1000 cells are scored, and so on up to a maximum count of 3000 cells.

4. ACKNOWLEDGMENTS

The authors want to thank the technical assistance of José Avilés and QFB Marta Luna. This work has been partially funded by PAPIIT 207196.

5. REFERENCES

Anwar, W.A., and M.P. Rosin, 1993. Reduction in chromosomal damage in schistosomiasis patients after treatment with praziquantel. Mutation Res., 298: 3, 179–185.

Aposhian, H.V., A. Arroyo, M.E. Cebrián, L.M. Del Razo, K.M. Hurlbut, R.C. Dart, D. González Ramírez, H. Kreppel, H. Speisky, A. Smith, M.E. Gonsebatt, P. Ostrosky-Wegman, and M. Aposhian, 1997. DMPS—Arsenic challenge test I: Increased urinary excretion of monomethylarsonic acid in humans given dimercaptopropane sulfonate, J. Pharmacol. Exp. Ther., 282: 192–200.

Backer, L.C., 1995. Pet dogs as sentinels for human exposure to environmental pollution, In: *Biomonitors and biomarkers as indicators of environmental change*, Butterworth, Corkum and Guzmán-Rincón (Eds.), Plenum Press, N. York, pp. 193–206.

Cairns, J., 1975. Mutational selection an the natural history of cancer, Nature (London), 255: 197–200.

Ballarin, C., F. Sarto, L. Giacomelli, G. Battista Bartolucci, and E. Clonfero, 1992. Micronucleated cells in nasal mucosa of formaldehyde-exposed workers, Mutation Res., 280: 1–7.

Calderón-Garcidueñas, L., N. Onaya-Brizuela, L. Ramírez-Martínez, and A. Villareal-Calderón, 1996. DNA strand breaks in human nasal respiratory epithelium are induced upon exposure to urban pollution. Environmental Health Perspectives, 104(2): 160–168.

Castillejos, M., D.R. Gold, and D. Dockery, 1992. Effects of ambient ozone on respiratory function and symptoms in Mexico City schoolchildren. Am. Rev. Respir Dis., 145: 276–282.

Cebrián, M.E., A. Albores, M. Aguilar, and E. Blakely, 1983. Chronic Arsenic Poisoning in the North of Mexico. Human Toxico., 2: 121–133.

Chen, C.J., Y.C. Chuang, T.M. Lin, and H.Y. Wu, 1985. Malignant neoplasms among residents of a Blackfoot disease-endemic area in Taiwan: high arsenic artesian well water and cancers, Cancer Res., 45: 5895–5899.

Cunningham, M.L., 1996. Role of increased DNA replication in the carcinogenic risk of nonmutagenic chemical carcinogens, Mutation Res., 365: 59–70.

Fontham, E., P. Correa, E. Rodriguez, and Y. Lin, 1986. Validation of smoking history with the micronuclei test, In: *Mechanism in Tobacco Carcinogenesis*, Hoffman, D., and Harris, C.C. (Eds.), Cold Spring Harbor, N. York, pp. 113–119.

Hsueh, Y.-M., G.-S. Cheng, M.-M. Wu, H.-S. Yu, T.-L. Kuo, and C.-J. Chen, 1995. Multiple risk factors associated with arsenic-induced skin cancers: effects of chronic liver disease and malnutritional status, British Journal of Cancer, 71: 109–114.

Fortoul, T.I., W. Lambert, M. Bliss, H. Bravo, P. Sánchez, I. Lopez, I. Sanchez, L. Villadermar, O. Serrano, and J. Samet, 1995. Acute changes in lung function associated with daily ozone exposures of children attending day camp in Mexico City. Am J. Crit. Care and Lung Physiol., 151: A496.

Gonsebatt, M.E., L. Vega, A.M. Salazar, R. Montero, P. Guzmán, J. Blas, L.M. Del Razo, G. García-Vargas, A. Albores, M.E. Cebrián, M. Kelsh, and P. Ostrosky-Wegman, 1997. Cytogenetic effects in human exposure to arsenic, Mutation Res., 386: 3, 219–228.

González Cid, M., D. Loria, M. Vilensky, J.L. Miotti, and E. Matos, 1991. Leather tanning workers: chromosomal aberrations in peripheral lymphocytes and micronuclei in exfoliated cells in urine, Mutation Res., 259: 197–201.

Hangmar, L., S. Bonassi, U. Strömberg, A. Brøgger, L.E. Knudsen, H. Norpa, C. Reuterwall, and the European Study Group on Cytogenetic Biomarkers and Health, Chromosomal Aberrations in Lymphocytes Predict Human Cancer: A report from the European Study Group on Cytogenetic Biomarkers and Health (ESCH), Cancer Res., 58: 4117–4121, 1998.

Livingston, G., R. Reed, B. Olson, and J. Lockey, 1990. Induction of nuclear aberrations by smokeless tobacco in epithelial cells of human oral mucosa, Environ. Mol. Mutagen., 15: 136–144.

Mandard, A.M., F. Duigou, J. Marnay, P. Masson, S.L. Qiu, J.S. Yi, P. Barrellier, and G. Lebigot, 1987. Analysis of the results of the micronucleus test in patients presenting upper digestive tract cancers and in non-cancerous subjects. Int. J. Cancer, 39: 442–444.

Mittelman, F., 1994. *Catalog of Chromosome Aberrations in Cancer*, 5th. Edn., Wiley, N. York.

Montero, R., L. Serrano G., and P. Ostrosky-Wegman, 1997. In vitro induction of micronuclei in lymphocytes: the use of bromodeoxyuridine as a proliferation marker, Mutation Res., 391: 135–141.

Moore, L.E., N. Titenko-Holland, P.J.E. Quintana, and M. Smith, 1993. Novel Biomarlers of genetic damage in humans: use of fluroescent in situ hybridization to detect aneuploidy and micronuclei in exfoliated cells, J. of Toxicology and Environ Health, 40: 349–357.

Moore, L.E., M.L. Warner, A.H. Smith, D. Kalman, and M.T. Smith, 1996. The use of fluorescent micronucleus assay to detect the genotoxic effects of radiation and arsenic exposure in exfoliated human epithelial cells, Environm. Mol. Mutagen, 27: 176–184.

Moore, L.E., A.H. Smith, C. Hopenhayn-Rich, M.L. Biggs, D.A. Kalman, and M.T. Smith, 1997. Decrease in bladder cell micronucleus prevalence after intervention to lower the concentration of arsenic in drinking water, Cancer Epidemiol Biomarkers Prev., 6: 1051–1056.

Müller, W.-U., and C. Streffer, 1994. Micronucleus Assays, In: *Advances in mutagenesis research*, Obe, G. (Ed.), Springer Verlag, Berlin, pp. 1–108.

Muñoz, N., M. Hayashi, J.B. Lu, J. Wahrendorf, M. Crespi, and F.X. Bosh, 1987. Effect of riboflavin, retinol, and zinc on micronuclei of buccal mucosa and of esophagus: A randomized double-blind intervention study in China, J. Natl. Cancer Inst., 79: 687–691.

Murad, T.M., and C. August, 1985. Radiation Induced Atypia, Diagnostic Cytopathol, 1: 137–152.

Naib, Z.M., 1996. *Cytophatology*, Little Brown & Co., Boston, pp. 1–14.

Norman, A., D. Bass, and D. Roe, 1985. Screening human populations for chromosome aberrations. Mutation Res., 143: 155–160.

Petras, M., M. Vrzoc, S. Pandragani, S. Ralph, and K. Perry, 1995. Biological monitoring of environmental genotoxicity in Southwestern Ontario, In: *Biomonitors and biomarkers as indicators of environmental change*, Butterworth, Corkum and Guzmán-Rincón (Eds.), Plenum Press, N. York, pp. 115–138.

Raafat, M., S. El-Gerzawi, and H.F. Stich, 1984. Detection of mutagenicity in urothelial cells of bilhardizial patients by the "micronucleus test", J. Egypt. Natl. Cancer Inst., 1: 63–67.

Reali, D., F. DiMarino, S. Bahrmandpour, A. Carducci, R. Barale, and N. Loprieno, 1987. Micronuclei in exfoliated urothelial cells and urine mutagenicity in smokers, Mutation Res., 192: 145–149.

Ribeiro, L.R., E.M.M. Cerqueira, D.M.F. Salvadori, H.S. Barbosa, and E.B. Whorton Jr., 1990. Monitoring of individuals occupationally exposed to aromatic amines In: *Mutation and the Environment*, Part C, Mendelsohn and Albertini (Eds.), Whiley-Liss, N. York, pp. 387–396.

Rosin, M.P., and J. German, 1985. Evidence for chromosome instability in vivo in Bloom syndrome: increased numbers of micronuclei in exfoliated cells, Hum. Genet., 71: 187–191.

Rosin, M.P., and H.D. Ochs, 1986. In vivo chromosomal instability in ataxia-telangiectasia homozygotes and heterozygotes, Hum. Genet., 74: 335–340.

Rosin, M.P., and A.M. Gilbert, 1990. Modulation of genotoxic effects in human, In: *Mutation and the Environment*, Part E, Mendelsohn and Albertini (Eds.), Wiley-Liss, N. York, pp. 351–359.

Rosin, M., 1992. Micronuclei as intermediate endpoints in intervention, In: *The biology and prevention of aerodigestive tract cancers*, Newell and Hong (Eds.), Plenum Press, N. York, pp. 95–104.

Rosin, M.P., H.D. Ochs, R.A. Gatti, and E. Boder, 1989. Heterogeneity of chromosomal breakage levels in epithelial tissue of ataxia telangiectasia homozygotes and heterozygotes, Hum. Genet., 83: 133–138.

Sarto, F., S. Finotto, L. Giacomelli, D. Mazzotti, R. Tomanin, and A.G. Levis, 1987. The micronucleus assay in exfoliated cells of the human buccal mucosa, Mutagenesis, 2: 11–17.

Sarto, F., R. Tomanin, L. Giacomelli, A. Canova, F. Raimondi, C. Ghiotto, and M.V. Fiorentino, 1990a. Evaluation of chromosomal aberrations in lymphocytes and micronuclei in lymphocytes, oral mucosa and hair root cells of patients under antiblastic therapy, Mutation Res., 228: 157–169.

Sarto, F., R. Tomanin, L. Giacomelli, G. Iannini, and A.R. Cupiraggi, 1990b. The micronucleus assay in human exfoliated cells of the nose and mouth: application to occupational exposures to chromic acid and ethylene oxide, Mutation Res., 244: 345–351.

Stich, H.F., J.R. Curtis, and B.B. Parida, 1982a. Application of the micronucleus test to exfoliated cells of high cancer risk groups: tobacco chewers. Int. J. Cancer, 30: 553–559.

Stich, H.F., W. Stich, and B.B. Parida, 1982b. Elevated frequency of micronucleated cells in the buccal mucosa of individuals at high risk for oral cancer: betel quid chewers, Cancer Lett., 17: 125–134.

Stich, H.F., B. Bohm, K. Chatterjee, and J. Saito, 1983. The role of saliva-borne mutagens and carcinogens in the etiology of oral and esophageal carcinomas of betel nut and tobacco chewers. In: *Carcinogens and mutagens in the environment*, Vol III, Naturally Occurring Compounds: Epidemiology and Distribution Stich, H.F. (Ed.), CRC Press, Boca Raton, pp. 43–58.

Stich, H.F., and M.P. Rosin, 1983. Quantitating the synergistic effect of smoking and alcohol consumption with the micronuclei test on human buccal mucosa cells, Int. J. Cancer, 31: 305–308.

Stich, H.F., and M.P. Rosin, 1984. Micronuclei in exfoliated human cells as a tool for studies in cancer risk and cancer intervention, Cancer Letters, 22: 241–253.

Stich, H.F., W. Stich, and M.P. Rosin, 1985. The micronuclei test on exfoliated human cells. Basic Life Sci., 34: 337–342.

Stich, H.F., 1987. Micronucleated cells as indicators for genotoxic damage and as markers in chemoprevention trials, Journal of Nutrition, Growth and Cancer, 4: 9–18.

Tolbert, P.E., C.M. Shy, and J.W. Allen, 1992. Micronuclei and other nuclear anomalies in buccal smears: methods development, Mutation Res., 271: 69–77.

Tomanin, R., F. Sarto, D. Mazzotti, L. Giacomelli, F. Raimondi, and C. Trevisan, 1990. Louis-Bar syndrome: spontaneous and induced chromosomal aberrations in lymphocytes and micronuclei in lymphocytes, oral mucosa and hair root cells, Hum. Genet., 85: 31–38.

Tucker, J.D., and R.J. Preston, 1996. Chromosome aberrations, micronuclei, sister chromatid exchanges, and cancer risk assessment, Mutation Res., 365: 147–160.

Williams, J.R., J. Russel, J.F. Dicello, and M.H. Mabry, 1996. The genotype of the human cancer cell: implications for risk analysis, Mutation Res., 365: 17–42.

18

p53 MUTATION LOAD

A Molecular Linkage to Carcinogen Exposure and Cancer

S. PERWEZ HUSSAIN AND CURTIS C. HARRIS

Laboratory of Human Carcinogenesis, National Cancer Institute, NIH, Bethesda, MD 20892

1. INTRODUCTION

Normal tissue homeostasis is maintained by the interplay of a number of different genes that control cellular proliferation and cell death. Many of these genes, for example, *ras*, *p53* and *Rb*, are altered in a variety of human cancers and are implicated in the natural history of human cancer. Several reports have described that each endogenous or exogenous mutagen induces a characteristic pattern of DNA alteration (Greenblatt et al., 1994; Patel et al., 1998), which can be displayed in a mutational spectrum that provides the type, location and frequency of alterations in a gene. The *p53* tumor suppressor gene is of particular interest because it is mutated in about half of all cancer types arising in a variety of tissues, and it manifests a high frequency of missense mutations (substitution of an amino acid in the encoded protein) (Hollstein et al., 1991; Levine et al., 1991; Ponder, 1988). The extensive study of *p53* mutation spectra has both provided information relevant to human cancer and generated several hypotheses. For example, a *p53* mutation spectrum can reflect the DNA damage of a particular carcinogen and aid in defining the biochemical mechanisms responsible for the genetic lesions in DNA that cause human cancer. The frequency and type of *p53* mutations can also act as a molecular dosimeter of carcinogen exposure and provide information about the molecular epidemiology of cancer risk. In this article, we will discuss how a characteristic mutation in the *p53* tumor suppressor gene can indicate the exposure to a particular mutagen or carcinogen in the environment. In addition, the *p53* gene can be a target to evaluate the mutagenicity of an unknown environmental pollutant, by using human cell cultures and a highly sensitive genotypic mutation assay (Aguilar et al., 1994; Aguilar et al., 1993; Hussain et al., 1994a; Hussain et al., 1994b).

Biomonitors and Biomarkers as Indicators of Environmental Change 2, Edited by Butterworth *et al.*
Kluwer Academic/Plenum Publishers, New York, 2000

2. THE *p53* TUMOR SUPPRESSOR GENE

p53 was first discovered by several independent groups of scientists as a cellular 53 KDa protein that was bound to the large T antigen in SV40 DNA tumor virus transformed cells and expressed at high levels in certain murine or human cancer cell lines (Lane and Crawford, 1979; Linzer and Levine, 1979; DeLeo et al., 1979; Kress et al., 1979). The *p53* gene was initially characterized as an oncogene because of its ability to efficiently induce neoplastic transformation of primary rodent cells when cotransfected with *RAS* and *E1A* (Eliyahu et al., 1984; Parada et al., 1984; Jenkins et al., 1984; Wolf et al., 1984). In fact, the *p53* cDNA constructs, isolated earlier from human or murine tumor cells, were all missense mutants. The wild-type gene isolated from normal cells failed to induce neoplastic transformation (Finlay et al., 1988; Eliyahu et al., 1988; Hinds et al., 1989), inhibited tumor cell growth and blocked neoplastic transformation of primary rodent cells by a mutated *p53* and *E1A* (Eliyahu et al., 1989; Finlay et al., 1989; Baker et al., 1990; Diller et al., 1990; Mercer et al., 1990; Chen et al., 1991). Then *p53* was found to be mutated in about 50% of all human cancers [reviewed in (Hollstein et al., 1991; Greenblatt et al., 1994)]. These facts then led to the reclassification of *p53* as a tumor suppressor gene.

The *p53* gene is well suited for mutational spectrum analyses for several reasons. First, *p53* mutations are common in many types of human cancer and a sizeable database of more than 10,000 entries has accrued, therefore, analyses from this large database can yield statistically valid conclusions (Hollstein et al., 1996; Hainaut et al., 1998), and the database is readily accessible on the world wide web (http://www.iarc.fr/p53/homepage.html). Second, the modest size of the *p53* gene (11 exons, 393 amino acids) permits the study of the entire coding region, and it is highly conserved in vertebrates, allowing the extrapolation of data from animal models (Soussi et al., 1990). Third, the missense mutations that alter *p53* function are distributed over a large region of the molecule, especially in the hydrophobic midportion of the protein (Hollstein et al., 1991; Levine et al., 1991; Greenblatt et al., 1994) (Figure 1). These numerous base substitutions alter *p53* conformation (Cho et al., 1994) and sequence-specific transactivation activity (Martinez et al., 1991; Michalovitz et al., 1990), thus correlations between distinct mutants and functional changes are possible. Frameshift (change in the reading frame of the entire subse-

FIGURE 1. Schematic of the p53 molecule. The p53 protein consists of 393 amino acids with functional domains and evolutionarily conserved domains containing regions designated as mutational hotspots. Functional domains include the transactivation region (amino acids 20–42, diagonal-striped block), sequence-specific DNA-binding region (amino acids 100–293), nuclear localization sequence (amino acids 316–325, vertical striped block), and oligomerization region (amino acids 319–360, horizontal-striped block). Cellular or oncoviral proteins bind to specific areas of the p53 protein. Evolutionarily conserved domains (amino acids 17–29, 97–292, and 324–352; black areas) were determined using the MACAW program. Seven mutational hotspot regions within the large conserved domain are identified: amino acids 130–142, 151–164, 171–181, 193–200, 213–223, 234–258, and 270–286 (checkered blocks). Vertical lines above the schematic, missense mutations.

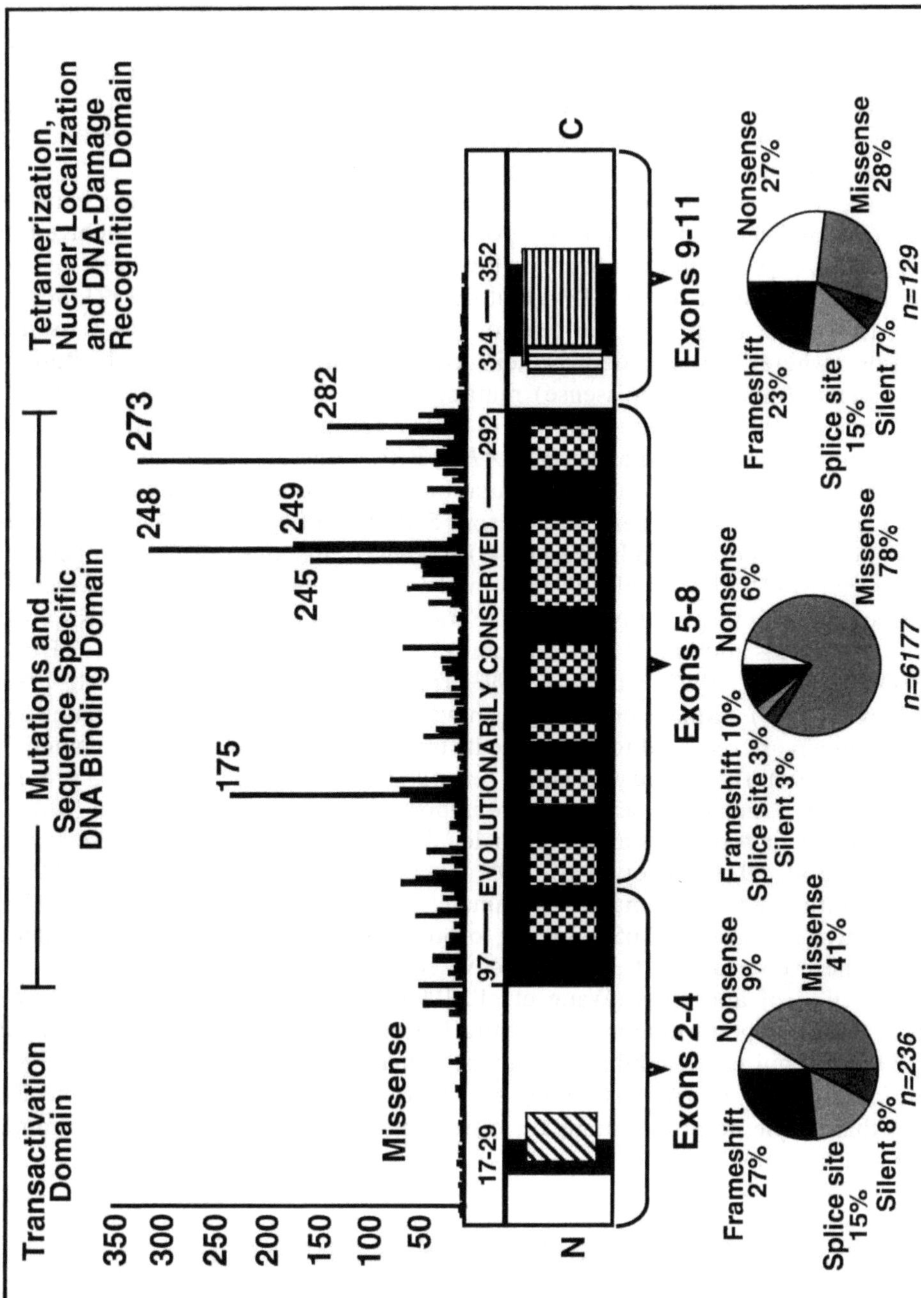
Transactivation Domain
Mutations and Sequence Specific DNA Binding Domain
Tetramerization, Nuclear Localization and DNA-Damage Recognition Domain
350
300
250
200
150
100
50
Missense
175
245
248
249
273
282
N
C
17-29
97
EVOLUTIONARILY CONSERVED
292
324
352
Exons 2-4
Exons 5-8
Exons 9-11
Frameshift 27%
Nonsense 9%
Missense 41%
Splice site 15%
Silent 8%
n=236
Frameshift 10%
Splice site 3%
Silent 3%
Nonsense 6%
Missense 78%
n=6177
Frameshift 23%
Nonsense 27%
Missense 28%
Splice site 15%
Silent 7%
n=129

quent coding sequence of the gene) and nonsense mutations (generation of a stop codon) that truncate the protein are generally located outside of these regions (Figure 1), and deletions and insertions are commonly found in exons 2–4 and 9–11 (Greenblatt et al., 1996), so evaluation of the entire DNA sequence yields relevant data. In contrast to *p53*, *ras* oncogenes possess transforming mutations that occur primarily in three codons, a few sequence-specific motifs and a critical functional domain (Park et al., 1989). Thus, the diversity of *p53* mutational events permits more extensive inferences for the mechanism of DNA damage involved.

Exogenous mutagenic agents and/or endogenous mutagenic mechanisms have been implicated in the induction of cancer. These mutations are archived in the spectrum of *p53* mutations found in human cancer (Hollstein et al., 1991; Levine et al., 1991; Greenblatt et al., 1994; Gottlieb and Oren, 1996; Harris, 1993; Soussi et al., 1994). The most common type of mutation found in *p53* is the missense mutation, where the encoded protein contains an amino acid substitution. Nonmissense (other than missense) mutations, e.g., deletions and insertions, are also found in the *p53* gene–but with a lower frequency–when compared with other tumor suppressor genes, such as *APC*, *BRCA* and *ATM*. In the distribution of missense mutations, these types of mutations occur more frequently in exons 2–4 (59%) and 9–11 (72%) rather than in exons 5–8 (22%) (Figure 1). The N-terminus of the p53 protein (encoded by exons 2–4) [reviewed in (Liu et al., 1993; Vogelstein and Kinzler, 1992; Thut et al., 1995; Lu and Levine, 1995)], has an abundance of acidic amino acids that are involved in the transcriptional function of *p53* (Raycroft et al., 1990; Fields and Jang, 1990), and binds to transcription factors such as TBP in TFIID (Seto et al., 1992; Liu et al., 1993; Truant et al., 1993; Martin et al., 1993; Mack et al., 1993). Experimental studies have shown that multiple point mutations in this domain are required to inactivate its transcriptional transactivation function (Lin et al., 1994). The carboxy-terminus (encoded by exons 9–11) of the p53 protein is enriched in basic amino acids that are important in: the oligomerization (formation of dimer or tetramer) and nuclear localization of the p53 protein [reviewed in (Clore et al., 1994; Jeffrey et al., 1995; Hupp and Lane, 1995; Lee et al., 1994)], the recognition of DNA damage (Jayaraman and Prives, 1995; Bakalkin et al., 1994), the negative regulation of p53 binding to promoter sequences of genes regulated by p53, the transcription of *p53*-transactivated genes (Horikoshi et al., 1995), and the induction of apoptosis (Wang et al., 1996). Laboratory studies have shown that at least two point mutations (substitution of a single base) in the N-terminus of *p53* are required to inhibit its transcriptional transactivity (Lin et al., 1994); therefore, deletions and insertions are more detrimental mutagenic mechanisms than single point mutations for disrupting these N-terminal and C-terminal functional domains.

3. MOLECULAR LINKAGE BETWEEN CARCINOGEN EXPOSURE AND CANCER

A number of specific *p53* mutational hotspots have been recognized in different types of human cancer (Figure 2). The occurrence of an identical mutation

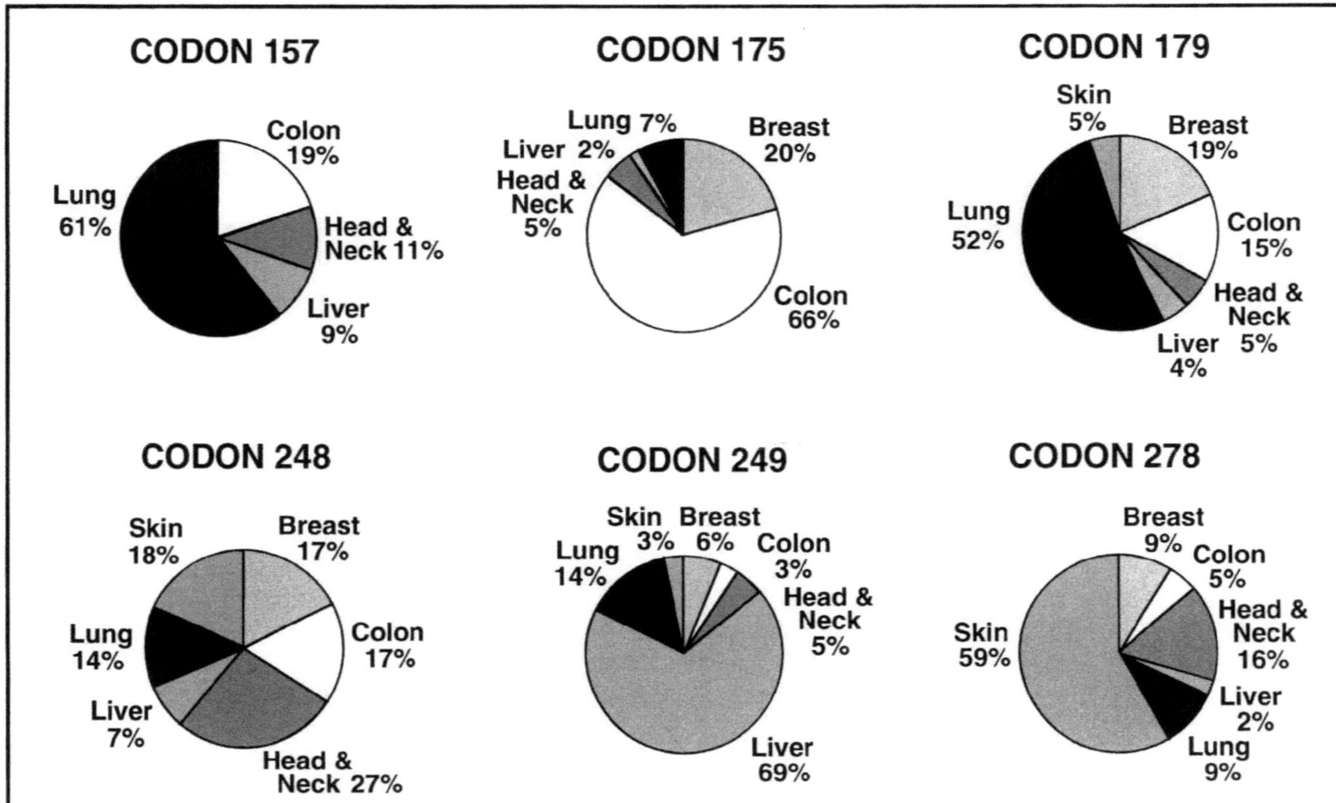

FIGURE 2. *p53* mutational hotspots in human cancers. Most types of human cancers show the domination of specific *p53* mutations at particular mutational hotspots. The characteristic patterns hypothesize molecular linkage between a particular cancer and a specific exogenous or endogenous carcinogen.

in human cancers, that is experimentally induced by a carcinogen, supports a causative role of a specific environmental carcinogen in certain tumor types [reviewed in (Hussain and Harris, 1998)]. Molecular linkages between exposure to carcinogens and cancer are best exemplified by the *p53* mutational spectrum of hepatocellular carcinoma, skin cancer and lung cancer.

The most common techniques used for *p53* mutation analysis include PCR-based assays like single strand conformational polymorphism (SSCP), denaturing gradient gel electrophoresis (DGGE), and DNA sequencing. Recently, another method based on yeast functional assays was developed to detect *p53* mutations (Scharer and Iggo, 1992; Ishioka et al., 1995; Moshinsky and Wogan, 1997). In this assay, the loss of DNA binding and transcriptional-transactivation function in mutant *p53* are detected by the colony color of yeast. Measuring the *p53* mutation load or the frequency of mutated alleles in nontumorous tissue may indicate previous carcinogen exposure and identify individuals at increased cancer risk. However, the detection of rare cells with mutations in a protooncogene or tumor suppressor gene in normal-appearing human tissue represents a challenging task. The average spontaneous mutation per base pair in human cells is estimated to be in the range of 10^{-8} to 10^{-10} and these frequencies increase only 10- to 1,000-fold upon exposure to a mutagen. Therefore, methods that allow the detection of a few altered DNA sequence from 10^5 to 10^{10} copies of corresponding wild-type sequences in the presence of large quantities of cellular DNA are required. The development of a highly sensitive genotypic assay (as described below) by Cerutti and coworkers has allowed the detection of low frequency mutations in normal-appearing human tissues as well as in cells exposed to an environmental carcinogen (Aguilar et al., 1994; Aguilar et al., 1993; Hussain et al., 1994a; Hussain et al., 1994b; Hussain et al., 1997). The detection of a particular mutation in normal-appearing tissue provides further support for the involvement of a specific carcinogen in a particular human cancer and may help identify individuals at increased cancer risk exposed to a particular pollutant in the environment.

3.1. Mutation Assay

The principal steps of this highly sensitive mutation assay are described in Figure 3. This assay allows the determination of the frequency of mutations which occur in restriction enzyme recognition sequences and render the mutated site resistant to cleavage by the corresponding endonuclease. The protocol includes the following steps.

The cellular DNA from 10^7–10^9 cells is enriched in the mutated sequence of the *p53* gene by exhaustive digestion with a chosen specific restriction enzyme (for example, MspI that recognizes hotspot codon 248). MspI restriction enzyme cleaves the wild-type CCGG recognition sequences, but leaves the mutated MspI site intact. A known copy number of the internal control (which we call mutant standard, MS), is added for later data calibration. The digested DNA sample is then loaded on a 2% agarose gel and a 350 bp to 550 bp fragment population is isolated by size fractionation. The isolated DNA fraction contains the predicted 462 bp

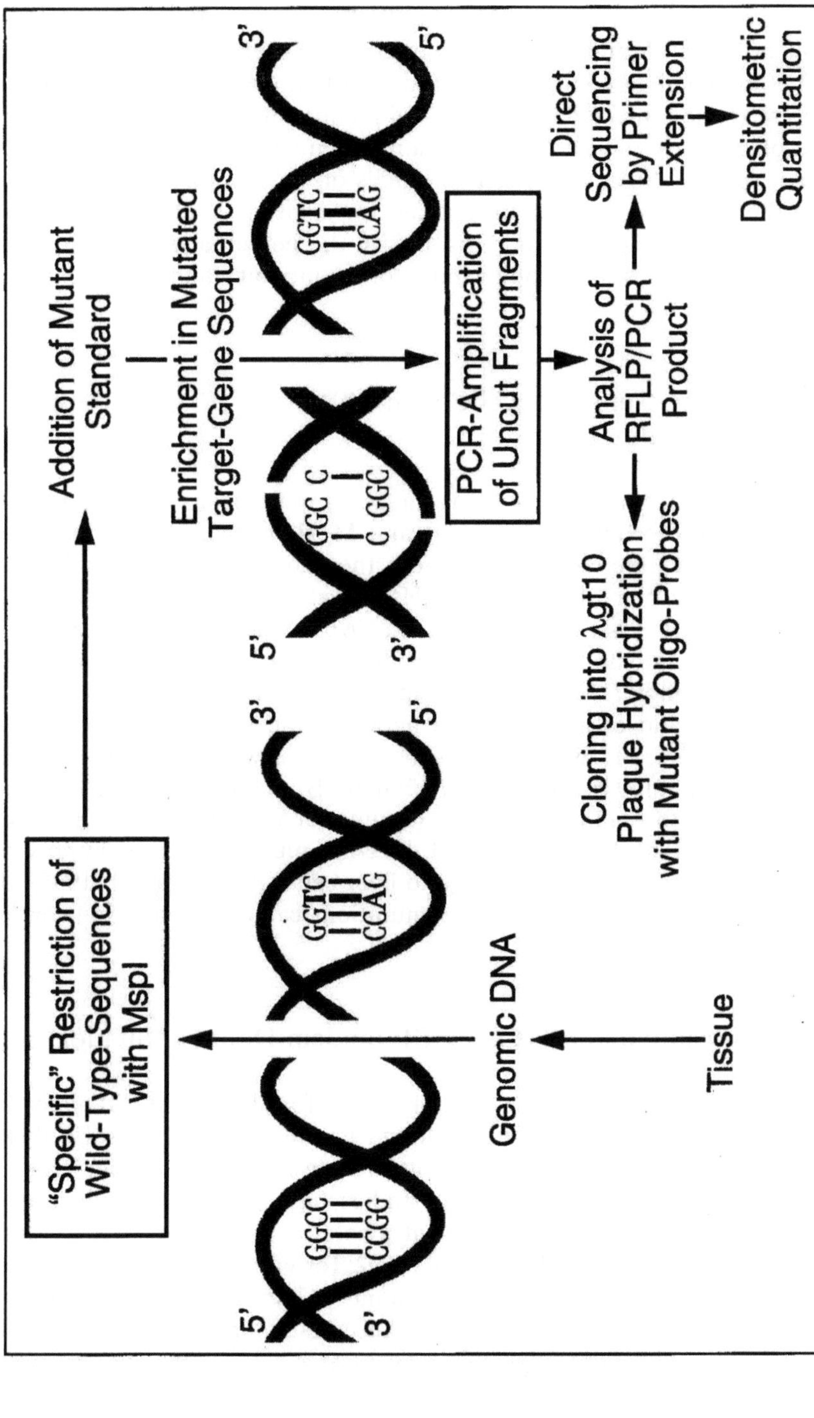

FIGURE 3. Schematic representation of the mutation assay.

mutated *p53* fragment and 468 bp MS. This enriched DNA preparation is then used as the template for DNA amplification, using the high fidelity enzyme *Pyrococcus furiosus* polymerase. The resulting PCR product is then again digested with MspI in order to eliminate the maximum possible elimination of wild-type sequence. The amplified DNA is then cloned into lambda phage gt10 arms. The cloned phage were then infected into *E.coli* and plaques are developed on LB-agar plates. These plaques are then lifted onto plaque membranes and hybridized with 19-mer mutant specific oligos. The positive signals on autoradiograph are counted to determine the percentage of mutant plaques for a specific mutation. The mutation frequency is determined by comparison with the percentage of mutant standard plaques arises from a known copy number of the mutant standard that is added to the DNA at the outset.

3.2. Aflatoxin B_1 Exposure and *p53* Codon 249^{ser} (AGG → AGT) Mutation

In liver tumors from persons living in geographic areas such as Qidong (Republic of China), Mozambique (Africa) and Mexico, where aflatoxin B_1 and hepatitis B virus (HBV) are cancer risk factors, the majority of p53 mutations are at the third nucleotide pair of codon 249 (Hsu et al., 1991; Bressac et al., 1991; Scorsone et al., 1992; Li et al., 1993; Soini et al., 1996). A dose-dependent relationship between dietary aflatoxin B_1 intake and codon 249^{ser} *p53* mutations is observed in hepatocellular carcinoma from Asia, Africa and North America [reviewed in (Harris, 1996; Montesano et al., 1997)]. In addition, the mutation load of 249^{ser} mutant cells in nontumorous liver is positively correlated with dietary aflatoxin B_1 exposure (Aguilar et al., 1994). Exposure of aflatoxin B_1 to human liver cells *in vitro* produces 249^{ser} (AGG to AGT) *p53* mutants (Aguilar et al., 1993; Mace et al., 1997). These results indicate that the expression of the 249^{ser} mutant p53 protein provides a specific growth and/or survival advantage to liver cells (Puisieux et al., 1995) and are consistent with the hypothesis that *p53* mutations can occur early in liver carcinogenesis.

3.3. Sunlight Exposure and *p53* CC → TT Tandem Double Mutation

Sunlight exposure is a well known risk factor for skin cancer. Tandem CC to TT transition mutations are frequently found in squamous and basal cell skin carcinoma (Brash et al., 1991), while they are rarely reported in other types of cancers (Greenblatt et al., 1994). In vitro studies have shown the induction of the characteristic CC to TT mutations by ultraviolet exposure (Hsia et al., 1989; Bredberg et al., 1986; Brash, 1988; Kress et al., 1992; Tornaletti et al., 1993). Sunlight-exposed normal and precancerous skin contain CC to TT tandem mutations (Nakazawa et al., 1994; Ziegler et al., 1994). These results indicate that CC to TT mutations induced by sunlight exposure may play a role in the occurrence of skin cancer.

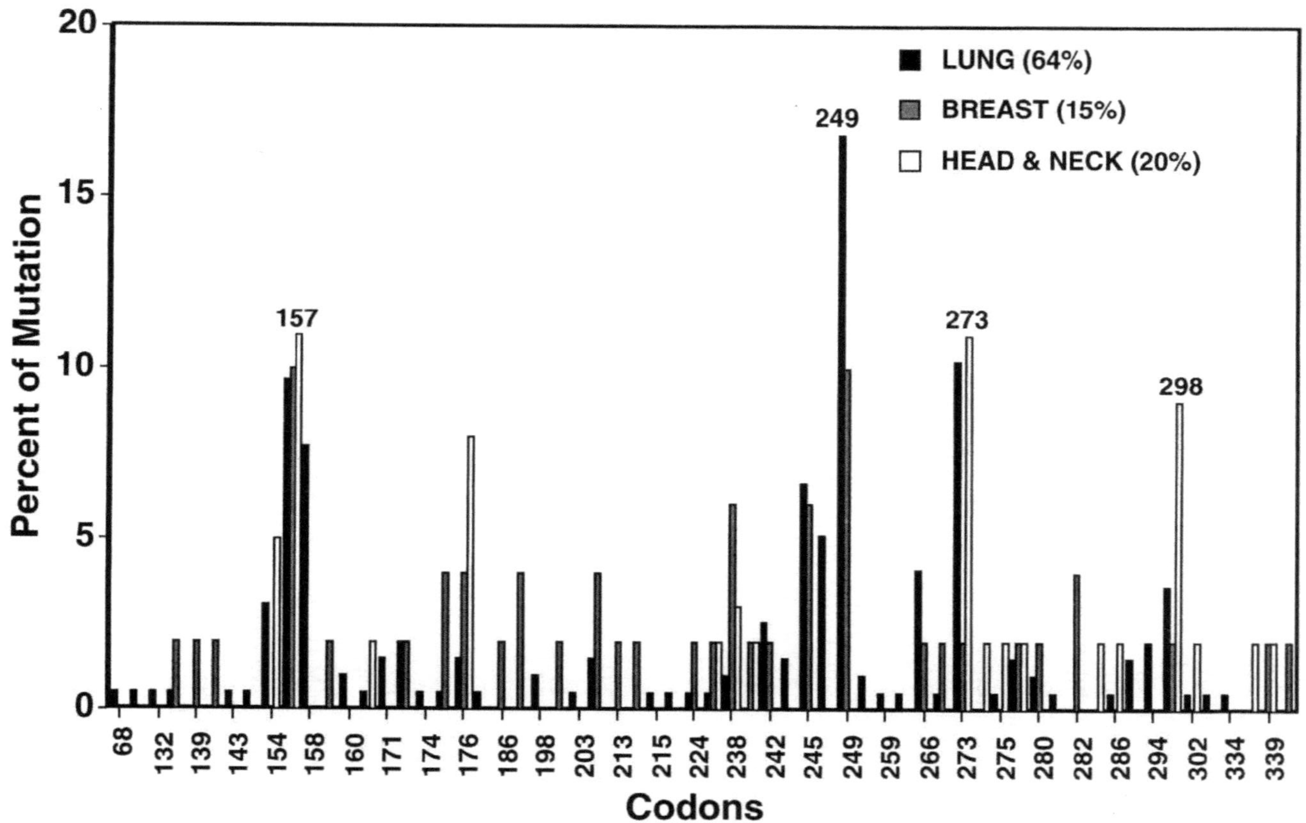

FIGURE 4. *p53* mutational hotspots of G to T transversions in human breast, head and neck, and lung cancers. Codon 157 is one of the mutational hotspots in lung cancer. A transversion of G to T at codon 157 in lung cancer is found more frequently among smokers than never-smokers and is a candidate early-marker for identifying individuals at higher cancer risk. The number in parentheses shows the prevalence of *p53* mutation in a particular cancer.

3.4. Benzo[a]pyrene Exposure and *p53* Mutations at Hotspot Codons 157, 248 and 273

Cigarette smoking has been established as a major risk factor for the incidence of lung cancer. Codons 157, 248 and 273 of the *p53* gene have been designated as mutational hotspots in lung cancer. The majority of mutations found at these codons are G to T transversions. Furthermore, in addition to lung cancer, codon 157 also constitutes one of the hotspots for G to T transversions in breast, and head and neck cancers (Figure 4). In smoking-associated lung cancer, the occurrence of G to T transversions has been linked to the presence of benzo(a)pyrene (BP) in cigarette smoke. Interestingly, codon 157 (GTC to TTC) mutations are not found in lung cancer from never-smokers (Greenblatt et al., 1994; Hernandez and Hainaut, 1998). A dose-dependent increase in *p53* G to T transversion mutations with cigarette smoking has been reported in lung cancer (Takeshima et al., 1993). Recently, it was shown that BP diol epoxide, the metabolically activated form of BP, binds to guanosine residues in codons 157, 248 and 273 (mutational hotspots in lung cancer) (Denissenko et al., 1996). Cigarette smoke condensate or BP also neoplastically transforms human bronchial epithelial cells *in vitro* (Klein-Szanto et al., 1992).

The occurrence of a characteristic *p53* mutation spectrum upon exposure to a particular carcinogen can add to the "Aweight of the evidence" for implicating an environmental pollutant/contaminant in the etiology of human cancer (Hussain and Harris, 1998).

4. ACKNOWLEDGMENT

We thank Dorothea Dudek for her editorial and graphic assistance.

5. REFERENCES:

Aguilar, F., S.P. Hussain, and P. Cerutti. 1993. Aflatoxin B1 induces the transversion of G → T in codon 249 of the p53 tumor suppressor gene in human hepatocytes. Proc. Natl. Acad. Sci. U. S. A 90: 8586–8590.

Aguilar, F., C.C. Harris, T. Sun, M. Hollstein, and P. Cerutti. 1994. Geographic variation of p53 mutational profile in nonmalignant human liver. Science 264: 1317–1319.

Bakalkin, G., T. Yakovleva, G. Selivanova, K.P. Magnusson, L. Szekely, E. Kiseleva, G. Klein, L. Terenius, and K.G. Wiman. 1994. p53 binds single-stranded DNA ends and catalyzes DNA renaturation and strand transfer. Proc. Natl. Acad. Sci. U. S. A 91: 413–417.

Baker, S.J., S. Markowitz, E.R. Fearon, J.K. Willson, and B. Vogelstein. 1990. Suppression of human colorectal carcinoma cell growth by wild-type p53. Science 249: 912–915.

Brash, D.E. 1988. UV mutagenic photoproducts in Escherichia coli and human cells: a molecular genetics perspective on human skin cancer. Photochem. Photobiol. 48: 59–66.

Brash, D.E., J.A. Rudolph, J.A. Simon, A. Lin, G.J. McKenna, H.P. Baden, A.J. Halperin, and J. Ponten. 1991. A role for sunlight in skin cancer: UV-induced p53 mutations in squamous cell carcinoma. Proc. Natl. Acad. Sci. U. S. A 88: 10124–10128.

Bredberg, A., K.H. Kraemer, and M.M. Seidman. 1986. Restricted ultraviolet mutational spectrum in a shuttle vector propagated in xeroderma pigmentosum cells. Proc. Natl. Acad. Sci. U. S. A 83: 8273–8277.

Bressac, B., M. Kew, J. Wands, and M. Ozturk. 1991. Selective G to T mutations of p53 gene in hepatocellular carcinoma from southern Africa. Nature 350: 429–431.

Chen, P.L., Y. Chen, R. Bookstein, and W.H. Lee. 1991. Genetic mechanisms of tumor suppression by the human p53 gene. Science 250: 1576–1580.

Cho, Y., S. Gorina, P. Jeffrey, and N.P. Pavletich. 1994. Crystal structure of a p53 tumor suppressor-DNA complex: A framework for understanding how mutations inactivate p53. Science 265: 346–355.

Clore, G.M., J.G. Omichinski, K. Sakaguchi, N. Zambrano, H. Sakamoto, E. Appella, and A.M. Gronenborn. 1994. High-resolution solution structure of the oligomerization domain of p53 by multi-dimensional NMR. Science 265: 386–391.

DeLeo, A.B., G. Jay, E. Appella, G.C. Dubois, L.W. Law, and L.J. Old. 1979. Detection of a transformation-related antigen in chemically induced sarcomas and other transformed cells of the mouse. Proc. Natl. Acad. Sci. U. S. A 76: 2420–2424.

Denissenko, M.F., A. Pao, M. Tang, and G.P. Pfeifer. 1996. Preferential formation of benzo[a]pyrene adducts at lung cancer mutational hotspots in P53. Science 274: 430–432.

Diller, L., J. Kassel, C.E. Nelson, M.A. Gryka, G. Litwak, M. Gebhardt, B. Bressac, M. Ozturk, S.J. Baker, B. Vogelstein, and et al. 1990. p53 functions as a cell cycle control protein in osteosarcomas. Mol. Cell Biol. 10: 5772–5781.

Eliyahu, D., A. Raz, P. Gruss, D. Givol, and M. Oren. 1984. Participation of p53 cellular tumour antigen in transformation of normal embryonic cells. Nature 312: 646–649.

Eliyahu, D., N. Goldfinger, O. Pinhasi-Kimhi, G. Shaulsky, Y. Skurnik, N. Arai, V. Rotter, and M. Oren. 1988. Meth A fibrosarcoma cells express two transforming mutant p53 species. Oncogene 3: 313–321.

Eliyahu, D., D. Michalovitz, S. Eliyahu, O. Pinhasi-Kimhi, and M. Oren. 1989. Wild-type p53 can inhibit oncogene-mediated focus formation. Proc. Natl. Acad. Sci. U. S. A 86: 8763–8767.

Fields, S., and S.K. Jang. 1990. Presence of a potent transcription activating sequence in the p53 protein. Science 249: 1046–1048.

Finlay, C.A., P.W. Hinds, T.H. Tan, D. Eliyahu, M. Oren, and A.J. Levine. 1988. Activating mutations for transformation by p53 produce a gene product that forms an hsc70-p53 complex with an altered half-life. Mol. Cell Biol. 8: 531–539.

Finlay, C.A., P.W. Hinds, and A.J. Levine. 1989. The p53 proto-oncogene can act as a suppressor of transformation. Cell 57: 1083–1093.

Gottlieb, T.M., and M. Oren. 1996. p53 in growth control and neoplasia. Biochim. Biophys. Acta. 1287: 77–102.

Greenblatt, M.S., W.P. Bennett, M. Hollstein, and C.C. Harris. 1994. Mutations in the p53 tumor suppressor gene: clues to cancer etiology and molecular pathogenesis. Cancer Res. 54: 4855–4878.

Greenblatt, M.S., A.P. Grollman, and C.C. Harris. 1996. Deletions and insertions in the p53 tumor suppressor gene in human cancers: confirmation of the DNA polymerase slippage/misalignment model. Cancer Res. 56: 2130–2136.

Hainaut, P., T. Hernandez, A. Robinson, P. Rodriguez-Tome, T. Flores, M. Hollstein, C.C. Harris, and R. Montesano. 1998. IARC Database of p53 gene mutations in human tumors and cell lines: updated compilation, revised formats and new visualisation tools. Nucleic Acids Res. 26: 205–213.

Harris, C.C. 1993. p53: at the crossroads of molecular carcinogenesis and cancer risk assessment. Science 262: 1980–1981.

Harris, C.C. 1996. The 1995 Walter Hubert Lecture—Molecular epidemiology of human cancer: insights from the mutational analysis of the p53 tumor suppressor gene. Br. J. Cancer 73: 261–269.

Hernandez, T.M., and P. Hainaut. 1998. A specific spectrum of p53 mutations in lung cancer from smokers: review of mutations compiled in the IARC p53 database. Environ. Health Perspect. 106: 385–391.

Hinds, P., C. Finlay, and A.J. Levine. 1989. Mutation is required to activate the p53 gene for cooperation with the ras oncogene and transformation. J. Virol. 63: 739–746.

Hollstein, M., D. Sidransky, B. Vogelstein, and C.C. Harris. 1991. p53 mutations in human cancers. Science 253: 49–53.

Hollstein, M., B. Shomer, M. Greenblatt, T. Soussi, E. Hovig, R. Montesano, and C.C. Harris. 1996. Somatic point mutations in the p53 gene of human tumors and cell lines: updated compilation. Nucleic Acids Res. 24: 141–146.

Horikoshi, N., A. Usheva, J. Chen, A.J. Levine, R. Weinmann, and T. Shenk. 1995. Two domains of p53 interact with the TATA-binding protein, and the adenovirus 13S E1A protein disrupts the association, relieving p53-mediated transcriptional repression. Mol. Cell Biol. 15: 227–234.

Hsia, H.C., J.S. Lebkowski, P.M. Leong, M.P. Calos, and J.H. Miller. 1989. Comparison of ultraviolet irradiation-induced mutagenesis of the lacI gene in Escherichia coli and in human 293 cells. J. Mol. Biol 205: 103–113.

Hsu, I.C., R.A. Metcalf, T. Sun, J.A. Welsh, N.J. Wang, and C.C. Harris. 1991. Mutational hotspot in the p53 gene in human hepatocellular carcinomas. Nature 350: 427–428.

Hupp, T.R., and D.P. Lane. 1995. Allosteric activation of latent p53 tetramers. Curr. Biol. 4: 865–875.

Hussain, S.P., F. Aguilar, P. Amstad, and P. Cerutti. 1994a. Oxy-radical induced mutagenesis of hotspot codons 248 and 249 of the human p53 gene. Oncogene 9: 2277–2281.

Hussain, S.P., F. Aguilar, and P. Cerutti. 1994b. Mutagenesis of codon 248 of the human p53 tumor suppressor gene by N-ethyl-N-nitrosourea. Oncogene 9: 13–18.

Hussain, S.P., C.H. Kennedy, P. Amstad, H. Lui, J.F. Lechner, and C.C. Harris. 1997. Radon and lung carcinogenesis: mutability of p53 codons 249 and 250 to 238Pu alpha-particles in human bronchial epithelial cells. Carcinogenesis 18: 121–125.

Hussain, S.P., and C.C. Harris. 1998. Molecular epidemiology of human cancer: contribution of mutation spectra studies of tumor suppressor genes. Cancer Res. 58: 4023–4037.

Ishioka, C., C. Englert, P. Winge, Y.X. Yan, M. Engelstein, and S.H. Friend. 1995. Mutational analysis of the carboxy-terminal portion of p53 using both yeast and mammalian cell assays *in vivo*. Oncogene 10: 1485–1492.

Jayaraman, L., and C. Prives. 1995. Activation of p53 sequence-speicific DNA binding by short single strands of DNA requires the p53 C-terminus. Cell 81: 1021–1029.

Jeffrey, P.D., S. Gorina, and N.P. Pavletich. 1995. Crystal structure of the tetramerization domain of the p53 tumor suppressor at 1.7 angstroms. Science 267: 1498–1502.

Jenkins, J.R., K. Rudge, and G.A. Currie. 1984. Cellular immortalization by a cDNA clone encoding the transformation-associated phosphoprotein p53. Nature 312: 651–654.

Klein-Szanto, A.J., T. Iizasa, S. Momiki, I. Garcia-Palazzo, J. Caamano, R. Metcalf, J. Welsh, and C.C. Harris. 1992. A tobacco-specific N-nitrosamine or cigarette smoke condensate causes neoplastic transformation of xenotransplanted human bronchial epithelial cells. Proc. Natl. Acad. Sci. U. S. A 89: 6693–6697.

Kress, M., E. May, R. Cassingena, and P. May. 1979. Simian virus 40-transformed cells express new species of proteins precipitable by anti-simian virus 40 tumor serum. J. Virol. 31: 472–483.

Kress, S., C. Sutter, P.T. Strickland, H. Mukhtar, J. Schweizer, and M. Schwarz. 1992. Carcinogen-specific mutational pattern in the p53 gene in ultraviolet B radiation-induced squamous cell carcinomas of mouse skin. Cancer Res. 52: 6400–6403.

Lane, D.P., and L.V. Crawford. 1979. T antigen is bound to a host protein in SV40-transformed cells. Nature 278: 261–263.

Lee, W., T.S. Harvey, Y. Yin, P. Yau, D. Litchfield, and C.H. Arrowsmith. 1994. Solution structure of the tetrameric minimum transforming domain of p53. Nature Struc. Biol. 1: 877–890.

Levine, A.J., J. Momand, and C.A. Finlay. 1991. The p53 tumour suppressor gene. Nature 351: 453–456.

Li, D., Y. Cao, L. He, N.J. Wang, and J. Gu. 1993. Aberrations of p53 gene in human hepatocellular carcinoma from China. Carcinogenesis 14: 169–173.

Lin, J., J. Chen, B. Elenbaas, and A.J. Levine. 1994. Several hydrophobic amino acids in the p53 amino-terminal domain are required for transcriptional activation, binding to mdm-2 and the adenovirus 5 E1B 55-kD protein. Genes Dev. 8: 1235–1246.

Linzer, D.I., and A.J. Levine. 1979. Characterization of a 54K dalton cellular SV40 tumor antigen present in SV40-transformed cells and uninfected embryonal carcinoma cells. Cell 17: 43–52.

Liu, X., C.W. Miller, P.H. Koeffler, and A.J. Berk. 1993. The p53 activation domain binds the TATA box-binding polypeptide in Holo-TFIID, and a neighboring p53 domain inhibits transcription. Mol. Cell Biol. 13: 3291–3300.

Lu, H., and A.J. Levine. 1995. Human TAFII31 protein is a transcriptional coactivator of the p53 protein. Proc. Natl. Acad. Sci. U. S. A 92: 5154–5158.

Mace, K., F. Aguilar, J.S. Wang, P. Vautravers, M. Gomez-Lechon, F.J. Gonzalea, J. Groopman, C.C. Harris, and A.M.A. Pfeifer. 1997. Aflatoxin B1 induced DNA adduct formation and p53 mutations in CYP450-expressing human liver cell lines. Carcinogenesis 18: 1291–1297.

Mack, D.H., J. Vartikar, J.M. Pipas, and L.A. Laimins. 1993. Specific repression of TATA-mediated but not initiator-mediated transcription by wild-type p53. Nature 363: 281–283.

Martin, D.W., R.M. Munoz, M.A. Subler, and S. Deb. 1993. p53 binds to the TATA-binding protein-TATA complex. J. Biol. Chem. 268: 13062–13067.

Martinez, J., I. Georgoff, and A.J. Levine. 1991. Cellular localization and cell cycle regulation by a temperature-sensitive p53 protein. Genes Dev. 5: 151–159.

Mercer, W.E., M.T. Shields, M. Amin, G.J. Sauve, E. Appella, J.W. Romano, and S.J. Ullrich. 1990. Negative growth regulation in a glioblastoma tumor cell line that conditionally expresses human wild-type p53. Proc. Natl. Acad. Sci. U. S. A 87: 6166–6170.

Michalovitz, D., O. Halevy, and M. Oren. 1990. Conditional inhibition of transformation and of cell proliferation by a temperature-sensitive mutant of p53. Cell 62: 671–680.

Montesano, R., P. Hainaut, and C.P. Wild. 1997. Hepatocellular carcinoma: from gene to public health. J. Natl. Cancer Inst. 89: 1844–1851.

Moshinsky, D.J., and G.N. Wogan. 1997. UV-induced mutagenesis of human p53 in a vector replicated in Saccharomyces cerevisiae. Proc. Natl. Acad. Sci. U.S. A 94: 2266–2271.

Nakazawa, H., D. English, P.L. Randell, K. Nakazawa, N. Martel, B.K. Armstrong, and H. Yamasaki. 1994. UV and skin cancer; Specific p53 gene mutation in normal skin as a biologically relevant exposure measurement. Proc. Natl. Acad. Sci. U. S. A 91: 360–364.

Parada, L.F., H. Land, R.A. Weinberg, D. Wolf, and V. Rotter. 1984. Cooperation between gene encoding p53 tumour antigen and ras in cellular transformation. Nature 312: 649–651.

Park, J.W., K.C. Cundy, and B.N. Ames. 1989. Detection of DNA adducts by high-performance liquid chromatography with electrochemical detection. Carcinogenesis 10: 827–832.

Patel, K.J., V.P.C.C. Yu, H. Lee, A. Corcoran, F.C. Thistlethwaite, M.J. Evans, W.H. Colledge, L.S. Friedman, B.A. Ponder, and A.R. Venkitaraman. 1998. Involvement of Brca2 in DNA repair. Molec. Cell 1: 347–357.

Ponder, B. 1988. Cancer. Gene losses in human tumours [news]. Nature 335: 400–402.

Puisieux, A., J. Ji, C. Guillot, Y. Legros, T. Soussi, K. Isselbacher, and M. Ozturk. 1995. p53-mediated cellular response to DNA damage in cells with replicative hepatitis B virus. Proc. Natl. Acad. Sci. U. S. A 92: 1342–1346.

Raycroft, L., H. Wu, and G. Lozano. 1990. Transcriptional activation by wild-type but not transforming mutants of the p53 anti-oncogene. Science 249: 1049–1051.

Scharer, E., and R. Iggo. 1992. Mammalian p53 can function as a transcription factor in yeast. Nucleic Acids Res. 20: 1539–1545.

Scorsone, K.A., Y.Z. Zhou, J.S. Butel, and B.L. Slagle. 1992. p53 mutations cluster at codon 249 in hepatitis B virus-positive hepatocellular carcinomas from China. Cancer Res. 52: 1635–1638.

Seto, E., A. Usheva, G.P. Zambetti, J. Momand, N. Horikoshi, R. Weinmann, A.J. Levine, and T. Shenk. 1992. Wild-type p53 binds to the TATA-binding protein and represses transcription. Proc. Natl. Acad. Sci. U. S. A 89: 12028–12032.

Soini, Y., S.C. Chia, W.P. Bennett, J.D. Groopman, J.S. Wang, V.M. DeBenedetti, H. Cawley, J.A. Welsh, C. Hansen, N.V. Bergasa, E.A. Jones, A.M. DiBisceglie, G.E. Trivers, C.A. Sandoval, I.E. Calderon, L.E. Munoz Espinosa, and C.C. Harris. 1996. An aflatoxin-associated mutational hotspot at codon 249 in the p53 tumor suppressor gene occurs in hepatocellular carcinomas from Mexico. Carcinogenesis 17: 1007–1012.

Soussi, T., C. Caron de Fromentel, and P. May. 1990. Structural aspects of the p53 protein in relation to gene evolution. Oncogene 5: 945–952.

Soussi, T., Y. Legros, R. Lubin, K. Ory, and B. Schlichtholz. 1994. Multifactorial analysis of p53 alteration in human cancer: a review. Int. J. Cancer 57: 1–9.

Takeshima, Y., T. Seyama, W.P. Bennett, M. Akiyama, S. Tokuoka, K. Inai, K. Mabuchi, C.E. Land, and C.C. Harris. 1993. p53 mutations in lung cancers from non-smoking atomic-bomb survivors. Lancet 342: 1520–1521.

Thut, C.J., J.-L. Chen, R. Klemm, and R. Tjian. 1995. p53 transcriptional activation mediated by coactivators TAFii40 and TAFii60. Science 267: 100–104.

Tornaletti, S., D. Rozek, and G.P. Pfeifer. 1993. The distribution of UV photoproducts along the human p53 gene and its relation to mutations in skin cancer [published erratum appears in Oncogene 1993 Dec;8(12):3469]. Oncogene 8: 2051–2057.

Truant, R., H. Xiao, C.J. Ingles, and J. Greenblatt. 1993. Direct interaction between the transcriptional activation domain of human p53 and the TATA box-binding protein. J. Biol. Chem. 268: 2284–2287.

Vogelstein, B., and K.W. Kinzler. 1992. p53 function and dysfunction. Cell 70: 523–526.

Wang, X.W., W. Vermeulen, J.D. Coursen, M. Gibson, S.E. Lupold, K. Forrester, G. Xu, L. Elmore, H. Yeh, J.H.J. Hoeijmakers, and C.C. Harris. 1996. The XPB and XPD helicases are components of the p53-mediated apoptosis pathway. Genes Dev. 10: 1219–1232.

Wolf, D., N. Harris, and V. Rotter. 1984. Reconstitution of p53 expression in a nonproducer Ab-MuLV-transformed cell line by transfection of a functional p53 gene. Cell 38: 119–126.

Ziegler, A., A.S. Jonason, D.J. Leffell, J.A. Simon, H.W. Sharma, J. Kimmelman, L. Remington, T. Jacks, and D.E. Brash. 1994. Sunburn and p53 in the onset of skin cancer. Nature 372: 773–776.

19

PLANT BIOMONITORS IN AQUATIC ENVIRONMENTS

Assessing Impairment Via Plant Performance

LESLEY LOVETT-DOUST AND JON LOVETT-DOUST

Department of Biological Sciences University of Windsor, Windsor, Ontario, N9B 3P4 Canada

1. INTRODUCTION

In this review we focus on use of the aquatic macrophyte, *Vallisneria americana*, as a biomonitor of overall environmental conditions in the Laurentian Great Lakes. An array of measures of plant performance have been investigated; estimates of the leaf-to-root surface area ratio have proved to be the most consistently effective and useful. The species has been used in many different ways to characterize plant response to single organochlorines and metals, PCB mixtures, and as a bioassay of sediment toxicity, in the lab and in the field, to evaluate designated Areas of Concern, and to focus upon individual microsites and point source impact zones.

1.1. Assessment and Biomonitoring

As the burden of persistent toxic compounds discharged into waterways continues to increase, there is a growing need for simple, relatively inexpensive methods to assess site quality in aquatic ecosystems, and to identify degraded microsites requiring remediation (Dolan and Hartig, 1996). Historically, much water quality assessment has been carried out by researchers with backgrounds in chemistry or engineering, hence chemical analysis was a dominant form of assessment. However, chemical analyses, particularly of such materials as organochlorines and polyaromatic hydrocarbons, can be very expensive, and local environmental factors may cause the actual exposure of an organism to be little correlated with chemical concentrations in the surrounding water or sediments. Furthermore, the utility of chemical analyses can be restricted by their detection limits. Often toxicants can cause adverse biological effects at concentrations below the detection capabilities of

Biomonitors and Biomarkers as Indicators of Environmental Change 2, Edited by Butterworth *et al.*
Kluwer Academic/Plenum Publishers, New York, 2000

analytical tests (Lovett-Doust et al., 1994a). To a large extent, toxicity testing has proceeded independently of environmental quality assessment in situ. Biological assessment of aquatic pollution has proven itself to be a very valuable, efficient and accurate tool.

Biotic assessment in its most basic form involves the simple tracking of mortality in exposed organisms. However, in most natural environments it is extended, chronic exposure to contaminants that has the most wide-ranging and significant repercussions—thus measures of sub-lethal impairment are favored. Biomonitoring has been defined as the use of organisms in situ to identify and quantify toxicants in an environment (Chaphekar, 1991). This procedure takes advantage of the capacity of living organisms to uptake and accumulate contaminants in their tissues through bioconcentration (uptake from the ambient environment) and biomagnification (uptake through the food chain) (see Gobas et al., 1991). In contrast to chemical analyses of abiotic samples that simply reflect the concentration of contaminants present in an area, the ability of biota to accumulate contaminants over time enables them to indicate the total pollutant *loadings* present in an environment (Lovett-Doust et al., 1994a,b).

The sampling of endemic organisms from naturally-occurring populations, and the subsequent chemical analysis of their tissues is referred to as *passive biomonitoring* (Chaphekar, 1991). Passive biomonitoring is perhaps the most frequently-utilized approach to biomonitoring and, if designed appropriately, such surveys of naturally-occurring organisms can yield valuable insights regarding the spatial distribution of the bioavailability of pollutants (Kauss and Hamdy, 1985). In contrast, *active biomonitoring* refers to the intentional introduction of well-defined organisms into field sites for a known period of time (Koutsandreas, 1980). Active biomonitoring may be especially useful in aquatic ecosystems (Lovett-Doust et al., 1993, 1994a).

1.2. Aquatic Plants As Biomonitors

Although they have been largely overlooked as biomonitoring candidates (but see Wang et al., 1997), there are significant benefits to using aquatic plants as biomonitors of environmental quality in aquatic ecosystems and several reasons why they are actually *more* appropriate than animals (Lovett-Doust et al., 1993, 1994a,b; Biernacki and Lovett-Doust, 1997; Powell, 1997; Anderson et al., 1997; Lewis and Wang, 1997; Puri et al., 1997). Since plants exist at the base of most food chains they will experience effects of toxic compounds released into the water sooner than organisms at higher trophic levels. In this respect, plants may be able to act as an "early warning signal" of impending contaminant impacts on other trophic levels in aquatic environments (Lovett-Doust et al., 1994a). Thus, correlations between concentrations of contaminants and observable impacts on plants that can be determined in toxicity tests should facilitate the prompt and timely detection of pollutants in aquatic environments. As largely sessile organisms there is no need to cage rooted macrophytes, unlike most animal biomonitors. Another benefit of being stationary is that macrophytes will directly reflect their local surroundings, in contrast to many fish, which are more mobile, or top predators such as birds, whose diets may not come entirely from the aquatic food chain (Lovett-Doust et al., 1994a).

Recent studies have also shown that aquatic plants are in many cases actually more sensitive to various contaminants than animal test species (Hughes, 1992). Some aquatic plants have been reported to be orders of magnitude more sensitive to herbicides (e.g. atrazine) than, for example, zooplankters or fish (Solomon et al., 1996). Until recently, the most frequently used aquatic macrophyte in toxicity testing (i.e. exposure of a test organism to a dilution series of a suspected toxicant) was *Lemna minor* (Hughes, 1992). This species is a free-floating miniaturized macrophyte that has been shown to be less susceptible to a number of pesticides compared to rooted macrophytes (see, e.g. Taraldsen and Norberg-King, 1990; Swanson et al., 1991). Similarly, various studies that measured the accumulation of metals by plants indicate that rooted-submerged plants tend to retain greater concentrations of metals, and do so more consistently than free-floating species (Outridge and Noller, 1991; Crowder, 1991). In comparison with seven other species of aquatic macrophyte (*Potamogeton perfoliatus*, *P. pectinatus*, *Lemna gibba*, *L. minor*, *Myriophyllum spicatum*, *Elodea canadensis*, and *Ceratophyllum demersum*), *Vallisneria americana* was reported to be the most sensitive to atrazine (Swanson et al., 1991; Solomon et al., 1996). Possibly because of inherent sensitivity and the influence of direct contact with both sediment-associated toxicants and waterborne pollutants, submersed rooted macrophytes may be especially suitable for monitoring aquatic environments, and testing contaminated sediments (see too Lewis, 1995; Boutin et al., 1995; Biernacki et al., 1997).

A limited number of studies have addressed the potential of aquatic plants as biomonitors. Many of these have used analytical methods to estimate directly the contaminant concentrations in tissues of aquatic plants (e.g. Mortimer, 1985; Chandra et al., 1993; St.-Cyr and Campbell, 1994) and filamentous algae (Jackson, 1985; Nyholm and Peterson, 1997). Although the use of plants as biomonitors in these studies likely provided a better estimate of bioavailability than would chemical analyses of water and/or sediment, they still employed expensive chemical analytic methods. An alternative perspective, from community ecology, considers the distribution of submersed aquatic macrophytes to assess water quality (e.g. Dennison et al., 1993; and see Hill, 1997). However, the study of Dennison et al. (1993) did not consider effects of sediment properties or anthropogenic contaminants. There are even fewer studies that have examined plant physiological and morphological responses, such as differential growth, reproduction and survival, as biomonitoring metrics of site quality. For example, sediment toxicity has been tested with the floating duckweed, *Lemna* sp., in terms of the number of fronds, chlorophyll production, root length, and carbon-14 uptake (Taraldsen and Norberg-King, 1990; Huebert and Shay, 1993). Root and shoot length, peroxidase and dehydrogenase activities, and chlorophyll production have been investigated as measures of sediment toxicity using the rooted submersed plant, *Hydrilla verticillata* (Klaine et al., 1990).

Species that are rooted in the sediments with leaves extending into the water column provide a very localized sampling device that can enable an investigator to assess the *relative* importance of contaminated sediments as opposed to effects due to the water column. Indeed, replicated individual genotypes of the rooted macrophyte, *Vallisneria americana*, have been placed into specific sites of concern and followed over time (Biernacki and Lovett-Doust, 1997). Either naturally-growing plants at an aquatic locality, or particular plants that have been placed in tubs in situ, and left to grow for a period of time (active biomonitors) can be used. Plant performance can be assessed in conjunction with contaminant levels in tissues. Evaluating contaminant concentrations in concert with plant behavior, may indicate the pathway of uptake.

1.3. *Vallisneria americana* and Passive Biomonitoring

Plants collected from the field have accumulated contaminants and responded to the quality of the environment over their entire lifespan; they therefore show integrated and cumulative effects of exposure rather than of particular, transient conditions. While some animals may be quite localized in terms of a home range, many migrate on a diurnal, annual or developmental basis. Plants, in contrast, are typically "rooted to the spot". This allows a spatial map to be constructed showing the intensity of environmental damage, based on the combined evidence gathered from multiple points in space (see Kauss and Hamdy, 1985). There are drawbacks to passive biomonitoring, however. Since only living material is collected from the field, individuals that have died (possibly because they contain high body burdens of contaminants or have suffered more environmental perturbation), remain unsampled. The method may therefore underestimate the severity of conditions at a site. Furthermore, it is likely that sites that have been disturbed or impaired for an extended time will contain biota resistant to the detrimental conditions. Such resistance may be a product of acclimation (an ongoing physiological response) or adaptation (a genetic response that is the result of selection for resistant phenotypes).

Our laboratory has compared the relative efficacy of many measures of plant morphology, growth patterns, and reproduction using the dioecious, rooted macrophyte *Vallisneria americana* (Catling et al., 1994; Lovett-Doust et al., 1994b; Biernacki et al., 1995a,b, 1996, 1997a,b). A particularly sensitive, easy-to-measure, and useful measure was the ratio of leaf-to-root surface area in *Vallisneria* plants. The subtle leaf-to-root surface area measures are very responsive to environmental parameters. The ratio increased with increasing concentrations of an array of organochlorines, and with increasing concentrations of heavy metal, as well as with estimates of general site impairment (Potter and Lovett-Doust, in press). Furthermore the response is detectable and comparable throughout the range of the Great Lakes, despite variation in the genetic makeup of *V. americana* populations (Lokker et al., 1994; Potter and Lovett-Doust, in press).

1.4. Active Biomonitors

We have developed an effective and reliable methodology for active biomonitoring in which most of the limitations of passive biomonitoring are obviated (e.g.

Biernacki et al., 1995a,b, 1996, 1997a,b). For example, plants are not locally adapted to new sites being investigated, so may show strong responses in terms of mortality, either in terms of their individual parts (leaves or shoots) or of whole plants. A particular advantage of active biomonitoring in sites where the test plant is also present naturally, is that performance of local plants can be directly compared with the transplants, brought in as active biomonitors. In fact, in order to control for the possible adverse effects of transplantation, *per se*, the appropriate methodology involves a "reciprocal transplant-replant" between and among sites. This enables transplanted individuals from a clean reference site to be compared with transplants that are native to the site under study. A relative "impairment index" can be calculated using measures of plant performance: shoot production, plant biomass, leaf-to-root surface area ratio, etc., to indicate the degree to which a particular *site* is responsible for plant performance (Lovett-Doust et al., 1994b).

1.5. Cloned Biomonitors

An important variable in active biomonitoring procedures is genetic variation among test individuals. Many biomonitor test protocols utilize cloned material that is deployed in different environments or in water and sediment samples from contrasting sites. For example, others have advocated using cloned organisms such as *Daphnia* (although this genus carries out both sexual and asexual life history phases, and it is therefore important to ensure that cultures consist of single asexual lineages) (see Lovett-Doust et al., 1993). Many bacterial tests (including the well-known Ames test for mutagenicity) also assume that the test organism is genetically uniform.

Vallisneria americana proliferates primarily through clonal growth, however sexual reproduction and establishment of new genotypes from seed is also possible (Kimber et al., 1995; Lokker et al., 1997). We have used cloned *Vallisneria americana* individuals—including some genotypes which are particularly tolerant to organochlorines. The responses of all plants, whatever their tolerance, were highly correlated with observed sediment toxicity and contaminant concentrations reported independently from these sites by other researchers (Biernacki and Lovett-Doust, 1997). Individual clones were represented by multiple ramets at each of five sites known to be contaminated. However, clonal lines differed in terms of their tolerance to environmental contaminants. Clones obtained originally from contaminated sites were more tolerant and lived longer at other contaminated sites than did plants originally collected from cleaner sites.

It seems likely that surviving plants in highly polluted areas may have evolved tolerance to at least some contaminants. In our study of the effects of local environment on survival, growth, and development in six genotypes of *Vallisneria americana* raised at five sites in the Huron-Erie corridor (Biernacki and Lovett-Doust, 1997), we concluded that tolerant genotypes are particularly useful, indeed essential in the most highly-contaminated sites, where non-tolerant genotypes die following exposure. However, non-tolerant genotypes were also useful, providing a graded assessment of sites with intermediate levels of contamination. Leaf-to-root surface area ratios in *V. americana* showed particular utility in measuring differences in environmental quality, providing a consistent measure of site quality unaffected by genotype, time, or the somewhat artificial conditions of life in a plastic tub. One concern may be that the use of organisms with evolved tolerance in biomonitoring studies may foster a false sense of security, and could underestimate environmental impairment due to contamination. This is avoided by using both tolerant and non-tolerant genotypes.

2. METHODOLOGY

At a crude level, the birth and death of whole individuals can be used to assess the toxicity or overall quality of an environment. However, such traditional measures as the LD_{50} or LC_{50} (the dose or concentration, respectively, which kills half of those in the study over a specified period of time) detect only the most severe and fast-acting of effects. Such experiments are rarely run with lower doses over longer times—the kinds of conditions more likely to occur in nature. It is unlikely that the response of biota to environmental change will be linear, so such studies have little predictive power for the kinds of problem we typically encounter in variable natural ecosystems. As indicated above, several different measures have been developed to record the growth response of plants to environmental change. Selection of a particular method can depend on such factors as the presence or absence of endemic plants at the site in question, the availability of replicate plants from the field, the expertise of the field researcher, and the laboratory facilities available.

2.1. Survivorship

We have shown that the simple survival of individual plants over one or two growing seasons is highly indicative of environmental conditions, in particular when non-tolerant plants from a clean reference site are placed in an area that is highly polluted or otherwise degraded. Tracking survival is relatively straightforward: in passive biomonitoring studies the number of individuals in marked quadrats can be recorded at regular intervals over the growing season. This will indicate not only overall environmental quality, but also any regional variation in plant phenology (the developmental timetable, including for example the time of germination of buried turions (the overwintering structures), the time of floral onset, fruit maturation in summer, and the appearance of new turions in fall. In the case of active biomonitoring, plant survival and individual performance over the season can be tracked by regularly surveying the tubs or flats of plants which have been placed in the site. It is particularly helpful to track survival (as opposed to the more subtle measures of plant performance described below) in situations where extraordinarily high levels of contaminant are being introduced either directly, at an "end-of-pipe" location, or in experiments where extreme concentrations of a contaminant are known to exist, in order to characterize the range of concentrations over which plants can survive (e.g. VanderWal and Lovett-Doust, 1998), and enable calibration of the sub-lethal responses characterized below.

2.2. Plant Growth Analysis

Our general procedure is based on two approaches: classical plant growth analysis (Evans, 1972), and the characterization of plant performance using demographic data about individual plant modules. The former calls for the measurement of the mass of major plant tissues: root, stem, leaf, and reproductive structures. Plant growth analysis was developed originally to characterize the performance of crops and weeds; the major interest was to understand how much energy and material was available (for subsequent harvesting) as either reproductive or vegetative parts, and to assess the relative importance of below-ground as opposed to above-ground tissues. The allocation of additional biomass to roots has been interpreted as a response to below-ground competition for nutrients and water, while distribution to stem and leaf tissues is seen as a result of the availability of light, and competition with neighboring individuals for access to light. Other indices of plant growth have been developed from these basic assessments of mass. For example, when both leaf mass and leaf area are known, various metrics may be determined (leaf area index [mass/area], specific leaf area [area/mass], root/shoot mass ratio). Leaves that are thin but large in area tend to result from low light conditions, whereas plants with thicker but smaller leaves typically result either when growing in full sunlight, or in drought; their structure should reduce water loss by reducing the surface-area-to-volume ratio.

2.3. Modular Demography

We have investigated module demography in a variety of settings (see e.g. Lovett-Doust and Eaton, 1982; Lovett-Doust et al., 1983; Lovett-Doust, 1993; Biernacki, 1996). The approach involves tracking the production ("birth"), life-span and death of individual modules of plant growth, and is based upon Goethe's early characterization of a plant as a "matriarchal tribe of units, each superimposed upon the other" (see Harper, 1977). Plants exist morphologically as structures made up of repeating units (like "lego bricks"), or more technically plant metamers (see Harper and White, 1974). These individual units are readily distinguished and often exist in plants at two levels. For example, in *Vallisneria americana* each genetic individual grows, iteratively, as a series of modular units, initially inter-connected by stolons that run in the sediment (see Figure 1). These modules, termed ramets, in turn are made up of individual leaves. It is therefore possible to monitor performance in this kind of plant in terms of the number and turnover of both ramets and leaves.

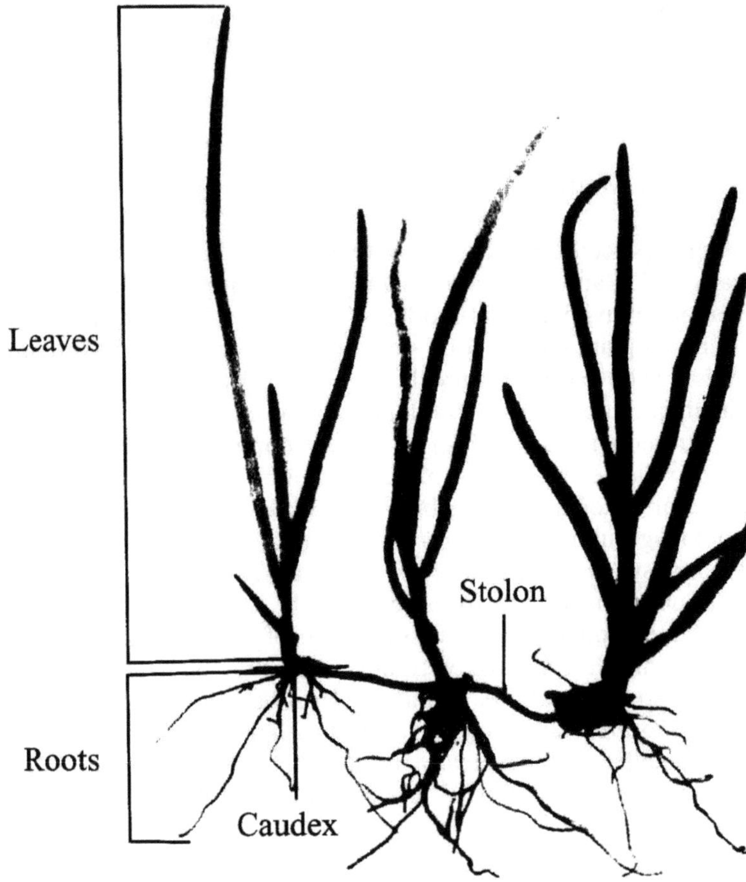

FIGURE 1. The structure of a plant of *Vallisneria americana*. Roots are embedded in sediment, while leaves and reproductive structures are in the water column. The original parent plant can produce daughter plants (ramets) on surface-to-subsurface stolons, produced in late summer (this plant is made up of three ramets). In winter, each plant dies back, losing its leaves, to a resilient overwintering turion (tuber-like structure). Each spring clonal regrowth occurs from turions forming new shoots. Turions are rich in nutrients and energy, and provide an important food resource for feeding ducks.

2.4. Leaf-To-Root Surface Area Ratio

As described above, this ratio is a particularly sensitive metric, and can show the effects of contaminants delivered both via the water column (simulating current discharges) and contaminants present in sediment (representing historic discharges that have accumulated in sediments). When *Vallisneria* plants are exposed to any of a variety of detrimental conditions, they show a response in terms of increasing the relative area of leaves as opposed to the area of root surface. In *Vallisneria*, we interpret this as a switch from nutrient uptake via roots to uptake that is increasingly across the submersed leaf surface. The response is consistent across an array of different kinds of chemical contaminants. Furthermore, it is evident in sites where light intensity is low (due both to high plant density as a result of conventional nutrient contamination, and to low light penetration due to turbidity). The ratio becomes elevated both in contaminated field sites and in experimental conditions with higher contaminant levels. Controlled lab studies of sediment toxicity tend to give lower values of the leaf-to-root surface area ratio than the values found in the field, but these sediment toxicity tests used clean water columns, adequate aeration and other conditions that should favor plant growth, and ran for relatively brief periods of exposure (Biernacki et al., 1997a). It is easy to cross-calibrate field and lab results; when this is done, the ranking of sites in terms of assessments in the two ways are consistently identical (see Biernacki et al., 1996).

In our initial studies, leaf and root surface areas were estimated manually as follows. Total length and mean width for all leaves sampled, and the length and mean diameter of all roots were measured for each ramet with electronic digital calipers (e.g. Mitutoyo Digimatic Caliper). Leaf surface area was calculated by doubling the product of the length and the average width (to account for both sides of the leaf) and totalling all leaves. Root surface area was calculated as for a cylinder (the roots are cylindrical in shape), by multiplying the length by the average diameter, by the value of pi.

More recently, we have mechanized leaf and root surface area determination using a desktop scanner to capture images of leaf and root outlines (VanderWal and Lovett-Doust, 1998). Following cleanup of the images (removing water spots, enhancing contrast, etc., using Correl Paint software), ImageTool (or other appropriate analytic software) is used to determine the surface area of each structure. Individual plant data is stored electronically and, using a statistical package such as SYSTAT (Systat, 1997), total surface areas and ratios are calculated. This reduced the time needed to process samples by a factor of approximately ten and had the advantage of capturing data ready for statistical analysis, eliminating the need for manual data transmission and possible errors. It is possible to teach undergraduate research assistants how to scan and clean up images in approximately three sessions of 30 minutes each. Further details of this protocol are available in VanderWal and Lovett-Doust (1998).

2.5. Sediment and Soil Assays

Sometimes a site needs to be assessed, though it may not have a natural *Vallisneria* population—either because conditions are unsuitable or simply because plants may not yet have established there. We have studied sediment from a number of sites that were inherently unlikely to allow *Vallisneria* growth—for example, sediments from deep harbor sites, rocky localities, or from channels where fast currents do not allow plant establishment.

In some instances we used plant biomonitors collected from "clean" sites, and subsequently grown in test sediments, some of which were severely impaired (Biernacki et al., 1995b, 1997a; Potter, 1998). It is important to note that only the sediment portion of the environment is being represented in these assays; the water column provided is essentially clean. This is likely to alter the partitioning of contaminants between the sediment pore water and the water column, with net movement out of the sediments into solution. Obviously, other (possibly synergistic) detrimental factors characteristic of the source site will also be absent (such as fast water movement, considerable water depth, or low light penetration). As a result such sediment toxicity tests are likely to underestimate the negative effects of growing at that particular site.

We have found that susceptible plants (collected from clean reference sites) respond in terms of such factors as total biomass, and rate of mortality, leaf turnover, etc., but that test plants collected from sites known to be impaired demonstrate apparent resistance to environmental degradation. Such tolerant plants are particularly useful for sediment toxicity testing, because they will show a response in terms not only of overall growth, but also in terms of the adjustment of leaf and root surface areas.

2.6. An Integrated Approach

What is presently most pressing for environmental managers is to locate hotspots of contamination within Areas of Concern (AOCs—some of which, like the Detroit River, or the Toronto waterfront, may be very extensive) and to focus remediation efforts where they are most needed (Dolan and Hartig, 1996). There has been considerable cross-validation of the leaf-to-root surface area ratio as a metric of plant impairment, and the presence of contaminated water and/or sediments (see above). Relatively untrained individuals can play a significant role in carrying out this assessment procedure, and it would seem ideal as a way of involving high school and college students in the local implementation of remedial action plans.

Done properly, plant biomonitoring is cheaper and superior to regular measurements of individual chemicals in water and sediment. A literature search for reports of contaminant levels in Canadian Areas of Concern indicated many factors making cross-comparisons difficult. For example, data have been gathered by many different investigators, across a number of different years; different kinds of instruments have been used, with different limits of detection, etc. However, despite this a meaningful, integrated evaluation of overall contamination at local microsites within the AOCs could be made. Potter and Lovett-Doust (in press) classified into five categories the individual contaminant concentrations reported in the literature for an AOC. Total microsite scores (summed across the particular contaminants measured at an AOC) were compared to the maximum possible score (determined by the number of contaminants reported for that locality); these were then calculated as an *overall* index of contamination at the AOC. Plant leaf-to-root surface area ratios and leaf-to-root mass ratios, using plant samples collected from each AOC, proved to be highly correlated to these indices, and suggested an overall response to contaminated sediments as well as a strong and significant response to metal contaminants in these sediments (Potter and Lovett-Doust, in press). It may therefore be unnecessary to repeatedly measure all contaminants in terms of their individual concentrations.

From the point of view of tracking remediation efforts and potential recovery of the ecosystem (e.g., following removal of contaminated sediment) it would be important to characterize the site *prior* to sediment removal, and then again, afterward. We suggest that the leaf-to-root surface area ratio, representing plant performance, provides an effective tracking protocol with much biological significance, that captures more contaminant exposure and integrates other environmental variables affecting the quality of life for aquatic biota. Plant biomonitoring protocols are more sensitive and cheaper by several orders of magnitude than the array of chemical assays necessary to give an accurate representation of contaminants in the aquatic environment.

2.6.1. Low Budget Approach

A study protocol can be developed to suit every budget (!) The most low-tech approach for a field survey (at local or regional scales) would be to excavate about 15 plants from each microsite, taking care to minimize any loss of roots. The plant and sediment can be shoveled up from the mud, and rinsed in a large bucked or plastic hamper. The investigator can then weigh the leaves and roots on the spot as fresh weight (after gently blotting the roots and leaves, but not squeezing them!) Alternatively tissues may be put in paper bags and dried to constant mass (a drying oven at 70°F for 4 days is sufficient). Dry mass is less variable than fresh mass, so is preferable for this purpose if facilities are available. While the leaf to root mass ratio is not quite as sensitive as the leaf to root surface area, it takes very little time to determine, and can use relatively untrained fieldworkers. Total plant mass is also a useful measure of growth, and if several different sites are being compared these data, with replicates of 15 plants per site, it should be possible to rank sites in terms of their relative impairment. In collaboration with D. C. Freeman at Wayne State University and R. Hough at Oakland University, this method has been used with schoolchildren, who have successfully surveyed sections of the Clinton River, an Area of Concern in Michigan that flows into Lake St. Clair.

2.6.2. Photocopier Approach

With this approach it is possible to determine leaf area and root area at a relatively inaccurate level, requiring access to a photocopier and plenty of inexpensive labor. It also requires some permissions from the users and/or owners of the photocopier. Approximately 10 plants per microsite should be excavated with a shovel from the sediments, and roots and leaves carefully recovered as they are rinsed off. Plants should be stored temporarily in a 5% formalin solution, in a screwtop glass jar. In the lab, leaves are placed on the surface of a photocopy machine, laid parallel and adjacent to each other to form a contiguous patch of tissue. The material can be photocopied onto acetate sheets, and the outline cut out and weighed. The mass of the acetate per unit area can be determined by photocopying a piece of cm graph paper, and a calibration curve drawn. Thus the mass of each outline can be converted into a outline area, to be multiplied by 2 for leaves, and π for roots. The mass of roots and leaves can also be determined.

3. FINAL COMMENTS

Impairment issues, assessment methods and, indeed, restoration options all appear to differ at different spatial and temporal scales (catchment, sub-watershed, reach, and site; plus between-season and year-to-year, etc. Most monitoring protocols involve site- or reach-level approaches. Thus, it is worth considering how these protocols can be applied at the sub-watershed scale, or higher. Indeed, since most restoration work is carried out at the site or reach level, perhaps one should ask whether the issue of scale is even relevant (or at least *how* it matters).

Many toxicity tests in the past have used plants such as lettuce or onion to evaluate sediments from the bottom of contaminated harbors (Fiskesjo, 1997). It seems inherently preferable to use an aquatic macrophyte, and indeed to use one that can contact with both the sediment and the water column—that is, a rooted submersed macrophyte such as *Vallisneria americana*. This provides much more meaningful information about the aquatic system than does a terrestrial plant, or even a floating aquatic like *Lemna*. The active biomonitoring studies that we have carried out involved deployment of planted ramets of *Vallisneria americana* growing in sediments from specific sources, and placed at the bottom of a river or lake area. For the most part these biomonitor placements survived well; they were staked down with very long (60 cm) rods, and the sediment in them kept them well weighted down. However, there was some "industrial vandalism" on tubs adjacent to chemical industries and plants in the St Clair River and the Detroit River. This raises an issue of concern; it is quite clear that so-called "surprise" or random visits by ministry and regulatory agency officials are announced ahead of time, so chemical discharge exceedances can be avoided at those times. The tendency recently is toward "self-monitoring," with companies deciding themselves when and where to sample waste discharges. A particular advantage of *in situ* active plant biomonitors is that they would not miss such "pulsed releases"; instead, they will show the evidence in terms of the accumulated level of contaminants, and in terms of the effect on plant growth. They are therefore silent, but effective sentinels of these periodic pulsed releases.

4. REFERENCES

Anderson, T.A., Hoylman, A.M., Edwards, N.T., and Walton, B. 1997. Uptake of polycyclic aromatic hydrocarbons by vegetation: a review of experimental methods. In Wang, W., Gorsuch, J.W., and Hughes, J.S. (Eds.): Plants for environmental studies, pp. 451–480. Lewis Publishers. New York.

Biernacki, M. 1996. The use of modular demography in the aquatic macrophyte *Vallisneria americana* to evaluate its potential as a biomonitor of organic contaminants. Ph. D. thesis. University of Windsor, Windsor, ON, Canada.

Biernacki, M., and Lovett-Doust, J. 1997. *Vallisneria americana* (Hydrocharitaceae) as a biomonitor of aquatic ecosystems: comparison of cloned genotypes. American Journal of Botany 84: 1743–1751.

Biernacki, M., Lovett-Doust, J., and Lovett-Doust, L. 1995a. The effects of trichloroethylene, plant sex and site of origin on modular demography in *Vallisneria americana*. Journal of Applied Ecology 32: 761–777.

Biernacki, M., Lovett-Doust, J., and Lovett-Doust, L. 1995b. Sediment effects on the uptake of trichloroethylene by roots and leaves in *Vallisneria americana*. In Munawar, M., Edsall, T., and

Leach, J. (Eds.): The Lake Huron ecosystem: ecology, fisheries and management, pp. 413–426. Ecovision World Monograph series, S.P.B. Academic Publishing, The Netherlands.

Biernacki, M., Lovett-Doust, J., and Lovett-Doust, L. 1996. *Vallisneria americana* as a biomonitor of aquatic ecosystems: leaf-to-root surface area ratios and organic contamination in the Huron-Erie corridor. Journal of Great Lakes Research 22: 289–303.

Biernacki, M., Lovett-Doust, J., and Lovett-Doust, L. 1997a. Laboratory assay of sediment phytotoxicity using the macrophyte *Vallisneria americana*. Environmental Toxicology and Chemistry 16: 472–478.

Biernacki, M., Lovett-Doust, J., and Lovett-Doust, L. 1997b. Temporal biomonitoring using wild celery, *Vallisneria americana*. Journal of Great Lakes Research 23: 97–107.

Boutin, C., Freemark, K.F., and Keddy, C.J. 1995. Overview and rationale for developing regulatory guidelines for nontarget plant testing with chemical pesticides. Environmental Toxicology and Chemistry 14: 1465–1475.

Catling, P.M., Spicer, K.W., Biernacki, M., and Lovett-Doust, J. 1994. The biology of Canadian weeds. 103. *Vallisneria americana* Michx. Canadian Journal of Plant Sciences 74: 883–897.

Chandra, P.R.D., Tripathi, U.N., Rai, S.S., and Garg, P. 1993. Biomonitoring and amelioration of non-point source pollution in some aquatic bodies. Water Science Technology 28: 323–326.

Chaphekar, S.B. 1991. An overview on bioindicators. Journal of Environmental Biology 12: 163–168.

Crowder, A. 1991. Acidification, metals and macrophytes. Environmental Pollution 71: 171–203.

Dennison, W.C., Orth, R.J., Moore, K.A., Stevenson, J.C., Carter, V., Kollar, S., Bergstrom, P.W., and Batiuk, R.A. 1993. Assessing water quality with submersed aquatic vegetation: habitat requirements as barometers of Chesapeake Bay health. BioScience 43: 86–94.

Dolan, D., and Hartig, J. 1996. Research managers seek innovative ways to deal with cutbacks. Focus 21: 3–4.

Evans, G.C. 1972. The quantitative analysis of plant growth. Blackwell Scientific Publications, Oxford.

Fiskesjo, G. 1997. Allium test for screening chemicals; evaluation of cytological parameters. In Wang, W., Gorsuch, J.W., and Hughes, J.S. (Eds.): Plants for environmental studies, pp. 307–333. Lewis Publishers. New York.

Gobas, F., McNeil, E.J., Lovett-Doust, L., and Haffner, G. 1991. Bioconcentration of chlorinated aromatic hydrocarbons in aquatic macrophytes. Environmental Science and Technology 25: 924–929.

Huebert, D.B., and Shay, J.M. 1993. Considerations in the assessment of toxicity using duckweeds. Environmental Toxicology and Chemistry 12: 481–483.

Harper, J.L. 1977. The population biology of plants. Academic Press, London.

Harper, J.L., and White, J. 1974. The demography of plants. Annual Review of Ecology and Systematics 5: 419–463.

Hill, B.H. 1997. Aquatic plant communities for impact monitoring and assessment. In Wang, W., Gorsuch, J.W., and Hughes, J.S. (Eds.): Plants for environmental studies, pp. 277–305. Lewis Publishers. New York.

Hughes, J.S. 1992. The use of aquatic plant toxicity tests in biomonitoring programs. pp. 169–174. Proceedings of the 18th Annual Aquatic Toxicity Workshop: Sept. 30–Oct. 3, 1991; Ottawa Canada. Canadian Technical Report of Fisheries and Aquatic Sciences No. 1863.

Jackson, M.B. 1985. The dominant attached filamentous algae of Lake Huron: field ecology and biomonitoring potential 1980. Aquatic Ecosystems Section, Water Resources Branch, Ontario Ministry of the Environment. 49pp.

Kauss, P.B., and Hamdy, Y.S. 1985. Biological monitoring of organochlorine contaminants in the St. Clair and Detroit Rivers using introduced clams, *Elliptio complanatus*. Journal of Great Lakes Research 11: 247–263.

Kimber, A., Korschgen, C.E., and van der Valk, A.G. 1995. The distribution of *Vallisneria americana* seeds and seedling light requirements in the Upper Mississippi River. Canadian Journal of Botany 73: 1966–1973.

Klaine, S.J., Brown, K., Byl, T., and Hinman, M.L. 1990. Phytotoxicity of contaminated sediments. Abstracts from the Annual Meeting of the Society of Environmental Toxicology and Chemistry. No. P092; Arlington, VA.

Koutsandreas, J.D. 1980. Overview and Assessment of Biological Sensing. Second Interagency Workshop on in situ Water Sensing: Biological Sensors. pp. 45–54. Pensacola Beach, FL.

Lewis, M. 1995. Use of freshwater plants for phytotoxicity testing: a review. Environmental Pollution 87: 319–336.

Lewis, M.A., and Wang, W. 1997. Water quality and aquatic plants. In Wang, W., Gorsuch, J.W., and Hughes, J.S. (Eds.): Plants for environmental studies, pp. 141–175. Lewis Publishers. New York.

Lokker, C., Lovett-Doust, L., and Lovett-Doust, J. 1997. Seed output and the seed bank in *Vallisneria americana* (Hydrocharitaceae). American Journal of Botany 84: 1420–1428.

Lokker, C.D., Susko, D., Lovett-Doust, L., and Lovett-Doust, J. 1994. Population genetic structure of *Vallisneria americana*, a dioecious clonal macrophyte. American Journal of Botany 81: 1004–1012.

Lovett-Doust, J. 1992. The influence of plant density on flower, fruit, and leaf demography in bush bean, *Phaseolus vulgaris*. Canadian Journal of Botany 70: 958–964.

Lovett-Doust, J., and Eaton, G.W. 1982. Demographic aspects of flower and fruit production in bean plants, *Phaseolus vulgaris* L. American Journal of Botany 69: 1156–1164.

Lovett-Doust, J., Lovett-Doust, L., and Eaton, G.W. 1983. Sequential yield component analysis and models of growth in bush bean (*Phaseolus vulgaris* L.) American Journal of Botany 70: 1063–1070.

Lovett-Doust, J., Lovett-Doust, L., Biernacki, M., Mal, T., and Lazar, R. 1997. Organic contaminants in submersed macrophytes drifting in the Detroit River. Canadian Journal of Fisheries and Aquatic Sciences 54: 2417–2427.

Lovett-Doust, J., Schmidt, M., and Lovett-Doust, L. 1994a. Biological assessment of aquatic pollution: a review, with emphasis on plants as biomonitors. Biological Reviews 69: 147–186.

Lovett-Doust, L., Lovett-Doust, J., and Biernacki, M. 1994b. American wildcelery, *Vallisneria americana*, as a biomonitor of organic contaminants in aquatic ecosystems. Journal of Great Lakes research 20: 333–354.

Lovett-Doust, L., Lovett-Doust, J., and Schmidt, M. 1993. In praise of plants as biomonitors—send in the clones! Functional Ecology 7: 754–758.

Mortimer, D.L. 1985. Freshwater aquatic macrophytes as heavy metal monitors—the Ottawa River experience. Environmental Monitoring and Assessment 5: 311–323.

Nyholm, N., and Peterson, H.G. 1997. Laboratory bioassays with microalgae. In Wang, W., Gorsuch, J.W., and Hughes, J.S. (Eds.): Plants for environmental studies, pp. 225–276. Lewis Publishers. New York.

Outridge, P.M., and Noller, B.N. 1991. Accumulation of toxic trace elements by freshwater vascular plants. Reviews of Environmental Contamination and Toxicology 121: 1–63.

Potter, K. 1998. The aquatic macrophyte *Vallisneria americana* as a biomonitor of site quality in Great Lakes Areas of Concern. M. Sc. Thesis. University of Windsor, Windsor, ON, Canada.

Potter, K., and Lovett-Doust, L. In press. Biomonitoring site quality in stressed aquatic ecosystems using *Vallisneria americana*. Ecological Applications.

Powell, R.L. 1997. The use of vascular plants as "field" biomonitors. In Wang, W., Gorsuch, J.W., and Hughes, J.S. (Eds.): Plants for environmental studies, pp. 335–365. Lewis Publishers. New York.

Puri, R.K., Ye, Q., Kapila, S., Lower, W.R., and Puri, V. 1997. Plant uptake and metabolism of polychlorinated biphenyls (PCBs). In Wang, W., Gorsuch, J.W., and Hughes, J.S. (Eds.): Plants for environmental studies, pp. 481–513. Lewis Publishers. New York.

Solomon, K. et al. 1996. Ecological risk assessment of atrazine in North American surface waters. Environmental Toxicology and Chemistry 15: 31–76.

St.-Cyr, L., and Campbell, P.G.C. 1994. Trace metals in submerged plants of the St. Lawrence River. Canadian Journal of Botany 72: 429–439.

Swanson, S., Rickard, C., Freemark, K., and MacQuarrie. 1991. Testing for pesticide toxicity in aquatic plants: Recommendations for test species. In Gorsuch, J.W., Lower, W.R., Wang, W., and Lewis, M.A. (Eds.): Plants for toxicity assessment, pp. 77–97; 2nd vol., STP 1115. American Society for Testing Materials, Philadelphia, PA.

SYSTAT. 1997. Systat 7.0 for Windows: Statistics. SPSS, Chicago, IL.

Taraldsen, J.E., and Norberg-King, T.J. 1990. New method for determining effluent toxicity using duckweed (*Lemna minor*). Environmental Toxicology and Chemistry 9: 761–767.

VanderWal, J., and Lovett-Doust, L. 1998. Assessment of plant response to contaminants using image analysis. Poster Presentation, IAGLR Conference, McMaster University, Canada.

Wang, W., Gorsuch, J.W., and Hughes, J.S. (Eds.) 1997. Plants for environmental studies. Lewis Publishers. New York. 563pp.

20

FISH CHROMOSOMES AS BIOMARKERS OF GENOTOXIC DAMAGE AND PROPOSAL FOR THE USE OF TROPICAL CATFISH SPECIES FOR SHORT-TERM SCREENING OF GENOTOXIC AGENTS

MANUEL URIBE-ALCOCER AND PÍNDARO DÍAZ-JAIMES

Instituto de Ciencias del Mar y Limnología, UNAM. Apdo. Postal 70-305, México, D.F. 10810, México. FAX: ++ (525) 616 0748, e-mail: muribe@mar.icmyl.unam.mx

1. INTRODUCTION

Aquatic systems are essential for life. Human subsistence and health are linked to these systems, as they are indispensable for food production through agriculture, livestock raising and fisheries and to industrial processes to satisfy society's needs. However, these systems have often been converted, as negative by-products of progress, into reservoirs of wastes expecting natural degradation or recycling, or as long-term or permanent storage for materials of great endurance. The unawareness of the threat of a faster accumulation of wastes than what the systems can recycle, has turned the aquatic environments into regions of accumulation and concentration of dangerous genotoxic agents with serious risks to the organisms that depend directly or indirectly on water, that is, all living beings.

Abreviations used: B(a)P: benzo(a)pyrene; BrdU: 5-bromodeoxyuridine; BrdU-PFG: 5-bromodeoxyuridine plus fluorescence and Giemsa; CA: chromosome aberration; CP: cyclophosphamide; EMS: ethyl methanesulfonate; MMC: mitomycin C; MMS: methyl methanesulfonate; MNNG: methyl-N'-nitroso-guanidine; MNT: micronucleus test; SCE: sister chromatid exchange.

Biomonitors and Biomarkers as Indicators of Environmental Change 2, Edited by Butterworth *et al.*
Kluwer Academic/Plenum Publishers, New York, 2000

Fishes of the class Osteichthyes constitute the oldest and one of the more extended classes of vertebrates. This class encompasses more than 20,000 species with an abundance and diversity not found among any other group of vertebrates. They have inhabited every aquatic niche in the planet and have become a prosperous and stable biological group.

Growing awareness of the threat posed by xenobionts in the aquatic environment have emphasized the need to apply to aquatic species the methodologies already developed to study genotoxicity in other organisms (Kligerman et al., 1973, Hooftman and Vink, 1980). Teleosts turned out to be suitable organisms to monitor the effects of pollutants present in the water column and in the sediments of aquatic systems due to their sensitivity to the mutagenic action of genotoxic agents and to the presence of an efficient microsomal system. Additional logistic advantages of the use of fishes in screening tests are the existence of time-tested methodologies and the relative low cost of their maintenance and breeding, particularly when the target species are used in aquaculture since many of these species have been selected for their fecundity and rapid development. Besides, limited space and equipment are required for their study.

Among the most commonly used methodologies to study DNA damage in fishes we can find the evaluation of number and types of chromosome aberrations (CAs), of sister chromatid exchanges (SCEs) and of micronuclei MNT) and, to a lesser extent, the study of variations in mitotic indexes. The endpoint of these tests is the DNA damage brought about in cells by carcinogens and/or mutagens. The measurement of chromosome alterations, and its eventual repair, allows to infer the presence and concentration of these substances in the aquatic environment.

In this chapter, studies on clastogenicity, having as end-points chromosome breakage, CAs and SCEs, using teleosts as biomarkers, will be revised, including a general overview of the methodologies employed, organisms and cell sources most commonly used, some of the substances tested, as well as their exposure routes. The review on the micronucleus test has already been ably done by Al-Sabti and Metcalf (1995) and recent papers (for instance, Zakhidov et al., 1996; Rao et al., 1997) do not alter the general assessment of the application of this methodology. In the second part of the chapter a proposal for the use of tropical catfishes chromosomes as biomonitors of water column and/or sediments contaminants, either of marine or brackish water, is made.

2. METHODOLOGIES

Investigation on fish chromosomes started in the thirties (Iriki, 1932; Katayama, 1937; Makino, 1934). The earliest trend in the study of fish chromosomes was the evaluation of the cytogenetic damage associated with abnormal development of fishes brought about by ionizing radiation (Soldberg, 1938) in search of applications to fish breeding and to aquaculture, mainly by Russian investigators (Belyaeva and Pokrovskaia, 1959; Migalovskaya, 1973). Since the publication of the methodologies to obtain fish chromosomes by McPhail and Jones (1966), papers dealing with fish cytogenetics increased significantly focusing mainly on evolution and/or taxonomy (Chiarelli and Capanna, 1973; Denton, 1973; Ojima et al., 1973;

Sola et al., 1981; Oliveira et al., 1988). Yet, some peculiarities of fish chromosomes like their small size and the difficulty to obtain neat and repeatable banding patterns have hindered high resolution comparisons as those obtained in other vertebrate groups, such as mammals, by means of the G, Q, or R banding patterns.

The many existing methods to obtain mitotic spreads are mainly based on the technique developed by McPhail and Jones (1966) and Lieppman and Hubbs (1969) to obtain fields from gill epithelium. The solid tissue technique of Kligerman and Bloom (1977) has also often been followed. The book of Denton (1974) and papers of Blaxhall (1977), Kligerman (1982), Uribe-Alcocer et al. (1983), Castorena-Sánchez et al. (1983), and Maldonado Monroy et al. (1985) present a number of procedures used in fish cytogenetics.

The techniques to detect SCEs are based mainly on the BrdU-FPG technique of Perry and Wolff (1974). Adaptation of this method to fish can be found in Kligerman and Bloom (1976) and suggested improvements in van der Kerkhoff and van der Gaag (1985), Maddock et al. (1986) and Wei and Lu (1987).

Early genotoxicity studies on fish relied on species with few and large chromosomes, mainly the central mudminnow *Umbra limi* (Kligerman and Bloom, 1976; Kligerman, 1979; Vigfusson et al., 1983) and the eastern mudminnow *Umbra pygmaea* (Alink et al., 1980; Poels et al., 1982; Hooftman and Vink, 1981) with 22 large metacentric and submetacentric chromosomes. This trend has been often followed, as shown by the studies on *Monopterus albus* (Di and Liu., 1992), species with 32 chromosomes (Kitada and Tagawa, 1972), on the killifish *Nothobranchius rachovii* (Wrisberg and van der Gaag, 1992) whose karyotype has 16 (Post, 1965) or 18 chromosomes (Scheel, 1972). Barker and Rackham (1979) used *Ameca splendens* transformed tetraploid cell cultures, whose specific diploid chromosome complement is formed by 26 large chromosomes (Uyeno and Miller, 1972). Unfortunately, the number of species with a low diploid number karyotype is very small (Table 1). The search for large chromosomes has also been considered important as shown by the paper by Krishnaja and Rege (1982) in which the mud-slipper species *Boleophthalmus dussumieri* with 46 fairly large acrocentric chromosomes was selected as experimental species.

While some authors have contended that the low diploid number is not an indispensable condition (Zajicek and Phillips, 1984), logistic needs have driven the search for other experimental species. The convenience of counting with a steady and sufficient supply of fishes to allow testing a large number of specimens at low cost is reflected in the number of studies in which cultured fish species have been used. The most noteworthy are the tilapias *Oreochromis mossambicus* (Manna and Mukherjee, 1989 and 1992; Manna and Biswas, 1992a and b; Manna and Sadhukhan, 1992a and b; Ouseph et al., 1994) and various types of carps: the common carp *Cyprinus carpio* (Al-Sabti, 1985a; Al-Sabti, 1986; Pankaj et al., 1990), the grass carp *Ctenopharyngodon idellus* (Al-Sabti, 1985a; Wei and Lu, 1987; Mao and Wang, 1990) as well as the tench *Tinca tinca* (Al-Sabti, 1985a), and the goldfish *Carassius auratus* (Ueda et al., 1992). The rainbow trout *Salmo gairdneri (= Oncorhynchus mikiss)* has also been used in genotoxic studies by Zajicek and Phillips (1984), Al-Sabti (1985b) and Kocan et al., (1985). Fishes cultured in smaller scale have also been used: the rose bitterling *Rhodeus ocellatus ocellatus*, the metropolitan bitterling *Tanakia tanago* (Ueda et al., 1992), as well as the tropical fish *Etroplus suratensis* (Das and

TABLE 1. Fish species with diploid numbers under 2N = 30.

Apteronotidae	*Apteronotus albifrons*		24	Howell, 1972
Belontiidae	*Sphaerichthys osphoromoides*		16	Calton and Denton, 1974 Almeida Toledo et al., 1981
Bothidae	*Citarichthys spilopterus*		28	LeGrande, 1975
Cyprinodontidae	*Aphyosemion bivittatum* (Rio Muni)		26	Scheel, 1972
	Aphyosemion cameronense		24	Scheel, 1972
			26	Scheel, 1972
			28	Scheel, 1972
	Aphyosemion celiae		20	Scheel, 1972
	Aphyosemion celiae		26	Scheel, 1972
	Aphyosemion christyi		18	Scheel, 1972
	Aphyosemion cognatum		26	Scheel, 1972
	Aphyosemion franzwerneri		22	Scheel, 1972
	Aphyosemion labarrei		28	Scheel, 1972
	Aphyosemion louenssense		20	Scheel, 1972
	Aphyosemion schoutedeni		22	Scheel, 1972
	Aphyosemion seymouri		28	Scheel 1972
	Notobranchius rachovii		16	Post, 1965
			18	Scheel, 1972
	Pterolebias longipinnis	10		Post, 1965
	Pterolebias peruensis		20	Post, 1965
Gadidae	*Gaidropsaurus mediterraneus*		28	Vasil'ev, 1978
Galaxiidae	*Galaxias maculatus*		22	Campos, 1972 Merrilees, 1975
Goodeidae	*Ameca splendens*		22	Miller and Fitzimons, 1971
	Characodon lateralis		24	Uyeno and Miller, 1972
Lebiasinidae	*Nanostomus trifasciatus* type C	12		Scheel, 1973
	Poecylobrycon unifasciatus	11		Scheel, 1973
Mugilidae	*Mugil curema*		28	LeGrande and Fitzimons, 1976
Sternopygidae	*Eigenmania sp.*		28	Falçao and Bertollo, 1985
Synbranchidae	*Fluta alba = Monopterus albus*		24	Kitada and Tagawa, 1972
Tetragonopterinae	*Hemigrammus unileatus* (G. neon)	14	22	Scheel, 1973
Umbridae	*Umbra limi*		22	Beamish et al., 1971
	Umbra pygmaea		22	Beamish et al., 1971

John, 1997). The local geographic component has also been an important factor in the selection of the fish species employed. In Russia, fishes of the genus *Alburnus* and *Alburnoides* (Sofradzija et al., 1980) and *Misgurus* (Pechkurenkov and Kostrov, 1982; Pokrovskaya and Pechkurenkov, 1984) have been employed; in India, the genera *Boleophthalmus* (Krishnaja and Rege, 1982), *Labeo, Catla*, *Cirrhinus* (Manna and Biswas, 1992a) and *Channa* (Mohanty and Prasad, 1982; Rishi and Grewal, 1995) whereas in the Southern Atlantic, off the coast of Spain, the genera *Liza* and *Rutilus* (Lobillo et al., 1991). In the United States, the genera *Fundulus* (Means et al., 1987; Perry et al., 1988), *Lepomis* (Kocan and Stark, 1991), Morone (Baksi, 1987) *Parophrys* (Stromberg et al., 1981), and *Opsanus* (Maddock et al., 1986) have been used.

TABLE 2. Screening tests performed by means of fish chromosome aberrations as end-points.

Organism	Tissue	Exposure route	Assayed factors	Reference
Coregonus peled	sperm	bath	dimethyl sulfate	
Salmo irideus			nitrosomethyl urea	Tsoi, 1970
Cyprinuys carpio L	sperm	bath	dimethyl sulfate	
			nitrosoethyl urea	
			1,4-bis-diazoacetyl butane	Tsoi et al., 1975
Umbra limi	gill	bath	Rhine river water	Prein et al., 1978
Alburnus alburnus	kidney		pesticide Unden	
Alburnoides bipunctatus			pesticide Simazin	
			detergent	Sofradzija et al., 1980
Notobranchius rachowi	gill	bath	B(a)P	
			EMS	Hooftman, 1981
Umbra pygmaea	gill		B(a)P	
	fin		EMS	
			Rhine river water	Hooftman and Vink, 1981
Boleophthalmus dussumieri	gill, fin	i.p. injection	MMC	
	kidney, scales	i.m. injection	chromium dichromate	
	intestine	bath	selenium dioxide	
	stomach		phenuyl mercury acetate	Krishnaja and Rege, 1982
Lebistes reticulatus		*in vivo*	gamma-irradiation	Panlaque, 1982
Misgurnus	eggs	bath	mercury chloride,	
			phosphorus 32	Pechkurenkov and Kostrov, 1982
	fish cell cultures	field conditions	sediments Puget Sound	Landolt and Kocan, 1984

TABLE 2. *Continued*

Organism	Tissue	Exposure route	Assayed factors	Reference
Misgurnus fossilis	eggs larvae		beta-particles	Pokrovskaya and Pechkurenkov, 1984
Cyprinus carpio *Tinca tinca* *Ctenopharyngodon idellus*	kidney gills	i.p. injection	aflatoxin B1, benzidine, B(a)P, 20-methyl-cholanthrene, Aroclor 1254	Al-Sabti, 1985a
Salmo gairdneri	gills kidney		phenol, decamethrine, malathion, neguvon crude oil	Al-Sabti, 1985b
Lepomis macrochirus *Salmo gairdneri*	gonadal cell cultures	*in vitro* exposure	B(a)P	
	cell cultures	*in vitro* exposure	MNNG polycyclic aromatic hydrocarbons, polychlorinated biphenyls chlorinated hydrocarbons heavy metals.	Kocan et al., 1985
Cyprinus carpio L	kidney	i.p. injection	aflatoxin B1, benzidine, B(a)P, 20-methyl-cholanthrene Aroclor 1254	Al-Sabti, 1986
Morone saxatilis *Cyprinodon variegatus*	eggs larvae embryos		genotoxic agents	Baksi, 1987
	embryo		trace metals, and chlorinated hydrocarbons, polycyclic aromatic hydrocarbons	Cross et al., 1987

Morone saxitilis	gills,		mutagen	
Cyprinodon variegatus	kidney			
Fundulus heteroclitus	intestinal cells			Means et al., 1987
Fundulus heteroclitus	embryos		methylmercury	Perry et al., 1988
Anabas testudineus		i.p. injection	*Pseudomonas aeruginosa*	Biswas and Manna, 1989
Umbra limi	Cell Line (ULF-23HU)	*in vitro*	N-nitroso-N-methyl urea	Park et al., 1989
Cyprinodon variegatus	embryos		produced waters (PW) Pass Fourchon and Timbalier Island	Daniels and Means, 1990
Cyprinus carpio	gill		Cadmium nitrate	Pankaj et al., 1990
Lepomis macrochirus	BF-2 cells	*in vitro*	MNNG	Kocan and Stark, 1991
Oreochromis mossambicus	gill	i.p. injection	spores of *Aspergillus niger*, *A. ochraceus* *Helminthosporium oryzae*	Manna and Sadhukhan, 1991
Orechromis mossambicus	gill	i.p. injection	*Bacillus subtilis*	Biswas and Manna, 1992
Oreochromis mossambicus	gill	i.p. injection	*Pseudomonas aeruginosa*	
Anabas testudineus				Manna and Biswas, 1992a
Oreochromis mossambicus	gill, kidney	i.p. injection	*Staphylococcus aureus*	
Anabas testudineus				Manna and Biswas, 1992b
Oreochromis mossambicus.	Gill	bath	insecticide Endrin	Manna and Mukherjee, 1992
Oreochromis mossambicus	gill	bath and i.p. injection	insecticide Aldrex 30	Manna and Sadhukhan, 1992a
Oreochromis mossambicus	gill testes	bath and i.p. injection	anisole (methyl phenyl ether)	Manna and Sadhukhan, 1992b
Rhodeus ocellatus o.	gill, kidney	i.p. injection	MMC	Ueda, et al., 1992
Channa punctatus	kidney	bath	insecticide Baytex 1000	Rishi and Grewal, 1993
	eggs		18 sites near and offshore	Klumpp and Von Westernhagen, 1995
Channa punctatus	kidney	bath	insecticide dichlorvos	Rishi and Grewal, 1995

TABLE 3. Screening tests performed by means of fish chromosome sister chromatid exchanges as end-points.

Organism	tissue	exposure route	assayed factors	reference
Umbra limi	scales	i.p. injection	BrdU	
	intestines			
	kidney			
	gills			Kligerman and Bloom, 1976
Ameca splendens	transformed cell culture	*in vitro* exposure	EMS	
			MMC	
			MNNG	
			MMS	Barker and Rackham, 1979
Umbra limi	gill	i.p. injection	MMS	
	intestines		CP	
	kidney		neutral red dye	Kligerman, 1979
Umbra pygmaea	gill	bath	Rhine river water	
	testis			Alink et al., 1980
Umbra pygmaea	gill		B(a)P	
	fin		EMS	
			Rhine river water	Hooftman and Vink, 1981
Umbra pygmae	gill, testis	bath	Rhine river water	Poels et al., 1982
Parophrys vetulus		bath	BrdUrd	
			polluted site in Duwamish	
			River, Seattle, WA	Stromberg et al., 1981
Notobranchius rachowi	gill	bath	EMS	
			CP	van der Hoeven et al., 1982
Channa punctatus	gills and kidney	*in vivo*	BrdU	Mohanty and Prasad, 1982
Notobranchius rachowi	gill	bath	Rhine river water	van der Gaag et al., 1983
Umbra limi	gill	bath	endrin	
			chlordane	
			diazino	
			guthion	Vigfusson et al., 1983

Salmo gairdneri	embryos	bath	MNNG	
			gamma radiation	Zajicek and Phillips, 1984
Opsanus tau	kidney		BrdUrd	Maddock et al., 1986
Ctenopharyngodon idellus	kidney	bath	MMC	
			MNNG	Wei and Lu, 1987
Fundulus heteroclitus	embryos		MeHG	Perry et al., 1988
Umbra limi	Cell Line	*in vitro*	N-nitroso-N-methyl urea	
	(ULF-23HU)			Park et al., 1989
Oreochromis mossambicus	gill	bath	sodium arsenite	Manna and Mukherjee, 1989
Ctenopharyngodon idellus	cell strain CZ-7901		As3+	
	kidney		Cd2+	
			Cr6+	
			HG2+	
			Pb2+	
			Carbofunam	
			Dipterex	
			Methamidophos	
			Tricyluzole	Mao and Wang, 1990
Rutilus alburnoides	kidney	i.p. injection	S. Atlantic off Spain	
Liza aurata				Lobillo et al., 1991
Monopterus albus	lymphocytes		CP	
	spleen		MMC	
	kidney			Di and Liu, 1992
Tanakia tanago	embryos	bath	MMC	
Nothobranchius rachowi	fin cell culture			Ueda et al., 1992
	gill	bath	chlorine dioxide	
			bleaching effluent	Wrisberg and van der Gaag, 1992
Oreochromis mossambicus	gill	bath	River Cooum, India	
	kidney			Ouseph et al., 1994
Etroplus suratensis	gill tissues	i.m. injections	MMS	
			CP	Das and John, 1997

The cell sources most extensively employed in genotoxic studies in fish are the gills and kidney (Tables 2 and 3). Intestinal epithelium, (Kligerman and Bloom, 1976; Means et al., 1987) and gonads (Alink et al., 1980; Manna and Sadhukhan, 1991) have also been used. For studies of *in vivo* exposure to clastogenic agents, metaphases derived from eggs have been favored (Pechkurenkov and Kostrov, 1982; Perry et al., 1988; Klumpp and Von Westernhagen, 1995), as well as from embryos (Perry et al., 1988; Ueda et al., 1992) and larvae (Pokrovskaya and Pechkurenkov, 1984; Baksi, 1987). In some instances metaphases from fish cell culture have been used (Barker and Rackham, 1979; Park et al., 1989; Mao and Wang, 1990; Kocan and Stark, 1991; Ueda et al., 1992) and, occasionally, the spleen (Di and Liu, 1992) and scales (Kligerman and Bloom, 1976) have been the source of metaphase spreads.

No agreement has been reached as to which tissue is more suitable to study genotoxic damage. According to Alink et al. (1980) no difference was found between the SCE rate in gill and testicular cells of fish exposed to the Rhine water. They suggest the direct contact of gills with the surrounding water does not make them more vulnerable to water-borne mutagens than the testis cells because no blood-testis barrier is present for these substances. Al Sabti (1985b) in his studies on the genotoxicity of four pesticides and of crude oil in rainbow trout, found that rates of CAs were always higher in gill cells than in kidney cells and Krishnaja and Rege (1982) in a comparative study on gill, kidney, intestine, scale, fin, and stomach tissues of *Boleophthalmus dussumieri* found concurrently that gill tissue yielded three times as many usable metaphases as did other tissues. Hooftman and Vink (1981) also found that gill tissue appeared to be the most suitable tissue for SCE analysis, although no significant difference was found in the production of CAs from fin and gill exposed to B(a)P and EMS.

On the other hand, Rishi and Grewal (1995) considered that the fish kidney CAs test is suitable to study the genotoxic effect of pesticides; and Biswas and Manna (1989) and Manna and Biswas (1992a) in their study of the genotoxicity of *Pseudomonas aeruginosa*, on the climbing perch *Anabas testudineus*, reported that, although the gill and kidney cells had qualitatively the same types of CAs, their frequencies differed indicating a higher sensitivity of kidney over gill cells. The testis tissue might also be suitable, but it seems to have the disadvantage of having only one short period of active cell division along the year (Hooftman and Vink, 1981).

Substances have been administered *in vivo* or *in vitro*. Administration *in vitro* has been done by exposing cell cultures to clastogenic agents. Barker and Rackham (1979) exposed transformed cell cultures derived from diploid embryonic tissues of *Ameca splendens* to EMS, MMC, MMS, and MNNG; Landolt and Kocan (1984) employed cell cultures derived from the gonads of *Salmo gardnieri* to assay Puget Sound sediment extracts. Kocan and Stark (1991) exposed BF-2 cells derived from the bluegill sunfish *Lepomis macrochirus* to MNNG. Mao and Wang (1990) studied five metal ions and four pesticides using as biomarker the cell strain ZC-7901 as well as kidney cells *in vivo* from grass carp (*Ctenopharyngodon idella*) juveniles.

The main advantages of using *in vitro* administration to cell cultures are: 1) availability of cells for tests; 2) possibility of duplicating the assays as many times as necessary; 3) elimination of individual cell variability; 4) absence of detoxifying and concentrating mechanisms; 5) greater accuracy of the estimated amount of the applied substances since fewer barriers to target molecules exist; 6) the cell-cycle kinetics studies are easier to perform; 7) tested methodologies for chromosome spreads can be used; 8) better quality and more mitotic spreads can be obtained.

However, most investigations have used *in vivo* administration because it allows the operation of the metabolic and detoxifying mechanism, as well as the potential activation of promutagens (Protic-Sabljic and Kurelec, 1983) and passage of the compounds across the blood testis barriers found also in the organisms for which the extrapolation of the results might be desired.

Clastogenic substances have been administered *in vivo* through direct or indirect exposure. The former include injection of the test substance into the fish, either intraperitoneally, the most usual form, (Tables 2 and 3) or, in a few particular cases, intramuscularly (Das and John, 1997; Krishnaja and Rege, 1982). Indirect exposure is produced either by bathing the target specimens in aqueous solutions of the test substance (for instance, Manna and Mukherjee, 1989; Wrisberg and van der Gaag, 1992; Ueda et al., 1992; Rishi and Grewal, 1995), by exposing them to waters extracted from test sites like the "produced waters" from Pass Fourchon and from Timbalier Island in the northwestern Gulf of Mexico (Daniels and Means,1990) or the Rhine River water (Alink et al., 1980; Poels et al., 1982), by actual exposure to water samples of interest or by collecting fish from the screening site, as done at the Puget Sound, Washington, and the Duwamish River, in Seattle, U.S.A. (Stromberg et al., 1981), in the Southern Atlantic off the coasts of Spain (Lobillo et al., 1991), in the River Cooum in India (Ouseph et al., 1994), or in nearshore and offshore locations on the Great Barrier Reef (Klumpp and Von Westernhagen, 1995).

The injection of samples extracted from the water bodies, either as extracts or as concentrates, permits to infer the amount of pollution found in those water bodies through the extent of the DNA and chromosomes damage. A comparison of these two types of exposure to heavy metals made by Krishnaja and Rege (1982) on *Boleophthalmus dussumieri*, showed that selenium had the strongest clastogenic potential followed by mercury and chromium in the direct-exposure series, whereas mercury induced a higher number of aberrations followed by selenium and chromium in the indirect exposure series. Differential pathways and barriers to these metals can probably explain their results.

A parallel line of contamination research studies has focused on differential activitation of metabolic enzimes, such as hepatic ethoxyresorufin-O-deethylase (EROD) or aryl hydrocarbon hydroxylase (AHH) in some fish species, as parameters related to the presence of environmental pollution in aquatic systems (Davies et al., 1984; Protic-Sabljic, 1984; Pangrekar and Sikka, 1992; Narbonne et al., 1994; Hasspieler et al., 1994; Rodríguez-Ariza et al., 1995; Otto and Moon, 1996; Moon et al., 1996; Eufemia et al., 1997; Stien et al., 1998; Au et al., 1999).

3. CHROMOSOMAL ABERRATION AND SISTER CHROMATID EXCHANGE TESTS

Relevant increments of sister chromatid exchanges have been brought about by most of the genetically-active mutagens or carcinogens when they were tested in *in vivo* animal systems or *in vitro* systems with activation (Carrano et al., 1978). Since the studies of Latt (1974) and Perry and Evans (1975) revealed that most repaired chromosomal alterations are not detected by means of the classical chromosome breakage methods, SCEs have been generally considered as a more sensitive method for the study of chromosome damage than the classical methods based on chromosome breakage. Kato and Shimada (1975) found that significant increases in SCEs were induced in cultured cells with a number of chemicals (MMC, EMS, MMS) at levels that produce little or no gross chromosome damage. Wrisberg and van der Gaag (1992) based on a study of SCEs in gill cells from the fresh water fish *Nothobranchius rachowi*, considered that the sensitivity of this test was comparable or even higher than the Ames test, as it was able to detect genotoxins in unconcentrated samples from a chlorine dioxide bleached effluent that the Ames test failed to detect.

Although the sensitivity of the SCEs in fishes is considered satisfactory to detect environmental genotoxic substances, a comparison of the SCE rate among some animal and plant systems showed that the mudminnow *Umbra limi* presented a low sensitivity (Kligerman and Bloom, 1976), surpassed only by the Chinese hamster cell line D-6 (Kato, 1974; Kato and Shimada, 1975) and by mouse spermatogonial cells (Morales-Ramírez et al., 1994). The differences between the SCE rates of *Vicia faba* and of *Umbra limi* were more than tenfold: *Vicia faba* showed between 28 and 32 exchanges/metaphase (Gómez-Arroyo and Villalobos-Pietrini, 1995a and b; Gómez Arroyo et al., 1997) vs. 2.59 exchanges/metaphase in *Umbra limi* (Kligerman and Bloom, 1976). Nevertheless, in a fish cell line (ULF-23HU), derived from fins of the central mudminnow, an increased spontaneous SCE rate of 5.3 per metaphase was obtained (Park et al., 1989).

In order to take some of the steps to validate short-term genotoxicity test on fishes, a number of studies have proved classical mutagens like CP (Kligerman, 1979; Di and Liu, 1992; Das and John, 1997), EMS (Barker and Rackham, 1979; Hooftman and Vink, 1981), B(a)P (Hooftman and Vink, 1981; Al-Sabti, 1985a; Kocan et al., 1985; Al-Sabti, 1986) or MMC (Kligerman, 1979; Barker and Rackham, 1979; Wei and Lu, 1987; Ueda, et al., 1992; Di and Liu, 1992; Das and John, 1997) and found that they did induce chromosome damage in these organisms, either like CAs, SCEs or by producing micronuclei.

Chromosome breakage produced by a genotoxic agent in one group of organisms produces the same effect in other groups due to the universality of the DNA. Chromatid gaps, breaks, centromeric gaps, acentric fragments and picnosis are reported to have been produced in *Channa punctatus* by the insecticides Dichlorovos (Rishi and Grewal, 1995) and Baytex 1000 (Rishi and Grewal, 1993). Injection of bacterial cultures of *Pseudomonas aeruginosa* in the climbing perch *Anabas testudineus* (Biswas and Manna, 1989) and of *Bacillus subtilis* in the tilapia *Oreochromis mossambicus* (Biswas and Manna, 1992) had the same effects. Al-Sabti (1985a) found in an assay with three carp species that Aroclor 1254 induced more breaks and fragments than ring and dicentric chromosomes, and that B(a)P enhanced fragment induction. Yet, in complex mixtures such as those found in 'produced waters' (waters used in oil extraction) differences have been found. Daniels and Means (1989) detected that all classes of aberrations were formed, especially chromatid and chromosome breaks, in embryos exposed to "produced waters" from the Timballier Island, whereas the embryos exposed in another site, Pass Fourchon, most frequently presented in changes in ploidy.

Studies bearing on several species treated with the same genotoxic substances are informative as to their differential sensitivities (Table 4). For instance, Al-Sabti (1985a) compared the response of kidney cells of the common carp, the tench, and

TABLE 4. Factors assayed by means of chromosomal mutations (CAs and SCEs).

Factors assayed mutagens	Organism	test	reference
aflatoxin B1	*Ctenopharyngodon idella*	CA	Al-Sabti, 1985a
aflatoxin B1	*Cyprinus carpio*	CA	Al-Sabti, 1985a
aflatoxin B1	*Tinca tinca*	CA	Al-Sabti, 1985a
aflatoxin Bl	*Cyprinus carpio*	CA	Al-Sabti, 1986
benzidine	*Ctenopharyngodon idella*	CA	Al-Sabti, 1985a
benzidine	*Cyprinus carpio*	CA	Al-Sabti, 1985a
benzidine	*Tinca tinca*	CA	Al-Sabti, 1985a
benzidine	*Cyprinus carpio*	CA	Al-Sabti, 1986
B(a)P	*Notobranchius rachowi*	CA	Hooftman, 1981
B(a)P	*Umbra pygmaea*	CA	Hooftman and Vink, 1981
B(a)P	*Umbra pygmaea*	SCE	Hooftman and Vink, 1981
B(a)P	*Ctenopharyngodon idella*	CA	Al-Sabti, 1985a
B(a)P	*Cyprinus carpio*	CA	Al-Sabti, 1985a
B(a)P	*Tinca tinca*	CA	Al-Sabti, 1985a
B(a)P	*Lepomis macrochirus*	CA	Kocan et al., 1985
B(a)P	*Salmo gardnieri*	CA	Kocan et al., 1985
B(a)P	*Cyprinus carpio*	CA	Al-Sabti, 1986
CP	*Umbra limi*	SCE	Kligerman, 1979
CP	*Notobranchius rachowi*	SCE	van der Hoeven et al., 1982
CP	*Monopterus albus*	SCE	Di and Liu 1992
CP	*Etroplus suratensis*	SCE	Das and John, 1997
dimethyl sulfate	*Coregonus peled*	CA	Tsoi, 1970
dimethyl sulfate	*Salmo irideus*	CA	Tsoi, 1970
dimethyl sulfate	*Cyprinus carpio L*	CA	Tsoi et al., 1975
EMS	*Ameca splendens*	SCE	Barker and Rackham, 1979
EMS	*Notobranchius rachowi*	SCE	Hooftman, 1981
EMS	*Umbra pygmaea*	CA	Hooftman and Vink, 1981
EMS	*Umbra pygmaea*	SCE	Hooftman and Vink, 1981
EMS	*Notobranchius rachowi*	SCE	van der Hoeven et al., 1982
20-methyl-cholanthrene,	*Ctenopharyngodon idella*	CA	Al-Sabti, 1985b
20-methyl-cholanthrene	*Cyprinus carpio L*	CA	Al-Sabti, 1985b
20-methyl-cholanthrene,	*Tinca tinca*	CA	Al-Sabti, 1985b
20-methyl-cholanthrene	*Cyprinus carpio L*	CA	Al-Sabti, 1986
MMS	*Ameca splendens*	SCE	Barker and Rackham, 1979
MNNG	*Ameca splendens*	SCE	Barker and Rackham, 1979
MNNG	*Salmo gairdneri*	SCE	Zajicek and Phillips, 1984
MNNG	*Salmo gardnieri*	CA	Kocan, 1985
MNNG	*Lepomis macrochirus*	CA	Kocan et al., 1985
MNNG	*Ctenopharyngodon idellus*	SCE	Wei and Lu, 1987
MNNG	*Lepomis macrochirus*	CA	Kocan and Stark, 1991

TABLE 4. *Continued*

Factors assayed mutagens	Organism	test	reference
MMC	*Ameca splendens*	SCE	Barker and Rackham, 1979
MMC	*Boleophthalmus dussumieri*	CA	Krishnaja and Rege, 1982
MMC	*Ctenopharyngodon idellus*	SCE	Wei and Lu, 1987
MMC	*Monopterus albus*	SCE	Di and Liu, 1992
MMC	*Rhodeus ocellatus ocellatus*	CA	Ueda et al., 1992
MMC	*Rhodeus ocellatus ocellatus*	SCE	Ueda et al., 1992
MMC	*Tanakia tanago*	SCE	Ueda et al., 1992
MMS	*Umbra limi*	SCE	Kligerman, 1979
MMS	*Etroplus suratensis*	SCE	Das and John, 1997
neutral red dye	*Umbra limi*	SCE	Kligerman, 1979
nitrosoethyl urea	*Cyprinuys carpio L*	CA	Tsoi et al., 1975
nitrosomethyl urea	*Coregonus peled*	CA	Tsoi, 1970
nitrosomethyl urea	*Salmo irideus*	CA	Tsoi, 1970
nitrosomethyl urea	*Umbra limi*	CA	Park et al., 1989
nitrosomethyl urea	*Umbra limi*	SCE	Park et al., 1989
phenol	*Salmo gairdneri*	CA	Al-Sabti, 1985b
1,4-bis diazacetyl butane	*Cyprinus carpio L*	CA	Tsoi et al., 1975
mutagens	*Morone saxitilis*	CA	Means et al., 1987
mutagens	*Cyprinodon variegatus*	CA	Means et al., 1987
mutagens	*Fundulus heteroclitus*	CA	Means et al., 1987
genotoxic agents	*Morone saxatilis*	CA	Baksi, 1987
genotoxic agents	*Cyprinodon variegatus*	CA	Baksi, 1987
pesticides			
Aldrex 30	*Oreochromis mossambicus*	CA	Manna and Sadhukhan, 1992a
Anisole (methyl phenyl ether)	*Oreochromis mossambicus*	CA	Manna and Sadhukhan, 1992b
Aroclor 1254	*Ctenopharyngodon idella*	CA	Al-Sabti, 1985a
Aroclor 1254	*Cyprinus carpio L*	CA	Al-Sabti, 1985a
Aroclor 1254	*Tinca tinca*	CA	Al-Sabti, 1985a
Aroclor 1254	*Cyprinus carpio L*	CA	Al-Sabti, 1986
Baytex 1000	*Channa punctatus*	CA	Rishi and Grewal, 1993
Carbofunam	*Ctenopharyngodon idellus*	SCE	Mao and Wang, 1990
Chlordane	*Umbra limi*	SCE	Vigfusson et al., 1983
Decamethrine	*Salmo gairdneri*	CA	Al-Sabti, 1985a
Diazino	*Umbra limi*	SCE	Vigfusson et al., 1983
Dichlorvos	*Channa punctatus*	CA	Rishi and Grewal, 1995
Dipterex	*Ctenopharyngodon idellus*	SCE	Mao and Wang, 1990
Endrin	*Oreochromis mossambicus*	CA	Manna and Mukherjee, 1992
Endrin	*Umbra limi*	SCE	Vigfusson et al., 1983
Guthion	*Umbra limi*	SCE	Vigfusson et al., 1983
Malathion	*Salmo gairdneri*	CA	Al-Sabti, 1985b
Methamidophos	*Ctenopharyngodon idellus*	SCE	Mao and Wang, 1990
Neguvon	*Salmo gairdneri*	CA	Al-Sabti, 1985b
Simazin	*Alburnus alburnus*	CA	Sofradzija et al., 1980
Simazin	*Alburnoides bipunctatus*	CA	Sofradzija et al., 1980
Tricyluzole	*Ctenopharyngodon idellus*	SCE	Mao and Wang, 1990
Unden	*Alburnus alburnus*	CA	Sofradzija et al., 1980
Unden	*Alburnoides bipunctatus*	CA	Sofradzija et al., 1980
detergent	*Alburnus alburnus*	CA	Sofradzija et al., 1980
detergent	*Alburnoides bipunctatus*	CA	Sofradzija et al., 1980

TABLE 4. *Continued*

Factors assayed mutagens	Organism	test	reference
metal ions			
As^{3+} sodium arsenite	*Oreochromis mossambicus*	SCE	Manna and Mukherjee, 1989
As^{3+}	*Ctenopharyngodon idellus*	SCE	Mao and Wang, 1990
Cd^{2+} Cadmium nitrate	*Cyprinus carpio*	CA	Pankaj et al., 1990
Cd^{2+}	*Ctenopharyngodon idellus*	SCE	Mao and Wang, 1990
Cr^{6+} chromium dichromate	*Boleophthalmus dussumieri*	CA	Krishnaja and Rege, 1982
Cr^{6+}	*Ctenopharyngodon idellus*	SCE	Mao and Wang, 1990
Hg^{2+} mercury chloride	*Misgurnus*	CA	Pechkurenkov and Kostrov, 1982
Hg^{2+}	*Ctenopharyngodon idellus*	SCE	Mao and Wang, 1990
Hg^{2+} phenyl mercury acetate	*Boleophthalmus dussumieri*	CA	Krishnaja and Rege, 1982
Hg^{2+} methylmercury	*Fundulus heteroclitus*	CA	Perry et al., 1988
Hg^{2+} methylmercury	*Fundulus heteroclitus*	SCE	Perry et al., 1988
Ph_{32}	*Misgurnus*	CA	Pechkurenkov and Kostrov, 1982
Pb^{2+}	*Ctenopharyngodon idellus*	SCE	Mao and Wang, 1990
Se selenium dioxide	*Boleophthalmus dussumieri*	CA	Krishnaja and Rege, 1982
heavy metals	*Salmo gairdnieri*	CA	Kocan et al., 1985
heavy metals	*Lepomis macrochirus*	CA	Kocan et al., 1985
trace metals		CA	Cross et al., 1987
radiations			
gamma-irradiation	*Lebistes reticulatus*	CA	Panlaque, 1982
gamma-irradiation	*Salmo gairdnieri*	SCE	Zajicek and Phillips, 1984
beta-particles	*Misgurnus fossilis*	CA	Pokrovskaya and Pechkurenkov, 1984
hydrocarbons			
crude oil	*Salmo gairdneri*	CA	Al-Sabti, 1985a
chlorinated hydrocarbons	*Salmo gairdnieri*	CA	Kocan et al., 1985
chlorinated hydrocarbons	*Lepomis macrochirus*	CA	Kocan et al., 1985
chlorinated hydrocarbons		CA	Cross et al., 1987
polycyclic aromatic hydrocarbons	*Salmo gairdnieri*	CA	Kocan et al., 1985
polycyclic aromatic hydrocarbons	*Lepomis macrochirus*	CA	Kocan et al., 1985
polycyclic aromatic hydrocarbons		CA	Cross et al., 1987
polychlorinated biphenyls	*Salmo gairdnieri*	CA	Kocan et al., 1985
polychlorinated biphenyls	*Lepomis macrochirus*	CA	Kocan et al., 1985
bacteria			
Bacillus subtilis	*Oreochromis mossambicus*	CA	Biswas and Manna, 1992
Pseudomonas aeruginosa	*Oreochromis mossambicus*	CA	Manna and Biswas, 1992a
Pseudomonas aeruginosa	*Anabas testudineus*	CA	Manna and Biswas, 1992a
Staphylococcus aureus	*Oreochromis mossambicus*	CA	Manna and Biswas, 1992b
Staphylococcus aureus	*Anabas testudineus*	CA	Manna and Biswas, 1992b
fungi			
Aspergillus niger	*Oreochromis mossambicus*	CA	Manna and Sadhukhan, 1991

TABLE 4. *Continued*

Factors assayed mutagens	Organism	test	reference
Aspergillus ochraceus	*Oreochromis mossambicus*	CA	Manna and Sadhukhan, 1991
Helminthosporium oryzae	*Oreochromis mossambicus*	CA	Manna and Sadhukhan, 1991
environmemental pollutants			
Atlantic S. off Spain	*Rutilus alburnoides*	SCE	Lobillo et al., 1991
Atlantic S. off Spain	*Liza aurata*	SCE	Lobillo et al., 1991
Puget Sound sediments		CA	Landolt and Kocan, 1984
Puget Sound sediment extracts	*Salmo gairdnieri*	CA	Kocan et al., 1985
Puget Sound sediment extracts	*Lepomis macrochirus*	CA	Kocan et al., 1985
produced waters (PW)			
Pass Fourchon	*Cyprinodon variegatus*	CA	Daniels and Means, 1990
Timbalier Island	*Cyprinodon variegatus*	CA	Daniels and Means, 1990
18 sites by Great Barrier		CA	Klumpp and Von Westernhagen, 1995
River Cooum, India	*Oreochromis mossambicus*	SCE	Ouseph et al., 1994
River Duwamish, USA	*Parophrys vetulus*	SCE	Stromberg et al., 1981
River Rhine water	*Umbra limi*	CA	Prein et al., 1978
River Rhine water	*Umbra pygmaea*	SCE	Alink et al., 1980
River Rhine water	*Umbra pygmaea*	SCE	Hooftman and Vink, 1981
River Rhine water	*Umbra pygmaea*	CA	Hooftman and Vink, 1981
River Rhine water	*Umbra pygmaea*	SCE	Poels et al., 1982
River Rhine water	*Notobranchis rachowi*	SCE	Van der Gaag et al., 1983
bleaching effluent	*Nothobranchius rachowi*	SCE	Wrisberg and van der Gaag, 1992
chlorine dioxide	*Nothobranchius rachowi*	SCE	Wrisberg and van der Gaag, 1992

the grass carp to known mutagens. In the common carp, the higher induction of CAs was produced by Aroclor 1254 and decreased from benzidine, aflatoxin B_1, and 2′-methylchloanthrene to B(a)P. In the tench, benzidine induced more CAs, whereas Aroclor 1254 induced a higher percentage of CA in the grass carp. He attributed these responses to differential concentrating potential of the kidney in the studied species or to the different length of the cell cycle in the species studied. Furthermore, the study of van der Hoeven et al. (1982) found that the killlifish *Notobrachius rachowi* and the eastern mudminnow *Umbra pygmaea* showed a similar sensitivity to EMS.

Assays of several genotoxic agents on the same species have shown differential responses. The studies of Al Sabti (1985b) in the rainbow trout showed that in this species the CAs most frequently found as a response to phenol were aneuploidy. Non-specific aberrations were increased by Neguvon and Malathion, although the latter also produced the highest rate of structural aberrations.

The studies of genotoxicity of cultures of the bacterium *Pseudomonas aeruginosa* injected intraperitoneally in the tropical tilapia *Oreochromis mossambicus* (Manna and Biswas, 1992a), showed that CAs seemed to be nonrandomly distributed, since the larger metacentric chromosome pair, characteristic of the tilapias, bore on the average 11 CAs, whereas the other 21 pairs showed only 12, against the expected ratio of 1 metacentric: 21 nonmetacentric were they randomly distributed.

The variables involved among different protocols preclude many meaningful comparisons such as the relative potency of mutagens in the same species, the differential inter- or intraspecific sensitivities, the thresholds, and/or the saturated concentrations for CAs or SCEs as well as whether or not sensitivities variability exists, among different karyotypes. The existing variations among fish species and in the chromosome test employed, in the way of administering the substances (*in vivo* either by injection of by exposure to solutions, or *in vitro*), in the length of the exposure to the mutagen, in the tissue in which the response has been measured, might bias conclusions drawn from comparisons between studies based on different protocols.

Few comparisons have been made, for instance Kligerman (1979) reported that on a microgram/gram basis, CP was found to be a 2–4 times more potent inducer of SCEs than MMS in the eastern mudminnow *Umbra limi*. Similarly van der Hoeven et al. (1982) detected that CP was almost 3 times more effective in inducing SCEs in *Notobranchius rachowi* than EMS on a molar basis. Furthermore, Barker and Rackhan (1979) found that, in tetraploid cell cultures of *Ameca splendens* the mutagens relative potency to induce SCEs was EMS < MMS < MNNG < MMC, similarly to results obtained in rodent and human cells (Perry and Evans, 1975 and Wolff et al., 1977). The work of Krishnaja and Rege (1982) showed that MMC chromosome-breaking efficiency is lower in the mud-skipper *Bolephthamus dussumieri* system than in *in vivo* mammalian systems. Thus, the performance of screening tests based on the same, or at least, similar protocols whenever possible is highly recommended in order to integrate the many data obtained from the fish chromosome studies.

The establishment of a causal relationship between the individual pollutants within the complex mixtures found in some aquatic environments and the toxic effects measured is often not possible. Several papers have emphasized the need to establish this relationship by suitable batteries of screening assays. Poels et al. (1982) argued for the need to integrate several types of tests: 1) long-term, such as embryo/larval tests, to appraise chronic genotoxicity; 2) short-term tests, like the SCEs and/or the Ames test to screen for short-term damage; and 3) chemical analysis, to know the specific mutagenic substances acting in particular assays. Weisburger and Williams (1991) have stressed the need to include several approaches, among which they mention CAs as well as SCEs tests, to study neoplasia in fish as a possible indicator of cancer risks to humans. To corroborate the mutagenic potential of the bacteria *Pseudomonas aeruginosa* and *Staphylococcus aureus* injected in fish species, Manna and Biswas (1992a and 1992b) used a battery of tests that included CAs in gill epithelia and first spermatocyte metaphase, MNT in peripheral blood, gill epithelia, and kidney cells, and lethal test for unfertilized eggs and death of embryos resulting from mating treated male parents with normal virgin females. The inclusion of variations to the MNT, like MN assays in binucleated erythrocytes (Al-Sabti, 1994) or in hepatic cells after treatment with allyl formate (Williams and Metcalfe, 1992; Rao et al., 1997) might help to attain a keener evaluation of the genotoxicity of the tested substances.

The variability found within teleosts is the result of a long evolutionary history. The use of the appellative "fish" encrypts many types of adaptive and physiological characters that determine differential responses to individual mutagenic agents reflected in the types and amounts of CAs, SCEs, MNT or in the alterations of mitotic indexes. These differences have determined the need to establish particular protocols, which in fact have been used in a number of grouped studies, such as those using fishes of the mudminnow genus *Umbra*, the killifish species *Nothobranchius rachowii*, the Indian carps, the tilapia *Oreochromis mossambicus*, etc. It is important then, to adapt suitable protocols for short-term genotoxicity tests in aquatic environments with the preferential use of local fish species for the ecotoxicological relevance of these organisms on regional studies to solve particular pollution problems.

4. CHROMOSOME ANALYSIS OF THREE TROPICAL CATFISHES

As seen earlier CAs and SCEs have been chosen as biomarkers to study genotoxicity in teleosts, and the target species have often been locally selected, mainly to screen for aquatic pollution *in situ*.

In this section of the chapter, we will describe the chromosome complement of three catfish species that might be useful biomarkers of pollution in marine and estuarine or fluvial habitats: *Cathorops fuerthi, C. aguadulce* (Ariidae) and *Istlarius balsanum* (Ictaluridae), respectively, to provide data-bases to evaluate possible chromosome damage brought about by environmental genotoxic substances in their respective biotopes.

Among the 35 Siluriform families, the ariids belong to one of the two marine families of the order (Nelson et al., 1994). Some species of this family, as *Cathorops fuerthi* or *C. aguadulce*, are adapted to estuarine and lagoonar habitats. Studies on the presence of genotoxic agents in these habitats are important, since many commercially relevant species of fish, shellfish, and crustaceans use them as spawning grounds, and it is crucial that any potential mutagenic and clastogenic risk to the organisms developing in these places or to the organisms that directly or indirectly feed on them, be accurately assessed.

Cathorops fuerthi and *C. aguadulce* display a wide distributional range and live along the coasts of the Mexican Pacific and of the Gulf of Mexico, respectively, where they play an important ecological and economic role. These species endure strong environmental pressures brought about by the sharp shifts in water salinity and temperature occurring in estuaries and lagoons. Particularly during the dry season, these conditions are extreme and only few species survive. The euryhalo-thermic adaptation potential of *C. fuerthi* and of *C. aguadulce* allows them to tolerate temperature ranges from 19 to 35 °C and of salinity from 0 to 40 ppm.

Istlarius balsanus, like other members of the Ictaluridae family, lives in rivers. This species was first described by Jordan and Snyder (1900) for the Balsas River Basin of Mexico. Little biological information about this catfish has since been gathered. Kato and Romo (1981) have determined it is a carnivorous fish that modifies its diet in accordance to the abundance of available invertebrates. With nocturnal habits and extended reproductive season, with its peak in April and May, it reaches reproductive age sooner than other ictalurids. Sexual dimorphism is not manifest and it has a higher survival rate than other members of its family. Maximum size record for this fish is 1.25 m long and 18 kg (Rosas, 1976).

4.1. Materials and Methods

Specimens were collected on the following sites:

a) In the Alvarado Lagoon (18.73° and 18.86°N, and 95.73° and 95.95°W) and in the town of Tlacotalpan (18.63°N and 95.66°W) on the margins of the Papaloapan River in the Mexican state of Veracruz, six males and three female specimens of *C. aguadulce*;
b) In the Tres Palos lagoon (between 16.71° and 16.82°N, and 99.65° and 99.77°W), NW of Acapulco in the state of Guerrero, four males and five female specimens of *C. fuerthi*;
c) In the River Huámito, (18.95°N and 101.87°W) in the state of Michoacán and in the river Amacuzac (18.52°N and 99.22°W), in the state of Guerrerro, four female and one male specimen, and three female and two male specimens, respectively, of *Istlarius balsanus*.

The specimens were collected by hook and line, and occasionally by castnets. Some specimens were processed *in situ* and other were transported alive to our Laboratory in Mexico City, where the process was performed. Specific determination of the specimens was made with the assistance of colleagues in charge of the Ichthyological Collection of the Instituto de Biología of the National University of Mexico, considering mainly the number of tooth patches on the palate as key feature to sort the different species as proposed by Taylor and Menezes (1978).

The cytogenetic methodology used to obtain mitotic fields has been described in Uribe Alcocer et al. (1983), Castorena-Sánchez et al. (1983) and Maldonado Monroy et al. (1985), as follows:

1. Five hours prior to processing, specimens were injected intraperitoneally with an aqueous 0.1% calcium chloride solution, according to the size of the specimen: from 5 to 10 cm length, 0.5 ml; from 10 to 15 cm length, 0.75 ml; and from 15 to 20 cm length, 1 ml.
2. After three hours, a 0.1% colchicine aqueous solution was injected in the anterodorsal muscles, at a dose of 0.1 ml/10 g of body weight.
3. Two hours later, fishes were killed and gills were dissected and immersed in a 0.75 M KCl solution so that a hypotonic shock was simultaneously applied. While the gill lamellae were kept in the hypotonic solution at 37 °C for 30 minutes, they were gently scrapped so that cells were shed and kept suspended. Cartilage particles were removed.

TABLE 5. Chromosome data of the catfish species reported in this paper.

species (site)	Organisms (males + females)	cells counted	modal diploid number	karyotype formula	fundamental number
Cathorops aguadulce (Alvarado Lagoon)	9 (6 + 3)	129	56	8m + 4sm + 13st + 3t	80
Cathorops fuerthi (Tres Palos Lagoon)	9 (4 + 5)	139	52	8m + 12sm + 5st + 1t	92
Istlarius balsanus (Amacuzac River)	5 (*)	115	58	9m + 13sm + 7st	102
Istlarius balsanus (Huámito River)	5 (*)	49	58	9m + 13sm + 7st	102

(*) not determined.

4. The cell-suspension was centrifuged 5 minutes at 800 rpm, the supernatant was discarded and the pellet was fixed in freshly made metanol-acetic acid mixture (3:1). It was resuspended after a 10-minutes rest. This was repeated three or four more times.
5. The pellet was resuspended again, and preparations were made by letting fall two or three drops of the cellular suspension on clean microscope slides. These were air-dried and stained with 2% aceto-orcein or Giemsa solution in phosphate buffer, pH 7.
6. The methods of Levan et al. (1964) were followed to classify chromosomes according to the centromere's position.

4.2. Results and Discussion

Table 5 shows the number and sex of analyzed specimens, number of cells counted to determine the diploid number, as well as the modal diploid, fundamental number, and the karyotype formula for every species. The karyotypes of the studied species are shown in Figure 1.

The chromosome diploid number of the ariid species *Arius felis* (LeGrande, 1980; García-Molina and Uribe-Alcocer, 1989) and *Bagre marinus* (Fitzimmons et al., 1988) have been determined as 2N = 54, whereas for it is *C. aguadulce* 2N = 56 and for *C. fuerthi*, 2N = 58. The fundamental number found in this work for

FIGURE 1. (a) Representative karyotype of *C. aguadulce*. The top row shows metacentric chromosomes, the second row, submetacentrics; third and fourth rows: subtelocentrics; and fifth row, acrocentric chromosomes. Bar = 10 µm. (b) Karyotype of *C. fuerthi*. First row shows metacentric chromosomes; second and third, submetacentrics, and fourth row shows the subtelocentrics as well as the acrocentric chromosomes. Bar = 10 µm. (c) Karyotype of *I. balsanus*. First and second rows show metacentric chromosomes; third and fourth rows, submetacentrics; and fifth row, the subtelocentric chromosomes. Bar = 10 µm.

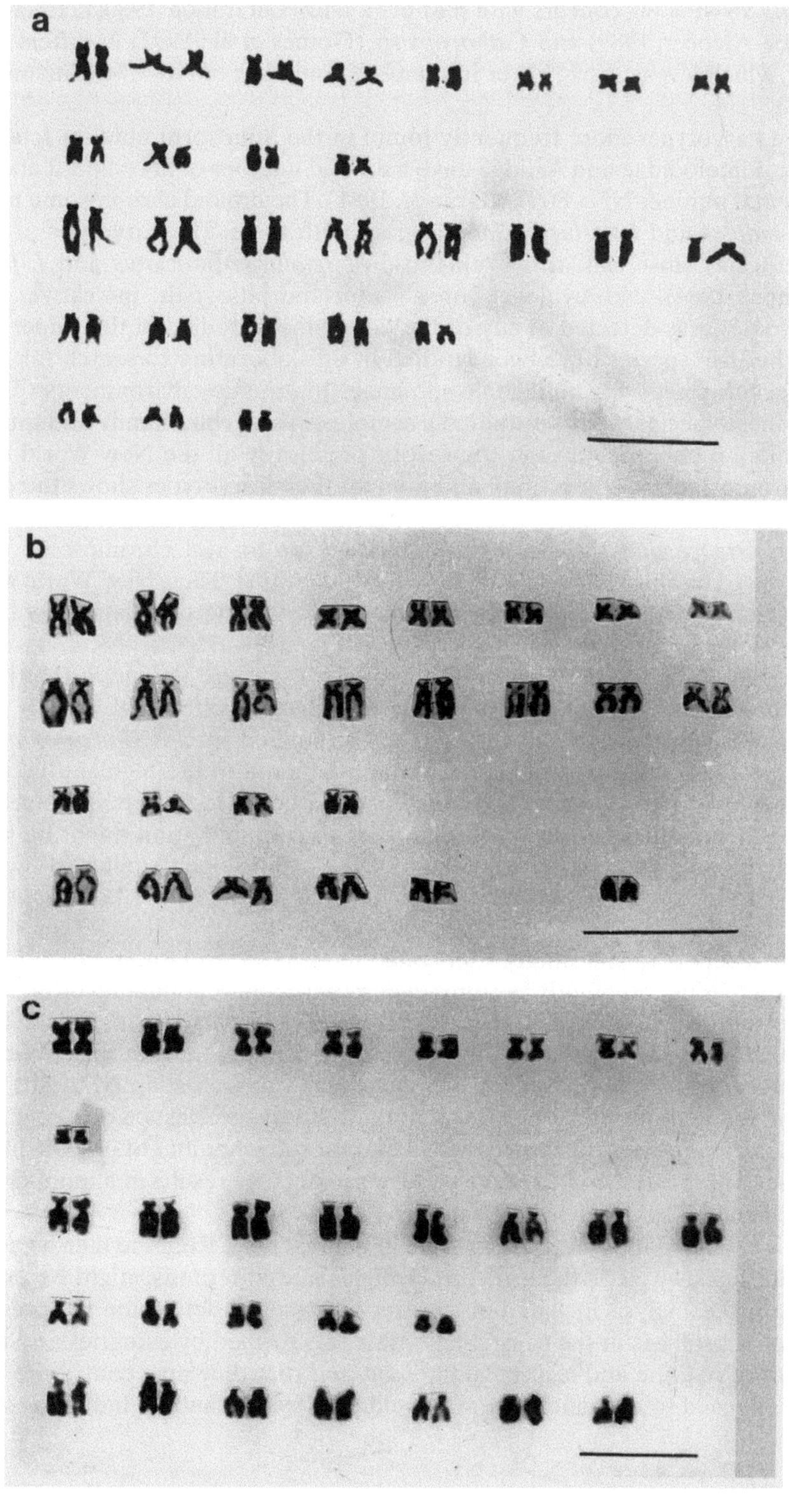
a
b
c

C. aguadulce, NF = 80, concurs with that of *A. felis* (LeGrande, 1980; García-Molina and Uribe-Alcocer, 1989) and *Cathorops* sp. (Gomes et al., 1992). It differs from *B. marinus*, which has a clearly lower fundamental number of NF = 74 (Fitzimmons et al., 1988).

The karyotypes more frequently found in the Siluriform families Ictaluridae, Bagridae, Pimelodidae and Ariidae have a diploid number of 2N = 56 ± 2 and a high fundamental number NF > 80 (Le Grande, 1981). The diploid chromosome numbers of *C. aguadulce* and *Istlarius balsanus* agrees with them. The karyotype of the latter species has close similarity with those of *Ictalurus punctatus* and *I. furcatus*. (LeGrande, 1984). No evidence of a heteromorphic pair indicative of sex-chromosomes, was detected in any of the populations studied in this paper.

Other fish species have been studied in our laboratory to search for suitable karyotypes to screen for mutagenic subtances in aquatic environments (Table 6). Among these species, we have studied a sample of the cichlid family and found that their diploid number is 48 as in the rest of practically all the New World cichlids. Their chromosome sizes are small and none of their karyotypes shows the conspicuous large marker pair of the old world cichlids, the tilapias and their allies exemplified in *Oreochromis mossambicus*, which seems to nest chromosome damage preferentially, as shown by Manna and Biswas (1992a). Thus, New World Cichlids chromosomes would not seem to be suitable to assess genotoxic damage with either the CAs or the SCEs end-point tests.

The atherinids of the Central Plateau of Mexico are unique in that they are among the few fresh-water atherinids and that they have followed closely the turbulent geological history of this region. The two studied species, *Chirostoma jordani* and *C. patzcuaro* form part of the typical lentic fauna of the highland lakes. They are abundant and sturdy fishes able to withstand the periodic harsh changes occurring in their habitat. Their use as biomarkers of contamination might be suitable, especially *C. patzcuaro*, which has a large subtelocentric pair similar to the largest pair of *Oreochromis mossambicus*, which could be likewise more sensitive to genotoxicants.

The sleepers of the family Eleotridae tend to spend large part of their time laying and feeding on muddy bottoms, where particulate pollutants often accumulate on estuaries and coastal lagoons (Lesko et al., 1996; Baumann et al., 1996). The use of benthic species to detect the presence of contaminants in this environment is advantageous for the pertinence of using organisms well adapted to this habitat. Although their chromosome number is not low, studies performed on a related goby species *Bolephthalmus dussumieri*, with a chromosome number of 2N = 46 like some of the eleotrid species studied, have yielded satisfactory results in a number of tests (Krishnaja and Rege, 1982).

It is expected that the chromosome data presented, from the marine and estuarine catfishes as well as those of the eleotrids and atherinids, might be useful for the metaphase analysis included in a battery of tests, to determine the presence of genotoxic substances in the coastal environments formed by estuaries and lagoons, as well as in riverine and lacustrine habitats, and that they may contribute to solve the threat posed to acquatic natural populations by human and industrial wastes.

TABLE 6. Chromosome studies on other fish species from Mexico.

	2N	chromosome formula	NF	author
Ariidae				
Arius Felis	54	16m + 12sm + 20st + 6a	82	García-Molina and Uribe-Alcocer, 1989
Atherinidae:				
Chirostoma patzcuaro	44	12st + 32a	44	in preparation
Chirostoma jordani	48	8m + 12sm + 10st + 18a	68	in preparation
Chirostoma	48	10sm + 22st + 16a	58	in preparation
Cichlidae				
Tilapia:	44	6m + 4sm + 34a	50	Castorena-Sánchez et al., 1983
Oreochromis mossambicus	44	8sm + 2st + 34	52	Uribe-Alcocer and Arreguín-Espinosa, 1989
O. urolepis hornorum	44	8m + 2st + 34a	52	Uribe-Alcocer and Arreguín-Espinosa, 1989
Cichlasoma ellioti	48	6sm + 42a	54	Uribe Alcocer et al., 1992
Cichlasoma trimaculatum	48	8sm + 40a	56	Uribe-Alcocer et al., 1992
Cichlasoma istlanum	48	8m + 8sm + 32a	56	Uribe-Alcocer et al., 1999
Eleotridae				
Dormitator latifrons	46	12m + 22sm + 10st + 2a	80	Uribe-Alcocer et al., 1983 Ramírez-Escamilla and Uribe-Alcocer, 1989
Dormitator maculatus	46	12m + 22sm + 10st + 2a	80	Maldonado-Monroy et al., 1985 Ramírez-Escamilla and Uribe-Alcocer, 1989
Gobiomorus dormitor	48	2m + 4sm + 42a	54	Maldonado-Monroy et al., 1985
Gobiomorus maculatus	46	12m + 10sm + 24a	68	Uribe-Alcocer et al., 1988
Eleotris pisonis	46	2m + 44a	48	Uribe-Alcocer et al., 1994
Eleotris picta	52	52a	52	Uribe-Alcocer and Díaz-Jaimes, 1996
Engraulidae				
Engraulis mordax	48	48a	48	Uribe-Alcocer, et al., 1996
Gobiidae				
Gobionellus microdon	56	4m + 6sm + 46a	66	Uribe-Alcocer and Díaz-Jaimes, 1996

5. CONCLUDING REMARKS

Fish chromosomes constitute efficient short-term biomarkers to screen the mutagenic potential of pollutants of the aquatic environment and to predict the impact of mutagens and/or carcinogens on populations of aquatic organisms. Analysis of CAs and SCE, as well as MNT reviewed elsewhere, appear to be sensitive indicators of chromosome damage.

Even though there might be techniques which under some circumstances may prove to be more sensitive and specific for the detection of genotoxic pollutants, the actual selection of a methodology to screen for noxious xenobionts depends on particular circumstances, such as the local fish fauna, estimated pollution level, available expertise, and on several budgetary aspects, such as actual cost of reagents on site, available equipment, possibility of acquiring new equipment, and its cost.

The establishment of suitable protocols for short-term genotoxicity tests in aquatic environments is a pressing endeavour, with the preferential use of local fish species for the ecotoxicological relevance of regional studies to solve the particular pollution problems.

6. REFERENCES

Alink, G., E.M.H. Frederix-Wolters, M.A. van der Gaag, J.F.J. van der Kerkhopff, and C.L.M. Poels, 1980. Induction of sister-chromatid exchange in fish exposed to Rhine water. Mutat. Res. 78: 269–374.

Almeida Toledo, L.F., F. Foresti, and S.A. Toledo Filho, 1981. Constitutive heterocromatin and nucleolus organizer regions in the knifefish *Apteronotus albifrons* (Pisces, Apteronotidae). Experientia 37: 953–954.

Al-Sabti, K., and C.D. Metcalfe, 1995. Fish micronuclei for assessing genotoxicity in water. Mutat. Res. 343(2–3): 121–135.

Al-Sabti, K., 1985a. Carcinogenic-mutagenic chemicals induced chromosomal aberrations in the kidney cells of three cyprinids. Comp. Biochem. Physiol. C. 82C(2): 489–493.

Al-Sabti, K., 1985b. Frequency of chromosomal aberrations in the rainbow trout, *Salmo gairdneri* Rich., exposed to five pollutants. J. Fish Biol. 26(1): 13–19.

Al-Sabti, K., 1986. Clastogenic effects of five carcinogenic mutagenic chemicals on the cells of the common carp, *Cyprinus carpio* L. Comp. Biochem. Physiol., C. 85C(1): 5–9.

Al-Sabti, K., 1994. Micronuclei induced by selenium, mercury, methylmercury and their mixtures in binucleated blocked fish erythrocyte cells. Mutation Res. 320: 157–163.

Au, D.W.T., R.S.S. Wu, B.S. Zhou, and P.K.S. Lam, 1999. Relationship between ultrastructural changes and EROD activities in liver of fish exposed to Benzo[a]pyrene. Environmental Pollution-London Then Barking 104(2): 235–248.

Baksi, S.M., 1987. Development of a mutagenicity assay on early life stages of *Morone saxatilis* and *Cyprinodon variegatus*. Diss. Abst. Int. Pt. B—sci. and eng. 47(10): 167 pp.

Barker, C.J., and B.C. Rackhan, 1979. The induction of sister chromatid exchanges in culture fish cells (*Ameca splendens*) by carcinogenic mutagens. Mutat. Res. 68: 381–387.

Baumann, P.C., I.R. Smith, and C.D. Metcalfe, 1996. Linkages between chemical contaminants and tumors in benthic Great Lakes fish. J. Great Lakes Res. 22(2): 131–152.

Beamish, R.J., M.J. Merrilees, and E.J. Crossman, 1971. Karyotypes and DNA values for members of suborder Esocoidei (Osteichthyes: Salmoniformes). Chromosoma 34: 436–447.

Belyaeva, V.N., and G.L. Pokrovskaia, 1959. Changes in the radiation sensitivity of loach spawn during the first embryonic mitosis. Dokl. Akad. Nauk SSST 125: 192–195 (Transl. From Russian, p. 632).

Biswas, S., and G.K. Manna, 1989. The mutagenic potentiality of the bacterium *Pseudomonas aeruginosa* tested on Climbing perch *Anabas testudineus*. Perspectives in Cytology and Genetics 6: 573–578.

Biswas, S., and G.K. Manna, 1992. The "Hay-Bacillus", *Baccillus subtilis* as genotoxic agent in treated fresh water Tilapia. Perspectives in Cytology and Genetics. 7: 945–952.

Blaxhall, P.C., 1975. Fish chromosome techniques—a review of selected literature. J. Fish. Biol. 7: 315–320.

Calton, M.S., and T.E. Denton, 1974. Chromosomes of the chocolate gourami: a cytogenetic anomaly. Science 185: 618–619.

Campos, H.H., 1972. Karyology of three galaxiid fishes *Galaxius maculatus, G. platei* and *Brachygalaxias bullocki*. Copeia 1972(2): 368–370.

Carrano, A.V., L.H. Thompson, P.A. Lindl, and J.L. Minklre, 1978. Sister chromatid exchanges as an indicator of mutagenesis. Nature 271: 551–553.

Castorena Sánchez, I., M. Uribe Alcocer, and J. Arreguín Espinosa, 1983. Estudio cromosómico de poblaciones del género *Tilapia* Smith (Pisces: Cichlidae) provenientes de tres regiones de México. Veterinaria México. Univ. Nal. Auton. México. 14(3): 137–145.

Cross, J.N., J.T. Hardy, J.E. Hose, G.P. Hershelman, L.D. Antrim, R.W. Gossett, and E.A. Crecelius, 1987. Contaminant concentrations and toxicity of sea-surface microlayer near Los Angeles, California. Mar. Environ. Res. 23(4): 307–323.

Chiarelli, A.B., and E. Capanna, 1973. *Cytotaxonomy and vertebrate evolution.* Academic Press. London and New York.

Daniels, C.B., and J.C. Means, 1990. Assessment of the genotoxicity of produced water discharges associated with oil and gas production using a fish embryo and larval test. Mar. Environ. Res. 28(1–4): 303–307.

Das, P., and G. John, 1997. In vivo induction of sister chromatid exchanges (SCE) in a tropical fish, *Etroplus suratensis* (Bloch). Acta Biol. Hung. 48(2): 167–172.

Davies, J.M., J.S. Bell, and C. Houghton, 1984. A comparison of the levels of hepatic aryl hydrocarbon hydroxylase in fish caught close to and distant from north sea oil fields. Mar. Environ. Res. 14: 1–4.

Denton, T.E., 1973. *Fish chromosome methodology.* Charles C. Thomas Publisher. Springfield, Illinois, USA.

Di, S., and L. Liu, 1992. The establishment of in vivo SCE detection system of chromosomes in *Monopterus albus*. Acta genet. Sin. 19(3): 212–220.

Eufemia, N.A., T.K. Collier, J.E. Stein, D.E. Watson, and R.T. Di Giulio, 1997. Biochemical responses to sediment-associated contaminants in brown bullhead (*Ameiurus nebulosus*) from the Niagara River ecosystem. Ecotoxicology 6(1): 13–34.

Falçao, J.N., and L.A.C. Bertollo, 1985. Chromosome characterization in Acestrorhyncinae and Cynopotaminae (Pisces, Characidae). J. Fish. Biol. 27: 603–610.

Fitzimmons, J.M., W.H. LeGrande, and J.W. Korth, 1988. Karyology of the marine catfish *Bagre marinus* (Ariidae) with an analysis of chromosome numbers among siluriform fishes. Jap. J. Icthyol. 35: 189–193.

García-Molina, F., and M. Uribe-Alcocer, 1989. Análisis cromosómico del Bagre Marino *Arius felis* (Ariidae: Siluriformes) de la Región de la Laguna de Términos Campeche. An. Inst. Cienc. del Mar y Limnol. Univ. Nal. Auton. México. 16: 69–74.

Gomes, V., V.N. Phan, M.J. de, and A.C.R. Passos, 1992. The karyotype of *Cathorops* sp., a marine catfish from Brazil. Bolm. Inst. Oceanogr., S. Paulo 40: 79–85.

Gómez-Arroyo, S., and R. Villalobos-Pietrini, 1995a. Chromosomal aberrations and sister chromatid exchanges in *Vicia faba* as genetic monitors of environmental pollutants. In: *Biomonitors and Biomarkers as Indicator of Environmental Change Eds.*: Butterworth, F.M., L.A. Corkum, and J. Guzmán-Rincón, Volume 50, Environmental Science Research, Series Editor: H.S. Rosenkranz, Plenum Press, NY.

Gómez-Arroyo, S., M.A. Armienta, J. Cortés-Eslava, and R. Villalobos-Pietrini, 1997. Sister chromatid exchanges in *Vicia faba* induced by arsenic-contaminated drinking water from Zimapan, Hidalgo, Mexico. Mutation Research 394: 1–7.

Gómez-Arroyo, S., M.E.Calderón-Segura, and R. Villalobos-Pietrini, 1995b. Sister chromatid exchange in human lymphocytes induced by propaxur following plant activation by *Vicia faba.* Environ. Mol. Mutagenesis 26: 324–330.

Hasspieler, B.M., J.V. Behar, D.B. Carlson, D.E. Watson, and R.T. Di Giulio, 1994. Susceptibility of channel catfish (*Ictalurus punctatus*) and brown bullhead (*Ameiurus nebulosus*) to oxidative stress: A comparative study. Aquat. Toxicol. 28(1–2): 53–64.

Hooftman, R.N., 1981. The induction of chromosome aberrations *in Nothobranchius rachowi* (Pisces: Cyprinodontidae) after treatment with ethyl methanesulfonate or benzo(a)pyrene. Mutation Res. 91: 347–352.

Hooftman, R.N., and G.J. Vink, 1980. The determination of toxic effects of pollutants with the marine polychaete worm *Ophryotrocha diadema.* Ecotoxicol. Environ. Saf. 4(3): 252–262.

Hooftman, R.N., and G.J. Vink, 1981. Cytogenetic effects on the eastern mudminnow, *Umbra pygmaea*, exposed to ethyl methanesulfonate, benzo(a)pyrene, and river water. Ecotoxicol. Environ. Saf. 5(3): 261–269.

Howell, W.M., 1972. Somatic chromosomes of the black ghost knifefish, *Apteronotus albifrons* (Pisces: Apteronotidae). Copeia 1972(1): 191–193.

Iriki, S., 1932. On the chromosomes of *Aplocheilus latipes*. Sci. Rep. Tokyo Bun. Dai. Sec. B. 1: 127–131.

Jordan, S., and J. Snyder, 1900. Notes on a collection of fishes from the rivers of Mexico, with description of twenty new species *U.S. Comission of fisheries*. Washington pp. 115–147.

Katayama, M., 1937. On the spermatogenesis of the teleost, *Oryzyas latipes*. Bull. Jap. Soc. Sci. Fish. Tokyo 5: 277–278.

Kato, E., and M. Romo, 1981. Algunos aspectos biológicos del bagre dulceacuícola *Istlarius balsanus* en el río Amacuzac, Morelos. Tesis Profesional. Escuela Nacional de Estudios Profesionales Iztacala, Universidad Nacional Autónoma de México.

Kato, H., 1974. Spontaneous sister chromatid exchanges detected by BudR-labelling method. Nature 251: 70–72.

Kato, H., and H. Shimada, 1975. Sister chromatid exchanges induced by mitomycin C: A new method of detecting DNA damage at chromosomal level. Mutation Res. 28: 459–464.

Kitada, J., and M. Tagawa, 1972. On the chromosomes of the ricefield eel (*Fluta alba = Monopterus albus*). La Kromosomo II (88–89): 2804–2807.

Kligerman, A.D., 1979. Induction of sister chromatid exchanges in the central mudminnow following in vivo exposure to mutagenic agents. Mutat. Res. 64: 205–217.

Kligerman, A.D., 1982. The use of cytogenetics to study genotoxic agents in fishes. In: *Cytogenetic assays of environmental mutagens* Ed: Hsu, R.C. Allanheld, Osmun. Totowa, N.J. pp. 161–181.

Kligerman, A.D., and S.E. Bloom, 1976. Sister chromatid differentiation and exchanges in adult mid-minnows (*Umbra limi*) after in vivo exposure to 5-bromodeoxyuridine. Chromosoma (Berl.) 56: 101–109.

Kligerman, A.D., and S.E. Bloom, 1977. Rapid chromosome preparations from solid tissues of fishes. J. Fish. Res. Board Can. 34: 266–269.

Kligerman, A.D., S.E. Bloom, and W.M. Howell, 1973. *Umbra limi*: a model for the study of chromosome aberration in fishes. Mutat. Res. 31: 225–233.

Klumpp, D.W., and H. Von Westernhagen, 1995. Biological effects of pollutants in Australian tropical coastal waters: Embryonic malformations and chromosomal aberrations in developing fish eggs. Mar. Pollut. Bull. 30(2): 158–165.

Kocan, R.M, K.M. Sabo, and M.L. Landolt, 1985. Cytotoxicity/genotoxicity: The application of cell culture techniques to the measurement of marine sediment pollution. Aquat. Toxicol. 6(3): 165–177.

Kocan, R.M., and K.L. Stark, 1991. Event-frequency relationship between two genotoxic end-points and cell death in cultured fish cells. *Comptes rendus du dix-septième colloque annuel sur la toxicologie aquatique*: 5–7 Novembre 1990, Vancouver, (C.B.). Vols. 1, 2. Eds.: Chapman, P., F. Bishay, E. Power, K. Hall, L. Harding, D. McLeay, M. Nassichuk, and W. Knapp. Can. Tech. Rep. Fish. Aquat. Sci. 1774(1,2): 961–971.

Krishnaja, A.P., and M.S. Rege, 1982. Induction of chromosomal aberrations in fish *Boleophthalmus dussumieri* after exposured in vitro to mitomycin C and heavy metals mercury, selenium and chromium. Mutat. Res. 102(1): 71–82.

Landolt, M.L., and R.M. Kocan, 1984. Lethal and sublethal effects of marine sediment extracts on fish cells and chromosomes. *Diseases of marine organisms*. Helgol. Meeresunters. 37(1–4): 479–491.

Latt, S.A., 1974. Sister chromatid exchanges, indices of human chromosome damage and repair: detection by fluorescens and induction by mitomycin C. Proc. Nat. Acad. Sci. (Wash) 71: 3162–3166.

LeGrande, W.H., and J.M. Fitzimons, 1976. Karyology of the mullets *Mugil curema* and *M. cephalus* (Perciformes: Mugilidae). from Louisiana. Copeia 1976(2): 388–391.

LeGrande, W.H., 1975. Karyology of six species of Louisiana flatfishs (Pleuronectiformes: Osteichthyes). Copeia 1975(3): 516–522.

LeGrande, W.H., 1980. The chromosome complement of *Arius felis* (Siluriformes, Ariidae). Japan J. Icthyol. 27: 82–84.

LeGrande, W.H., 1981. Chromosomal evolution in North American catfishes (Siluriformes: Ictaluridae), with particular emphasis on the madtoms, Noturus. Copeia 1981(1): 33–52.

LeGrande, W.H., 1984. Karyology of three species of catfishes (Ictaluridae: *Ictalurus*) and four hybrid combiantions. Copeia 1984(4): 873–878.

Lesko, L.T., S.B. Smith, and M.A. Blouin, 1996. The effect of contaminated sediments on fecundity of the brown bullhead in three Lake Erie tributaries. J. Great Lakes Res. 22(4): 830–837.

Levan, A.K., A. Fredga, and R. Sandberg, 1964. Nomenclature for centromeric position of chromosomes. Hereditas 52: 201–220.

Liepmann, M., and C. Hubbs, 1969. A karyological analysis of two cyprinid fishes, *Notemigonus chrysoleucas* and *Notropis lutrensis*. Tex. Rep. Biol. And Med. 27(2): 427–435.

Lobillo, J., J.V. Delgado, and A. Rodero, 1991. Sister chromatid exchange test detection of toxin-induced damage in cultured fish. Genet. Sel. Evol. 23(sup. 1): 160S–162S.

Maddock, M.B., H. Northrup, and T.J. Ellingham, 1986. Induction of sister-chromatid exchanges and chromosomal aberrations in hematopoietic tissue of a marine fish following in vivo exposure to genetoxic carcinogens. Mutat. Res. 172(2): 165–175.

Makino, S., 1934. The chromosomes of the sticklebacks *Pungitis tymensis* and *P. Pungitius*. Cytologia 5: 155–168.

Maldonado-Monroy, M.C., M. Uribe-Alcocer, J. Arreguín-Espinosa, and A. Castro-Pérez, 1985. Karyotypical studies on *Dormitator maculatus Bloch* and *Gobiomorus dormitor* Lacépède (Gobiidae: Perciformes). Cytologia (Tokyo) 50: 15–21.

Manna, G.K., and A. Sadhukhan, 1991. Genotoxic potential of the spores of three species of fungi experimented on cichlid fish, *Oreochromis mossambicus*. La Kromosomo II, (63–64): 2129–2134.

Manna, G.K., and A. Sadhukhan, 1992a. Cytogenetical assays of fish, *Oreochromis mossambicus* treated with an organochlorine insecticide, Aldrex 30. La Kromosomo II (65): 2173–2182.

Manna, G.K., and A. Sadhukhan, 1992b. Mutagenic potential of anisole (methyl phenyl ether) on *Oreochromis mossambicus*. J. inland fish. Soc. India 24(1): 40–49.

Manna, G.K., and P.K. Mukherjee, 1989. A study of the genotoxic potentiality of the inorganic weedicide, sodium arsenite in the experimentally treated tilapia fish. J. Freshwater Biol. 1(2): 147–159.

Manna, G.K., and P.K. Mukherjee, 1992. Genotoxicity assays of the insecticide, Endrin in experimentally treated cichlid fish, *Oreochromis mossambicus*. J. Inland fish. Soc. India 24(2): 1–9.

Manna, G.K., and S. Biswas, 1992a. Cytogenetic assays of the mutagenic potential of the bacterium, *Pseudomonas aeruginosa* in five species of experimentally treated fish. Cytologia 57(4): 427–436.

Manna, G.K., and S. Biswas, 1992b. The bacterium *Staphylococcus aureus* as genotoxic agent in experimentally treated five species of fresh water teleosts. La Kromosomo II (67–68): 2286–2297.

Mao, S., and C. Wang, 1990. The effect of some pollutants on SCE of grass carp (*Ctenopharyngodon idellus*) cells. Oceanol. Limnol. Sin. 21(3): 205–211.

McPhail, J.D., and R.L. Jones, 1966. A simple technique for obtaining chromosomes from teleost fishes. J. Fish. Res. Bd. Canada 23(5): 767–768.

Means, J.C., C.B. Daniels, and S.M. Baksi, 1987. Development of in vivo genotoxicity tests in estuarine fish and their application to aquatic toxicology. Mar. Environ. Res. 24(1–4): 327–331.

Merrilees, M.J., 1975. Karyotype of *Galaxias maculatus* from New Zealand. Copeia 1975(1): 176–178.

Migalovskaya, V.N., 1973. Effect of X-irradiation on the gametes and embryonal cells of the Atlantic salmon. In: Effect of ionizing radiation on the organism. Ed.: Sorokin, B.P. Washington D.C. USAEC: 89–99.

Miller, R.R., and J.M. Fitzimons, 1971. *Ameca splendens*, a new genus and species of Goodeid fish from western Mexico, with remarks on the classification of the Goodeidae. Copeia 1971(1): 1–13.

Mohanty, L., and R. Prasad, 1982. Sister chromatid exchanges in a live fish *Channa punctatus*. Nucleus (Calcutta) 25(3): 161–164.

Moon, T.W., D.M.E. Otto, D.M. Arquette, and J.K. Buttner, 1996. Detection of contaminant exposure in fish by activation of detoxication enzymes. Great lakes res. Rev. 2(2): 36–40.

Morales-Ramírez, P., M.T. Mendiola-Cruz, T. Vallarino-Kelly, and R. Rodríguez-Reyes, 1994. Comparison of sister-chromatid induction in murine germinal and somatic cells by gamma radiation exposure in vivo. Environ. Mol. Mutagenesis 24: 89–95.

Narbonne, J.F., D. Ribera, X. Michel, C. Raoux, P. Garrigues, J.L. Monod, P. Lemaire, F. Galgani, M. Roméo, J.P. Salaün, and M. Lafaurie, 199. Indicateurs biochimiques de contamination de l'environnement marin: étude comparative en mer Méditerranée. Oceanis 17(3): 57–27.

Nelson, S.J., 1994. *Fishes of the World*. 3d. John Wiley and Sons, Inc. New York.

Ojima, Y., K. Ueno, and M. Hayashi, 1973. A review of the chromosome numbers in fishes. La Kromosomo II (1): 19–47.

Oliveira, C., L.F. Almeida-Toledo, F. Foresti, H.A. Britski, and S.A. Toledo Filho, 1988. Chromosome formulae of Neotropical freshwater fishes. Rev. Brasil. Genet. 11(3): 577–624.

Otto, D.M.E., and T.W. Moon, 1996. Phase I and II enzymes and antioxidant responses in different tissues of brown bullheads from relatively polluted and non-polluted systems. Arch. Environ. Contam. Toxicol. 31(1): 141–147.

Ouseph, A., D. Sudarsanam, P. Gandheeswari, and T. Ambrose, 1994. Influence of physico-chemical parameters of River Cooum on mitotic index in tissues of *Oreochromis mossambicus* (Peters). J. Ecotoxicol. Environ. Monit. 4(2): 105–108.

Pangrekar, J., and H.C. Sikka, 1992. Xenobiotic metabolizing enzyme activity in the liver and kidney of the brown bullhead (*Ictalurus nebulosus*). Mar. Environ. Res. 34(1–4): 287–291.

Pankaj, G., A.H. Dholakia, and M. Gadhia, 1990. Cadmium nitrate induced chromosomal aberrations in a common carp *Cyprinus carpio*. [A kadmium-nitrat okozta kromoszomaelvaltozasok a pontynal.-Izmeneniya kromosom kharpa, vyzbannye nitratom kadmiya]. Aquacult. Hung. 6: 19–23.

Panlaque, C.A., 1982. Observations on the radiosensitivity of guppy (*Lebistes reticulatus* Peters). Thesis (M.Sc.). Gregorio Araneta Univ. Philipines. 62 pp.

Park, E.H., J.S. Lee, A.K. Yi, and H. Etoh, 1989. Fish cell line (ULF-23HU) derived from the fin of the central mudminnow (*Umbra limi*): Suitable characteristics for clastogenicity assay. In Vitro: Cell. Dev. Biol. 25(11): 987–994.

Pechkurenkov, V.L., and B.P. Kostrov, 1982. On the combined effect of mercury chloride and phosphorus 32 on the developing *Misgurnus* eggs. Radiobiologiya. 22(1): 70–7.

Perry, D.M., J.S. Weis, and P. Weis, 1988. Cytogenetic effects of methylmercury in embryos of the killifish, *Fundulus heteroclitus*. Arch. Environ. Contam. Toxicol. 17(5): 569–574.

Perry, P., and H.J. Evans, 1975. Cytological detection of mutagen-carcinogen exposure by sister chromatid exchange. Nature (London) 258: 21–125.

Perry, P., and S. Wolff, 1974. New Giemsa method for the differential staining of sister chromatids. Nature (lond.) 251: 156–158.

Poels, C.L.M., M.A. van der Gaag, and A. Noordsij, 1982. Methodology of research concerning micropollutants: Biological tests. *Micropollutants in the environment.* Wat. Sci. Tech. 14(12): 143–152.

Pokrovskaya, G.L., and V.L. Pechkurenkov, 1984. Effect of pH of the strontium-90-yttrium-90 solutions on egg incubation in loach *Misgurnus fossilis* (L.) (Cobitidae). [Vliyanie pH rastvorov strontsiya-90-ittriya-90 na inkubatsiyu ikry v'yuna. *Misgurnus fossilis* (L.) (Cobitidae)]. Vopr. Ikhtiol. 24(1): 146–150.

Post, A., 1965. Vergleichende Untersuchungfen der Chromosomenzahlen bei Süss-wasser-Teleosteern. Z. Zool. Syst. Evol. 3: 47–9.

Prein, A.E., G.M. Thie, G.M. Alink, J.H. Koeman, and C.L.M. Poels, 1978. Cytogenetic changes in fish exposed to water of the river Rhine. Sci. Total Environ. 9: 287–291.

Protic-Sabljic, M., and B. Kurelec, 1983. High mutagenic potency of several polycyclic aromatic hydrocarbons induced by liver postmitochondrial fractions from control and xenobiotic treated immature carp. Mutation Res. 118: 177–189.

Protic-Sabljic, M., 1984. Metabolism of carcinogens in the liver of common carp: Induction of drug-metabolizing enzymes and formation of mutagenic metabolites. Mar. Environ. Res. 14(1–4): 437–438.

Ramírez-Escamilla, A., and M. Uribe-Alcocer, 1989. Comparación citogenética entre las especies del Género *Dormitator* (Pisces: Gobiidae). An. Inst. Cienc. del Mar y Limnol. Univ. Nal. Auton. México. 16: 75–80.

Rodríguez-Ariza, A., F.M. Díaz-Méndez, J.I. Navas, C. Pueyo, and J. López-Bare, 1995. Metabolic activation of carcinogenic aromatic amines by fish exposed to environmental pollutants. Environ. Mol. Mutagenesis 25: 50–57.

Rao, S.S., T. Neheli, J.H. Carey, and V.W. Cairns, 1997. Fish hepatic micronuclei as an indication of exposure to genotoxic environmental contaminants. Environ. Toxicol. Water Qual. 12(3): 217–222.

Rishi, K.K., and S. Grewal, 1993. Cytogenetic screening of effect of a Mosquito larvicide on kidney cell chromosomes of *Channa punctatus*. Abstr. In Eight All India Congress of Cytology and Genetics, pp. 87.

Rishi, K.K., and S. Grewal, 1995. Chromosome aberration test for the insecticide, dichlorvos, on fish chromosomes. Mutat. Res.-genet. Toxicol. 344(1–2): 1–4.

Rosas, M., 1976. Sobre la existencia de un nemátodo parásito de *Tilapia nilotica* (*Goezia* sp. Goeziidae), de la presa Adolfo López Mateos (Infiernillo, Mich.). Memorias del Simposio sobre Pesquerías en Aguas Continentales. Tuxtla Gutiérrez, Chiapas. Instituto Nacional de Pesca. Secretaría de Industria y Comercio, México. 239–270.

Scheel, J.J., 1972. Rivuline karyotypes and their evolution (Rivulinae, Cyprinodontidae, Pices) Z. zool. syst. Evol. 180–209.

Scheel, J.J., 1973. Fish chromosomes and their evolution. Interval Report of Danmarks Akvarium, Charlottenlund, Denmark.

Sofradzija, A., T. Vukovic, and R. Hadziselimovic, 1980. Effects of Some Pesticides and Detergents on the Fish Chromosome Sets—Results of the Preliminary Investigations. [Efekti Nekin Pesticida i Deterdzenata u Hromosomskim Garniturama riba -Rezultati Preliminarnih Istrazivanja]. Acta biol. Iugosl. (e ichthyol.) 12(1): 71–76.

Sola, L., S. Cataudella, and E. Capanna, 1981. New developments in vertebrate cytotaxonomy. Genetica 54: 285–328.

Soldberg, A.N., 1938. The susceptibility of *Fundulus heteroclitus* embryos to X-radiation. Jour. Exp. Zool. 78: 441–465.

Stien, X., P. Percic, M. Gnassia-Barelli, M. Roméo, and M. Lafaurie, 1998. Evaluation of biomarkers in caged fishes and mussels to assess the quality of waters in a bay of the NW Mediterranean Sea. Environ. Pollut 99(3): 339–345.

Stromberg, P.T., M.L. Landolt, and R.M. Kocan, 1981. Alterations in the frequency of sister chromatid exchanges in flatfish from Puget Sound, Washington, following experimental and natural exposure to mutagenic chemicals. NOAA Tech. Memo., NOAA/OMPA, Boulder, Co (Usa), 50 pp. NOAA/OMPA, Boulder, Co (USA).

Taylor, W.R., and N.A. Menezes, 1978. Ariid. *In: FAO species identification sheets for fishery purposes. Western Central Atlantic (Fishing area 31)* Ed.: Fischer, W. Vol. II.

Tsoi, R.M., A.I. Men'shova, and Y.F. Golodov, 1975. Specificity of the influence of chemical mutagens on spermatozoids of *Cyprinus carpio* L. Soviet Genetics 10: 190–193.

Tsoi, R.M., 1970. Effect of nitrosomethyl urea and dimethyl sulfate on sperm of rainbow trout (*Salmo irideus* Gibb) and peled (*Coregonus peled* Gmel.) Dokl. Adad. Nauk SSST 189: 849–851.

Ueda, T., M. Hayashi, Y. Ohtsuka, T. Nakamura, J. Kobayashi, and T. Sofuni, 1992. A preliminary study of the micronucleus test by acridine orange fluorescent staining compared with chromosomal aberration test using fish erythropoietic and embryonic cells. *Hazard assessment and control of environmental contaminants in water*. Wat. Sci. Technol. 25(11): 235–240.

Uribe-Alcocer, M., and J. Arreguín-Espinosa, 1989. Los cromosomas de los peces *Oreochromis urolepis hornorum* y *Oreochromis mossambicus* (Pisces: Cichlidae). An. Inst. Cienc. del Mar y Limnol. Univ. Nal. Autón. México. 16: 199–206.

Uribe-Alcocer, M., and P. Díaz-Jaimes, 1996. Chromosome complements of *Gobionellus microdon* (Gilbert, 1891) and *Eleotris picta* Kner Steindachner (Gobioidea, Perciformes) collected in Mexico. J. Fish. Biol. 48: 796–798.

Uribe-Alcocer, M., B.L. Náder-García, and N. Valdés-Morales, 1992. The Chromosomes of two cichlids from Mexico *Cichlasoma ellioti* and *C. trimaculatum*. Jap. J. Ichthyol. 39(2): 174–177.

Uribe-Alcocer, M., C. Téllez-Vargas, and P. Díaz Jaimes, 1999. Chromosomes of *Cichlasoma istlanum* (Perciformes: Cichlidae) and karyotype comparison of two presumed subspecies. In press. Rev. Biol. Tropical.

Uribe-Alcocer, M., J. Arreguín-Espinosa, and S. Rojas Romero, 1988. The karyotype of a Gobiid, *Gobiomorus maculatus*, from Mexico. Jap. J. Ichthyol. 34(4): 509–511.

Uribe-Alcocer, M., J. Arreguín-Espinosa, A. Torres-Padilla, and A. Castro-Pérez, 1983. Los cromosomas de *Dormitator latifrons* (Pisces: Gobiidae). An. Inst. Cienc. del Mar y Limnol. Univ. Nal. Auton. México. 10(1): 23–30.

Uribe-Alcocer, M., N. Valdés-Morales, P. Díaz Jaimes, Y. Hornelas Orozco, and V. Arenas, 1996. Comparación de los cariotipos de las poblaciones central y sureña de la anchoveta *Engraulis mordax*, Girard 1854 (Engraulidae, Pisces). Ciencias Marinas 22(3): 361–376.

Uribe-Alcocer, M., R. Montes-Pérez, and P. Díaz-Jaimes, 1994. The chromosome complement of *Eleotris pisonis* (Gobiidae; Perciformes) from Mexico. A new case of heteromorphic sex chromosomes in fishes. Cytobios 77: 183–187.

Uyeno, T., and R.R. Miller, 1972. Second discovery of multiple sex chromosomes among fishes. Experientia 28: 223–225.

Van der Gaag, M.A., G.M. Alink, P. Hack, J.C.M. van der Hoeven, and J.F.J. van der Kerkhoff, 1983. Genotoxicological study of Rhine water with the killifish *Notobrachius rachowi*. Mutation Res. 113: 311.

Van der Hoeven, J.C.M., I.M. Bruggeman, G.M. Alink, and J.H. Koeman, 1982. The killifish *Nothobranchius rachowi*, a new animal in genetic toxicology. Mutation Res. 97: 35–42.

Van der Kerkhoff, J.F.J., and M.A. van der Gaag, 1985. Some factors affecting optimal differential staining of sister-chromatids in vivo in the fish *Nothobranchius rachowi*. Mutat. Res. 143(1–2): 39–43.

Vasil'ev, V.B. 1978. Karyotypes of 5 species of fishes (Pisces) from the Black Sea. Tsitologiya 20(9): 1092–1094.

Vifgusson, N.V., E.R. Vyse, C.A. Pernsteiner, and R.J. Dawson, 1983. In vivo induction of sister-chromatid exchange in *Umbra limi* by the insecticides endrin, chlordane, diazinon and guthion. Mutation Res. 118(1–2): 61–68.

Wei, Y., and R. Lu, 1987. Preliminary studies on frequencies of sister chromatid differential (SCD) and exchanges (SCE) in renal cells of grasscarp (*Ctenopharyngodon idellus*) *in vivo*. Acta hydrobiol. Sin. 11(1): 29–33.

Weisburger, J.H., and G.M. Williams, 1991. Critical effective methods to detect genotoxic carcinogens and neoplasm-promoting agents. Environ. Health perspect. 90: 121–126.

Williams, R.C., and C.D. Metcalfe, 1992. Development of an in vivo hepatic micronucleus assay with rainbow trout. Aquatic toxicology 23: 193–202.

Wolff, S., B. Rodin, and J.E. Cleaver, 1977. Sister chromatid exchanges induced by mutagenic carcinogens in normal and xeroderma pigmentosum cells. Nature (London) 265: 347–349.

Wrisberg, M.N., and M.A. van der Gaag, 1992. In vivo detection of genotoxicity in waste water from a wheat and rye straw paper pulp factory. Sci. Total environ. 121: 95–108.

Zajicek, K.D., and R.B. Phillips, 1984, Mitotic inhibition and anaphase aberrations in rainbow trout embryos treated with MNNG and gamma radiation. 1984 Joint Meeting of the Genetics Society of America and the Genetics Society of Canada, Vancouver, B.C. (Canada), 12–15 Aug. 1984. Genetics, 107(3, pt.2): s117.

Zakhidov, S.T., V. Chebotareva Iu, K.A. Savvaitova, and V.A. Maksimov, 1996. [A cytogenetic study of hematopoietic cells in fish from the reservoirs of the Noril'sk-Pyasina water system (Taimyr)]. Izv. Akad. Nauk. Ser. Biol. ISS 1, P10–5.

21

THE *TETRAMITUS* ASSAY

ROBERT L. JAFFE

Environmental Toxicology Laboratory LLC 45-10 Court St., Long Island City NY 11101

1. INTRODUCTION

A simple test for measuring cytotoxic agents using the flagellate phenotype of *Tetramitus rostratus* has been developed. The test measures dose-dependent inhibition of cell division by DNA-damaging agents and other toxicants. The *Tetramitus* Assay is rapid, cost effective, exhibits a high level of statistical reliability and requires no animals or animal byproducts. An additional attribute of the assay which is useful for exposure monitoring and risk assessment is the ability to measure whole particle toxicity without the need for prior extraction and solvent-substitution procedures. The assay is five to ten times more sensitive than standard Whole Effluent Toxicity tests (WET Tests such as *Ceriodaphnia* and fat head minnow) and can be performed on non-sterile environmental samples. The test method is robust and is suitable for international acceptance. A detailed protocol with standard operating procedures is described; and, evaluation of growing cell populations in seed flasks prior to the actual performance of the test is predictive of test performance, thus avoiding the loss of valuable samples.

The *Tetramitus* Assay affords the regulatory community the opportunity to develop extensive data bases on complex mixtures found in drinking water, industrial effluents, air, and soil that previously have been difficult to develop because of time and cost constraints. The *Tetramitus* Assay will allow for frequent testing thus permitting development of more accurate hazard assessments and comprehensive exposure models.

Tetramitus rostratus is a unicellular organism which can exist as three distinct phenotypes: flagellate, ameba, or cyst. *Tetramitus* is estimated to have originated 1.0 to 1.2 billion years ago (Sagan, 1967). Single flagellates can be isolated and grown in liquid culture with bacteria as the only food source. Flagellates are quite stable (no amebae have been observed in more than 10,000 subcultures of flagellate populations reaching densities of up to 3×10^7 cells/mL). Because *Tetramitus* flagellates are particle feeders, the organism is useful for assessing whole particle cytotoxicity. The flagellate has a rigid cytoskeleton, four flagellae, and a gullet which starts from the ventral depression and extends into the body of the cell (Figure 1).

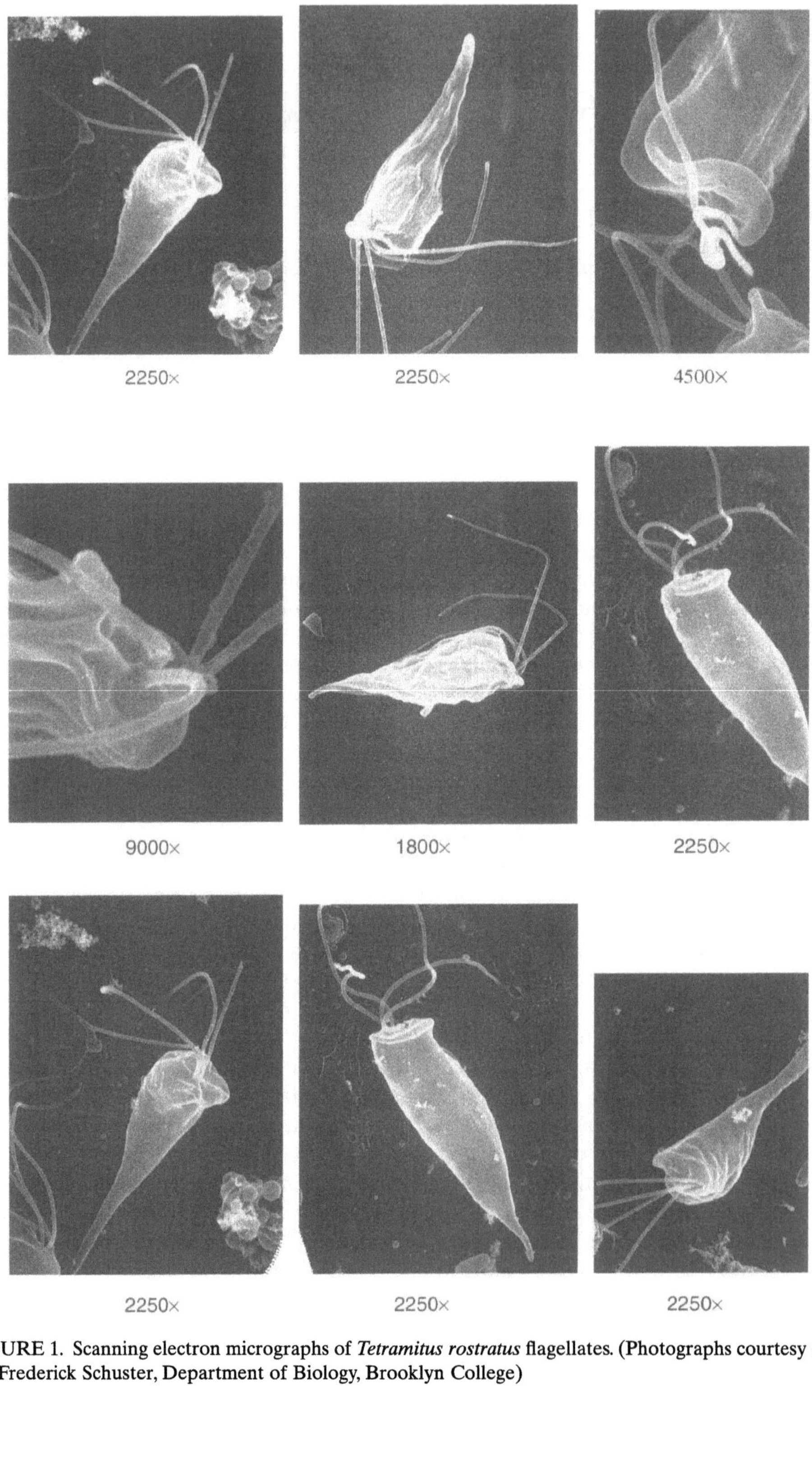

FIGURE 1. Scanning electron micrographs of *Tetramitus rostratus* flagellates. (Photographs courtesy of Dr. Frederick Schuster, Department of Biology, Brooklyn College)

The assay measures dose-dependent inhibition of cell division by agents which cause DNA-damage. Recent studies have demonstrated the existence in *Tetramitus* of a 21.4 kb extrachromosomal DNA plasmid (rDNA) which codes for the ribosomal RNA (Clark and Cross, 1988). Minor sequence differences have been shown to cause drastic changes in the growth rate of *E. coli* cells harboring mutant rDNA plasmids (Steen et al., 1986). Walsh (1996) has estimated the existence of 4,000 copies of rDNA amounting to 17% of the total cell DNA in single *Tetramitus* amebae. The action of DNA-damaging agents on flagellates causes both decreased rates of cell division and a decrease in cell size.

2. METHODS

The original description of methods for culturing flagellates and suggested test protocols has been published (Jaffe, 1995). The following text describes the latest methods' revisions.

2.1. Original Stock Cultures

Tetramitus flagellates are maintained in association with *Klebsiella pneumoniae* in YPP medium (0.05% Difco yeast extract and 0.05% Difco proteose peptone in distilled water) and grown in Corning 16 × 125 mm sterile polystyrene tissue culture tubes (25200). Flagellates inoculated from YPP medium into bacteria-buffer cultures usually take 5–6 subcultures before optimal growth conditions were observed (mean division time of 4 hrs at 30 °C).

2.2. Standard Bacteria-Buffer Maintenance Cultures

Tetramitus flagellates are grown in MS-1 buffer containing a dense suspension of *Klebsiella pneumoniae* (Kp). MS-1 contains 0.1 mM KCl, 0.3 mM $CaCl_2$, 0.3 mM NaH_2PO4, 0.0008% phenol red (pH indicator), and 1.4 mM $NaHCO_3$. The bicarbonate is added separately after autoclaving (Fulton, 1970). The original formulation of MS-1 called for inclusion of EDTA, which is now omitted in order not to interfere with toxicity testing of heavy metals.

Cultures of *Klebsiella* are grown overnight in a shaking water-bath at 35 °C in 300 ml of 2.5% Oxoid #2 nutrient Broth (Unipath-CM67) per 2,800 mL Fernbach nephlo flask, or 90 mL per 500 mL flask (Bellco Glass 2581-14135). Teflon lined screw caps are used instead of the non-toxic rubber caps. Kp growth is monitored by determining the turbidity in a Klett-Somerson nephelometer using a red filter. The Kp are harvested by centrifugation at 2,500 RCF for 10 mins in Corning 50 mL sterile polypropylene tubes (25330-50). The Oxoid #2 supernatant is decanted, and the pellets are resuspened in 40 mL of MS-1 by vigorous mixing with a Vortex Genie mixer. The Kp suspensions are recentrifuged as above, the MS-1 supernatants are decanted and the pellets are resuspended in fresh MS-1 (30 mL for each original

50 mL of Kp culture). The final suspension is referred to as Kp "soup" and 90 mL volumes (Kp suspension from 3 tubes) are incubated in 300 mL baffled DeLong® flasks (Bellco 2510-00500) in a shaking water bath at 26 °C @ 180 rpm. We have found the New Brunswick Innova 3,000 model to be very reliable; all our units are in use continuously without interruption. Our first unit has been in operation 24 hours a day for eight years without a single malfunction.

Standard flagellate cultures were incubated in 125 ml baffled DeLong® flasks (Bellco 2510-00125) in 10 mL of Kp "soup" at 30 °C @180 rpm. Specially designed 125 mL flasks with 38 mm necks (Bellco Glass, special order as described in ETL/Bellco draft-specifications) are used for seed cultures in order to facilitate rapid pipeting of 50 μL aliquots into individual tubes (see below). Seed cultures which are in log phase for at least three division cycles are optimal for toxicity testing. Cultures which are about to enter into early stationary phase should not be used for testing; the dose-growth curves can exhibit regression data with r^2 values below 0.97. Use of the statistics program for the Multisizer IIe allows for determination of the mean diameter of the cell population. Flagellates in log phase have an average diameter of 9.2–9.4 μ. As the population enters stationary phase, the mean cell diameter decreases and any subculture derived from stationary phase cultures will exhibit an increase in cell diameter corresponding to the slope at that time point, reflecting as well the growth status of the culture as it re-enters log phase (Figure 2). The mean cell diameter of any culture can be used to ascertain if that culture is in log phase, thus, serving as an objective quality control indicator for a given seed flask.

2.3. Counting

Cell concentrations were determined with the use of either a model ZM Coulter Counter or a Multisizer IIe (Beckman-Coulter Electronics, Miami, Florida) using a 100 μ aperture tube. 0.2 mL aliquots were transferred to Folin-Wu tubes containing 30 mL of electrolyte (0.4% NaCl [w/v] in distilled water). The volume was adjusted to 35.0 mL by adding saline from a plastic wash bottle to the etched 35 mL volume line of the Folin-Wu tubes. The contents of each tube were agitated using a Vortex-Genie mixer, aliquots were transferred to Coulter disposable counting cuvettes and 2 counts were determined at threshold settings of 10–99.9; current, 400 mA; attenuation, 4; preset gain, 1; and manometer selection, 500 μL for the ZM. The narrow channel option with lower channel = 6.03 μ and upper channel = 15.03 μ settings are used for the Multisizer. We have found this method of counting to be extremely reliable; the correlation coefficients of the growth curves are usually 0.998 or higher. One correlation coefficient of a four-point growth curve was 0.999999. *The precision of the Coulter Counters, both the Multisizer IIe and the ZM, over the course of 10 years of research experience has been a constant ingredient in the production of a database containing control and dose growth regression curves with high correlation coefficients.*

Standard hemacytometer counting methods can be employed by pipetting a 100 μL of sample into a 12 × 75 mm polystyrene test tube (Fisher 14-961-10) containing 10 μL of Lugol's iodine. After vortex-mixing, aliquots are transferred to a

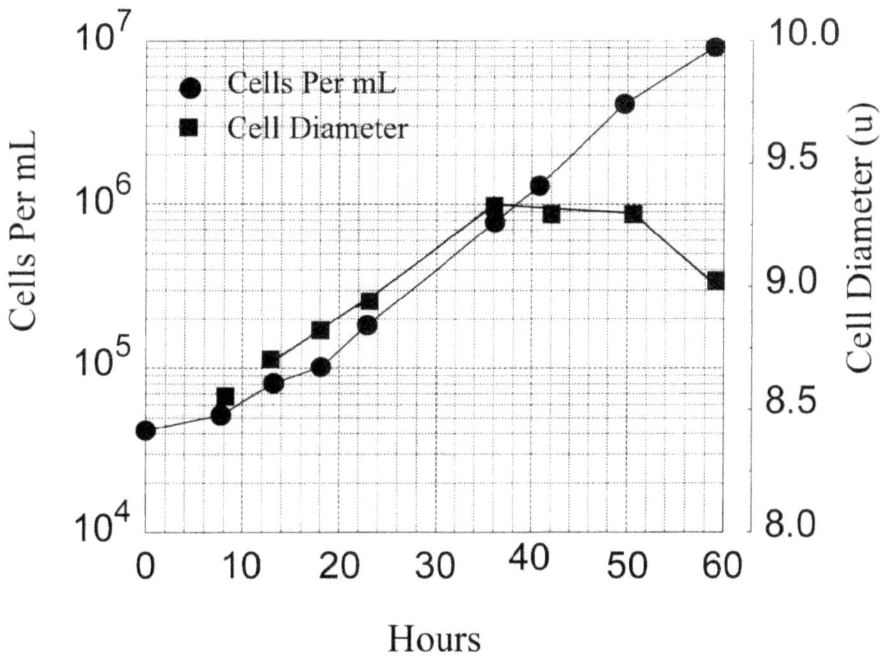

FIGURE 2. Growth of *Tetramitus* flagellates. The mean cell diameters are potted at the times indicated.

counting chamber. Four counts of 100+ are obtained, and the cell concentration is calculated by multiplying the average count by 1.1 (to compensate for the 10 μL of Lugol's iodine). The larger area of the hemacytometer chamber can be employed using the dilution multiplier of 1.1×10^3. The lowest flagellate concentration which can be measured by this method would be 1×10^5 cells/mL. For calculation of cell concentrations below this level, counts totaling less than 100 would be accepted, bearing in mind the decreased reliability of the count.

2.4. Water Particle Counting

The practical lower size limit for water particles is 2 μ using the 100 μ aperture tube. Use of the 50 or 70 μ aperture tubes resulted in clogging. This is due to the high aspect ratio (length to width) of particles usually found in water samples. The lower threshold on the ZM is set at 1.0. Particles are adhesive and stick to the glass Folin-Wu dilution tubes. In order to eleminate this problem, particle dilutions are made in Corning 50 mL disposable centrifuge tubes (see above). 100 μL aliquots of sample in 25 mL of 0.4 M NaCl are used for particle suspensions. To convert CC Counts (using the manometer setting of 500 μL) to particles/mL multiply by 500. For neat water samples, 1000 uL aliquots are diluted in 25 mL and the conversion factor is 50.

TABLE 1. Lotus 123™ spreadsheet.

Test 1 Control Dilution Factor: 1/350			
	Initial	1	2
Time of reading	13:21:89	4:38:31	14:07:33
Elapsed Time	0.00	15.28	22.70
Coulter Counts			
Reading 1	43	948	10,966
Reading 2		975	11,360
Average	43	961.5	11,163
Log Cells/mL	4.1795	5.5270	6.5918
Cells/mL	1.512×10^4	3.365×10^5	3.907×10^6

EFFECT OF CONCENTRATION (Summary Sheet)

Concentration	Slope	r^2	Slope Ratio
MS-1 (Control)	0.0869	0.9999	1
20%	0.0811	0.9931	0.933
50%	0.0749	0.9939	0.862
90%	0.0645	0.9969	0.740

2.5. Data Management

Cell counts were entered into a Lotus™ spreadsheet (Table 1) which was modified to list the time of sampling, elapsed time, 2–4 Coulter Counter determinations, average cell concentration, the log of the cell concentration, and regression calculations. Templates for toxicity tests employing 4, 5, or 6 doses also contained macros which provided summary sheets listing the slope-ratio calculations for each dose. The data from the Lotus™ spreadsheets were transferred to Psi-Plot (Poly Software, Salt Lake City, UT) spreadsheets in order to produce growth and dose-response graphics. Complete spread sheets containing raw data of the dose growth curves for one sample are presented in the report to the Agency for Toxic Substances and Disease Registry. The ATSDR Report is available @ ***http://www.envirolab.com.***

2.6. Cultures for Toxicological Studies

2.6.1. Individual Toxicants

Toxicological studies are carried out in 17 × 100 mm Falcon (35-2057) sterile, disposable polystyrene tubes; final volumes are 1.0 mL/tube. For organic toxicants, 10 μL aliquots of serial dilutions of toxicant dissolved in dimethylsulfoxide (DMSO) are added to 990 μL aliquots of flagellate cultures in order to obtain a series of dose-culture tubes. 10 μL of DMSO is used for the 0-dose or control tube. Inorganic

TABLE 2. Schedule of components for testing either whole water samples or whole particle suspensions. The units of measurement for whole water dilutions are % Effluent; and P-MEQ/mL (particle milliliter equivalents per mL) for particle suspensions.

Whole Effluent Whole Particle	MS-1	Flagellates (Seed Flask)	Kp Suspension (20X)
0 μL	900 μL	50 μL	50 μL
100 μL	800 μL	50 μL	50 μL
200 μL	700 μL	50 μL	50 μL
500 μL	400 μL	50 μL	50 μL
900 μL	—	50 μL	50 μL

toxicants are dissolved and diluted in MS-1. Some organics which are not soluble in DMSO, such as benzo[a]pyrene are dissolved in cyclohexane.

2.6.2. Environmental Samples

Whole effluent and whole particle testing are conducted in 1.0 mL final volumes according to the dilution matrix described in Table 2. The protocol has been modified in order to permit allocation of 90% of the volume of the test cultures for delivery of the sample. In order to reconstitute the water samples in MS-1 buffer; 60 μL of solution A, 30 μL of 0.1 M $CaCl_2.2H_2O$, and 100 μL of 0.1 M $NaHCO_3$ are added to 10 mL of neat water sample. Solution A contains 10 mL of 0.1 M $NaHPO_4$, 10 mL of 0.5% phenol red solution (Sigma P-0290), and 3 mL of 0.1 M KCl.

2.6.3. Whole Particle Preparation

Concentrated suspensions are obtained from water samples collected according to the following method: cold water taps are turned on at a moderate flow rate (10 L/min) and allowed to run for 2 min or longer. Two Corning 250 mL plug seal, disposable, sterile centrifuge tubes (25350-250) are then filled to the top, closed and transported back to the laboratory. On days where the ambient temperature is above 75°F, samples are placed in a cooler chest with ice. The centrifuge tubes are either stored at 5–10 °C or centrifuged in an IEC 2V centrifuge at 2,400 RCF (Relative Centrifugal Force) in a 266 rotor for 8 hrs. The temperature of the centrifuge was maintained at 16–18 °C by air-cooling through a three inch air hose connected to a Carrier 28,000 BTU air conditioner. Higher speed, refrigerated centrifuges can reduce the processing time to 30 mins.

The supernatant was removed by decantation (first 150 mL) and then by aspiration of the remaining 110 mL in order not to disturb the flocculent orange-brown pellet. The pellets from two tubes were resuspended in a total volume of 1.6 mL of MS-1 buffer by vortex mixing, and brought to 2.0 mL in pyrex 2 mL volumetric test tubes (Corning 5640-2). Aliquots of 0.9, 0.5, 0.2 and 0.1 mL are delivered to 14 mL sterile disposable polystyrene tissue culture tubes (Falcon 2057) for whole particle testing in the *Tetramitus* Assay.

2.6.4. 20X Kp Preparation and Seed Cultures

The 20X Kp (*Klebsiella pneumonia*) suspension (see Table 2) is obtained by recentrifuging the Kp suspension (see above) and resuspending the Kp in 1/20 of the original volume (e.g. resuspend the pellet obtained from 100 mL of Kp soup in 5.0 mL of MS-1). The starting flagellate concentration in the 1.0 mL of test culture will be 1/20 of the seed culture concentration. This revised protocol allows for testing a higher dose of particles or whole effluent. The same dilution matrix is used for testing whole water toxicity—this is used where water samples have high levels of toxicants which would produce toxicity data without the need for concentration procedures. Aliquots of whole water samples also may be filtered through membrane filters (small volumes obtained with Swinnex filters) of known pore size in order to determine the toxicity of filtrates which are selected for exclusion of particles of specific sizes.

The Sequence for the Test is:

1) Set up the seed flask culture to contain 2.0×10^5 cells per mL (572 counts determined by Multisizer or ZM) at anticipated time of delivery. 50 μL delivered to each tube would result in a starting concentration of 1×10^4 cells per mL. The slope of the dose response curves for individual toxicants, whole effluents, and water concentrates is steeper at lower starting flagellate concentrations (Figure 3). Consequently all tests are standardized for this starting concentration. The usual slopes of log phase cultures range between 0.082–0.093 (log cell concentration per hour). The mean division time is about 4 hrs and the time required to grow 1 log is 12 hrs (e.g. $1 \times 10^4 - 1 \times 10^5$). Seed cultures can be diluted with Kp soup several hours before tests commence to ensure the 1×10^4 per mL starting concentration.
2) Test components are delivered into each tube in the following sequence: MS-1, particles or whole effluent, 20 X Kp, vortex and place in water bath.
3) Count cells in seed flask and then deliver 50 μL aliquots as rapidly as possible. Use the recorded time of seed flask determination and designate the starting flagellate concentration as 1/20 of the recorded seed flask count value. Delivery of 10×5 aliquots (10 tests with 4 dilutions + control) usually takes 12–15 mins.
4) Set up Test 1 spreadsheet (separate file label) with recorded time and calculated starting Coulter Multisizer counts for the control culture, then copy these values to the 4 dose-culture spread sheet cells; save the file. Exit the spreadsheet and copy this file to 9 separate files, assigning different file labels for different tests. The spreadsheet can be formatted any time prior to the first sampling time.

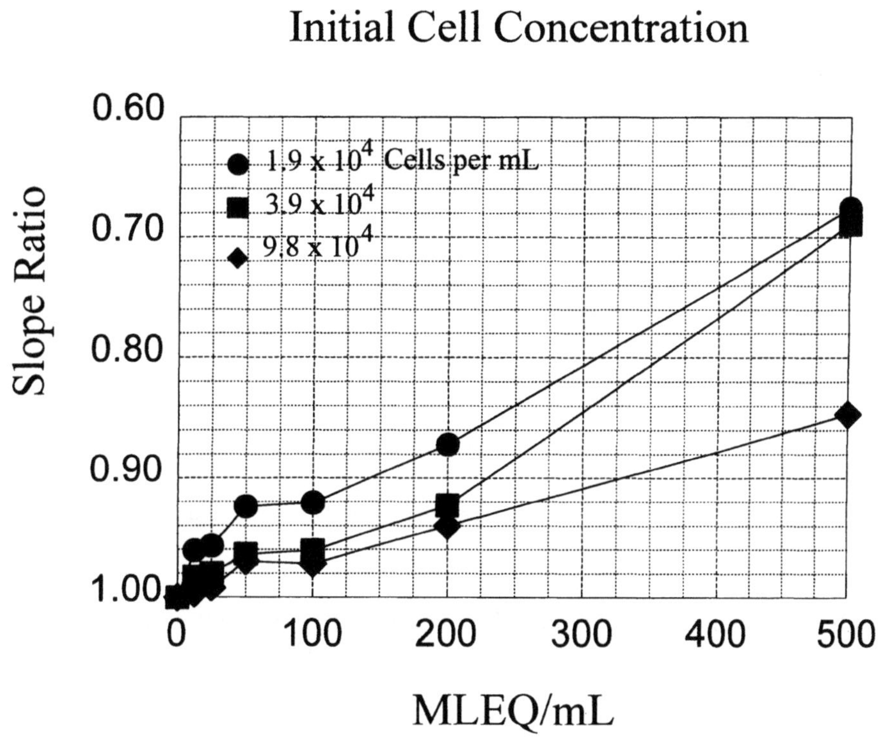

FIGURE 3. The effect of initial cell concentration on the dose response curve of a Netherlands water concentrate.

5) Count each test series 5 tubes at a time at two subsequent sampling times (usually 12–15 hrs and 17–20 hrs). The summary page will automatically list the calculated values of the slopes, r^2 values and the slope-ratios for each dilution. These values can be exported to a statistics program (PSI-Plot, Excel, Statmost etc.) to produce graphics of the dose-response curves.

Determination of the storage time stability of particles both in whole water and in suspension after centrifugation, will be valuable for scheduling the elapsed times after sample collection for optimal performance of tests.

2.6.5. Calculation of Whole Particle Toxicity

Figure 4 is a particle dose response curve of three tap water samples. The toxicity value is obtained by reading the P-MLEQ/mL intercept for a slope ratio value of 0.90; obtaining the reciprocal and multiplying by 1,000. For WI 43 the $SR_{0.90} = 84$ P-MLEQ/mL; the toxicity would be 11.9 ($1/84 \times 1{,}000$).

There are variations in the shapes of the dose response curves obtained from different samples. The ETL (Environmental Toxicology Laboratory) particle dose response curve exhibits a flattening of the dose response just after 50 P-MLEQ/mL and the extent of the departure from linearity is more pronounced above 100 P-

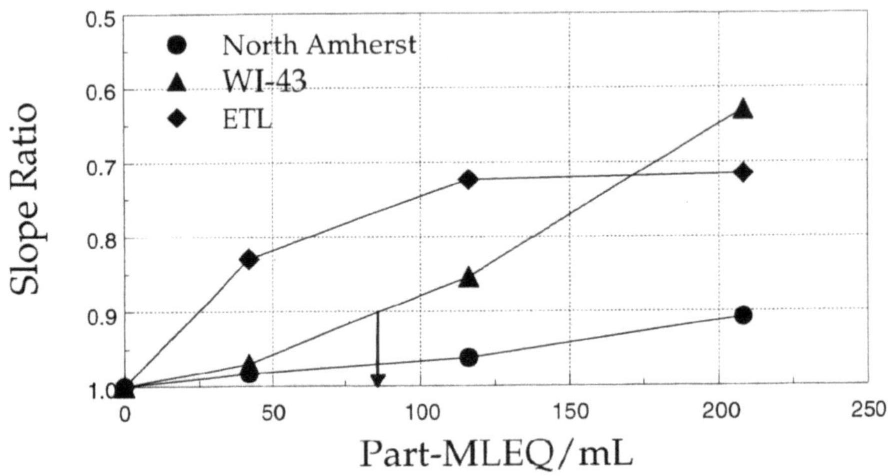

FIGURE 4. Particle Dose Response curves. The dropped-line for WI-43 indicates $SR_{0.90}$ MLEQ/mL value.

MLEQ/mL. We have encountered this in several effluent samples taken from industrial sources. The total particle concentration may influence the dose response after ingestion by the flagellates, possibly by the particles readsorbing particle-released toxicants at higher particle concentrations. Another possible explanation is that at higher particle concentrations particle aggregates are formed and limit release of toxicants. This effect can be minimized by testing several dose intervals below 50 P-MLEQ/mL. Brief ultrasonic treatment of the particle suspensions prior to delivery to the tubes may break up some of the larger particle aggregates and create more homogeneous suspensions. This is currently under investigation. Particle toxicity is a complex measurement; model experiments with added known toxicants will serve to define some of the interactions. There are probably several different classes of particle toxicities.

2.6.6. Validation of Particle Assay Method

Triplicate samples were taken from the kitchen cold water faucet of NYC-1 and centrifuged in three separate 150 mL bottles. The final resuspension volumes varied slightly because the residual water over each pellet after decantation was not equal. The water drawn from the bathtub was included as a fourth sample for comparison. The particle dose response curves were plotted and 11/12 values for the three kitchen samples and the one bathtub sample produced similar regression lines (Figure 5). Although the dose response curves were biphasic, all four samples displayed similar responses. As noted above, the shapes of the dose response curves vary from sample to sample. The coefficients of variance for each group of dose points are listed in Table 3. The average CV for all the groups of data points excluding Sample B, 49–54 MLEQ/mL is 3.46%. These data demonstrate the statistical reliability of the particle assay, even when testing involves a complex, biphasic dose response curve.

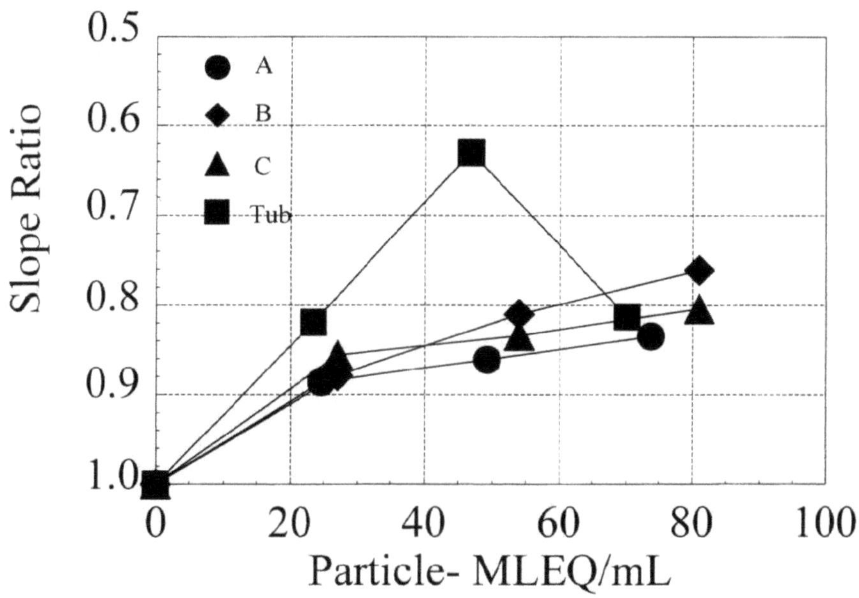

FIGURE 5. Particle toxicity method validation. Three kitchen tap water samples and one bath tub tap were centrifuged in separate bottles; each pellet was resuspended and tested for particle toxicity.

TABLE 3. Mean Slope Ratio Values for given Dose Ranges.

	25–27	49–54	49–54	74–81
Slope Ratio A	0.886	0.861	0.861	0.835
Slope Ratio B	0.819	0.630	—	0.813
Slope Ratio C	0.856	0.831	0.831	0.804
" Tub	0.879	0.810	0.810	0.761
MEAN	0.860	0.784	0.835	0.803
CV	3.51	13.3	3.06	3.86

The average coefficient of variance for the three dose ranges is **6.89%** (all points included). The average CV for all points minus the Slope Ratio—B value at dose range 49–54 is **3.48%**.

2.7 Drinking Water Concentrates

2.7.1. Collection:

Cold water taps are turned on a moderate flow rate (10 L/min) and allowed to run for 2 mins or longer. 3×1 gal samples are collected in Poland Spring polyethylene containers. The containers originally contained Poland Spring distilled water which is used for all media formulation in our laboratory. This water was analyzed by Massachusetts DEP (Oscar Pancorbo, Ph.D.-Director) and was found to have extremely low levels of all metals. Each 3 gallon sample was transported to the laboratory in cooler chests containing ice if the ambient temperature was above 75°F.

2.7.2. XAD Concentration

One gal. aliquots are filtered through a 90 mm Whatman Glass microfiber filter (934-AH) using a Kontes ultra-ware filtration apparatus (K953840-4090). One filter was used for 3 × 1 gal aliquots and the filters were removed and stored in 100 mm petri dishes. Dried filters were arranged and labelled in 8 × 10 acetate sheets and photographed as permanent records of sediment deposition for each house sampled. The filtered water was transferred to a 2.5 gal polycarbonate carboy with a spigot (Nagle #2317-0020; actual capacity was 3.2 gal) and connected to the XAD column with silicone tubing and a male ground glass joint which had glass tubulation on one end. Ten L of filtered water were decanted into the carboy and passed through thc XAD columns. The bottom ends of the columns were connected by silicone tubing to a floor drain, thus eliminating a labor-intensive step of collecting and discarding column effluents. When the volume of water in the carboy fell below the level of the outlet spigot, the connection between the carboy and the column was disengaged, a 1 L spherical glass reservoir with a 24/40 male ground glass joint was connected to the column and the remaining water in the carboy (600–1,000 mL) was added for the final amount to be passed through the column. Two gal rectangular carboys (Fisher 02-963-10A) will accommodate 10 L volumes, occupy less shelf space, and because the spigot is closer to the bottom, will eliminate the need to transfer the sample to the spherical glass reservoir. The columns were packed sequentially with a 3 cm section of silicone-treated glass wool (Alltech 4037), 8 cm of purified XAD-2 resin (Alltech 17311), 8 cm of purified XAD-8 resin (Alltech 17318), and a 3 cm glass wool plug. The prepared columns were covered by a top layer of distilled water and the columns were monitored to ensure that the tops did not run dry due to absence of water flow through the column. After 10 L were passed through, the columns were rinsed with 200–300 mL of distilled water, the tubing disconnected and the columns transferred to a supporting frame in a fume hood. The excess water was forced through the columns with a brief wash of prepurified N_2 (a male ground glass joint, with a glass tubulation at one end, was directly connected to the nitrogen tank by silicone tubing) and a 10 mL aliquot of ethyl acetate was forced into the column with another brief N_2 wash. Semivolatile organics were eluted from the XAD columns with 50 mL of ethyl acetate (Patterson et al., 1995) and collected in 60 mL separatory funnels (Teflon stopcocks). The lower water layer was drawn off and the ethyl acetate was transferred to 40 × 130 mm glass tubes (Corning 9856-40). Groups of six tubes were placed in a Labconco RapidVap® rotor (79065) and the ethyl acetate was evaporated under vacuum at 38 °C. The dried concentrates were reconstituted with 1.0 mL of 25% DMSO/MS-1 buffer and serially diluted for testing.

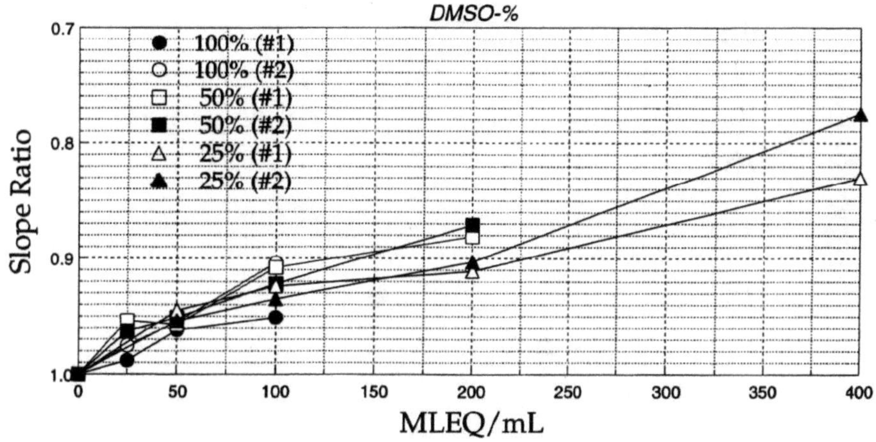

FIGURE 6. Dose Response Curve for ETL water concentrates prepared with 100, 50, and 25% DMSO solutions.

2.7.3. Validation of Concentrate Method

The final concentration step is reconstitution of dried concentrate in 1 mL of dimethylsulfoxide (DMSO) in 40 × 100 mm tubes. Smaller reconstitution volumes may result in greater variability. The optimal final concentration of DMSO in each of the 1 mL test cultures was determined to be 1%. DMSO at higher concentrations was found to be toxic to flagellates. Reconstitution solutions of 50 and 25% DMSO/MS-1 buffer were used to reconstitute duplicate dried concentrates. This permitted delivery of larger aliquot volumes to each 1 mL of test culture. The highest dose value with 25% DMSO now would be 400 MLEQ/mL (milliliter-equivalents) compared to 100 MLEQ/mL using 100% DMSO as reconstituting solution. Lowering the DMSO concentration to 25% did not effect the total toxicity recovered. Figure 6 describes the dose response curve for duplicate concentrates at each DMSO dilution. The r^2 for the dose response regression curve derived from combining the data from all DMSO concentrations was –0.947 (negative because the slope ratio axis is inverted from $1 \rightarrow 0$).

3. RESULTS

3.1. Otis Air National Guard Monitoring Well, Ann Arbor Industrial Effluents, and Flushing and Warwick, N.Y. Whole Drinking Water Toxicity

The first whole effluent toxicity test using the *Tetramitus* Assay was performed in 1997 on samples drawn from monitoring wells just outside Otis Air National Guard Base in Sandwich, Massachusetts. The original test protocol was conducted

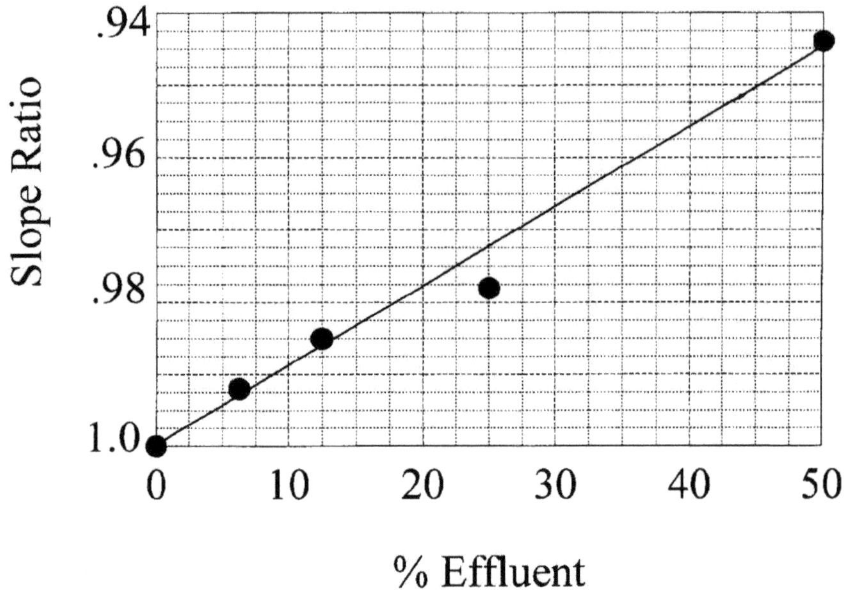

FIGURE 7. *Tetramitus* whole water toxicity test of Sample FS 343-79 collected at Otis Air National Guard Base on September 10, 1997 with Denis LeBlanc, USGS.

on 1.0mL samples which were added to 1.0mL of *Tetramitus* seed culture (final volumes = 2.0mL). Consequently, the highest concentration was 50% effluent. Use of the new protocol allows for testing of samples at 90% effluent (see above). At 50% effluent the slope ratio for the Otis ANG sample was 0.944 (Figure 7). Table 4 lists the regression data used to construct the graphic. Although the slope ratios for the 4 sample dilutions are somewhat lower than those observed for industrial effluents (see below), the r^2 values for each of the dose-growth curves, including the control, are all greater than 0.998 (see Table 4). This provides data with high statistical reliability. The r^2 value for the dose response curve itself is –0.991 (4 experimental points and the measured 0-dose point). This value is negative because of the inversion of the slope ratio axis, with a starting value of 1.0 and subsequent decreasing values. Acceptance of dose response data are contingent on linearity of the dose-response regression curve. These data pass the most stringent acceptance criteria.

A subsequent study comparing the *Tetramitus* Assay to standard EPA whole effluent assays (WET tests) using the micro crustacean *Ceriodaphia* and the fat head minnow were conducted on 2^0 sewage, pharmaceutical plant, and automobile plant effluents (Jaffe, RL, Sweet, LI, and Meier, PG—manuscript in preparation). Table 5 illustrates that the *Tetramitus* Assay was at least five times more sensitive than the standard WET test. Only the pharmaceutical plant produced a NOEC (no observable effect concentration) of 60% effluent using standard WET testing; both the 2^0 sewage and auto manufacturing plant effluents had NOEC's of 100%. The NOEC's

TABLE 4. Dose Response Data Summary for Graphic presented in Figure 7.

% Effluent	Slope (Growth Curve)	r^2	Slope-Ratio
0	0.0823	0.9996	1.00
6.25	0.0817	0.9987	0.992
12.5	0.0811	0.9999	0.985
25	0.0805	0.9998	0.978
50	0.0777	0.9999	0.944

for the *Tetramitus* assays were 20–40%. All the assays were conducted with starting flagellate concentrations of 1.0×10^5 cells per mL. Figure 8 is a graphic comparing WET Tests performed on different effluents. One series using a starting cell concentration of 1×10^4 produced a dose response curve which was 2.5× more sensitive (based on comparison of the slope-ratio intercept of 0.90). Similar increased sensitivity also was observed on water concentrates (Figure 9). Whole effluent samples also may be filtered through 0.45 μ pore size membrane filters. The filtered whole effluents can be tested in order to observe the effect of particle contribution (greater than 0.45 μ) on the unfiltered whole effluent dose response curves. In addition, particle toxicity can be directly measured by testing concentrated particle suspensions (obtained by centrifugation).

3.1.1. Particle Toxicity in Effluent Samples

Particle toxicity interactions at higher particle concentrations varies with the source and time of sampling. Several outcomes are possible, demonstrating the effect of higher particle concentrations on toxicity measurements:

TABLE 5. Comparison of WET Tests using *Tetramitus*, *Ceriodaphnia*, and Fathead minnow.

Source of Effluent	WET Test Organism	NOEC
Pharmaceutical Plant	*Tetramitus*	20% effluent
	Ceriodaphnia	60% effluent
	Fathead minnow	60% effluent
Auto-Manufacturing Plant	*Tetramitus*	20% effluent
	Ceriodaphnia	100% effluent
	Fathead minnow	100% effluent
2° Sewage Plant	*Tetramitus*	20% effluent
	Ceriodaphnia	100% effluent
	Fathead minnow	100% effluent

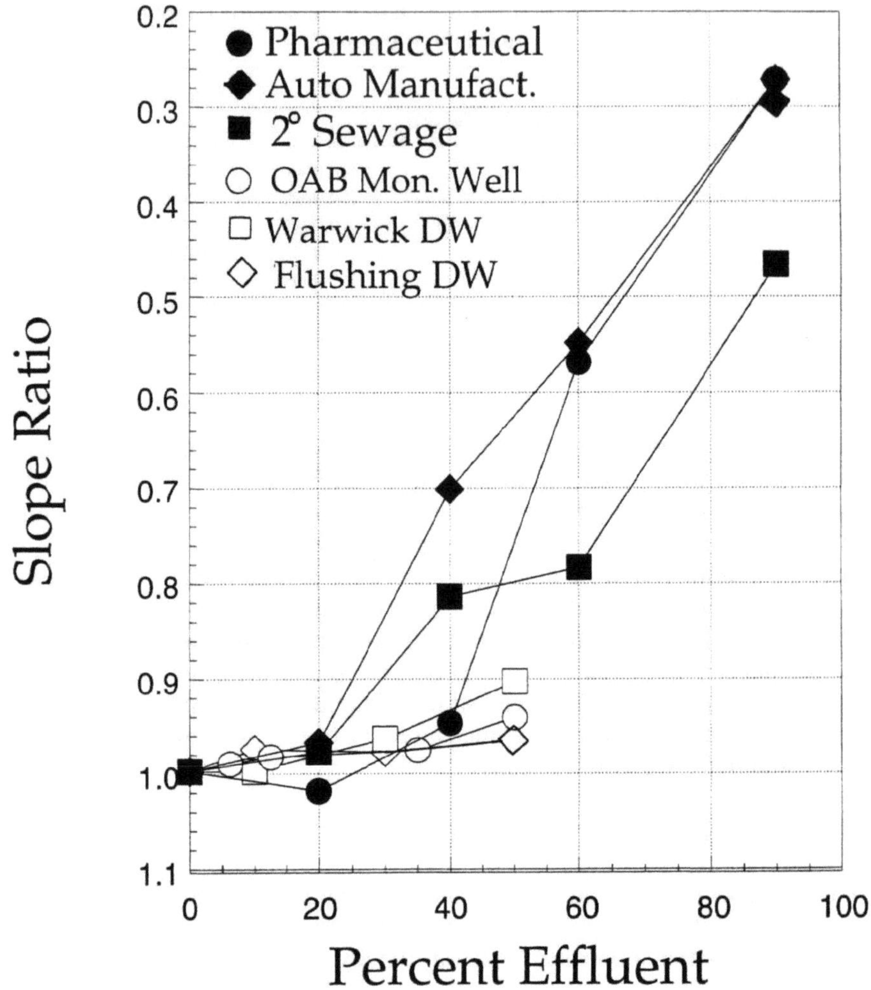

FIGURE 8. Michigan Split Sample Study, Otis ANG, Warwick and Flushing Drinking Water.

- No effect of particles on whole effluent toxicity value. Filtered whole water toxicity is lower than whole effluent value and the difference can be accounted for by the *addition* of a whole particle toxicity value (Figure 9).
- Both Particle Toxicity and Whole Effluent *dose response curves flatten* out at higher concentrations (Part-MLEQ/mL and % effluent), while the curve for Filtered Whole Water Toxicity rises as a linear function of increased dose. At higher whole water concentrations the particles are interfering with the dose response signal of both Whole Particle and Whole Water tests (Figure 10).
- Whole Effluent and Filtered Whole Effluent curves are the same (Figure 11).

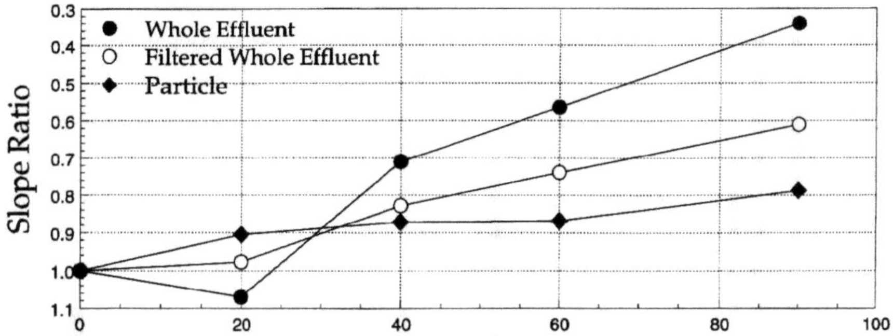

FIGURE 9. Dose response curves of components in an effluent from an auto manufacturing plant. The whole water toxicity is the sum of the filtrate + particulate toxicity.

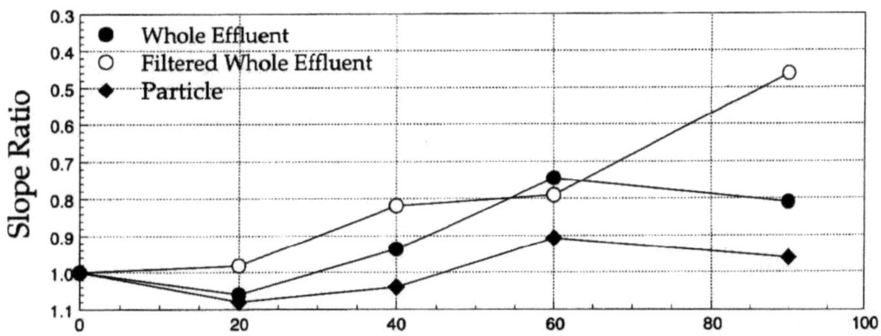

FIGURE 10. Dose response curves of components from a secondary sewage treatment plant effluent. Particle interactions above 60% effluent and 60 Part-MLEO/mL cause flattening of both whole effluent and particle dose response curves while the filtered whole effluent curve remains linear.

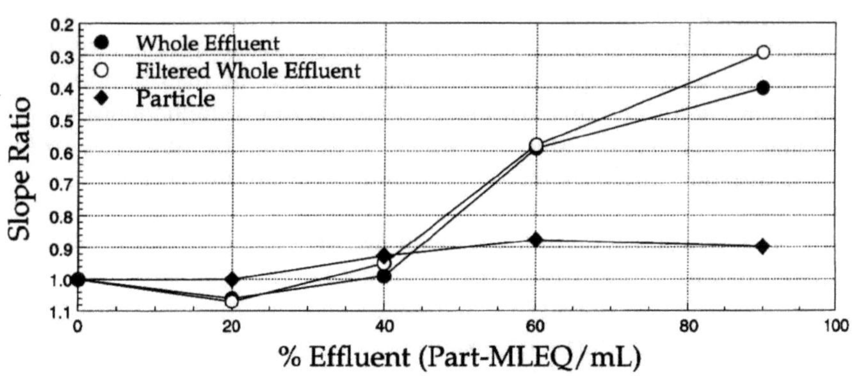

FIGURE 11. Dose response curves of components from a pharmaceutical plant effluent. Whole Effluent and Filtered Whole Effluent curves are similar.

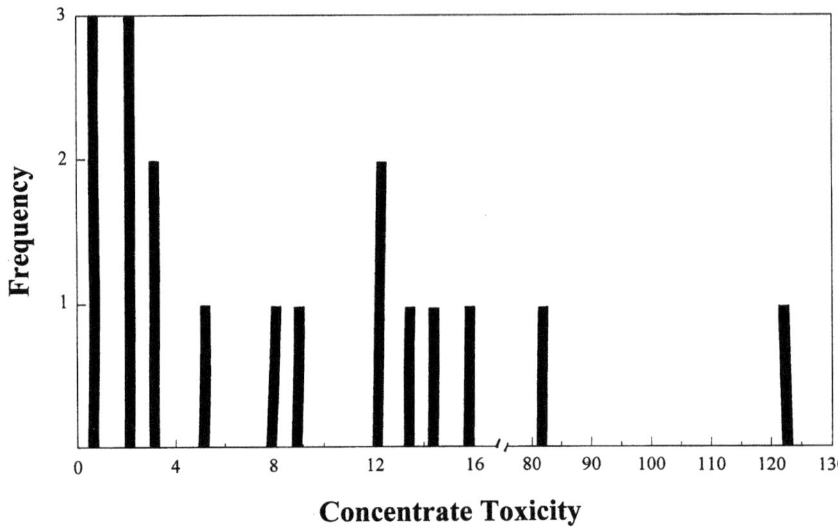

FIGURE 12. Concentrate toxicity in West Islip tap water samples. 10/28 homes had no measureable toxicity and are not indicated in the graphic.

3.2. West Islip Tap Water Toxicity

3.2.1. Drinking Water Concentrates

Prompted by the search for risk factors associated with higher incidence of breast cancer on Long Island, a study has been started to examine the distribution of toxicity using the *Tetramitus* Assay in drinking water obtained from household cold water taps in West Islip. In order to plan a population-based study relating drinking water toxicity to breast cancer incidence, the distribution of toxicity values and the coefficient of variance of triplicate measurements can serve to derive a power estimate of the test and indicate the appropriate sample size for this study. twenty eight tap water samples were concentrated and the distribution of toxicity is presented in Figure 12. For 26 of the samples, the low to high values vary over a 10-fold range. There are two "outliers" which are 80- and 126- times the low/normal values. The question raised by these preliminary measurements is:

> Are women who reside in homes with higher toxicity drinking water at greater risk for breast cancer (or other disease experiences)?

3.2.2. Drinking Water Particle Toxicity

Preparation of the drinking water for concentration necessitates a filtration step to prevent particles from clogging the XAD columns. Deposition of sediment

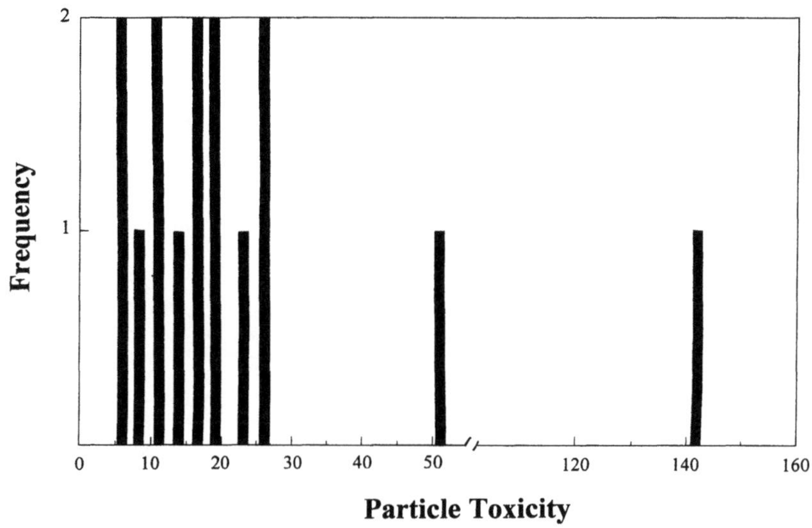

FIGURE 13. Whole Particle Toxicity in West Islip tap water samples. 21/36 homes had no measurable toxicity and are not indicated in the graphic.

on the 90mm glass fiber filters can indicate the relative amount of particles in that water sample and serve as a visual record for that house. There appears to be a seasonal increase in drinking water sediment from mid-May to early October. The particle toxicity also increases. The distribution of particle toxicity amongst 36 homes in West Islip (Figure 13) seems to be similar to water concentrate toxicity in the spread of values. Here again the issue of particle toxicity and potential risk emerges.

3.3. Levittown Data

3.3.1. Whole Water Toxicity

On September 26, 1998 six homes were sampled on Ramble Lane in Levittown, N.Y., Zip Code 11756. Whole water toxicity dose response curves for each tap water sample (LI80-85) are shown in Figure 14. Tap water from LI 85 was more toxic than effluents obtained from a pharmaceutical, auto manufacturing, and secondary sewage treatment facility in the Michigan Split Sample Study (Figure 8). LI 81, 82, and 84 are almost as toxic. Approximation of the NOEC dose-intercept of the dose response curves for LI81, 82, and 84 is at 50% effluent, 20% for LI 85, and 0% for LI 80. The dose-intercept is a graphic estimation of the NOEC (no observable effect concentration) used in evaluating EPA WET Test data (see above). Inhibition of the rate of cell division is a quantitative change and, to the extent that linearity of the dose response regression curve is demonstrated,

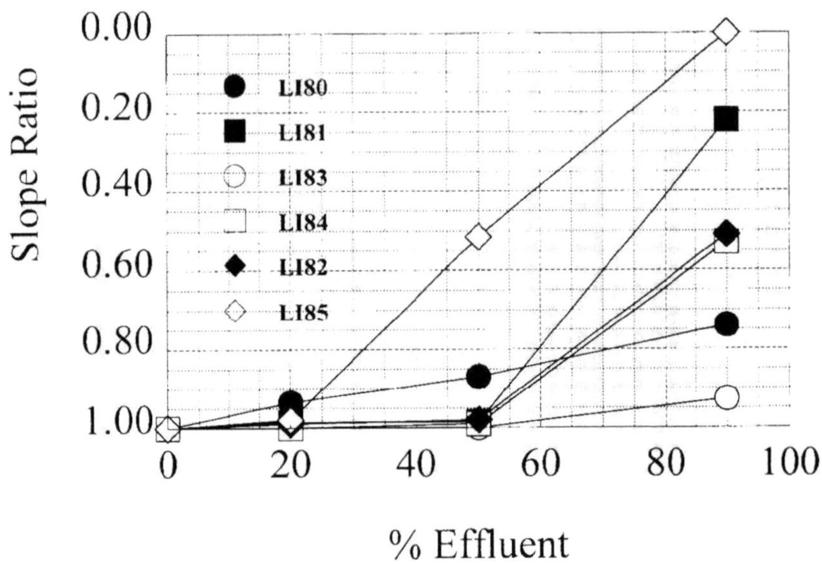

FIGURE 14. Whole water toxicity of 6 tap water samples obtained from homes on Ramble Lane in Levittown, L.I.

comparison of toxic potencies of different samples can be achieved by reading the dose intercept value for a given slope ratio.

Further work was planned in order to obtain data points at more frequent dose intervals and to repeat these observations. However, during the week of October 5, 1998 The Hempstead Water Authority performed a purging of the entire water system by opening all fire hydrants and flushing through large volumes of water. Subsequent water sampling revealed that the sediment was almost completely eliminated (Figure 15). Three liter aliquots were filtered through 47 mm glass fiber filters in order to obtain a visual record of the sediment attributable to particles greater than 1.5 μ. Prior to the Levittown samples, sediment visuals were obtained using 3 gal aliquots filtered through 90 mm filters and required processing at the laboratory. Use of 47 mm filters with a 4 L vacuum flask and a portable vacuum pump (Fisher 13-875-220) permitted on-site preparation of sediment-on-filters. This was useful because the homeowner felt more connected to the testing process with an immediate visual record of the sediment. A duplicate filter was offered as a record of the sampling activity for that day.

Whole water toxicity for samples LI 110-121 and LI 124-130 was 0 except for LI 117 and LI118 (Figure 16). The criteria for a positive toxicity dose response curve is that three points on the dose response curve must increase as a function of increasing dose with a r^2 for that dose response linear regression curve above 0.90. We had planned to perform repeat measurements with a focused dose response matrix (dose points of 60, 70, 80, and 90%). The purging of the sediment and concomitant

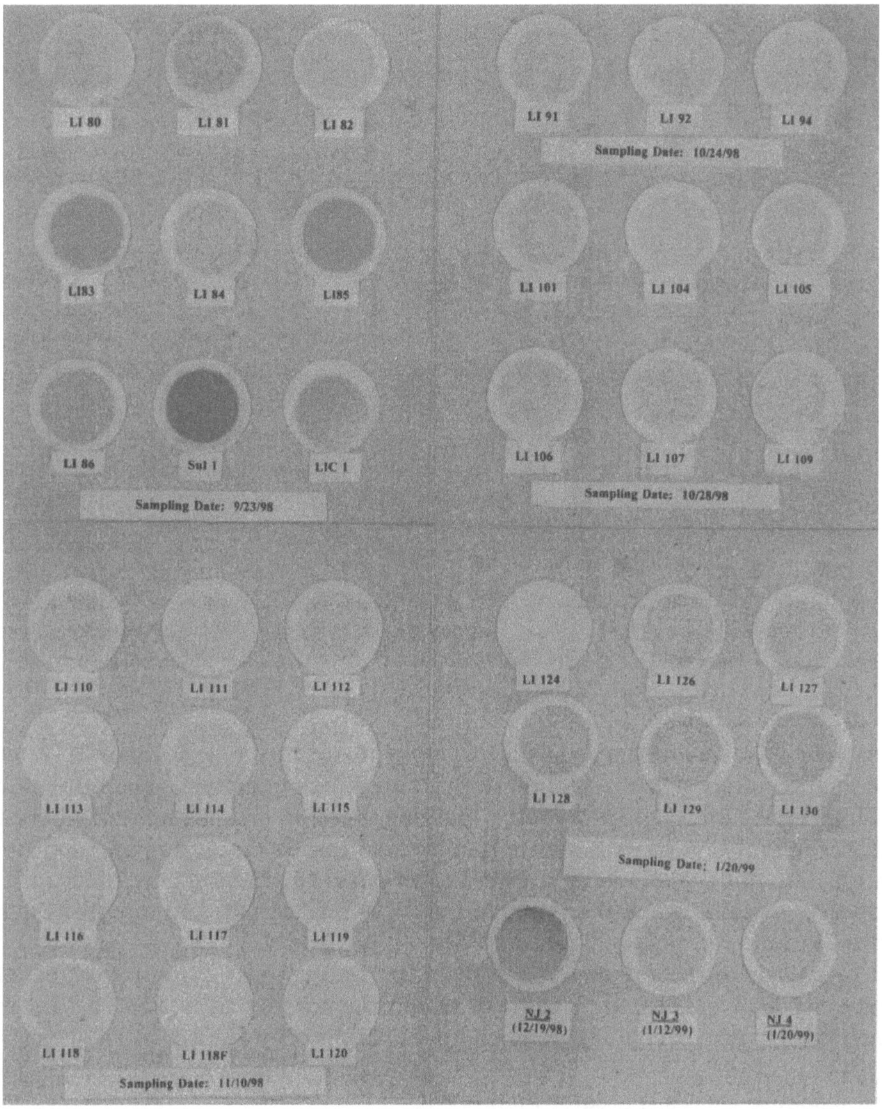

FIGURE 15. Photograph of sediment collected on 47 mm filters obtained by filtration of 3 liter samples. (For a color representation of this, see figure facing page 412).

reduction of whole water toxicity was observed for the time period up to January 20, 1999. Therefore, this study awaits further sample analyses over a longer time span in order to confirm seasonal variations in drinking water toxicity levels.

The reduction of sediment and toxicity by flushing the system illustrates the utility of the *Tetramitus* Test. The ability of the organism to ingest particles allows for monitoring of whole particle toxicity in water samples. Development of a data

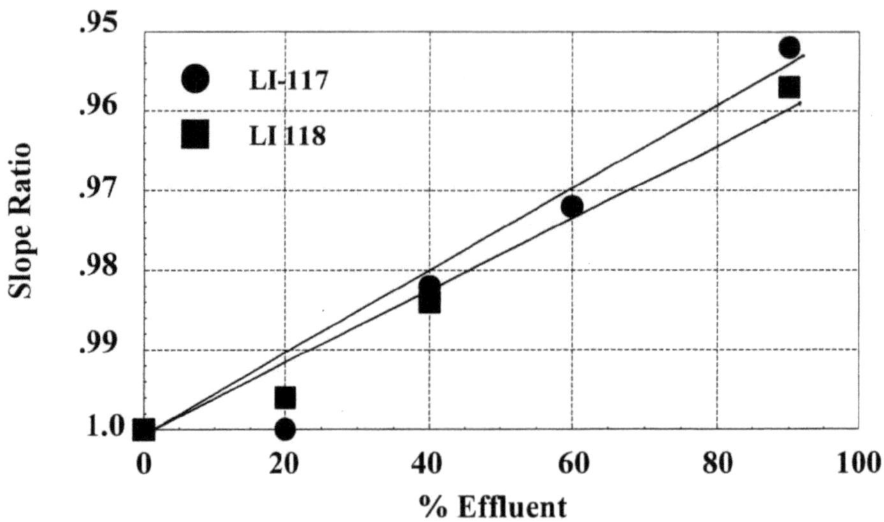

FIGURE 16. Whole water toxicity in samples LI-117 and LI-118.

base which demonstrates seasonal variation in toxicity levels points to a simple remediation strategy for reduction of drinking water toxicity; i.e., monthly purging of the system instead of semi-annual purging. Specific local situations, where toxic releases impinge on drinking water quality, also can be monitored and appropriate remediation effected when indicated. This exposure reduction may reduce the risk of specific diseases when the putative link between toxicity and disease is demonstrated in future population-based studies.

Analysis of the regression variables for whole water toxicity tests (Table 6) shows the ***high statistical reliability*** of data generated from LI 110-30 samples which display no toxicity. The coefficients of variance for the control slopes (LI 80-85, LI110-121, LI 124-130) range from 1.05–2.48%; the mean r^2 for these control dose response curves is greater than 0.998.

The coefficient of variance (CV) for the dose-growth curves of LI 110-121 is 2.54% (48 tests); for LI 124-130 the CV is 2.27% (31 tests). These data show that no false positives were obtained from samples which had no whole water toxicity. The average r^2 for regression data generated from 85 growth curves (controls included) was greater than 0.998.

3.3.2. Whole Particle Toxicity

Whole particle suspensions were prepared as described above and three doses were tested (Figure 17). Three homes (LI 82, 83, and 85) registered whole particle toxicity values for two doses, one home (LI 84) displayed positive toxicity for one

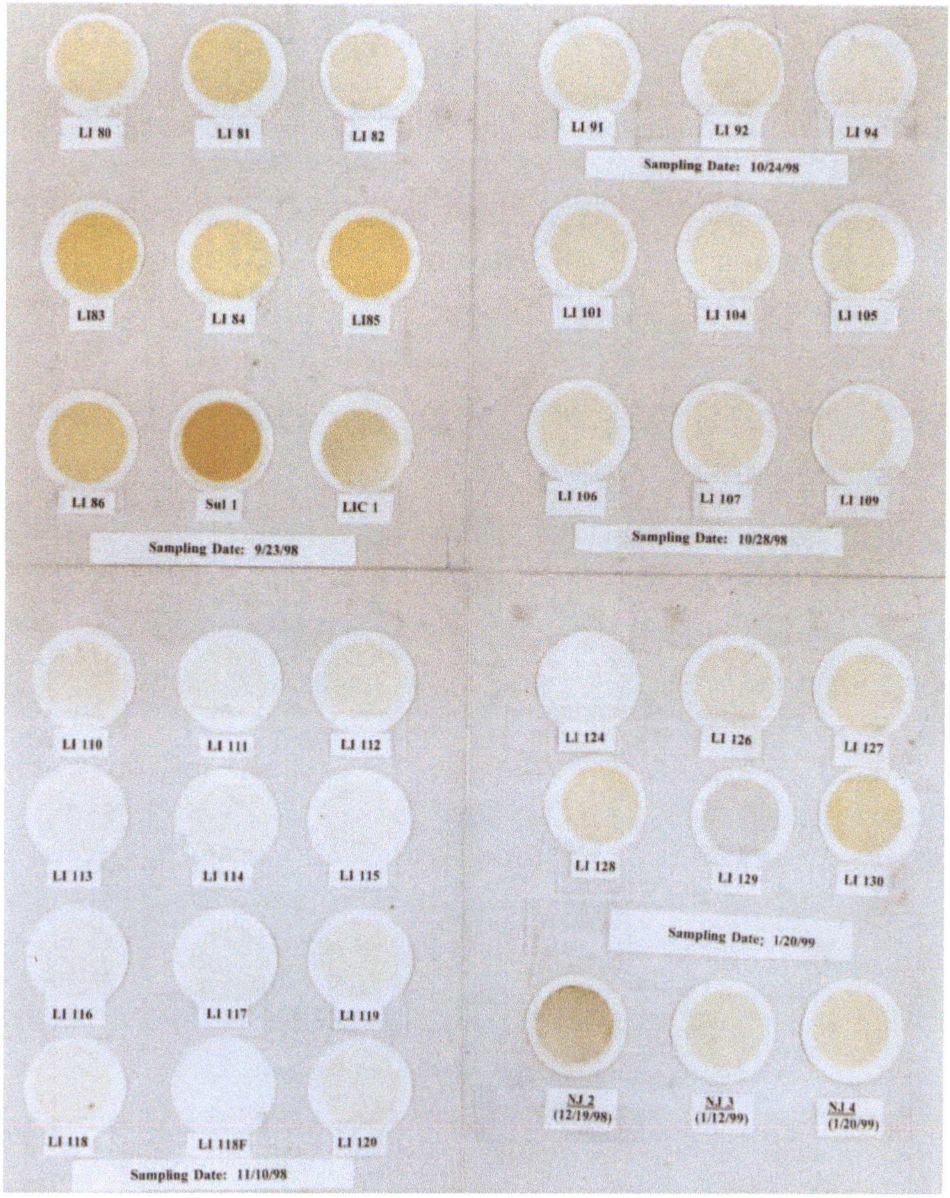

FIGURE 15. Photographs of sediment collected on 47 mm filters obtained by filtration of 3 liter samples.

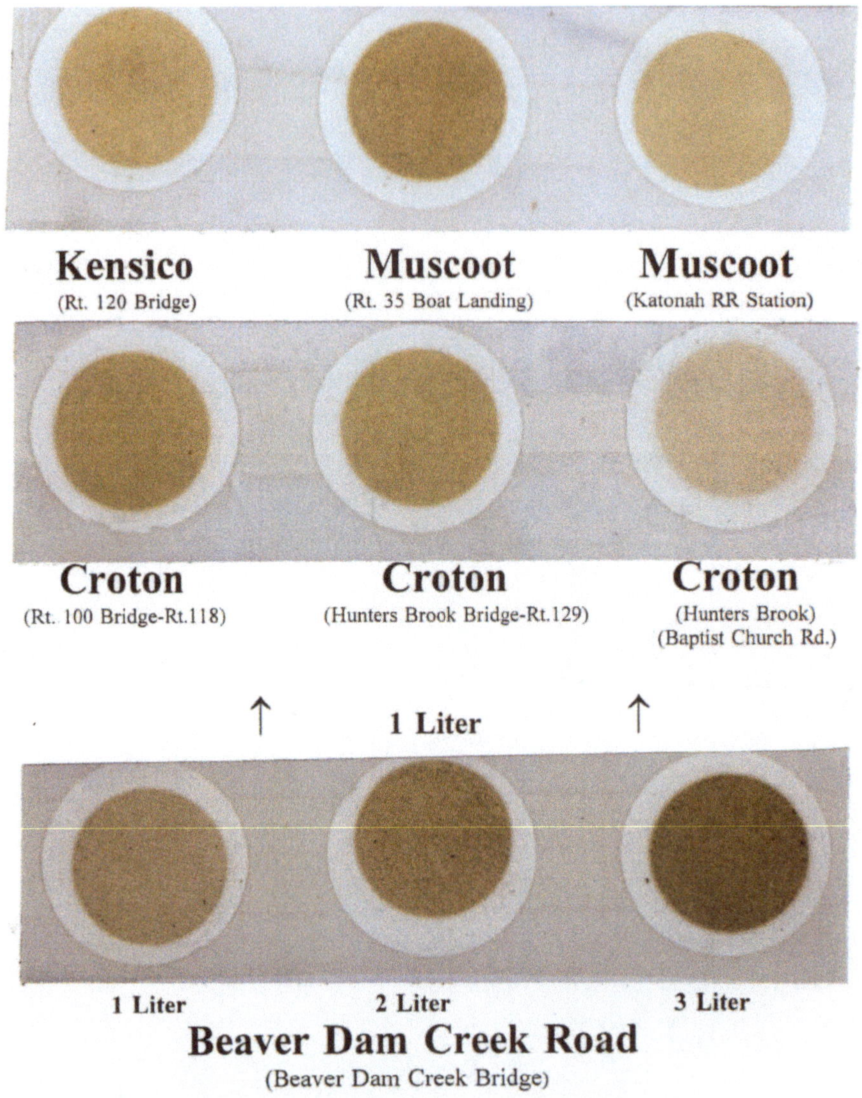

FIGURE 19. Photographs of sediment collected on 47 mm filters obtained by filtration of samples taken from N.Y.C. reservoirs on February 26, 1999.

TABLE 6. Summary of Regression Data for Whole Water Toxicity Tests.

Test Group	Controls				Sample Dose-Growth Curves					
	Slope Mean	Slope CV	R^2 Mean	R^2 CV	Slope Mean	Slope CV	R^2 Mean	R^2 CV	Slope Ratio Mean	Slope Ratio CV
LI 80–85	0.0866	1.05%	0.999	0	—	—	—	—	—	—
LI 110–121	0.0769	2.48%	0.998	0.117%	0.0764	2.54%	0.999	0.095%	0.991	2.53%
LI 124–130, NJ3	0.0785	1.30%	0.999	0	0.0817	2.27%	0.998	0.229%	1.04	2.36%

CV = Coefficient of Variance.

dose, while LI 80, 81 showed no particle toxicity. As evidenced by particle recovery calculations for the NYC Reservoir samples (see Table 7, below) the currently used low speed centrifugation methods yielded low particle recoveries which varied with each sample.

Even with low recovery, particles (probably 1% recovery) from LI 82 equal to 243 mL caused complete growth inhibition and cell death of *Tetramitus* flagellates after 18 hours of incubation. This dose of water particles was 1/8 of the daily intake of drinking water (assuming 2 L/day as the average dose of drinking water). If particle recovery were 100%, then particles from 2.4 mL of LI 82 would cause flagellate death. Determinations of particle concentration for these samples were subject to the uncertainties of not knowing the contribution of "sticky" particles on the glass

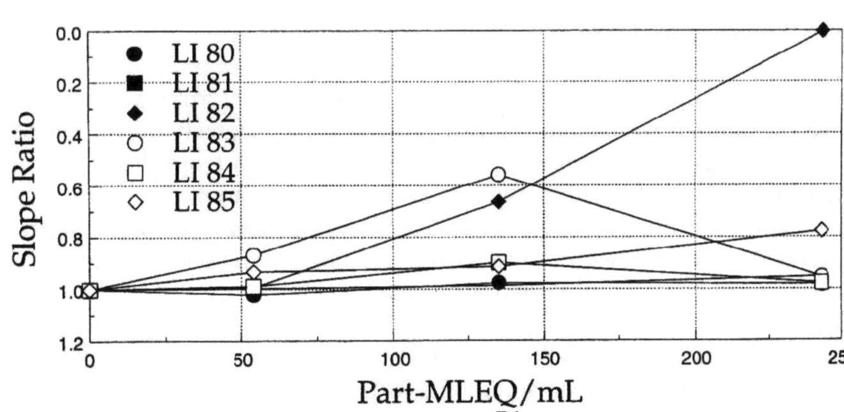

FIGURE 17. Whole particle toxicity in drinking water obtained from 6 homes on Ramble Lane in Levittown, N.Y.

TABLE 7. Total Sediment Dry Weight and water particles above 2 μ. Recovery Percent is determined by dividing the particle suspension concentration by 180 to give the "unconcentrated" value for the centrifuged/reconstituted preparation. This value divided by the original particle concentration will give % Recovery of particles in the centrifuged preparations.

	Whole Sample		Particle Suspension			
	TSP*	Par/mL	Par/mL	Concen. Factor	Part. Per mL (Calc.)	% Recovery
Kensico Rt. 120	1.3	3.8×10^4	1.2×10^5	180	6.7×10^2	1.8
Muscoot Rt. 35	1.7	2.6×10^4	2.3×10^5	180	1.3×10^3	5.0
Muscoot Katonah	0.83	1.0×10^4	2.8×10^5	180	1.6×10^3	16
Croton Rt. 100	1.5	2.2×10^4	1.9×10^5	180	1.1×10^3	5.0
Croton Rt. 129	2.3	4.0×10^4	3.5×10^5	180	1.9×10^3	4.8
Croton Hun Brk	0.86	2.2×10^4	5.9×10^4	180	3.3×10^2	1.5

* TSP = Total suspended particle matter—dry mg per liter (sediment from 1 L collected on 47 mm Whatman 934-AH Filter, pore size = 1.5 μ).

surfaces of the Folin-Wu dilution tubes. This problem was not perceived until the NYC Reservoir samples were analyzed (see below).

3.4. NYC Reservoir Data

Three gal samples were collected on February 26, 1999 from the sites indicated in Figure 18. Samples were collected in a 17 × 51 cm high-density polyethylene pipet jar (Nalgene 5242-0040) which had two holes drilled 1 cm from the top to accommodate a 50 ft length of braided nylon rope. In order to sample from the middle of each reservoir the sampling device (2 gal capacity) was lowered from bridges spanning each reservoir. Two swipes were required to fill 3 × 1 gallon containers; an initial aliquot was retrieved and discarded as a rinse.

Figure 19 shows the sediment collected on filters retrieved from 1 L aliquots from each reservoir. The particle concentrations and total dry weight of each sediment are listed in Table 7. The sediments in the 1 L reservoir samples is substantially greater than the sediment present in the three L aliquots used to prepare the Levittown filters (Figure 15).

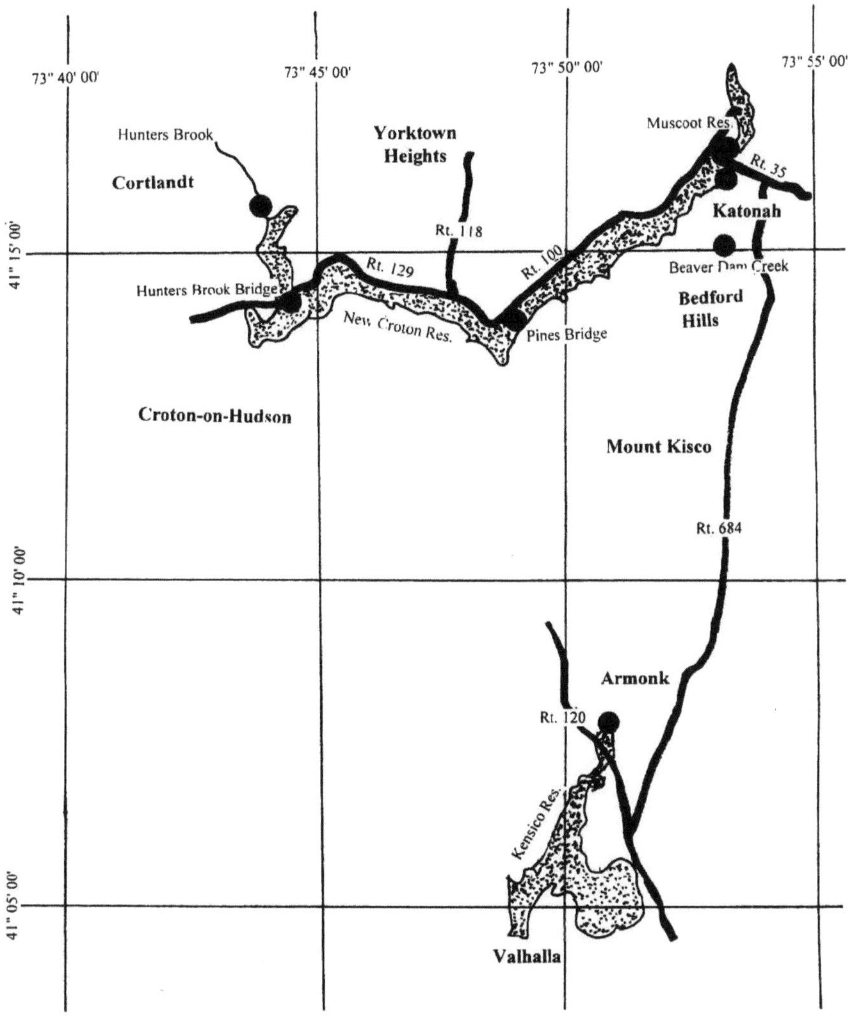

FIGURE 18. Map of Northern Westchester County. Closed circles (•) indicate sites of collection. (Modified with permission from Hagstrom Map Company, Inc.)

3.4.1. Whole Water Toxicity

No whole water toxicity was observed in any of the seven samples. Ten mL aliquots of each sample were pressure-filtered through 25 mm Swinnex filters containing Millipore HA filters ($0.45\,\mu$). Some filtrates prepared in this manner from effluents tested in the Michigan split sample study displayed greater toxicity than unfiltered samples (see below). Because of the high sediment content of the water the possible matrix effect of particles was examined. The filtered whole water samples also were not toxic.

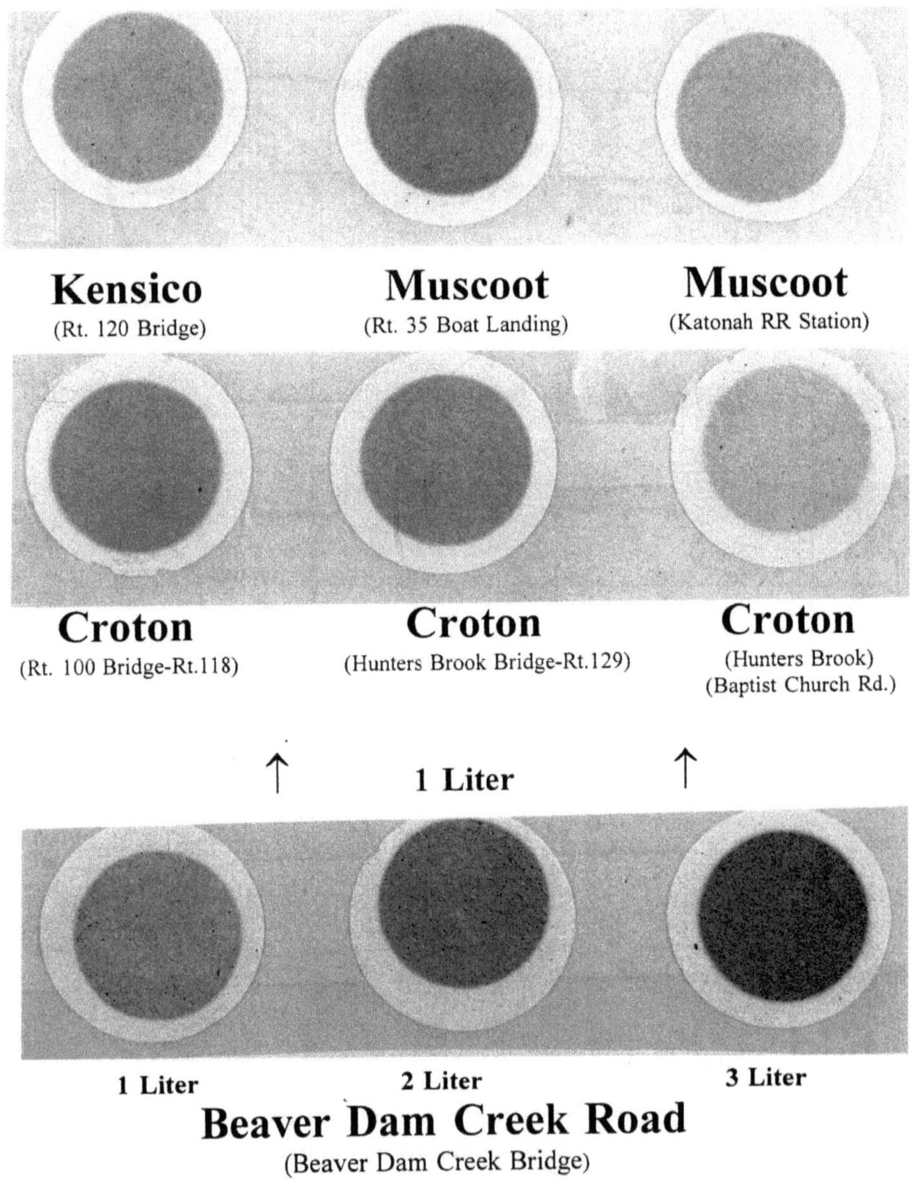

FIGURE 19. Photographs of sediment collected on 47 mm filters obtained by filtration of samples taken from N.Y.C. Reserviors on February 26, 1999. (For a color representation of this, see figure facing page 413).

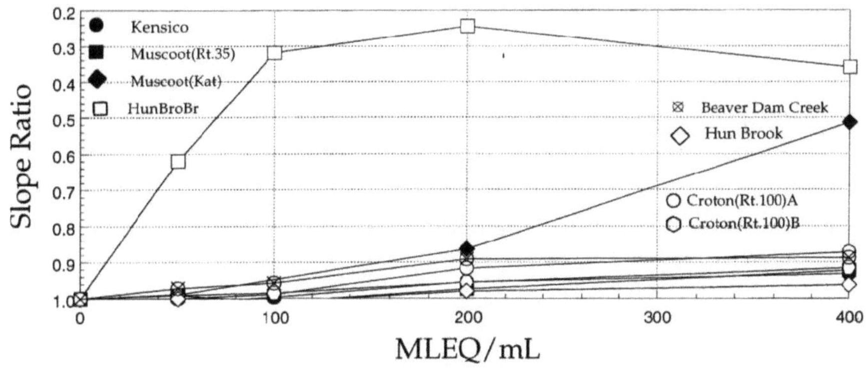

FIGURE 20. Concentrate toxicity obtained from 6 N.Y.C. Reservoirs and Beaver Dam Creek.

3.4.2. Concentrate Toxicity

The sample obtained from the Croton Reservoir, Hunters Brook Bridge, Route 129 displayed the highest toxicity; the $SR_{0.90}$ value was 71 (1/14 × 1,000). Figure 20 shows the concentrate dose response curves for the seven samples. The second highest concentrate toxicity was observed in the Muscoot Reservoir at the Katonah railroad station. The $SR_{0.90}$ value for that sample was 6.5 (1/155 × 1,000). A duplicate concentrate was prepared for the sample obtained from the Croton Reservoir, Route 100 (Pines Bridge) in order to compare a replicate concentrate dose response curve for one sample.

Hunters Brook is a suspected point of origin for possible Croton Reservoir contamination, but water sampled from the bridge on Baptist Church Road where Hunters Brook feeds into the Croton Reservoir had the lowest concentrate toxicity. Beaver Dam Creek, which feeds into the Muscoot Reservoir at Katonah, did display some concentrate toxicity. Tracking these concentrate toxicity levels over time should provide a data base which could help determine which human activities are contributing to the accumulation of toxic materials in the reservoirs. Separation of the components of these mixtures using the *Tetramitus* Assay as a guide could shed light on the source and the chemical identity of these toxicants.

3.4.3. Particle Toxicity

Again, the Hunters Brook Bridge (Route 129) sample displayed the greatest toxicity (Figure 21). The rise in the dose response curve occurred between 36–72 Particle-MLEQ/mL No actual data points were observed for the interval between Slope Ratio = 1 (no toxicity) and 0.49. The Pines Bridge (Croton, Route 100) and the Muscoot Reservoir (Route 35) samples displayed toxicity dose response curves with respective Slope Ratio values of 0.470 and 0.579 for the 162 P-MLEQ/mL dose point. The Muscoot Reservoir sample at Katonah displayed a 4-point linear dose

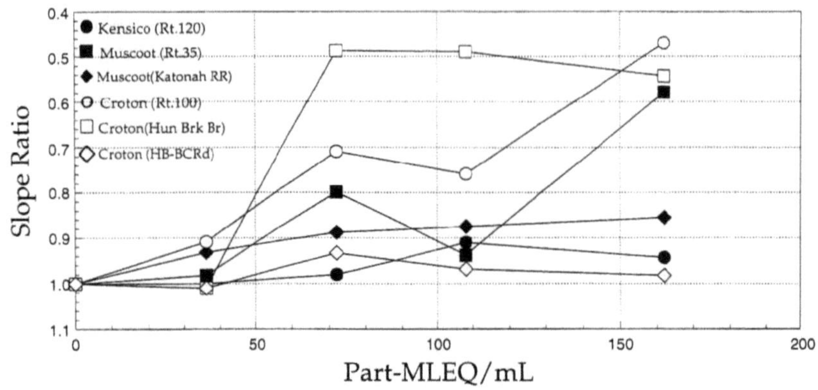

FIGURE 21. Particle toxicity in 6 N.Y.C. Reservoirs.

response curve with the 162 P-MLEQ/mL dose point Slope Ratio value = 0.855. The lowest particle toxicity was found in the Hunters Brook sample taken from the Baptipst Church Road Bridge.

Three reservoirs permitted an accurate estimate of $SR_{0.90}$ for both concentrates and particles. For these samples the P/C Tox Ratios (Table 8) ranged from 6.33 to 86. As sediments adsorb toxicants over time, the P/C Tox Ratio should reflect this process. Use of the *Tetramitus* Assay in conjunction with separation of individual toxicants can help to identify individual toxic components found in the original water matrix. Reconstitution of separated individual toxicants in various combinations can confirm possible synergistic effects of toxicant mixtures.

The toxicity data developed for the NYC Reservoirs were gathered using the most current protocols. Whole water and filtered whole water dilutions (up to 90% effluent) displayed no measurable toxicity. Concentration of soluble (XAD) and particulate (centrifugation) components did, however, confirm the presence of toxicants. The Croton Reservoir at Hunters Brook Bridge (Rt. 129) displayed the highest toxicity levels and signals a location which should be followed on a weekly basis. Mapping of the toxicity levels at specific locations distant to the original sampling point could provide evidence for the point of origin of these toxicants. The other reservoirs should be followed on a monthly schedule. Increases in toxicity levels at specific locations should trigger the reassignment of that location for weekly testing.

3.5. Use of Other Flagellate Species

Different species of flagellates produce populations of cells with discrete size distributions (Figure 22). Different sub-populations of particles may display different particle toxicity responses based on the average size of populations of flagellates. Furthermore, differences in gullet morphology and ingestion physiology also

TABLE 8. Calculation of Particle/Concentrate Toxicity Ratio.

Sample	Concentrate $SR_{0.90}$ Dose(1) (MLEQ/mL)	Particle $SR_{0.90}$ Dose(2) (MLEQ/mL)	Particle Recovery(3)	Adjusted Particle $SR_{0.90}$ Dose(4)	Particle/ Concentrate Tox. Ratio(5)	Total Suspended Particles(6)	Particles/mL(7)
Croton (Rt. 129)	14	42	4.8 % (21X)	2.0	**7.0**	2.3	4.0×10^4
Muscoot (Katonah)	150	60	16 % (6.3X)	9.5	**6.3**	0.83	1.0×10^4
Croton (Rt. 100)	310	72	5.0 % (20X)	3.6	**86**	1.5	2.2×10^4

(1) Dose of water concentrate (MLEQ/mL) corresponding to a Slope Ratio of 0.90 (that dose which causes a 10% reduction in the rate of cell division).
(2) Dose of particles (Particle-MLEQ/mL) corresponding to a Slope Ratio of 0.90 (that dose which causes a 10% reduction in the rate of cell division).
(3) Particle recovery (see Table 7); Reciprocal of % recovery provides conversion factor for expected toxicity value for 100% recovery.
(4) Lower MLEQ/mL values correspond to higher toxicity. For the Croton (Rt. 100) sample with a 5% recovery (20X less than expected for 100% recovery) 3.6 MLEQ/mL is 20 X more toxic than 72 MLEQ/mL.
(5) Concentrate $SR_{0.90}$ Dose/ Adjusted-Particle SR.90 Dose.
(6) mG Dry weight of sediment collected on filter after filtration of 1 Liter aliquot. (7) Particles greater than 2 μ determined by Coulter Counter analysis.
(7) Determined by Coulter Counter Analysis.
* Because of lower toxicity, the other three reservoir samples required dose values for $SR_{0.90}$ calculation greater than 400 MLEQ/mL.

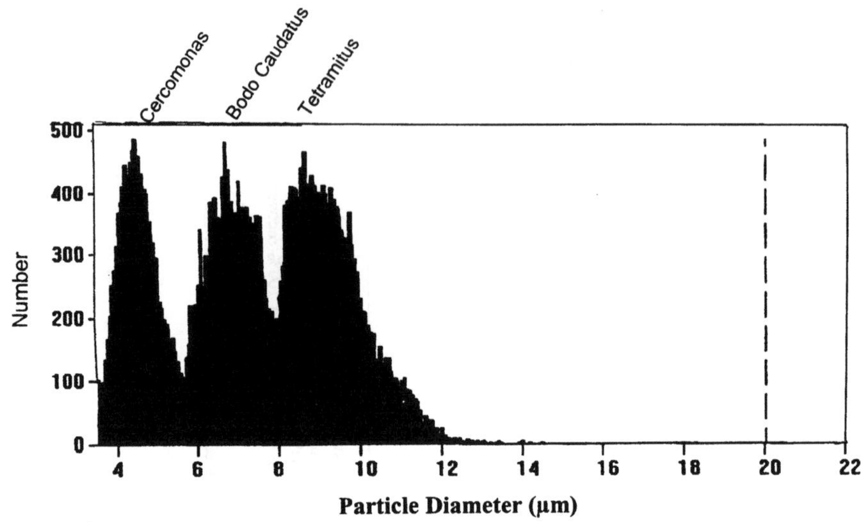

FIGURE 22. The size distribution of 3 different flagellate species.

may effect the dose response of different flagellate species. A test battery of different species could be used to develop particle toxicity fingerprints (PTFs) for particle sub-populations. Although this is a more costly analysis of particle toxicity, deployment of this technology could be applied to toxic tracking of heterogenous particle mixtures.

Both *Bodo caudatus* and *Cercomonas longicauda* grow under the same conditions as *Tetramitus* flagellates. *Bodo* divides every 2 h and *Cercomonas* 4.5 h (Figure 23). *Cercomonas* is more sensitive to cadmium than is *Tetramitus*; the dose per cell required to cause $SR_{0.90}$ is smaller. *Cercomonas* is 1/6 the volume of *Tetramitus* and this probably accounts for the greater sensitivity. Comparitive studies of reference toxicants as well as particle suspensions are planned for the three species. *Tetramitus* still appears to be the species of choice for developing a comphresive data base because it is the largest species and has the largest gullet.

3.6. Use of Size Distributions and Abnormal Movements as Early Warning Signals

In response to an RFP from the Batelle/EPA, "Advanced Monitoring Pilot", which specified the need for a real-time monitoring instrument which could detect shell fish and fin fish spoilage, a method utilizing the decrease in size of *Tetramitus* flagellates was developed. One gram pieces of Turbot (obtained from a local salad bar) were incubated at 37 °C for three days in order to increase spoilage. The aged pieces were homogenized in a 40 mL Kontes- Dounce glass homogenizer (K885300-0040) in 20 mL of MS-1. The slurry was centrifuged and the supernatant was

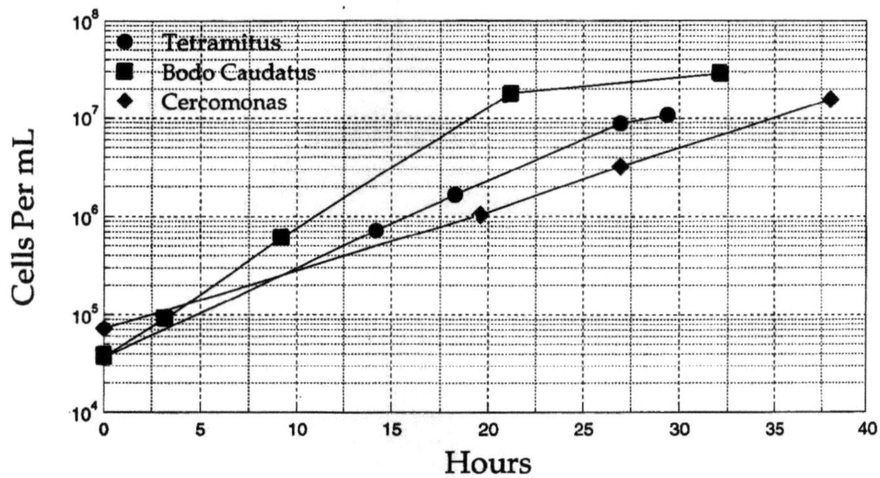

FIGURE 23. Growth of three species of flagellates in Kp/MS-1 at 30°C @ 180 RPM.

sterilized by pressure-filtration through a Swinnex assembly containing an HA Millipore filter. A 50% dilution of the extract produced cessation of flagellate growth in 10h followed by cell death. The size distribution of flagellates was monitored at short intervals after initial exposure to the spoiled Turbot extract (Figure 24). At 2h there is a marked decrease in the average flagellate diameter, at 5.45h the population decreases. In fact, there is a linear kill curve starting at 2–3h. The extract then can be analyzed at lower concentrations and % dilution dose response curves can be constructed. The initial size decrease analysis can be performed in a mobile laboratory for on site inspection and detailed dose response data can be generated at a home base laboratory. The newer Coulter Z1 and Z2 models no longer use mercury-filled glass manometers, thus the risk of mercury accidents is eliminated in field studies.

The use of *Bodo* for early detection applications might reduce the early warning times because of their faster growth rate. *Cercomonas*, which is smaller and more sensitive, may also provide earlier signals. Incubation of *Tetramitus* flagellates with asbestos produces abnormal swimming patterns. Many of the cells spin in circles, or do not exhibit the usual rhytmic straight ahead "boring" pattern. Similar patterns were observed in soil testing studies. ETL plans to create a film clip library of abnormal flagellate swimming patterns in response to reference toxicants and environmental mixtures. This library will be published on our website; ***http://www.envirolab.com*** We also plan to maintain a monthly fact sheet describing recent protocol revisions and newsworthy bulletins. The rapid transfer of information will aid in reducing interlaboratory variability. A system of performance ratings can be developed for laboratories which employ the *Tetramitus* Assay on a regular basis and the establishment of an information network will be essential for development of regional and national data bases.

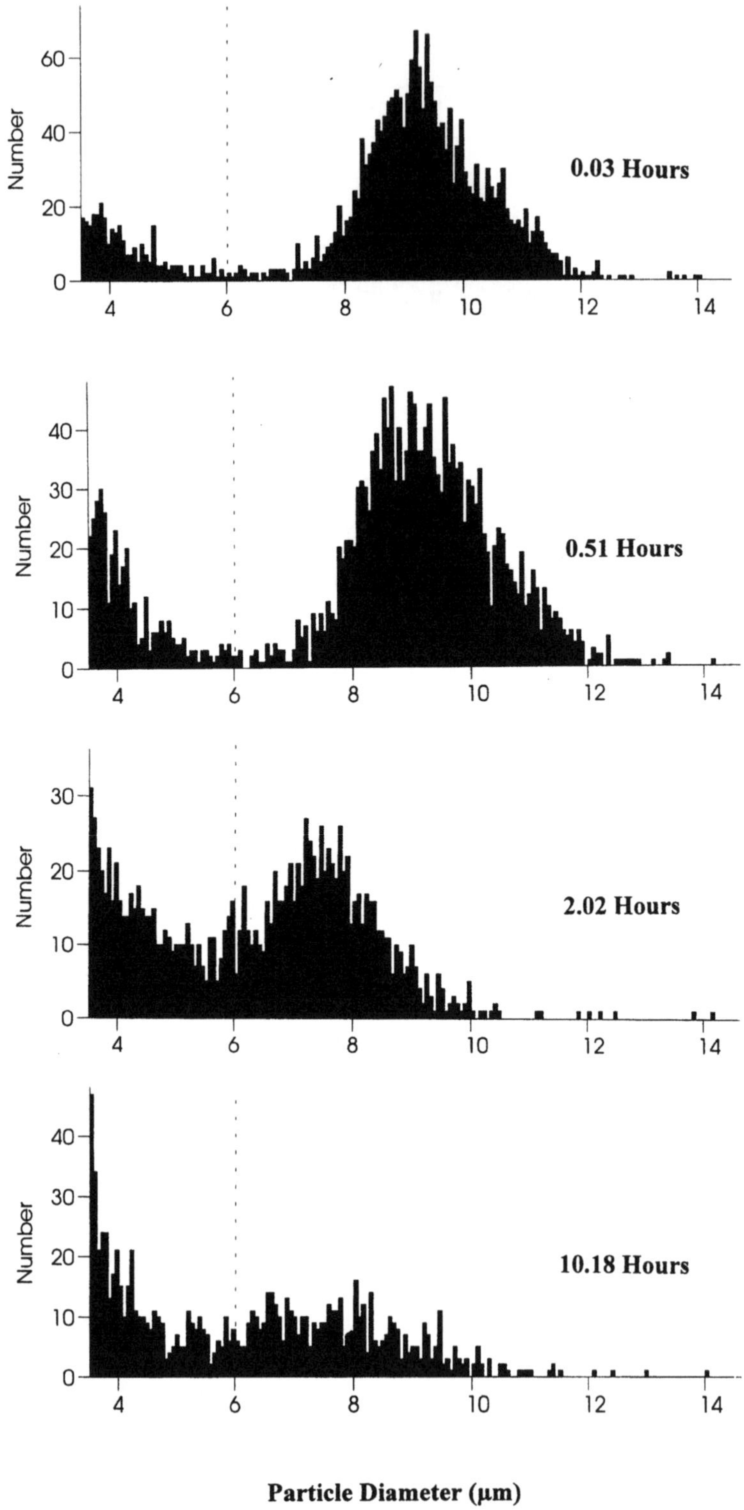

FIGURE 24. Cell size distribution after incubation with MS-1 extract of 3 day "spoiled" turbot. *Tetramitus* flagellates were incubated with a 50% extract, aliquots were withdrawn at indicated times, and cell size distributions were determined by counting in a Coulter Multisizer IIE.

This technology has applications for food inspection, site monitoring, toxic tracking, and early warning for use in **chemical weapons detection**. The threat of chemical terrorism is a growing concern with well-financed, hostile groups developing means to poison our drinking water supply.

3.7. Biological Plausibility

The issue of the biological plausibility of *Tetramitus* toxicity being linked to human disease experience has been raised by the N.Y. State Department of Health and the ATSDR. The following points argue in favor of possible linkage:

- The genes which regulate cell division in lower eukaryotes (yeast, protists, insects) have been conserved over hundreds of millions of years of evolution and bear close DNA sequence homology to mammalian and human genes.
- The functional and structural components of signal transduction path ways also display homology across a broad spectrum of eukarotic species. This includes MAPK (mitogen-activated protein kinase) signaling cascade (Seger and Krebs, 1995). The MAPK pathway provides a simple unifying explanation for the mechanism of action of most, if not all, nonnuclear oncogenes. It includes growth factor signaling and pathways initiated by phorbal esters, ionophors, and ligands for seven transmembrane receptors. Altered regulation caused by these factors often leads to cellular transformation or altered proliferation.
- Proto-oncogenes which function in the regulation of normal cell growth and/or differentiation may be converted to oncogenes by single step mutations. Activation of these oncogenes can transform a cell (Perantoni, 1998).
- *Tetramitus* flagellates exhibit a dose-dependent cytotoxic response to coal tar pitch (CTP) condensate and benzo[a]pyrene (BAP) used in a NIOSH mouse tumor study. The **tumorogenic response** of CTP and BAP in mice, on a per weight basis, were quite similar as were the **cytotoxic responses** in *Tetramitus* (Jaffe, 1995b).
- Preneoplastic cells exhibit **selective growth advantages** over normal cells in toxic environments (Perantoni, 1998).
- *Tetramitus* flagellates exhibit a dose-dependent cytotoxic response to 22 mutagens as well as chrysotile asbestos (Jaffe, 1995b) (a human, non-genotoxic, carcinogen).

An expanded summary of known proto-oncogenes and their homologies to human oncogenes is presented in the 2nd Edition Report to the ATSDR (Jaffe, 2000).

4. DISCUSSION

White, et al. (1996) have examined the genotoxicity of dichloromethane extracts from 50 effluent samples collected from 42 industries and have concluded: "Expression of potency values ... revealed that *effluent particulate matter* is on average, *almost four orders of magnitude more potent than aqueous filtrates.*" Generation of whole particle toxicity data using the *Tetramitus* Assay can reveal the extent to which particle toxicity is the major component in environmental matrices. The limited observations of particle toxicity in the N.Y.C. Reservoirs underscores the need to examine this issue. Scant data exists on risk assessment of toxic particles in drinking water. Toxic particle interaction with epithelial linings may change exposure parameters which, in turn, would impact on risk assessment estimates.

The results presented in this report demonstrate the statistical reliability of whole water toxicity testing using the *Tetramitus* Assay as well as validation of the particle and XAD concentrate methods. A study released by the General Accounting Office found that 45 states did not evaluate all components of their public water systems recommended by the Federal Environmental Protection Agency. The study concluded that federal financing for state water quality programs was a fraction of the amount required (Hinds, 1993). Utilization of the *Tetramitus* Assay would resolve these cost issues and provide a monitoring instrument for small drinking water purveyors who are granted regulatory concessions because of their size. The creation of a drinking water and surface water *Tetramitus* Toxicity database could be correlated to disease experience and would suggest further studies relating toxicity to disease.

5. ACKNOWLEDGMENTS

Funding for this project was provided by grants from the Beverly B. Perry Foundation, the West Islip Breast Cancer Coalition, and generous equity investment in ETL by Irving Weisman and Dr. Robert Jaros. Equipment for this study was provided by Beckman Coulter, Labconco, Brinkman Industries, New Brunswick Scientific Company, and Bellco Glass. Technical assistance was provided by Michael R. Jaffe and water sample collection was assisted by Valerie Josephson (Mt. Sinai School of Medicine), Virginia Regnante and Freddie O'Connor (West Islip Breast Cancer Coalition), and Mark Sullivan (Riverkeepers). ETL's website and e-mail at info @envirolab.com was managed by Christopher Norman and James Brooks. Legal consultation was provided by Frederick Mandel and Dr. Marina Larson (Patent and Licensing). A major portion of this manuscript was included in a report submitted to the ATSDR (Jaffe, 1999); the writing of this report was funded by this agency.

6. REFERENCES

- Clark, C.G., and G.A.M. Cross (1988) Circular Ribosomal RNA Genes Are A General Feature of Schizopyrenid Amoebae. J. Protozool. *35*: 326–329.
- Fulton, C. (1970) Amebo-flagellates as research partners: the laboratory biology of *Naegleria* and *Tetramitus*. Methods Cell Physiol. *4*: 341–476.
- Hinds, Michael deCourcy (1993) Survey Finds Flaws in States' Water Inspections. N.Y. Times, April 15 p.A14.

** Italics not in original text.

- Jaffe, R.L. (1995a) Rapid Assay of Cytotoxicity using *Tetramitus* Flagellates. Toxicology and Industrial Health *11*: 543–558.
- Jaffe, R.L. (1995b) Detection of Cytotoxic Agents using *Tetramitus Rostratus*. US Patent Number 5, 387, 508 (can be accessed @ ***http:/www.envirolab.com***).
- Jaffe, R.L. (1999) Utility and Reliability of the *Tetramitus* Assay as a Monitoring Instrument for Toxicity Reduction and Risk Assessment. Report submitted to the Agency for Toxic Substances and Disease Registry, April 16, 1999.
- Jaffe, R.L. (2000) 2nd Edition, Report to ATSDR, August 24, 2000. ***http:/www.envirolab.com***.
- Patterson, K.S., B.W. Lykins Jr., and S.D. Richardson (1995) Mutagenicity of drinking water following disinfection. J. Water SRT-Aqua. *44*: 1–9.
- Perantoni, A.O. Cancer-associated genes in McKinnell, R.G., R.E. Parchment, A.O. Perantoni, and G.B. Pierce (1998) The Biological Basis of Cancer, Cambridge University Press, 133–161.
- Sagan, L. (1967) On The Origin of Mitosing Cells. J. Theoret. Biol. *14*: 224–274.
- Seger, R., and E.G. Krebs (1995) The MAPK signaling cascade. FASEB J. *9*: 726–735.
- Steen, R., D.K. Jemiolo, R.M. Skinner, J.J. Dunn, and A.E. Dahlberg (1986) Expression of Plasmid-Coded Mutant Ribosomal RNA in *E. coli*: Choice of Plasmid Vectors and Gene Expression Systems. Progress in Nucleic Acid Research and Molecular Biology *33*: 1–18.
- Walsh, C. (1996) personal communication.
- White, P.A., J.B. Rasmussen, and C. Blaise (1996) Comparing the Presence, Potency and Potential Hazard of Genotoxins Extracted From A Broad Range of Industrial Effluents. Env. Molec. Mutagenesis, *27*: 116–139.

Epilogue

The Saga of *Tetramitus*

There once were amebae *T. rostratus*
who would not transform in most apparatus
but, give them the heat, in a manner quite neat,
and flagellates ye shall have *gratis*.
(Brandeis University—1969)

Then one day from their flask,
understanding their task
they exclaimed: Test us, we are sensitive to asbestos,
and if you inquire, we have millions to hire
to work for some scientific investos.
(Mount Sinai School of Medicine—1985)

T'was the year before the millenium and all through the land
T. rostratus flagellates, focused on particles, said let us expand.
To our brothers and sisters please honor this request
if ye are larger or smaller please help us to test-
different size particles in water, air and in soil,
Come *Bodo*, *Cercomonas*, *Chilodinella*, et al.
And they now can be heard chanting their call
Happy monitoring and good regulation
in our new protist-testing mall.
(Environmental Toxicology Laboratory—1999)

22

THE USE OF AQUATIC INVERTEBRATE TOXICITY TESTS AND INVERTEBRATE ENZYME BIOMARKERS TO ASSESS TOXICITY IN THE STATES OF AGUASCALIENTES AND JALISCO, MEXICO

ROBERTO RICO-MARTÍNEZ,
CÉSAR ALBERTO VELÁZQUEZ-ROJAS,
IGNACIO ALEJANDRO PÉREZ-LEGASPI, AND
GUSTAVO EMILIO SANTOS-MEDRANO

Universidad Autónoma de Aguascalientes, Department of Chemistry, Av. Universidad 940, Aguascalientes, Ags., C.P. 20100, Mexico

1. INTRODUCTION

This[1] paper describes a series of techniques using several aquatic invertebrates as indicators of adverse effects in natural and man-made aquatic environments. The information obtained in these techniques allowed the monitoring of the municipal drinking water system of the city of Aguascalientes, Mexico and the detection of toxicity in streams suspected of dumping of agricultural and industrial discharges in the states of Aguascalientes and Jalisco, in Mexico.

Aquatic organisms are exposed to a great variety of factors that cause them adverse effects. One of these factors is the toxicity due to anthropogenic activities (Snell et al., 1993). To characterize the toxicity of running water and reservoirs, we

---

[1] Abbrevations: DI = Deionized water; LC50 = Lethal Concentration where 50% of tests organisms die; LOEC = Lowest Observable Effect Concentration; NOEC = No Observable Effect Concentration.

Biomonitors and Biomarkers as Indicators of Environmental Change 2, Edited by Butterworth *et al.*
Kluwer Academic/Plenum Publishers, New York, 2000

need to determine the intensity and duration of toxic exposure to a series of test organisms (Rand and Petrocelli, 1985).

The use of aquatic invertebrates to detect the effect of toxic substances in an aquatic ecosystem is validated for the importance of these organisms as primary consumers in these ecosystems, as well as the great sensitivity of these organisms to a broad variety of toxic substances. Such sensibility, especially in the case of rotifers is higher when compared with other species traditionally used in aquatic toxicology (Burbank and Snell, 1994). Although more than one hundred species of aquatic invertebrates have been used for toxicity testing, only 39 have some kind of approval, and *Daphnia magna* has been specifically mentioned as approved by official agencies of several developed countries (Hueck-van der Plas, 1978).

The State of Aguascalientes, like the rest of Mexico is not indifferent to problems related to toxicity (Limón et al., 1989; Flores-Tena and Silva-Briano, 1995). Aguascalientes is one of the provinces, where zooplankton has been well-characterized (Silva-Briano, 1992; Rico-Martínez and Silva-Briano, 1993; Dodson and Silva-Briano, 1996). However, this knowledge has been rarely put to use in studies of aquatic toxicology or general ecology. The city of Aguascalientes like most cities of Mexico, lacks a specific water toxicity monitoring network that uses toxicological tests to identify adverse effects on aquatic communities. The actual monitoring network relies on microbiological, chemical, and physical parameters that do not measure directly the acute or chronic effect on the communities inhabiting these water reservoirs. This network based on these parameters only determines if the parameters found fulfill with the values allowed according to the official ecological criteria (Criterios Ecológicos de Calidad del Agua, Diario Oficial de la Federación, December 13th, 1989). However, even records of precise and accurate chemical concentrations in a particular ecosystem are not enough to protect the biota, for many mechanisms of toxicity are poorly understood (Baudo, 1987). For this reason, toxicity tests are preferred as tools for monitoring. Besides, toxicity tests once they are established are less expensive and allow us to make a better assessment of the intensity and duration of toxic exposure, thanks to the development of protocols like the Toxicity Identification Evaluation (TIE) which helps in the determination of the precise origin and identity of the toxic substance responsable for toxicity in a given area (Snell et al., 1993).

The present contribution includes a one year report on the use of *D. magna* and other cladoceran species acute toxicity tests and *L. quadridentata* biomarker tests to monitor acute and chronic toxicity in the municipal drinking water system of the city of Aguascalientes, and a few streams in the states of Aguascalientes and Jalisco, Mexico.

2. MATERIALS AND METHODS

2.1. *Daphnia magna* Acute Test

The city of Aguascalientes has a population of 582,628 inhabitants and it is located at an altitude of 1870 m (INEGI, 1995). Heavy extraction of water from wells had caused severe problems in the city. Geological faults and the necessity of deep digging to obtain water (due to aquifer overdraft) are potential factors than cause deterioration of groundwater, which is the main source of water for the city of Aguascalientes. Although the *D. magna* acute tests has been traditionally used to analyse wastewater discharges, the possibility that water in the municipal system could be contaminated justified our project. A total of 136 sampling sites in the city of Aguascalientes (which include wells, drinking water tanks, water pipes from homes, etc.) were established for a one year period from June 1997 to June 1998 (Figure 1) to monitor the presence of acute toxicity using the *D. magna* acute toxicity protocol detailed in the Mexican Norm (NOM-074-ECOL-1994). Detailed descriptions of each sampling site have been already published (Rico-Martínez et al., 1997). Additional data from each sampling point regarding dissolved oxygen, BOD, COD, pH, temperature, conductivity, calcium, magnesium, total alkalinity, chlorine, total solids, total dissolved solids, and total coliforms, were available to us from the municipal water board (CAPAMA). Samples came from surface water (from homes, water pipes, and tanks), and from wells. Samples were collected in bottles that were sequentially cleaned with detergent, tap water, nitric acid, DI water (Labconco Water Pro System), acetone and EPA medium 24–48 h prior to sample collection. Samples coming out of surface water were aereated for 24 h to eliminate chlorine.The *D. magna* acute test technique consisted in the 48 hr exposure of 24 hr-old neonates of *D. magna* to a control and five different toxicant concentrations determined through a range toxicity test (Rand and Petrocelli, 1985). In the control, ten neonates are placed in each of three vials with 100 ml of EPA medium (USEPA, 1985). The same is done for each treatment except that besides the EPA medium the vials contain the corresponding toxic concentration. No food is given to the neonates during this 48 hr period in an incubator with a 18:6 L:D schedule, light intensity was kept between 400 to 1,000 lux as determined by an illuminometer (Kyoritan Electrical Instruments), and temperature was kept at 20 ± 1 °C as determined by a microcomputer thermometer (Hanna Instruments). Lack of movement was the endpoint used to score the animals. We obtained LC_{50} values for cadmium, chromium, and toluene using chemicals of reference of the highest purity available (Sigma) in five replicate batches of three cladoceran species; *Daphnia magna*, *D. pulex*, and *Simocephalus vetulus*. Only nominal concentrations were used as no actual determination of chemicals were performed on the different toxicant concentrations. The only modification to this protocol consisted in the use of 250 ml polystyrene vessels instead of the 150 ml glass vessels recommended in the protocol. We have made several comparisons between both kind of containers for cadmium and we found no signifficant differences ($P > 0.05$, $n = 5$).

Statistical analyses were performed with the program Statistica 4.5 (Statsof Inc., 1993), and LC_{50} values were determined using the French program DL-50 (S.B.I.-I.R.C.T., Montpellier, 1987). This program calculates the LC_{50} values and test the linearity of the adjustment by using the chi-square statistic ($P < 0.05$).

2.2. Esterase and Phospholipase A2 Biomarkers in the Rotifer *Lecane quadridentata*

We used the technique developed by Burbank and Snell (1994) with slight modifications. This technique consists in the measurement of the *in vivo* inhibition

FIGURE 1. Localization of the 136 sampling sites in the city of Aguascalientes (scale 1:5,000). Dark circles numerated from 1 to 136 represent wells that provide water to the city. Dark cicles surronded by an additional black and white rings represent water tanks. Sanples from residential homes and water pipes around a well were also taken at each sampling date.

of the enzymes esterases and phospholipase A2 in the freshwater littoral rotifer *Lecane quadridentata*. We used neonate females hatching from asexual eggs instead of neonates hatching from cysts as in Burbank and Snell (1994). Hatching was performed at 25 °C in EPA medium, 18 h prior to the start of the tests. Parthenogenetic females were routinely grown in EPA medium and fed the green alga *Nannochloris oculata* originally purchased from UTEX Collection. A complete life-table analysis of *L. quadridentata* that includes culture techniques have been already developed at our lab (Pérez-Legaspi and Rico-Martínez, 1998). The reference chemicals used (cadmium, copper, and lead) were diluted in EPA medium. For each reference chemical we performed a range toxicity test and then a toxicity test that included five toxicity concentrations, and a negative control was performed.

The neonate rotifers were exposed to the correspondent fluorogenic substrates to determine *in vivo* enzyme activity. The esterase substrate is cFDAam (5-carboxyfluorescein diacetate acetoxymethylester), and the phospholipase A2 substrate is PLA2 (2-[6-(7-nitrobenz-2-oxa-1,3-diazol-4-yl) amino(-hexanoyl-1-hexadecanoyl-sn-glycero-3-phospholcholine). Both substrates were from Molecular Probes, Inc. (Oregon, USA). The test started with addition of 750 µL per well of the chemical of reference diluted in EPA medium, or EPA medium alone (negative control) to a 24-well plate (Corning Inc.). Thirty individuals were transferred for each well using a plastic micropipette. The transfer was done under a dissecting scope with up to 200X (Olympus). The individuals were then exposed to the different toxicant concentrations in the dark at 24 °C for 30 min. After toxicant exposure, 2 µL of cFDA and 1 µL or PLA2 were added to each well and the plate was gently shaken. The final concentration of cFDA in each well was 5.0 µM, and the PLA2 concentration was 17.2 µM. Then, the plate was incubated in the dark at a 24 °C for 15 min. Then, 150 µL of pentobarbital sodium 630 mg/L, was added to each well to anesthesize the rotifers. Two strands of tape are placed at each edge of a slide to allow support of the coverslide without damaging the rotifers. Rotifers were transfered to the slide and covered with the coverslide and observed under the compound microscope at 200X. Fluorescence was quantified with a emission spectrum of 450–490 nm and with a 515 nm filter. The intensity of the fluorescence was quantified using the image analysis software Harmony (Bioscan Inc.). Statistical analysis was done with the program Statistica using ANOVA and Duncan's multiple range tests to determine the points NOEC, and LOEC.

2.3. Ingestion Rate as a Biomarker of Toxicity in *Daphnia magna*

We used the protocol of Juchkelka and Snell (1995) with slight modifications. This protocol uses suspension feeding of this cladoceran species as endpoint for toxicity assessment. We used synthetic dyed beads of 1.9 µm of diameter (Bangs Laboratories, USA) instead of fluorescein labelled beads as in Juchelka and Snell (1995). We measured the percentage of the gut that is cover by beads as an index of ingestion rate, instead of measuring the intensity of fluorescence. Copper was

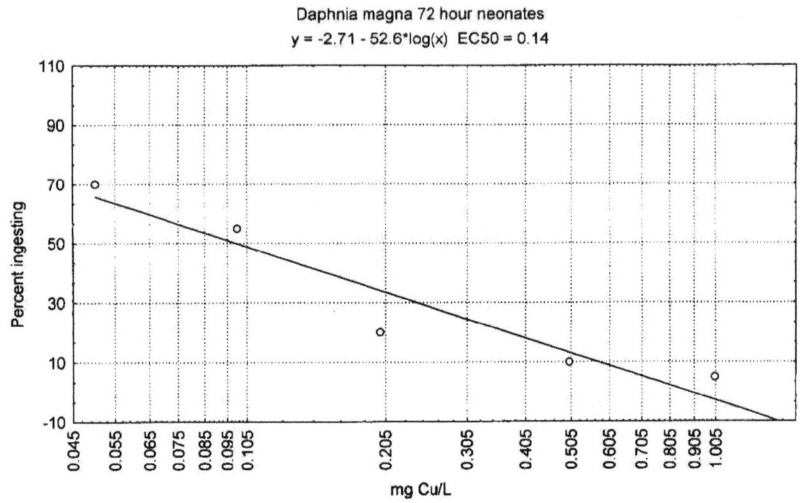

FIGURE 2. Calculation of the copper EC_{50} ingestion rate value for 72-hour neonates of *D. magna*. Cooper concentration was plotted in a long scale while the percentage of dyed bead ingestion was plotted in an arimethic scale. The formula of the regression and the EC_{50} inhibition of the ingestion rate value are also indicated in the figure.

used to standarize the technique and EC_{50}, NOEC and LOEC values were obtained for this chemical of reference, for 72-h *D. magna* neonates (Figure 2).

We are currently developing the same technique using 1.76 μm and 4.20 μm in diameter fluorescence-labelled polystyrene microspheres from Bangs Laboratories, instead of the dyed beads. This change in protocol would allow us to automate the technique in the near future.

3. RESULTS

3.1. Monitoring of the Drinking Water System of the City of Aguascalientes

In this one year period we have completed for each sampling site at least two rounds of sampling in 127 of the 136 proposed sampling sites. Nine sampling sites (14, 21, 29, 32, 70, 80, 100, 122 and 128) are wells that have not been monitored due to several reasons; seven wells (21, 29, 32, 70, 80, 100, and 128) have cancelled operations, one (122) is a private well and the municipal water authority (CAPAMA) is not allowed to monitor this well. The remaining well (14) is not actually in operation due to lack of pumping equipment. Out of these 127 sampling sites only three times (sampling sites 105, 108 and 119) acute toxicity was suspected. All three sampling sites correspond to wells. Sampling site 105 is a well at the Ciudad Industrial (the area of the city where factories and heavy industries are located). The sampling crew suspected of a malfunction of the pump (which is operated by oil) and the

TABLE 1. *Daphnia magna*, *D. pulex* and *Simocephalus vetulus* LC50's. Comparison with data published in the literature for related species. All values are in mg/l.

Compound	*D. magna* LC_{50} value	*D. pulex* LC_{50} value	*S. vetulus* LC_{50} value	Values reported in literature	References
Cadmium	0.19	0.18	0.03	0.13 for *D. magna*	Lewis and Weber, 1985
Chromium	3.35	1.76	3.37		
Toluene	58.97	36.59	42.50		

entrance of oil in the sample collected that day (August 28th, 1997). Later, several samples were collected a few days after the first sample at the well and surrounding areas, and later (two times in 1998), and in all cases no acute toxicity was found. In the case of sampling site 108, this is a well at Fraccionamiento Bosques del Prado that it is not in operation right now. CAPAMA has determined that several criteria including a *D. magna* acute toxicity test have to be met before this well can be allowed to operate. In the case of sampling site 119, this site corresponds to a well at Fraccionamiento La Herradura. We suspected entrance of chlorine to this sample. Aereation was provided to this sample to eliminate the chlorine and no toxicity was found. The sampling crew took a sample of this site one week later (July 18th, 1997) and we have sampled this well twice in 1998 and in all three samples no acute toxicity was found.

Results of the *D. magna*, *D. pulex*, and *Simocephalus vetulus* LC_{50} values we obtained with cadmium, chromium and toluene are presented in Table 1. These tests have been used as our positive toxicity controls. The use of other cladocerans species is recommended since *D. magna* is an European species introduced in North America, that has never been reported as a natural inhabitant of Mexican reservoirs. Both *D. pulex* and *S. vetulus* were collected from the pond at the Botanical Garden of the Universidad Autónoma de Aguascalientes, and have been cultured in bioclimatic chambers with the same conditions than *D. magna* for more than two years.

3.2. Monitoring of Los Gavilanes River

This small river located in Jalisco near the border of the states of Aguascalientes and Jalisco (for localization of this river see Rico-Martínez and Silva-Briano, 1993) was sampled two times (November 14th and 29th, 1998) at three sampling sites at the community of Los Gavilanes, Jalisco. We used *D. magna* acute tests to identify the presence of toxicity in elutriates and surface water. On both dates the tests identified the station where the discharge of agricultural products were suspected. Only a concentrated sample of the elutriate was recorded as having acute toxicity on both dates. A series of dilutions of 50, 25, 12.5 and 6.25% of this elutriate showed no acute toxicity. The other two sampling sites located at 100 and 200 meters downstream showed no acute toxicity at both dates. Dissolved oxygen values were higher than 4 ppm in all cases. Samples of surface water showed no acute toxicity, and more than ten species of planktonic invertebrates and three benthonic rotifers have been reported in three different points in three different dates in this river (Rico-Martínez and Silva-Briano, 1993; Rico-Martínez unpublished data), and therefore these facts might suggest that toxicity is present in the sediments. Agricultural products are suspected of causing the toxicity as no known industrial discharges are going into the river near the sampling area.

3.3. Monitoring of Industrial Discharges in the City of Aguascalientes

On March 21st, 1998, we took samples of industrial discharges from Parque Industrial del Valle de Aguascalientes (PIVA), Parque Industrial de San Francisco de Los Romo, and from the Ciudad Industrial of Aguascalientes (near sampling sites 103, 104, and 105 of Figure 1), the three sites near the city where heavy industries are located. These discharges eventually end up at the San Pedro River. Samples were collected in clean bottles according to the Mexican Norm for performing acute tests (NOM-074-ECOL-1994) and *D. magna* acute tests and *D. magna* ingestion tests (Juchelka and Snell, 1995) were performed to assess the presence of toxicity in each sample. The sample was fractioned as part of a Toxicity Identification Evaluation (TIE) sheme to obtain some data about the characteristics of the toxicants present in each sample. Columns of amberlite, C18, XAD-2, and zeolite, were used to characterize reductions in toxicity due to binding of toxicants to each column (Snell et al., 1993). Acute toxicity was found in the three samples from industrial discharges even with dilutions of 6.25% of the sample. Toxicity was then investigated using the fractionation scheme with the columns enumerated above and the ingestion rate as endpoint. Figure 2 shows the effect of copper on the ingestion rate of 72-h *D. magna* neonates. The column of XAD-2 removed substantial amounts of toxicity from the indutrial discharges of PIVA and Ciudad Industrial (80% and 100%, respectively). The amberlite column removed all the toxicity from the Parque Industrial de San Francisco de Los Romo industrial discharge.

3.4. Use of Enzyme Biomarkers on the Rotifer *Lecane quadridentata*

Figure 3 shows a photograph of neonate females of *L. quadridentata* degrading the fluorescein labelled-substrate of estereases. Important decreases in fluorescence can be seen after comparing the control (3A) and neonates exposed for 30 min to 0.0005 (3B) and 0.001 mg/L of copper (3C). The results of the esterase and phospholipase A2 biomarker tests are shown in Table 2. Only NOEC and LOEC values are presented here for three metals; cadmium, copper, and lead for both enzymes. These tests are currently used to assess the presence of toxicity in rivers near the city of Tamazunchale, S.L.P., Mexico (Rico-Martínez unpublished data).

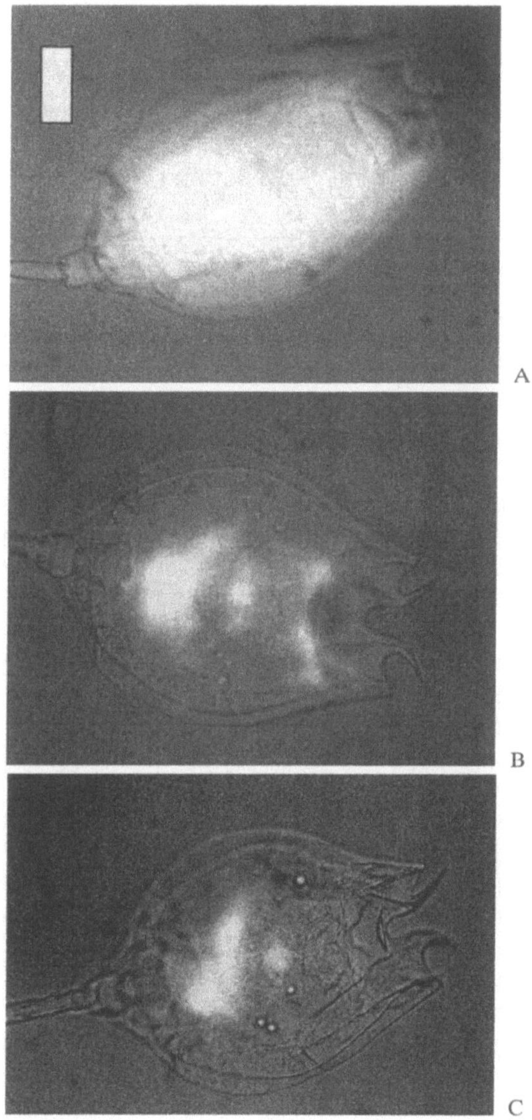

FIGURE 3. Photographs of *Lecane quadridentata* neonates degrading the esterase substrate. A) A control organism showing abundant fluorescence as large amounts of the esterase substrate are being degraded. B) A organism exposed to 0.0005 mg/L of copper for 30 min shows a significant decrease in esterase activity. C) This organism was exposed to 0.001 mg/L of copper for 30 min. The white bar at the bottom of 3A is the scale used for all photographs and represents 50 μm.

TABLE 2. Results of the esterase and phospholipase A2 "*In vivo*" biomarker tests with the freshwater littoral rotifer *Lecane quadridentata*. All values are in mg/l.

Compound	Esterase NOEC	Esterase LOEC	PLA2 NOEC	PLA2 LOEC	Values reported in literature	References
Cadmium	0.100	0.500	0.075	0.100	0.025* 0.050**	Burbank and Snell, 1994
Copper	0.00005	0.00010	0.00050	0.00100	0.025* 0.050**	Burbank and Snell, 1994
Lead	0.010	0.025	0.250	0.500		

* Esterase NOEC for *Brachionus calyciflorus*.
** Phospholipase A2 NOEC for *Brachionus calyciflorus*.

4. DISCUSSION AND CONCLUSION

The use of acute toxicity tests to monitor the drinking water system of the city of Aguascalientes has revealed an interesting fact. Most of the urban wells, water pipes, and water concetration tanks are free of acute toxicity. Unfortunately, the same can not be said about rural wells. As part of the cooperation between our lab and CAPAMA, we analyzed samples coming out of rural wells around the city of Aguascalientes, and we found acute toxicity in about 5% of the wells we examined. Although toxicity tests are used in developed countries mostly to set rules for discharges (Buikema et al., 1982), the use of the tests to monitor the drinking water system of the city of Aguascalientes is justified by the fact that Aguascalientes is a city that presentes several potential toxicity problems aggravated by the depletion of groundwater (CAPAMA, 1998, pers. com.).

Our LC_{50} values for *D. magna* for cadmium are found in the range reported by several authors (Lewis and Weber, 1985; Baudo, 1987). Our results show that for three toxicants tested *D. magna* is not the most sensitive species, in fact it is the less sensitive species in two cases (see Table 1). This result is aggravated by the fact that *D. magna* is not present in any reservoir of Mexico. Although the *D. magna* test is perhaps the most standarized and accepted acute toxicity test worldwide, in Mexico some precautions have to be taken to use this test in toxicity assessment. We recommend performing additional tests using native species or cosmopolitan species inhabiting Mexican reservoirs to determine the ecological relevance of the *D. magna* tests for Mexican reservoirs. The combination of both types of tests is suggested to account for the possibility that native species may have become less sensitive to local toxicants.

The use of *D. magna* acute tests has allowed us to assess toxicity from both agricultural (Los Gavilanes River) and industrial sources (industrial discharges near the city of Aguascalientes). The fractionation scheme used here demonstrated that most of the toxicity in the industrial discharges is due to organics since the XAD-2 column has a strong affinity for these compounds, and passage through this column removed most of the toxicity at two sites. The amberlite column removed cations that appear to account for all the toxicity found at Parque Industrial de San Francisco de Los Romo. Further studies combining fractionation schemes and chemical identification are necessary to completely assess the identity and toxicity of the three industrial discharges. However, this preliminary assessment of these discharges indicates that strong acute toxicity is being released into the municipal sewage system and that these discharges eventually end up at the San Pedro River or other streams near the city of Aguascalientes.

The results of the quantification of the enzymes, esterases and phospholipase A2, *in L. quadridentata* showed that esterases are more sensitive than the phospholipase A2 for copper and lead, but not for cadmium. In the case of copper our results showed that *L. quadridentata* had a remarkably higher sensitivity than *B. calyciflorus* (a zooplanktonic rotifer species), for both enzymes (Table 2). However, *L. quadridentata* is less sensitive to cadmiun than *B. calyciflorus*. These results suggest that there are important differences among rotifers in their sensitivity to metals, and these differences can be related to the types of habitats that a particular species inhabits. In this case *L. quadridentata*, a littoral species, seems a better model for sediment toxicity than *B. calyciflorus* as the primary consumer in a potential battery of organisms used to assess toxicity in sediments.

5. ACKNOWLEDGMENTS

We thank Fernando Martínez-Jeronimo of ENCB-IPN for donating a few organisms of the EPA strain of *D. magna*. We also thank CAPAMA for providing most of the samples and additional information on biological, chemical and physical parameters of the samples we analyzed. We thank Comisión Nacional del Agua Gerencia Estatal Aguascalientes for providing samples of the three industrial discharges near the city of Aguascalientes used in this work. We thank Jesús Gerardo and Victor Eduardo Estrada-Aguilera for Collecting and Processing of samples from the Gavilanes River. This work was supported by grant RN-20/96 of CONACyT-SIHGO to R.R.

6. REFERENCES

Baudo, R. 1987. Ecotoxicological testing with *Daphnia*. pp. 461–482. In: R.H. Peters, and R. de Bernardi (Eds.) *Daphnia*, Mem. Ist. Ital. Idrobiol., 45. Consiglio Nazionale delle Richerche, Verbania Pallanza, Italy.

Buikema, A.L., Jr., B.R. Niederlehner, and J. Cairns. 1982. Biological monitoring. Part IV—Toxicity Testing. Water Res. 16: 239–262.

Burbank, S., and T.W. Snell. 1994. Rapid toxicity assessment using esterase biomarkers in *Brachionus calyciflorus* (Rotifera). Environmental Toxicology and Water Quality 9: 171–178.

Dodson, S.I., and M. Silva-Briano. 1996. Crustacean zooplankton species richness and associations in reservoirs and ponds of Aguascalientes State, Mexico. Hydrobiologia 325: 163–172.

Flores-Tena, F.J., and M. Silva-Briano. 1995. A note on El Niagara, a polluted reservoir in Aguascalientes, Mexico. Hydrobiologia 308: 235–241.

Hueck-van der Plas, E.H. 1978. Experiences with an inventory of ecological tests based on an enquiry by the OECD chemicals group. pp. 63–73. In: Tests for the ecological effects of chemicals. Proc. Research Seminar, 7–9. Berlin. Erich Schmidt Verlag, Berlin.

Juchelka, C.M., and T.W. Snell. 1995. Rapid toxicity assessment using ingestion rate of cladocerans and ciliates. Arch. Environ. Contam. Toxicol. 28: 508–512.

Lewis, P.A., and C.I. Weber. 1985. A study of the reliability of *Daphnia* acute toxicity tests. pp. 73–86. In: R.D. Cardwell, R. Purdy, and R.C. Bahner (eds.). Aquatic toxicology and hazzard assessment seventh symposium. ASTM STP 854, American Society for Testing and Materials. Philadelphia, USA.

Limón, J.G.M., O.T. Lind, D.S. Vodopich, R. Doyle, and B.G. Trotter. 1989. Long- and short-term variation in the physical and chemical limnology of a large, shallow, turbid tropical lake (Lake Chapala, Mexico). Arch. Hydrobiol. Suppl. 83(1): 57–81.

Pérez-Legaspi, I.A., and R. Rico-Martínez. 1998. Effect of temperature and food concentration in two species of littoral rotifers. Hydrobiologia 387/388: 341–348.

Rand, G.M., and S.R. Petrocelli. 1985. Fundamentals of aquatic toxicology. Washington Hemisphere Publishing Corporation. Washington D.C. 670 pp.

Rico-Martínez, R., and M., Silva-Briano. 1993. Contribution to the knowledge of the Rotifera of Mexico. Hydrobiologia 255/256: 467–474.

Rico-Martínez, R., A.M. Jímenez-Rodríguez, C.A. Velázquez-Rojas, and I.A. Pérez-Legaspi. 1997. Desarrollo de bioensayos toxicológicos y su aplicación en programas de monitoreo de la calidad de pozos y tomas de agua del Municipio de Aguascalientes. Memoria del Cuarto Simposio Estatal: La investigación y el desarrollo tecnológico en Aguascalientes, Aguascalientes, Mexico: 150–154.

Snell, T.W., D. Dusenbery, L. Dunn, and N. Walls. 1993. Biomarkers for managing water resources. Georgia Institute of Technology. Environmental Resources Center. ERC 02-93 Publication, Atlanta, Georgia, USA. 43 pp.

Silva-Briano, M. 1992. Preliminary study of the zooplankton of Mexico. End of course report of the International training course "Lake Zooplankton: A tool in Lake Management." State University of Ghent. Ghent, Belgium: 12 pp.

U.S. Environmental Protection Agency. 1985. Methods for measuring the acute toxicity of effluents to freshwater and marine organisms. EPA-600/4-85-013, USA. Environmental Protection Agency, Washington D.C., USA.

BIOMONITORING OF PESTICIDES BY PLANT METABOLISM: AN ASSAY BASED ON THE INDUCTION OF SISTER-CHROMATID EXCHANGES IN HUMAN LYMPHOCYTE CULTURES BY PROMUTAGEN ACTIVATION OF *Vicia faba*

SANDRA GÓMEZ-ARROYO,[1]
MARÍA ELENA CALDERÓN-SEGURA,[1] AND
RAFAEL VILLALOBOS-PIETRINI[2]

Laboratorios de Citogenética[1] y Mutagénesis[2] Ambientales, Centro de Ciencias de la Atmósfera Universidad Nacional Autonóma de México, Ciudad Universitaria, Coyoacán 04510 D.F., México

1. INTRODUCTION

Experimental evidence has shown that some chemical agents, involving pesticides, induce DNA impairment. For many years it was not known that plants have enzymatic systems suitable to perform metabolic transformation, however, it has recently been shown that several compounds including pesticides can be activated by plants.

Plewa (1978) developed a method to evaluate mutagenic responses of several pesticides in the presence of a plant enzymatic system and coined the term "plant activation" to define the process by which a promutagen was activated into a

Biomonitors and Biomarkers as Indicators of Environmental Change 2, Edited by Butterworth *et al.*
Kluwer Academic/Plenum Publishers, New York, 2000

mutagen by some kind of biological action of the plant system (Plewa and Gentile, 1982). The chemical agents that are not mutagens by themselves, but need previous metabolic activation (of plants or animals) to produce such effect are called promutagens or indirect-acting mutagens (Plewa, 1978; Plewa and Gentile, 1982).

The formation of toxic metabolic products in plants depends upon the kind of compound, the physicochemical properties, concentrations, metabolic routes, etc. Toxicity can be increased, decreased or not affected after activation (Zilkah and Gressel, 1977; Hess, 1985, 1987; Sterling, 1994).

There are two general processes to study the plant promutagen activation, *in vivo* and *in vitro*. In the former, the chemical agent is introduced to a whole plant in the same conditions of the agricultural fields and in the last, the agent is incubated with a plant tissue homogenate or with homogenates from plant cell cultures, with or without cofactors, and finally the biochemical and genetic analysis of the metabolites from the plant tissues or cell cultures are carried out (Plewa, 1978; Wildeman and Nazar, 1982; Plewa et al., 1984, 1988). In both, the action of plant metabolism on a potential mutagen can be quickly assayed with an indicator microorganism as *Salmonella typhimurium* (Plewa et al., 1988), *Escherichia coli, Aspergillus nidulans* (Scott et al., 1978), *Saccharomyces cerevisae* (Loprieno and Adler, 1980), etc. or mammalian cells (Takehisa et al., 1982; Takehisa and Kanaya, 1983) in order to determine its genetic activity and to evaluate their genotoxic properties. With these procedures several pesticides can be activated by plant enzymatic systems to specific mutagenic compounds, such as aldrin, diallat, 2-aminofluorene, 2-acetylaminofluorene, benzidine, maleic hydrazide, phenyl urea, etc. using extracts and/or homogenates of seedlings of soybean, corn and barley, roots of broad bean and onion, tulip bulbs, and tobacco callus, like metabolizers (Plewa, 1978; Scott et al., 1978; Constantin and Nilan, 1982; Takehisa et al., 1982; Wildeman and Nazar, 1982; Takehisa and Kanaya, 1983; Takehisa, 1986; Higashi et al., 1985; Higashi, 1988; Seo et al., 1993).

These useful methods of plant activation were based on the extraction of mutagenic agents activated by whole plants (Plewa, 1978), homogenates of plant tissues (Plewa and Gentile, 1982; Rasquinha et al., 1988) or microsomal fractions of plant cells (Gentile et al., 1986). The test in which the suspension culture of plant cells is employed as the activator system and microorganisms as indicators of genetic damage, was called plant cell/microbe coincubation assay (Plewa et al., 1988; Plewa and Wagner, 1993). *In vitro* studies used *Salmonella typhimurium* to evaluate the mutagenic damage or cell-free homogenates of several plants or the microsomal fraction that has the enzymatic activity needed for the metabolism (Higashi et al., 1981; Higashi, 1988; Rueff et al., 1996). The extracts have been obtained mainly from wheat (Lai et al., 1978; Rasquinha et al., 1988), corn (Plewa, 1978), *Tradescantia* (Scott et al., 1978), potato tubers (Loprieno and Adler, 1980), tobacco (Gentile et al., 1982, 1985) and *Persea americana* (Rueff et al., 1996).

1.1. Promutagen Activation by *Vicia faba*

Vicia faba is a very sensitive plant to the effect of pesticides (Gómez-Arroyo et al., 1992, 1995; Gómez-Arroyo and Villalobos-Pietrini, 1995) and it is considered metabolically active because it contains the enzymatic fraction S10 (Takehisa et al., 1982, 1988; Takehisa and Kanaya, 1983; Takehisa, 1986; Gómez-Arroyo et al., 1995).

Several compounds have been activated through *Vicia faba* and used as indicators of cytogenetic damage measured as SCE in CHO cells (Takehisa et al., 1982; Takehisa and Kanaya, 1983) and SCE in human lymphocyte cultures (Gómez-Arroyo et al., 1995).

Plant activation of promutagens was studied by Takehisa et al. (1988) using *Vicia faba* S10 (*in vitro* activation) and the extracts were prepared from promutagen-treated roots of *Vicia faba* (*in vivo* activation). The induction of sister chromatid exchanges in Chinese hamster ovary cells was used as endpoint to evaluate the cytogenetic effects of promutagens activated by *Vicia faba*. Cyclophosphamide and ethyl alcohol were activated both by *Vicia* S10 and by *Vicia* extracts, and their activation resulted in an increase in SCE. Benzo(a)pyrene, 2-aminofluorene, and maleic hydrazide were not activated. Aniline although activated, had no effect on the induction of SCEs and inhibited mitotic progression.

Gómez-Arroyo et al. (1995) carried out experiments in which metabolic activation was done by *Vicia faba* roots, and SCE in human lymphocytes *in vitro* was used to assess the cytogenetic damage. Several concentrations of the carbamate insecticide propoxur were applied for 4 h to the roots of *Vicia faba*. Extracts prepared from this treatment were added to the lymphocyte cultures and a significant increase of SCE frequencies with a concentration-response relationship could be detected. The lymphocyte proliferation kinetics and the proliferation rate index (PRI), were also analyzed, and no effect was observed, except at the highest concentration (2,000/ppm). This general behaviour was in agreement with the presence of an enzymatic system (S10 fraction) in *Vicia* roots capable of metabolizing or activating the propoxur. At the concentration mentioned, cell necrosis was produced in *Vicia*; therefore, this extract did not induce SCE in lymphocytes. However, lymphocyte proliferation kinetics was delayed and PRI was significantly decreased. Ethanol, a promutagen activated by this plant (Takehisa et al., 1982, 1988; Takehisa and Kanaya, 1983), was applied directly to the lymphocyte cultures and the response was negative. On the other hand, the extracts of roots treated with ethanol increased the SCE more than twice that of the negative control, but lymphocyte proliferation kinetics and PRI were not affected. Ethanol is metabolized to acetaldehyde by the action of the cytosolic enzyme alcohol dehydrogenase (ADH) that depends upon nicotinamide adenine dinucleotide (NAD) as cofactor. This reaction occurred mainly in mitochondria and cytoplasm. The acetaldehyde is the main SCE inductor in human lymphocyte *in vitro* in the presence of ADH and NAD (Obe and Ristow, 1970; Obe et al., 1979, 1986; Obe and Anderson, 1987).

1.2. Plant Metabolism

The pesticides transformation in higher plants enclose primary (phase I), secondary (phase II) and tertiary reactions (phase III). The former were mainly non synthetic processes such as oxidations, reductions and hydrolysis, while the second were synthetic or conjugation-forming metabolites with low or null phytotoxicity. Most of the synthesized compounds were highly polar and water soluble, had limited mobility and were predisposed to phase III enzymes. In the reactions of phase III, intermediates were primary or secondary conjugates of insoluble residuals in water, that could be incorporated into cell organels, cytoplasmic constituents or bound to lignin biopolymers, that were not phytotoxic but had an unknown identity such as the final fate and the toxic activity (Shimabukuro et al., 1981, 1982).

In plants, as well as in animals, many enzymes involved in the xenobiotic metabolism are localized in the endoplasmic reticulum (Menn, 1978), however plant enzymes were also detected in the cytoplasm and in the cell wall region (Sandermann, 1988). Besides enzymatic reactions observed in animals, plants also include secondary conjugation enzymes, forming water insoluble metabolites which were, in general, biologically inactive and might be deposited in specific compartments of the cells (Sandermann, 1988).

The reactions of the oxidative metabolism were catalyzed by mixed-function oxygenases (MFO) which insert an oxygen atom from O_2 to produce water. MFO was cytochrome P-450 (cyt P-450) with terminal oxidase activity detected in microsomes of more than 30 plant species (Higashi et al., 1983; Higashi, 1985). These microsomes were considered as cyt P-450 analogs from the rat liver cyt P-450 (Rich and Bendall, 1975; Higashi, 1985, 1988). Higher plant cyt P-450 has increased specificity to substrates, but is less heterogeneous than in mammals. This enzyme has been involved in the induction of secondary metabolites that were mutagenic and originated from a promutagen. That was the case for benzo(a)pyrene and 2-aminofluorene. Reduced phosphorylated nicotinamide dinucleotide (NADPH) was the best source of electrons to reduce cyt P-450, while in animals it was the nicotinamide dinucleotide (NAD) (Frear et al., 1969; Higashi et al., 1981).

Two types of peroxidases were involved in the reactions. The classic used H_2O_2 and the other used molecular oxygen (Lamoureux and Frear, 1979) and originated free radical intermediates that interacted with DNA and those that increased mutagenicity of some chemicals (Yamasaki, 1974; Lamoureux and Frear, 1979; Donh and Krieger, 1981; Plewa et al., 1991). The hydrolysis was catalyzed by some hydrolases, esterases, amidases, O-alkylhydrolases, etc., and represented a general degradation mechanism of xenobiotics in animals, microorganisms and plants (Menn, 1978; Shimabukuro et al., 1981, 1982).

Conjugation represented the reactions of endogenous substrates with the production of xenobiotic metabolism in plants, formed by the plant's phase I system. The final products of these reactions were glucosides (conjugation with glucose), glutathione, D-glucose malonic acid or amino acid (Shimabukuro et al., 1982; Lamoureux and Rusness, 1987) or with lignin (Sandermann et al., 1983) or pectines and other polyssaccharides (Langenbach et al., 1986). This enzymatic pathway also actived the plant promutagens (Shimabukuro et al., 1982).

In animals, as well as in plants, the last step of molecular biotransformation was conjugation (Higashi, 1988). The conjugates were easily excreted by animals, but as excretory systems do not exist in plants they were polymerized and/or incorporated into their structural components in such a way that the initial products, the reactive intermediates and/or active oxygen species and the final products could produce firstly damage to the plant itself, secondly conjugates and/or could be stored in the same plant until they were set free in the gastrointestinal tract or in the animal's organs when they were eaten. The last event could specifically occur as a consequence of pesticides added consumable plants (Sandermann, 1988).

The fact that plants were considered as healthy, fresh and innocuous food must be reviewed because they contain pesticide residuals, as well as its metabolites, stored in several structures and, therefore, become a potential risk when they are ingested by animals or humans as they might impair organs, cause physiological damage, mutations, cancer and death. Based on these facts, the interest to investigate problems produced by such chemical agents and their metabolites has been increased.

2. *In vivo* AND *In vitro* PROMUTAGEN ACTIVATION OF THIOCARBAMATE HERBICIDES BY *Vicia faba*

Thiocarbamate herbicides such as butylate and molinate had a wide range of applications in agricultural soils and specially on rice, corn and other cereals. Most of the carbamic herbicides at low concentrations disturb several biochemical process like photosynthesis, respiration, lipid biosynthesis (waxes, cutin and suberin) and also modify lipases, isoprenoids (in particular kaurene, a gibberellin precursor) and flavonoids (anthocyanins) in roots, stems and leaves (Mann et al., 1965).

Taking in account that human and animal health are at risk because of pesticides as they could be related to direct-acting or indirect-acting mutagens, Calderón-Segura et al. (1999) analyzed the influence of *Vicia faba* metabolism on biotransformation of thiocarbamic herbicides: molinate and butylate through the damage produced to the plant itself using SCE analysis and by *in vivo* activation, in which the roots were exposed to the herbicides and its extracts were applied to human lymphocyte cultures. The activation systems containing the enzymatic S10 fraction of the plant roots *in vitro* were also added to human lymphocyte cultures in several concentrations of both herbicides in order to evaluate the cytogenetic effects by scoring the SCE.

The results obtained with molinate and butylate treatments in *Vicia faba* root tip meristems, showed that both herbicides produced significant increases in SCE when compared to the control. A concentration-response relationship was observed. This fact suggested that some metabolic products induced damage in the plant itself. Metaphase cells were not found at 100 ppm of molinate and 300 ppm of butylate (Table 1). As it has been demonstrated in *Vicia faba*, molinate is an S-dependent agent (Gómez-Arroyo et al., 1992) and if SCE is a S-dependent event (Wolff et al., 1974; Painter, 1982), then it would be expected that this type of agent should be an efficient SCE-inducer. Although there were no data on the production of chromosome aberrations by butylate in this study, this herbicide produced SCE and was less toxic than molinate.

The results obtained with direct treatments with both herbicides showed that they were not direct mutagens in cultured human lymphocytes. However both were cytotoxic, since 300/ppm of molinate and 200/ppm of butylate produced cell death. When the extracts of the roots treated for 4 h with molinate and butylate (*in vivo* activation) were applied to lymphocyte cultures, a positive response was

TABLE 1. Sister chromatid exchanges induced by molinate and butylate in *V. faba*[a] (from Calderón-Segura et al., 1999).

Control	$\overline{X} \pm SE$
	30.32 ± 2.97
Molinate [ppm]	
25	50.21* ± 2.96
50	66.71* ± 3.50
75	82.42* ± 2.41
100	Metaphase cells were not observed
Butylate [ppm]	
25	37.47* ± 2.10
50	57.17* ± 3.44
75	62.11* ± 1.20
100	69.50* ± 2.23
200	106.72* ± 3.08
300	Metaphase cells were not observed

[a] n = 50 metaphases in two experiments.
* Significant differences among control and each concentration were obtained by analysis of variance. $F_{molinate} = 13.11$, $F_{butylate} = 67.34$, in both cases P value was <0.0001, and therefore the Newman-Keuls multiple comparisons test was applied, $P < 0.001$.

obtained. The SCE frequencies increased significantly from 100 to 300 ppm of molinate, but with 400 ppm, decreased to control values. Meanwhile, butylate increased SCE in a concentration-response relationship. It was not phytotoxic for the roots, but the extracts were cytotoxic for lymphocytes producing cell death at 500 ppm (Table 2).

In the *in vitro* assays with and without S10 Vicia metabolic fraction in which molinate and butylate were added at 48 h to lymphocyte cultures for 4 h, treatment alone or in combination with the S10 mixture, showed negative response when both herbicides were applied directly. Cellular death was produced by 300 ppm of molinate and 200 ppm of butylate. However, in the treatments where the S10 metabolic fraction was added, the SCE frequencies were significantly higher than those of the control and a concentration-response relationship was not observed with molinate, since at 50 ppm, significant differences of SCE appeared. This result was asymptotic up to 100 ppm and decreased to control values at 300 ppm, and it did not affect the lymphocytes, because mitotic inhibition did not occur. Direct treatments and 300 ppm induced cell death, but when S10 mix was added, cell death did not appear. With butylate, the SCE frequencies also increased at 50 ppm, a concentration-response relationship was noted and the same detoxification behavior was observed. When butylate was applied without S10 mix, cellular death happened at 200 ppm; and with S10 mix addition, cellular death was induced at 300 ppm (Table 3).

Comparing both plant activation systems, *in vivo* SCE frequencies were higher than *in vitro*. These differences could be related to the exposure time of the lymphocytes, or/and to active metabolites. In the *in vivo* experiments, the treatment lasted 48 h and *in vitro*, the exposure lasted only 4 h. The concentration was lower,

TABLE 2. Sister chromatid exchanges in human lymphocytes cultures induced by the extracts of *V. faba* treated with molinate and butylate (in vivo activation)[a] (from Calderón-Segura et al., 1999).

$\overline{X} \pm SE$	Control		$\overline{X} \pm SE$
5.36 ± 0.14			5.30 ± 0.15
	Lymphocyte cultures treated directly with		
	Molinate [ppm]	**Butylate** [ppm]	
6.64 ± 0.20	25	6.25	5.26 ± 0.14
6.18 ± 0.22	50	12.50	5.48 ± 0.12
6.68 ± 0.21	75	50	5.14 ± 0.24
5.40 ± 0.19	100	75	6.22 ± 0.25
5.32 ± 0.14	200	100	6.56 ± 0.26
Cellular death[b]	300	200	Cellular death[b]
4.84 ± 0.33	Ethanol (3,600 ppm)[c] directly added		4.84 ± 0.33
6.44 ± 0.22	Negative control Lymphocyte cultures + *V. faba* extract without treatment		6.44 ± 0.22
11.00* ± 0.50	Positive control Lymphocyte cultures + *V. faba* extracts from treatment with ethanol (3,600 ppm)[c]		11.68* ± 0.50
	Lymphocyte cultures + *V. faba* extracts from treatment with		
	Molinate [ppm]	**Butylate** [ppm]	
8.02* ± 0.30	100	100	8.48* ± 0.58
12.40* ± 0.18	200	200	14.16* ± 0.46
17.78* ± 0.23	300	300	17.00* ± 0.84
4.46 ± 0.20	400	400	24.23* ± 1.20
6.12 ± 0.13	500	500	Cellular death[b]
5.48 ± 0.11	750		
6.34 ± 0.24	1,000		
6.00 ± 0.21	1,500		
5.45 ± 0.16	2,000		

[a] n = 50 metaphase in two experiments.
[b] Stimulated cells were not observed.
[c] Corresponding to 1×10^{-1} M.
* Significant differences among controls and each treated group were obtained by analysis of variance $F_{molinate} = 28.80$, $F_{butylate} = 129.92$, in both cases P value is <0.0001, and therefore the Newman-Keuls multiple comparisons test was applied, $P < 0.001$.

due to the fact that herbicides were in direct contact with the S10 enzymatic fraction.

These data showed that both herbicides needed the *Vicia faba* metabolism to produce damage to the DNA, meaning that the ezymatic system of the broad bean root could transform both herbicides into mutagenic metabolites or produced reactive intermediates that might have bound covalently to DNA. Thiocarbamates were

TABLE 3. Sister chromatid exchanges in human lymphocytes cultures with molinate and butylate with and without *Vicia* S10 mix (*in vitro* activation)[a] (from Calderón-Segura et al., 1999).

$\overline{X} \pm SE$	Control		$\overline{X} \pm SE$
4.96 ± 0.59			5.06 ± 0.41
	Lymphocyte cultures treated directly with		
	Molinate [ppm]	**Butylate** [ppm]	
5.10 ± 0.40	25	25	4.16 ± 0.29
4.12 ± 0.24	50	50	4.34 ± 0.38
4.78 ± 0.33	75	75	3.72 ± 0.28
5.26 ± 0.34	100	100	4.16 ± 0.25
5.42 ± 0.41	200	200	Cellular death[b]
Cellular death[b]	300		
4.60 ± 0.28	Ethanol (3,600/ppm)[c] directly added		4.60 ± 0.28
5.28 ± 0.39	Negative control Lymphocyte cultures + *V. faba* extract without treatment		5.28 ± 0.39
8.78* ± 0.28	Positive control Lymphocyte cultures + *V. faba* extracts from treatment with ethanol (3,600 ppm)[c]		8.78* ± 0.28
	Lymphocyte cultures + *V. faba* S10 mix from treated with		
	Molinate [ppm]	**Butylate** [ppm]	
5.74 ± 0.36	25	25	5.05 ± 0.58
7.06* ± 0.37	50	50	6.70* ± 0.40
8.30* ± 0.56	75	75	6.88* ± 0.59
7.36* ± 0.54	100	100	7.82* ± 0.50
7.60* ± 0.41	200	200	8.20* ± 0.48
5.58 ± 0.34	300	300	Cellular death[b]

[a] n = 50 metaphase in two experiments.
[b] Stimulated cells were not observed.
[c] Corresponding to 1×10^{-1} M.
* Significant differences among controls and each treated group were obtained by analysis of variance $F_{molinate} = 13.107$, $F_{butylate} = 17.692$, in both cases P value is <0.0001, and therefore the Newman-Keuls multiple comparisons test was applied, $P < 0.001$.

initially metabolized, in plants and mammals, by sulfoxidation and N-dealkylation, reactions depending upon the monooxygenase system of cytochrome P-450 know as mixed function oxidases to produce sulfoxides and N-alkylating metabolites (Hubbell and Casida, 1977). Molinate and butylate sulfoxide might be involved in the genetic damage to the lymphocytes in culture, because these metabolites behave as alkylating agents, the active forms of these herbicides, and are the potential candidates to interact with DNA.

To be certain about the metabolic capacity of the *Vicia faba* roots *in vivo* as well as *in vitro*, ethanol was used as positive control. As shown in Tables 2 and 3, the presence of ethanol doubled the SCE values in the lymphocytes. The data agreed well with that obtained by Takehisa and Kanaya (1983) and by Takehisa et al. (1988) in Chinese hamster ovary cells and in human lymphocytes by Gómez-Arroyo et al. (1995) with the same plant activation system.

The protein concentration used in the S10 mix was constant (*in vitro* activation) and, when the extracts obtained of roots treated with both herbicides and the positive control were applied to lymphocyte culture, the protein concentrations were not different from the negative control with the exception of the highest concentration of molinate that produced a decrease in protein synthesis, a secondary effect induced by most thiocarbamics (Mann et al., 1965; Moreland et al., 1969; Fuerst, 1987).

The mitotic (MI) and replication (RI) indexes and cell proliferation kinetics (CPK) of human lymphocyte cultures (HLC) were determined through the evaluation of the effects of the thiocarbamic herbicides butylate and molinate with and without metabolic activation by *Vicia faba in vivo* and *in vitro* (Calderón-Segura and Espinosa-Ramírez, 1998). Both herbicides when applied directly to HLC for 4h, stimulated CPK, increased M3 cells frequency and inhibited mitosis at higher concentrations. RI was not affected by any of the pesticides, but MI was diminished to control values by molinate. The two herbicides with *in vitro* activation showed similar effects for CPK, MI, IR, except for 75 ppm molinate that, when applied directly increased M3, 300 ppm had no action on the cell cycle (Table 4).

Direct treatment with butylate and molinate on HLC for 48h, delayed CPK and increased the M1 cell frequencies, higher concentrations produced cellular death. RI and MI decreased in a concentration-related maner. When the root extracts (*in vivo* activation) treated with butylate were added to HLC for 48h CPK was stimulated, and RI was not modified, while MI decreased. Molinate showed an opposite effect on CPK, MI and RI decreased and 400 ppm had no effect on these parameters (Table 5).

On the other hand, RI and MI were not affected by ethanol with and without metabolic activation *in vivo*, but had a positive response in *in vitro* assays, stimulating CPK and increasing M3 cells frequency. The *Vicia faba* extracts alone and S10 mixture had no effect on CPK, RI and MI (Tables 4 and 5).

Thiocarbamic herbicides butylate and molinate with and without plant activation *in vivo* and *in vitro* by *Vicia faba* roots showed altered CPK, MI, and RI. These results suggested that CPK, RI and MI measurements were useful to screen potential cytotoxic and cytostatic activities of pesticides.

3. PROTOCOL OF PLANT PROMUTAGEN ACTIVATION BY *Vicia faba*

3.1. *Vicia faba*/SCE Protocol

Vicia faba (var. minor) seeds were germinated between two cotton layers soaked with tap water. Primary roots reaching 2–3 cm were introduced into a solution containing 100 µM 5-bromo-2′-deoxyuridine (BrdU), 0.1 µM 5-fluorodeoxyuridine (FdU) and 5 µM uridine (Urd) for 20 h. Afterwards, they were

TABLE 4. Effects on cell kinetics, proliferation rate index (PRI) and mitotic index (MI) in human lymphocytes cultures treated for 4 h with molinate and butylate with and without *Vicia* S10 mix (in vitro activation)[a] (modified from Calderón-Segura and Espinosa-Ramírez, 1998).

	Treated directly					*In vitro* activation				
	M1	M2	M3	%RI[b]	%MI[c]	M1	M2	M3	%RI[b]	%MI[c]
Control	35.5	30.5	34.0	2.0	3.2	35.0	30.5	34.5	2.0	3.2
Molinate [ppm]										
25	28.0	43.5	28.5	2.0	2.3	18.0	31.5	50.5	2.3	4.1
50	20.5	29.0	50.5	2.3	1.8*	21.0	37.5	41.5	2.2	3.3
75	19.0	33.0	48.0	2.3	1.9*	19.5	34.5	46.0	2.3	5.5*
100	17.0*	30.5*	52.5*	2.4	2.1	26.5	31.0	42.5	2.2	3.9
200	15.0*	25.5*	59.5*	2.5	1.9*	19.5	32.5	48.0	2.3	3.8
300	Cellular death[d]					19.0*	28.5*	52.5*	2.3	3.0
Butylate [ppm]										
25	28.0	27.5	44.5	2.2	3.1	26.0	30.0	44.0	2.2	3.0
50	25.0	26.0	49.0	2.0	2.3	24.0	43.0	33.0	2.1	3.1
75	14.0*	34.0*	52.0*	2.4	3.2	20.0	33.5	46.5	2.3	2.6
100	10.0*	30.0*	60.0*	2.8	3.1	10.5	23.0	66.5	2.3	3.0
200	Cellular death[d]					11.5*	15.5*	73.0*	2.6	2.5
300						Cellular death[d]				
S10 Mix						16.0	49.0	35.0	2.2	4.0
Ethanol (3,600 ppm)[e]	21.0*	17.0*	62.0*	2.4	3.1	15.5*	20.0*	64.5*	2.5	2.6

[a] Average from two experiments.
[b] Replication Index.
[c] Mitotic Index, n = 2,000 cells.
[d] Stimulated cells were not observed.
[e] Corresponding to 1×10^{-1} M.
* Significant differences among controls and each treated group were obtained by X^2, in both cases $P < 0.05$.

treated for 4 h with molinate (S-ethyl-N,N-hexamethylenethiocarbamate) dissolved in distilled water at 25, 50, 75 and 100 ppm and butylate (S-ethyl-diisobutylthiocarbamate) also dissolved in distilled water at 25, 50, 75, 100, 200 and 300 ppm. Both herbicides with technical grade of 72%, were obtained from Quimica Lucava of Mexico. Controls of distilled water alone followed the same procedure as above.

Fresh solutions of BrdU, FdU and Urd were applied for a second replicative cyle (20 h). The treatments were carried out in the dark at 20 °C. Two experiments were run for each concentration.

The root tips were cut and treated with colchicine (0.05%) for 3 h and stained using the Feulgen differential technique described by Tempelaar et al. (1982), modified as follows: cuttings were fixed with glacial acetic acid for 1 h, then put in ethanol-acetic acid (3:1) for two days at –2 °C, and later in ethanol 70% for 15 min

TABLE 5. Effects on cell kinetics, proliferation rate index (PRI) and mitotic index (MI) in human lymphocyte cultures by the extracts of *Vicia faba* treated during 48h with butylate and molinate (*in vivo* activation)[a] (modified from Calderón-Segura and Espinosa-Ramírez, 1998).

	Treated directly					*In vivo* activation				
	M1	M2	M3	%RI[b]	%MI[c]	M1	M2	M3	%RI[b]	%MI[c]
Control	38.5	30.0	32.5	2.0	4.7	35.5	30.0	34.5	2.0	4.7
Molinate [ppm]										
25	31.1	38.5	30.4	2.0	4.0					
50	30.5	32.5	37.0	2.0	2.8					
75	47.5*	32.5*	20.0*	1.7*	2.0*					
100	65.5*	25.0*	9.5*	1.4*	1.0*	35.5	34.5	30.0	1.9	2.9
200	72.0*	25.0*	3.0*	1.3*	0.8*	51.1*	35.0*	14.0*	1.6*	2.6*
300	Cellular death[d]					68.0*	25.0*	7.0*	1.4*	1.9*
400						36.0	36.5	28.0	1.9	3.2
Butylate [ppm]										
6.25	30.0	32.5	37.5	2.1	3.4					
12.50	32.0	34.0	34.0	2.0	3.0					
50	41.0*	40.0*	19.0*	1.8*	2.4*					
75	46.0*	38.0*	16.0*	1.7*	1.3*					
100	79.5*	11.5*	9.0*	1.3*	0.3*	42.5	34.5	23.0	1.8	3.1
200	Cellular death[d]					24.0	32.5	43.5	2.2	2.8
300						18.0*	25.0*	57.0*	2.4	1.4*
400						12.0*	31.0*	57.0*	2.5	1.2*
500						Cellular death[d]				
S10 fraction						35.0	35.0	30.0	2.0	4.6
Ethanol (3,600 ppm)[e]	33.0	36.0	31.0	2.0	3.2	33.0	30.0	37.0	2.0	4.0

[a] Average from two experiments.
[b] Replication Index.
[c] Mitotic Index, n = 2,000 cells.
[d] Stimulated cells were not observed.
[e] Corresponding to 1×10^{-1} M.
* Significant differences among controls and each treated group were obtained by X^2, in both cases $P < 0.05$.

and hydrolysed in 5N HCl for 80 min at 20 °C. Then, they were washed three times with distilled water and stained with the Schiff reagent (Feulgen staining) for 12 min in the dark. Cuttings were treated with pectinase 2%, dissolved in 0.01 M citrate buffer, pH 4.7, for 15 min at 28 °C, followed by acetic acid 45% for 10 min and were finally transferred to cold ethanol for 30 min.

The cell squash was prepared in acetic acid 45%. Slides were made permanent by a dry-ice technique (Conger and Firchild, 1953), dehydrated by two absolute butanol changes, and then mounted in Canada balsam. Slides were coded and SCE was scored in 25 metaphase cells for each concentration in each one of the replicate experiments.

3.2. *In vivo* Activation by *Vicia faba*

Pimary roots of 4–6 cm length (seven days old) were immersed in molinate at 100, 200, 300, 400, 500, 750, 1,000, 1,500 and 2,000 ppm, butylate at 100, 200, 300, 400 and 500 ppm, ethanol (positive control) at 3,600 ppm (1×10^{-1} M, Takehisa et al., 1988) or distilled water (negative control) for 4 h in the dark at 20 °C. After treatment, primary roots were washed and the 2 cm tip of the main roots were macerated and homogenized at 4 °C in 0.1 M Na-phosphate buffer, pH 7.4. The ratio of buffer solution volume in ml (2.0–2.5) to fresh weight (2.0–2.5) of root cuttings in grams was 1:1 (Takehisa et al., 1988). The homogenate was centrifuged for 15 min at 10,000 × g at 0 °C. The supernatant was sterilized using Millipore filter (pore size 0.45 μm) and immediately used to treat the human lymphocytes in culture.

Total protein content of *Vicia faba* was determined by the Bio-Rad protein assay (Bradford, 1976). Bovine serum albumin was used as standard for the protein curve. Protein concentration was fairly constant from preparation to preparation, with values between 4.0–4.96 μg/μL except with molinate 500 to 2,000 ppm, in which the protein concentrations were reduced considerably from 2.61 to 1.1 μg/μL.

3.3. Human Lymphocyte Cultures

With an heparinized syringe, 5 mL of healthy donor's blood were extracted by venipuncture. Eight drops of blood were added to 3 mL of RPMI medium 1640 with L-glutamine (Gibco) plus 0.2 mL of phytohemagglutinin (Gibco). The cultures were incubated at 37 °C for 72 h. After 24 h, BrdU was added to the culture medium at a final concentration of 5 μg/mL.

Due to the fact that SCE and cell proliferation kinetics can be influenced by inter-individual variability in response to mitogens, culture conditions or to the blood samples themselves (Morgan and Crossen, 1981; Lamberti et al., 1983; Speit et al., 1986) different donors were used in each experiment to avoid inter-individual variability.

3.4. Direct Treatment with Molinate and Butylate of Human Lymphocytes in Culture and Treatment with *Vicia faba* Extracts

Molinate at 25, 50, 75, 100, 200 and 300 ppm and butylate at 6.25, 12.50, 50, 75, 100 and 200 ppm were dissolved in distilled water and applied directly to human lymphocyte cultures. Other lymphocytes cultures were treated with 100 μL of *Vicia faba* extracts previously exposed to different concentrations of molinate and butylate. Two experiments were carried out for each concentration. No changes in pH values were detected in cultures after the addition of *Vicia faba* roots extracts treated or untreated with molinate and butylate. Both types of cultures were incubated for an additional 48 h period at 37 °C, and colchicine (100 μL) was added 2 h prior to the harvest. Metaphase cells were harvested by centrifugation, treated with 0.075 M KCl and fixed in methanol-acetic acid (3 : 1). Slides were stained by the fluorescence-plus-Giemsa technique (Perry and Wolff, 1974). Twenty-five metaphases were scored for each culture.

3.5. *In vitro* activation

3.5.1. *Vicia faba* S10 Fraction

Two cm of the primary root tips of *Vicia faba* were macerated and homogeneized in 0.1 M Na-phosphate buffer (pH 7.4) containing 1 mM dithiothreitol, 1 mM EDTA and 0.6 M mannitol. The ratio of the volume of the buffer solution in ml to the fresh weight of root cuttings in grams was 1:1, unless otherwise stated. Immediately before homogeneization, polyvinylpolyrrolidone was added to the buffer at a ratio of 10% of the roots' fresh weight. The homogenate was centrifuged for 15 min at 10,000 × g at 4 °C (Takehisa et al., 1988). The supernatant was sterilized using a Millipore filter (pore size 0.45 µm) and immediately used for *in vitro* activation of molinate and butylate. Protein concentration used was 10 mg/500 µL of S10 reaction mixture.

3.5.2. Lymphocyte Treatment with and without *in vitro* Metabolic Activation

Lymphocyte cultures of 48 h were exposed to molinate at 25, 50, 75, 100, 200 and 300 ppm, and butylate at 25, 50, 75, 100 and 200 ppm for 4 h (the last 2 h shaking) in the dark at 37 °C with or without metabolic activation. After treatment, the cells were washed twice in 0.9% sodium chloride and reincubated for 24 h with RPMI medium containing 100 µL of BrdU. Colchicine (100 µL) was added 71 h after the beginning of the culture and the same procedure mentioned above for harvesting and staining was followed.

The S10 mixture was prepared from microsomal S10 fraction at the rate 1:9 (v/v) with the following compounds: 8×10^{-3} M MgCl2, 3.3×10^{-3} M KCl, 5×10^{-3} M glucose 6-phosphate, 4×10^{-3} M NADP and NAD and 10^{-1} M Na_2HPO_4- NaH_2PO_4 at pH 7.4. The 48 h cultures were incubated with 500 µL of the activation system and several concentrations of both herbicides. As positive control, 3,600 ppm of ethanol (1×10^{-1} M) was used.

3.6. Statistical Analysis

SCEs were assessed statistically using analysis of variance (ANOVA) to determine significant differences among the treated groups. When a significant F value was found ($P < 0.05$), the Newman-Keuls multiple comparison test was used to identify groups showing evidence of significant differences at $P < 0.001$ when compared with controls.

4. CONCLUDING REMARKS

The capacity of the microsomal system S10 of the root of *Vicia faba* to activate pesticides *in vivo* and the mutagenic expression of the metabolites contained in the extracts through sister chromatid exchanges (SCE) in human lymphocytes in culture has been shown.

The S10 enzymatic system of *Vicia faba in vitro* is a quick method to detect genotoxic metabolites when pesticides are activated.

The significant SCE increments in lymphocyte cultures and in *Vicia faba* root macerates show that SCE is a dependable and objective cytogenetic test to detect the activation of environmental promutagens.

It was defined that some chemicals such as the carbamic insecticide propoxur and the thiocarbamic herbicides molinate and butylate need the *Vicia faba* metabolic activation to induce sister chromatid exchanges in human lymphocyte cultures.

The replication index permitted the evaluation of cell kinetics proliferation and the mitotic index afforded data on pesticide cytotoxicity. Both indexes are worthwhile when directly applied to lymphocyte cultures, as well as to identify the effect of the metabolites following the plant metabolism.

The capacity of the plants to metabolically transform promutagen pesticides to mutagen should be taken into account because the role played by plants in the trophic chain and in the human diet might mean an actual health risk.

5. ACKNOWLEDGMENTS

We are grateful to Dr. Bettina Sommer, Instituto Nacional de Enfermedades Respiratorias for her kind review of the English translation.

6. REFERENCES

Bradford, M.M. 1976. A rapid and sensitive method for the quantitation of microgram quantities of protein utilizing the principle of protein-dye binding. Anal. Biochem. 72: 248–254.

Calderón-Segura, M.E., and M. Espinosa-Ramírez. 1998. Efecto del butilate y molinate sobre la división de los linfocitos en cultivo con y sin activaciòn metabólica *in vivo* e *in vitro* por *Vicia faba*. Rev. Int. Contam. Ambient. 14: 39–47.

Calderón-Segura, M.E., S. Gómez-Arroyo, R. Villalobos-Pietrini, and M. Espinosa-Ramírez. 1999. *In vivo* and *in vitro* promutagen activation by *Vicia faba* of thiocarbamate herbicides molinate and butylate to products inducing sister chromatid exchanges in human lymphocyte cultures. Mutat. Res. 438: 81–88.

Conger, A.D., and L.M. Fairchild. 1953. A quick-freeze method for making smear slides permanent. Stain Technol. 28: 281–283.

Constantin, M.J., and R.A. Nilan. 1982. Arylamine N-oxidation by microsomal fraction of germinating pea seedling (*Pisum sativum*). J. Agric. Food Chem. 31: 1276–1282.

Dohn, D.R., and R.I. Krieger. 1981. Oxidative metabolism of foreign compounds by higher plants. Drug Metab. Rev. 12: 119–157.

Frear, D.S., H.R. Swanson, and F.S. Tanaka. 1969. N-demethylation of substituted 3-(phenyl)-1-methylureas: isolation and characterization of a microsomal mixed function oxidase from cotton. Phytochemistry 8: 2157–2169.

Fuerst, E.P. 1987. Understanding the mode of the action of the chloroacetamide and thiocarbamate herbicides. Weed Technol. 1: 270–277.

Gentile, J.M., G.J. Gentile, J. Bultman, R. Sechriest, E.D. Wagner, and M.J. Plewa. 1982. An evaluation of the genotoxic properties of insecticides following plant and animal activation. Mutat. Res. 101: 19–29.

Gentile, J.M., G.J. Gentile, S. Townsend, and M.J. Plewa. 1985. *In vitro* enhancement of the mutagenicity of 4-nitro-0-phenylenediamine by plant S9. Environ. Mutagen. 7: 73–85.

Gentile, J.M., G.J. Gentile, and M.J. Plewa. 1986. *In vitro* activation of chemicals by plants: a comparison of techniques. Mutat. Res. 164: 53–58.

Gómez-Arroyo, S., L. Rodríguez-Madrid, and R. Villalobos-Pietrini. 1992. Chromosomal alterations induced by the thiocarbamate herbicide molinate (Ordram) in *Vicia faba*. Rev. Int. Contam. Ambient. 8: 77–80.

Gómez-Arroyo, S., and R. Villalobos-Pietrini. 1995. Chromosomal aberrations and sister chromatid exchanges in *Vicia faba* as genetic monitors of environmental pollutants. In: Biomonitors and Biomarkers as Indicators of Environmental Change. Eds.: Butterworth F.M., L.D. Corkum, and J. Guzmán-Rincón, Plenum Press, New York, pp. 95–113.

Gómez-Arroyo, S., M.E. Calderón-Segura, and R. Villalobos-Pietrini. 1995. Sister chromatid exchange in human lymphocytes induced by propoxur following plant activation by *Vicia faba*. Environ. Mol. Mutagen. 26: 324–330.

Hess, F.D. 1985. Herbicide absorption and translocation and their relationship to plant tolerances and susceptibility. In: Weed Physiology. Vol 2. Herbicide Physiology. Ed.: Duke S.O. CRC Press, Boca Ratón, Florida, pp. 191–214.

Hess, F.D. 1987. Herbicide effects on the cell cycle meristematic plant cells. Rev. Weed Sci. 3: 183–203.

Higashi, K., K. Nakashima, Y. Karasaki, M. Fukunaga, and Y. Mizugudu. 1981. Activation of benzo (a) pyrene by microsomes of higher plant tissues and their mutagenicity. Biochem. Int. 2: 373–380.

Higashi, K., K. Ikeuchi, Y. Karasaki, and M. Obara. 1983. Isolation of immunochemically distinct form of cytochrome from microsomes of tulip bulbs. Biochem. Biophys. Res. Commun. 115: 46–52.

Higashi, K. 1985. Microsomal cytochrome P-450 in higher plants. In: P-450 and Chemical Carcinogenesis. Eds.: Tagashira, Y., and T. Omura. Gann Monogr. on Cancer Res. Vol. 30, pp. 49–66.

Higashi, K., K. Ikeuchi, M. Obara, Y. Karasaki, H. Hirano, S. Gotoh, and Y. Koga. 1985. Purification of a single major form of microsomal cytochrome P-450 from tulips bulbs (*Tulipa gesneriana L.*). Agric. Biol. Chem. 49: 2399–2405.

Higashi, K. 1988. Metabolic activation of environmental chemicals by microsomal enzymes of higher plants. Mutat. Res. 197: 273–288.

Hubbell, J.P., and J.E. Casida. 1977. Metabolic fate of the N,N-dialkylcarbamoyl moiety of thiocarbamate herbicides in rats and corn. J. Agric. Food Chem. 25: 404–413.

Lai, C., B.J. Dabney, and C.R. Shaw. 1978. Inhibition of *in vitro* metabolic activation of carcinogens by wheat sprout extracts. Nutr. Cancer 1: 27–30.

Lamberti, L., P.P. Bigatti, and G. Ardito. 1983. Cell kinetics and sister-chromatid-exchange frequency in human lymphocytes. Mutat. Res. 120: 193–199.

Lamoureux, G.L., and D.S. Frear. 1979. Pesticide metabolism in higher plants. *In vitro* enzyme studies. In: Xenobiotic Metabolism: *In vitro* Methods. Eds.: Paulson G.D., D.S. Frear, and E.P. Marks. Symp. Series 97. Amer. Chem. Soc. Washington, D.C., pp. 77–128.

Lamoureux, G.L., and D.G. Rusness. 1987. EPTC metabolism in corn, cotton, and soybean: identification of a novel metabolite derived from metabolism of a glutathione conjugate. J. Agric. Food Chem. 35: 1–7.

Langenbach, R., S. Leavitt, C. Hix, Y. Sharief, and J.W. Allen. 1986. Rat and hamster hepatocyte-mediated induction of SCEs and mutation in V79 cells and mutation of Salmonella by aminofluorene and dimethylnitrosamine. Mutat. Res. 161: 29–37.

Loprieno, N., and I.D. Adler. 1980. Cooperative programme of the EEC on short-term assays for mutagenicity. In: Molecular and Cellular Aspects of Carcinogen Screening Tests. Eds.: Montesano R., H. Bartsch, and L. Tomatis. IARC Publ. No. 27 Lyon, pp. 331–341.

Mann, J.D., S.L. Jordan, and B.E. Day. 1965. A survey of herbicides for their effects upon protein synthesis. Plant Physiol. 40: 840–843.

Menn, J.J. 1978. Comparative aspects of pesticide metabolism in plants and animals. Environ. Health Perpect. 27: 113–124.

Moreland, D.E., S.S. Malhotra, R.D. Gruenhagen, and T.E. Shokrah. 1969. Effects of herbicides on RNA and protein synthesis. Weed Sci. 17: 556–563.

Morgan, W.F., and P.E. Crossen. 1981. Factors influencing sister-chromatid exchange rate in cultured human lymphocytes. Mutat. Res. 81: 395–402.

Obe, G., and H. Ristow. 1970. Acetaldehyde, but not ethanol, induces chromatid exchanges in Chinese hamster cells *in vitro*. Mutat. Res. 56: 269–272.

Obe, G., A.T. Natarajan, A. Meyers, and A. Den Hertog. 1979. Induction of chromosomal aberrations in peripheral lymphocytes of human blood *in vitro* by ethanol and its metabolite acetaldehyde. Mutat. Res. 68: 291–294.

Obe, G., R. Jonas, and S. Schmidt. 1986. Metabolism of ethanol *in vitro* produces a compound which induces sister-chromatid exchanges in human peripheral lymphocytes *in vitro*: acetaldehyde not ethanol is mutagenic. Mutat. Res. 174: 47–51.

Obe, G., and D. Anderson. 1987. Genetic effects of ethanol. Mutat. Res. 186: 177–200.

Painter, R.B. 1982. A replication model for sister-chromatid exchange. Mutat. Res. 70: 337–341.

Perry, P., and S. Wolff. 1974. New Giemsa method for the differential staining of sister chromatid. Nature (London) 251: 156–158.

Plewa, M.J. 1978. Activation of chemical into mutagens by green plants. A preliminary discussion. Environ. Health Perspect. 27: 45–50.

Plewa, M.J., and J.M. Gentile. 1982. The activation of chemicals into mutagens by green plants. In: Chemical Mutagens. Principles and Methods for their Detection. Eds.: De Serres F.J., and A. Hollaender. Plenum Press, New York, Vol. 7, pp. 401–420.

Plewa, M.J., E.D. Wagner, G.J. Gentile, and J.M. Gentile. 1984. An evaluation of the genotoxic properties of herbicides following plant and animal activation. Mutat. Res. 136: 233–245.

Plewa, M.J., E.D. Wagner, and J.M. Gentile. 1988. The plant cell/microbe coincubation assay for the analysis of plant-activated promutagens. Mutat. Res. 197: 207–219.

Plewa, M.J., S.R. Smith, and E.D. Wagner. 1991. Diethyldithiocarbamate suppresses the plant activation of aromatic amines into mutagens by inhibiting tobacco cell peroxidase. Mutat. Res. 247: 57–64.

Plewa, M.J., and E.D. Wagner. 1993. Activation of promutagens by green plants. Annu. Rev. Genet. 27: 93–113.

Rasquinha, I.A., A.G. Wildeman, and R.N. Nazar. 1988. Studies on the use of plant extracts in assessing the effects of the plant metabolism on the mutagenicity and toxicity of pesticides. Mutat. Res. 197: 261–272.

Rich, P.R., and D.S. Bendall. 1975. Cytochrome components of plant microsomes. Eur. J. Biochem. 55: 333–341.

Rueff, J., C. Chiapella, J.K. Chipman, F. Darroudi, I. Duarte Silva, M. Duverger-van Bogaert, E. Fonti, H.R. Glatt, P. Isern, A. Laires, A. Leonard, M. Llagostera, P. Mossesso, A.T. Natarajan, F. Palitti, A.S. Rodrigues, A. Schinoppi, G. Turchi, and G. Werle-Schneider. 1996. Development and validation of alternative metabolic systems for mutagenicity testing in short-term assays. Mutat. Res. 353: 151–176.

Sandermann, H., D. Scheel, and T.V.D. Trenck. 1983. Metabolism of environmental chemicals by plants-copolymerization into lignin. J. Appl. Polymer Sci. Appl. Polymer Symp. 37: 407–420.

Sandermann, H. 1988. Mutagenic activation of xenobiotics by plant enzymes. Mutat. Res. 197: 183–194.

Scott, B.R., A.H. Sparrow, S.S. Schwemmer, and L.A. Schairer. 1978. Plant metabolic activation of 1,2-dibromoethane (DBE) to a mutagen of greater potency. Mutat. Res. 49: 203–212.

Seo, K.Y., J. Riley, D. Cortez, E.D. Wagner, and M.J. Plewa. 1993. Characterization of stable high molecular weight mutagenic product(s) of plant-activated m-phenylenediamine. Mutat. Res. 299: 111–120.

Shimabukuro, R.H., G.L. Lamoureux, and D.S. Frear. 1981. Pesticide metabolism in plants: principles and mechanisms. In: Biological Degradation of Pesticides. Ed.: Matsumura F. Plenum Press, New York, pp. 123–145.

Shimabukuro, R.H., G.L. Lamoureux, and D.S. Frear. 1982. Pesticide metabolism in plants reactions and mechanisms. In: Biodegradation of Pesticides. Eds.: Matsumura F., and C.R.K. Murti. Plenum Press, New York, pp. 21–66.

Speit, G., R. Düring, and K. Mehnert. 1986. Variation in the frequency of sister chromatid exchanges in repeated human lymphocyte cultures. Hum. Genet. 72: 179–181.

Sterling, T.M. 1994. Mechanisms of herbicide absorption across plant membranes and accumulation in plant cells. Weed Sci. 42: 263–276.

Takehisa, S., N. Kanaya, and R. Rieger. 1982. Induction of SCEs in CHO cells by extracts from *Vicia faba* roots exposed to ethanol. Mutat. Res. 105: 169–174.

Takehisa, S., and N. Kanaya. 1983. A comparison of *Vicia faba*-root S10 and rat-liver S9 activation of ethanol, maleic hydrazide and cyclophosphamide as measured by sister-chromatid exchange induction in Chinese hamster ovary cells. Mutat. Res. 124: 145–151.

Takehisa, S. 1986. Pro-mutagen activation by *Vicia faba*: measurement with SCE induction in CHO cells. Biol. Zbl. 105: 37–40.

Takehisa, S., N. Kanaya, and R. Rieger. 1988. Promutagen activation by *Vicia faba*: an assay based on sister chromatid exchanges in Chinese hamster ovary cells. Mutat. Res. 197: 195–205.

Tempelaar, M.J., M.T.J. de Both, and J.E.G Versteegh. 1982. Measurement of SCE frequencies in plants: a simple Feulgen staining procedure for *Vicia faba*. Mutat. Res. 103: 321–326.

Wildeman, A.G., and R.N. Nazar. 1982. Significance of plant metabolism in the mutagenicity and toxicity of pesticides. Can. J. Genet. Cytol. 24: 437–449.

Wolff, S., J. Bodycote, and R.B. Painter. 1974. Sister chromatid exchanges induced in Chinese hamster cells by UV irradiation of different stages of the cell cycle: the necessity for cells to pass through S. Mutat. Res. 25: 73–81.

Yamasaki, I. 1974. Peroxidase. In: Molecular Mechanism of Oxigen Activation. Ed.: Hayaish O. Academic Press, New York, pp. 5335–5558.

Zilkah, S., and J. Gressel. 1977. Cell cultures vs. whole plants for measuring phytotoxicity. III. Correlations between phytotoxicity in cell suspension cultures and seedling. Plant Cell Physiol. 18: 815–820.

24

GENETIC MONITORING OF AIRBORNE PARTICLES

RAFAEL VILLALOBOS-PIETRINI, SANDRA GÓMEZ-ARROYO, AND OMAR AMADOR-MUÑOZ

Laboratorios de Mutagénesis y Citogenética Ambientales. Centro de Ciencias de la Atmósfera, UNAM, Ciudad Universitaria, Coyoacán 04510 D.F. México

1. INTRODUCTION

Environmental problems in huge cities have been related to a high demographic density and diverse industrial activities. Gasoline and diesel exhaust, incinerators and industries's stacks involved in combustion processes and forest and agriculture fires are some of the main air pollution sources (Westerholm et al., 1992). Many organic, inorganic and radioactive chemicals, identified as hazardous air pollutants (Kelly et al., 1994), are present in the urban atmospheres in a wide variety of compounds (Natusch, 1978).

Samet (1992) and Dockery et al. (1993), among others, considered air pollution as one of the main causes of lung cancer in man. The amount of total suspended particles (TSP) has been established as standard because they have been shown to be one of the best air pollution indicators (Dams et al., 1975). The Mexican norm adopted the permissible limit of 260 $\mu g/m^3$ in a 24 h period (SEDUE, 1987), the same standard that have been followed in USA since 1971 (Hileman, 1981). However, in 1987 it was substituted by the PM10 standard because it involved particles $\leq 10\,\mu m$. The PM10 permissible limits were 150 $\mu g/m^3$ in a 24 h period. This standard was closely related to health because these particles reach the bronchioalveolar spaces of the respiratory system and may impair cell functions (Hileman, 1981; Rodes and Evans, 1985; Seemayer et al., 1987) and was considered among the air pollutants currently identified by USEPA (1987). An expert panel on Air Quality Standards of the UK recommended 50 $\mu g/m^3$ as the permissible limit for PM10 in a 24 h period (EPAQS, 1995). Schwartz and Marcus (1990) suggested that these particles are the most toxic air pollutants among all the complex mixtures and are therefore the most significant contributors to human mortality even when compared to soil derived coarse particles (Ozkaynak and Thurston, 1987).

Biomonitors and Biomarkers as Indicators of Environmental Change 2, Edited by Butterworth *et al.*
Kluwer Academic/Plenum Publishers, New York, 2000

Recently, the US Environmental Protection Agency incorporated a new standard called PM2.5 considering that particles ≤2.5 µm were associated with human mortality and morbility. This standard has a permissible limit of 50 µg/m^3 in a 24 h period (Utell and Samet, 1993; USEPA, 1996). Ostro and Chestnut (1998) considered the existence of a safe threshold level for airborne particulate matter uncertain.

Based on the previous statements, this paper deals with particulate pollution, solid or liquid, emitted into the air with organic matter associated to it, composition of which changes from place to place. These particles were generated by combustion, condensation and conversion of gas to particles (Lioy et al., 1990) and their formation and fates were closely associated with the production of ozone and other oxidants (Finlayson-Pitts and Pitts, 1997).

TSP and PM10 have been significantly associated with daily mortality and morbidity (Fairley, 1990; Schwartz, 1991a,b; Dockery et al., 1992, 1993; Pope et al., 1992, 1995; Schwartz and Dockery, 1992a,b; Ostro, 1993; Li and Roth, 1995; Moolgolvkar et al., 1995; Anderson et al., 1996; Moolgolvkar and Luebeck, 1996; Choudhury et al., 1997). An increase of 100 µg/m^3 in daily TSP concentration has been related with about 6% increase in mortality (Schwartz, 1991a). However, other studies have related the increase in 10 µg/m^3 of PM10 with 1% of premature mortality (Dockery et al., 1993; Schwartz et al., 1996) or with 1% of daily mortality (Ostro, 1993; Dockery and Pope, 1994; Ostro and Chestnut, 1998); it was associated also with 3.4% in mortality due to respiratory causes and 1.8% in mortality due to cardiovascular failure (Pope et al., 1992). An increase of 10 µg/m^3 in PM2.5 increases the total daily mortality in 1.5% (Schwartz et al., 1996).

The relationship between TSP and PM10 was obtained by calculating the PM10/TSP ratio. In eight locations in the USA (Rodes and Evans, 1985) and four sites of Mexico City (Villalobos-Pietrini et al., 1995), PM10 constitutes between 37 and 58% of TSP, much lower than the 70% described for the Netherlands by van der Meulen et al. (1987). The disagreement could be influenced by the high relative humidity.

In samplings made in Philadelphia, Burton et al. (1996) noted that PM10 contained 75% of PM2.5. Chow et al. (1996a) observed that in California, PM2.5 constituted from 30 to 70% of PM10 and Suh et al. (1997) observed that PM2.5 comprised 77% of PM10 in Washington D.C. In the winter months in Birmingham, UK, PM2.5 comprised 80% of PM10 and during summer it accounted only 50% (Harrison et al., 1997).

There is evidence that health effects of airborne particles depended upon their size and chemical composition (Butler et al., 1985; Dideren et al., 1985). Causal relations between increases in airborne particles (≤10 µm in diameter) and elevation in prevalence and symptoms of respiratory diseases have been established by several epidemiologic studies (Dockery and Pope, 1994; Pope et al., 1995). The inhalation of particles >2.5 µm in the European urban environments seems to have no effect on human health, because these particles are rather scant. Health effects are more closely correlated with the number of ultrafine particles than with their mass concentrations because larger particles account for most of the mass concentration (Tuch et al., 1997). Churg and Brauer (1997) found that particles <2.5 µm were effectively retained by the human lung parenchyma and the PM2.5 range was related to the effects of chronic pollutants.

The polycyclic aromatic hydrocarbons (PAHs) are widely spread environmental pollutants produced by natural and anthropogenic sources and their concentration in the atmosphere varies mainly with the type of emitters, the combustion and meteorological conditions (Masclet et al., 1988; Baek et al., 1991a; Wallace, 1991).

More than 100 PAHs have been identified in the organic fraction associated to airborne particles of urban areas and some of them have been used as standards of reference (Baek et al., 1991a). The PAHs with five rings or more were present in particulate form, while compounds with three or four rings were in both particulate and gaseous forms. Those with two rings were only found in gaseous forms (Lee and Shuetzle, 1983). Cautreels and van Cauwenberghe (1978) found PAHs with five rings in the gaseous forms also. It is important to mention that the concentration of PAHs in gas phase increases with the temperature (Yamasaki et al., 1982; McVeety and Hites, 1988).

Atmospheric PAHs, as benzo(a)pyrene (B(a)P), have been considered as indicators of carcinogenesis and the accumulative risk of cancer had a linear correlation with B(a)P concentration (Redmon et al., 1976). Helmes et al. (1982) considered PAHs and their derivatives as one of the main chemical classes to indicate pollution resulting from combustion sources. Cederlöf (1978) described that an increment of one nanogram per cubic meter of B(a)P increased 5% the mortality rate of lung cancer in non-smoking males. In Mexico City in 1993, the malignant tumors were the second cause of mortality and, among them, lung cancer was the first one (Ponciano, 1996).

Research of the carcinogenic effects of organic matter of airborne particles has been expensive and still time consuming to be used routinely. Due to the fact that around 90% of carcinogens were also mutagens (McCann et al., 1975; McCann and Ames, 1976; Commoner et al., 1978), mutagenic tests with microorganisms have been used (Ames et al., 1973, 1975; Maron and Ames, 1983) to detect the genetic damage produced by organic complex mixtures associated to air particles.

2. SAMPLE TREATMENTS

2.1. Collection

Many kind of filters have been used to sample airborne particles. The great majority of publications are based on the employment of glass fiber filters because they are resistant, highly effective and relatively unexpensive (van Vaeck et al., 1979; Baek et al., 1991a,b), although teflon filters were considered as the best for sampling tools due to their high chemical inertness and efficiency (Shaw and Stevens, 1980; Grosjean, 1983; Lee and Schuetzle, 1983; Monn et al., 1997). Unfortunately, they are expensive and commercially restricted in size.

In order to stabilize the glass fiber filters, Lee and Schuetzle (1983) set them into a box containing drierite with relative humidity ≤45% during 24 h. Later they were weighed in a special balance, and maintained in the box for another 24 h period. Then, they were weighed again and if the weight did not vary in ±0.01 mg, the filter was ready for the sampling. Thrane and Mikalsen (1981) heated the filters at 400 °C for 4 h before they weighed them. EPA (1990) recommends the heating of the glass fiber filters at 600 °C for 5 h before exposing them. By this procedure it is possible to decrease the B(a)P or others PAHs contents are decrease to a level below 10 ng, then the state of the filters were considered as adequate for sampling.

The usual collection period for suspended particulate matter has been 24 h, but some researchers tried shorter collection times: two (Kado et al., 1986; Viras et al., 1990) and three hours (Pitts et al., 1982; Villalobos-Pietrini et al., 1995) or longer periods, up to some weeks (König et al., 1980). This was mainly done because PAH concentration in air is very low and high volume samplers are employed to obtain enough organic mass to make the studies. However, PAHs on filters could get lost by air streams, ambient temperature, volatilization, reactions on the filter surface and sampling time (Teranishi et al., 1977; Pitts et al., 1978b; Lewtas, 1980; Kolber et al., 1981; Chuang et al., 1987; McLachlan et al., 1990; Baek et al., 1991a,b). Grimmer (1983) noted that PAHs with boiling point below 400 °C such as fluoranthene, pyrene, benzofluorenes and their methylderivatives reevaporated after they had already been collected.

The air flow rate through the high volume samplers is controlled and maintained. For TSP, the lowest air flow speed is 1.0 std m^3/min and the highest 1.8 std m^3/min (USEPA, 1982). For PM10 the air flow speed is 1.13 m^3/min (±10%) (USEPA, 1987).

Recently, low volume samplers have been used with several kinds of inlets to collect particles of different sizes (Allen et al., 1996; Burton et al., 1996; Chan et al., 1997; Harrison et al., 1997; Sheu et al., 1997; Ohta et al., 1998; Sweet and Gatz, 1998) with different air flow speeds, i.e. 16.67 L/min used as reference (USEPA, 1996).

In mid latitude cities, collection of airborne particles is generally taken during each of the seasons that are clearly differentiated. However, in the tropics the climate is governed by different air masses. One is the dry westerly current that prevails from November to April associated with clear skies and cool temperatures. During the rest of the year (May–October), the westerlies give way to the moist unstable trade winds (Mosiño and García, 1973) that are linked to convective rain showers in the Mexico City basin (and most of Mexico) located at 2,250 mosl in the tropics (20° latitude N) and surrounded by mountains. Therefore, ventilation of air pollutants is usually poor, especially during the dry season when anticyclonic conditions prevail favoring the formation of surface based radiation inversions (Jauregui, 1986, 1997).

2.2. Extraction and Concentration

A very wide spectra of solvents has been used to extract organic matter from air particles. Depending upon the extraction method, different solvents could remove several organic compounds and therefore, different results were obtained for mutagenicity testing.

The two usual extractions were made in a soxhlet apparatus or in ultrasound agitation. Both proved to be suitable (Van Houdt et al., 1989; Matsushita et al., 1992). The soxhlet extraction during 12 h permitted the recovery of more than 95% of the extractable material (Lee and Schuetzle, 1983). Using cyclohexane, Golden and Sawicki (1975) extracted by sonication 15% more PAHs than with soxhletion. Also Krishna et al. (1983b) obtained higher mutagenic responses with sonication than with soxhletion. Grimmer (1983) and Lee and Schuetzle (1983) recommended the extraction to be made in absense of ultraviolet light or fluorescent lamps used for laboratory ilumination in order to avoid PAHs decomposition.

The extraction procedures of organic material from the air particles is normally done using a single organic solvent, a sequence of solvents increasing in polarity (Daisey et al., 1979, Butler et al., 1985) or intercalating organic, acid and basic solvents (Lee et al., 1976; Teranishi et al., 1978; Kolber et al., 1981; Lee and Schuetzle, 1983).

The efficiency of the extraction procedure and the analytical methods depended upon the kind of matrix and on the physical and chemical nature of the particles (Fitch and Smith, 1979; Griest et al., 1980; Lee and Schuetzle, 1983). The quantity of extractable organic matter (EOM) might be determined by weighing the filter before and after extraction or weighing the extract after evaporation of the solvents (Lee and Schuetzle, 1983).

The range of solvents employed are non-polar such as n-hexane, cyclohexane, benzene and toluene, moderately polar such as dichloromethane and chloroform and polar solvents such as acetone, ether and methanol mainly and also a mixture of the solvents.

According to Möller and Löfroth (1982) and Sideropoulos and Specht (1994), acetone and dichloromethane were the most suitable solvents for a complete extraction of the organic compounds associated to airborne particles. Jungers et al. (1981), Löfroth (1981) and Krishna et al. (1983a) found that, among different solvents, the acetone extracts produced the highest mutagenic activity and Lee et al. (1991) observed that acetone extracts were several times more effective than dichloromethane extracts to induce mutagenicity. Nielsen (1992) stated that dichloromethane extracts were the best for fractionation, chemical analysis and extraction of complex environmental mixtures. Hitzfeld et al. (1997) made the organic matter extraction from filters with ultrapurified water, obtaining an aqueous fraction, followed by another extraction with dichloromethane and Biran et al. (1996) prewetted the filters with ethanol and then used ultrapurified water to make the extractions.

After soxhlet or ultrasound and solvents, the extracts were evaporated almost until dryness using a rotatory evaporator or the Kuderna-Danish system and then transferred to vials. In order to dry extracts and to determine the extractable organic matter, the most frequently used was the high purity nitrogen.

When the extracts were not used immediately, they could be stored at –80 °C for short periods without modifications in the mutagenic potency (Matsushita et al., 1992). However Butler et al. (1985) showed that extracts stored during 6 to 15 months could lose ≈40% of the mutagenic activity per μg of extract when stored at ≈–30 °C.

In all works related to mutagenicity, after extraction and concentration, the extracts were redissolved with dimethyl sulfoxide (DMSO) to expose them to the biological test systems such as *Salmonella* strains, employing 0.5 ml as maximal (Ames, 1975) and non-toxic concentration (Salamone et al., 1979).

2.3. Fractions

The open column chromatography technique is one of the easiest ways to separate the compounds contained in the organic environmental complex mixtures of airborne particles according to its polarity. High performance liquid chomatography (HPLC) is another technique often used, which has high resolution power to separate compounds.

Although Pyysalo et al. (1987) noted the loss of mutagenicity due to fraction procedure on silica gel open column, which might degrade some unstable components, the synergistic potential effects between various fractions could not be totally excluded.

Cautreels and van Cauwenberghe (1976) fractionated the organic matter of the airborne particles in acidic, basic and neutral portions and in the last they found PAHs and nitrogen polyaromatics among other compounds. Teranishi et al. (1978) observed the mutagenicity of airborne particulate matter in the acidic, polyaromatic and oxygenated fractions. In the organic matter extracted by Moriske et al. (1985), the neutral fraction was further separated into aliphatic, PAHs and polar neutral compounds by thin layer chromatography and the whole organic extract showed lower mutagenic activity than the polar neutral compounds.

Sasaki et al. (1987) noted higher mutagenicity of TA98 and TA100 *Salmonella* strains and more incidence of lung tumors in mice with the neutral fraction than with acidic or basic fractions. While Chrisp and Fisher (1980) found the most potent mutagens in the acidic fraction of urban air, Crebelli et al. (1991) noted that mutagenicity of airborne particulate matter collected in downtown Rome depended mainly on direct-acting acidic and neutral compounds and less from basic promutagens detected when metabolic activation was involved.

2.4. PAHs

To determine the polycyclic aromatic hydrocarbons in the complex mixtures of airborne particles, several analytical techniques as thin layer chromatography (TLC) (with fluorescence detection), high performance liquid chromatography (HPLC) (with fluorescence or ultraviolet detector or mass spectrometry) and nuclear magnetic resonance technique (NMR) have been used, but the most common was the gas chromatography (GC) because it allowed the detection of substances at very low concentrations (ppm, ppb or ppt) and smaller amounts of samples were needed (generally, 1 or 2 µL), was relatively inexpensive, high numbers of compounds could be identified and quantified and it could be coupled with other techniques such as mass spectrometry and infrared spectrometry.

According to their vapor pressure, amount of fine particles, ambient temperature, concentration, affinity for the organic fraction of the particles and adsorption surface available, the distribution of PAHs in the atmosphere has been characterized to be between gas and particle phases (Natusch and Tomkins, 1978; Yamasaki et al., 1982; van Vaeck et al., 1984; Coutant et al., 1988; Tuominen et al., 1988; Baker and Eisenreich, 1990). Thus, particulate matter was considered as the most abundant and the most hazardous atmospheric pollutant to human health and because of this, most of the studies have been focused on particles (Seinfeld, 1989; Baek et al., 1991a).

Van Vaeck and van Cauwenberghe (1978) and Baek et al. (1991a), found above 95% of PAHs, heterocyclic polyaromatic, carboxylic acids and aliphatic hydrocarbons in particles <3 μm and between 60 to 70% in particles <1 μm (van Vaeck et al., 1979, 1984). Then, Baek et al. (1991a) estimated that 63 to 80% of the PAHs was contained in particles <1.1 μm and Pistikopoulos et al. (1990) observed that nonvolatile PAHs were in particles <1 μm and volatile PAHs adhered mainly to particles greater than this diameter, except fluoranthene.

Allen et al. (1996) showed that the distribution of PAHs with similar molecular weights was the same in the particles with similar size (0.14 to 1.9 μm). This was the case for benzo-fluoranthenes, benzo(a)pyrene, B(a)P and perylene (MW = 252). In the urban aerosols, the concentration of PAHs in particles with different sizes varied inversely proportional with their molecular weights. So there was a preferential accumulation of lower molecular weight PAHs (≤228) in larger particles (0.5 μm ≤ D ≤ 2.0 μm) (i.e. phenanthrene and pyrene) compared with higher molecular weight PAHs which were in particles of <0.5 μm diameter (i.e. indeno(1,2,3-cd)pyrene and coronene). This did not occur in rural aerosols, where coarse particles contained the larger fraction of higher molecular weight PAHs.

Generally, PAHs suffered several and complicated physical and chemical processes and, according with their properties, these facts modified their concentration and distribution in air. Other factors were also involved as wet and dry deposition, transport by mass air, turbulence and convection, reactions with other pollutants and exchanges between gas and particulate phases (U.S. National Academy of Science, 1972; Masclet et al., 1986).

Several urban pollution studies have used B(a)P as reference of airborne PAHs levels (Funcke et al., 1981; Hadnagy et al., 1986; Seemayer et al., 1987; Viras et al., 1990; DeMarini et al., 1994), but Katz and Chan (1980) considered B(a)P as a poor index for cities where motor vehicle traffic was one of the major sources of air pollution. B(a)P was not a major indirect acting mutagen in the air samples of four cities of Taiwan (Chou and Lee, 1990). In Los Angeles, Hamilton, Toronto and two other Ontario cities the dominant PAHs were benzo(ghi)perylene and coronene and the common activity for these cities was the emission of vehicle exhaust (Gordon, 1976; Katz et al., 1978; Masclet et al., 1986; Tuominen et al., 1988). For Barale et al. (1991) concentrations of cyclopenta(cd)pyrene correlated very well with mutagenicity. Takada et al. (1990) considered that the majority of particles containing PAHs were discharged by mobile sources. It is supposed that pyrene, cyclopenta(cd)pyrene, benzo(ghi)perylene and coronene were produced by the vehicle traffic (Gordon and Bryant, 1973; Jeltes, 1974). Coronene was one of the most stable PAHs in air (Falk et al., 1956).

The chemical decomposition of the PAHs depended on the kind of the adsorption substrate (Korfmacher et al., 1980). However, light is not such an important degradative factor as evaporative or oxidative reactions with gaseous pollutants (Korfmacher et al., 1979, 1981).

3. TESTER STRAINS

The TA strains used to detect mutagenicity in the organic extracts from airborne particles were derived from *Salmonella typhimurium* LT2 having the following genotypes *gal*, Δ*(chl, uvrB, bio) rfa* (Maron and Ames, 1983). The Ames test (Ames et al., 1975) was based on the bacteria requirements of histidine (his^-) in the plate incorporation assay. The original mutation consisted in the reversion from his^- (auxotrophy) to his^+ (prototrophy). In the strain TA100, the mutation hisG46, that is a base-pair substitution in the first structural gene of the histidine operon, blocked the histidine synthesis. While in the TA98 strain, the frameshift mutation hisD3052 in the last structural gene of the same operon codified for a defective histidinol dehydrogenase, and the last step in the histidine synthesis was therefore inhibited.

Other strains of *Salmonella typhimurium* lacking the pKM101 episome have also been used. TA1537 and TA1538 are involved in a frameshift mutation and strain TA1535 is related to a base-pair substitution (Ames et al., 1975).

Kado et al. (1986) used the TA98 strain with the standard plate incorporation and with the microsuspension or preincubation procedure (based on the addition of ten times more bacteria). The last one resulted more sensitive than the former, but Goto et al. (1992) tested the complex mixtures of airborne particles using TA98 and TA100 strains with the two methods mentioned above and the overall results showed similar values, good correlations and no significant differences between both assays. Schuetzle and Lewtas (1986) combined chemical fractions with a biological test simplifying the chemical characterization of the complex mixtures and verifying the mutagenicity of each fraction by means of a "bioassay-directed" analysis. Using this method, Nishioka et al. (1988) and Lewtas (1990) found that the highest mutagenic fractions contained more polar compounds than the nitro-PAHs.

The tester strains set also has been used in the presence or absence of mammalian metabolic activation, S9 mix. When mutagenicity was observed in the absence of metabolic activation, it indicated that direct-acting mutagens were present. But when metabolic activation increased the mutation frequency, it implied that the observed activity was due to an indirect-acting mutagen or promutagen.

TA98 showed greater sensitivity to the organic complex mixtures of air particles than TA100 (Teranishi et al., 1977; Tokiwa et al., 1977, 1980; Salamone et al., 1979; Alfheim and Möller, 1980; Talcott and Harger, 1980; Fukino et al., 1982; Tuominen et al., 1988; Barale et al., 1991; Nardini and Clonfero, 1992; Villalobos-Pietrini et al., 1995; Al-Khodairy and Hannan, 1997). However, De Flora et al. (1989a) detected TA100 as more sensitive than TA98 in metabolic activated fractions and greater sensitivity of TA98 in unfractionated samples. De Wiest et al. (1982) noted that TA100 was more sensitive to the airborne extracts than TA98.

DeMarini et al. (1994) found higher mutagenic activity in TA100 than in TA98 accounted in the neutral/base fraction from bioassay-directed chemical of organic extracts of air particles.

Other *Salmonella* strains used for mutagenicity tests of air particles were YG1020 (equivalent to TA98) and YG1025 (equivalent to TA100), variants that had multicopies of another plasmid cloning vector (pBR322) carrying the genes for the nitroreductase enzyme (YG1021 and YG1026) or for the acetyltransferase enzyme (YG1024 and YG1029). These strains were particularly sensitive in the absence of the mammalian metabolic fraction (Legzdins et al., 1995). Espinosa-Aguirre et al. (1993) found YG1024 as the most sensitive strain to mutagenic effects of Mexico City particulate matter, followed by YG1021 and YG1020, while the TA98NR, nitroreductase-deficient strain resisted the genotoxic effects of the same complex mixtures. These data suggest the presence of nitroarenes.

The nitropyrenes mutagenicity is due to active nitroreductases naturally present in enterobacteria (McCann and Ames, 1977; Rosenkranz and Mermelstein, 1980; Mermelstein et al., 1981).

A forward mutation in *Salmonella typhimurium* strain SV50 from arabinose-sensitive to arabinose-resistant was useful to detect the activity of environmental complex mixtures (Whong et al., 1983; Warner et al., 1991).

Siak et al. (1985) and Sato et al.(1995) observed the mutagenic response of TA98 to the organic matter extracted from TSP and the important decrease of mutagenicity in TA98NR (resistant to 1-nitropyrene and nitroreductase-deficient) and $TA98DNP_6$ (resistant to 1, 6 dinitropyrene and deficient in nitroreductase and O-acetyltransferase), showing the significant contribution of nitroarenes. In other urban air particulates, the lower mutagenicity of TA98NR when compared to TA98, meant higher contribution of direct-acting mutagenicity of nitroaromatic compounds (Tuominen et al., 1988; Nardini and Clonfero, 1992).

When no appreciable differences were found in strains TA98NR and $TA98DNP_6$ compared with the TA98 parental strain, it could mean that no significant amounts of direct-acting nitroderivatives were present (Barale et al., 1991).

DeMarini et al. (1993) found that 90% of the revertants induced in the TA98 strain contained a hot spot deletion of GC or CG within the sequence CGCGCGCG in the unfractionated mixture and in the neutral/base fraction of the air samples.

4. OTHER BIOLOGICAL TEST SYSTEMS

The increase of sister chromatid exchanges (SCE) was observed in bone marrow and spleen primary cells of mouse in culture but *in vivo*, this effect was not induced by intraperitoneal or oral administration of extracts from airborne particles (Krishna et al., 1986). The extracts of airborne particles from an industrialized city produced SCE, chromosomal aberrations and cell cycle delay in human lymphocytes in culture also in concentration-dependent way (Krishna et al., 1984; Hadnagy et al., 1986; Hadnagy and Seemayer, 1987).

SCE, chromosomal aberrations and mitotic arrest were induced in cultures of lung cells line V79 of Chinese hamster by airborne particle extracts of urban and non-urban locations (Alink et al., 1983; Seemayer et al., 1987; Motykiewics et al., 1988, 1991). Point mutations were also found at the HGPRT-locus in V79 cells, and malignant transformation in Syrian hamster kidney cells in culture was induced (Seemayer et al., 1987).

The extracts of diesel exhaust particles obtained with both dichloromethane and a pulmonary surfactant (dipalmitoyl phosphatidyl choline) yielded similar positive results of mutagenicity in *Salmonella* TA98 and SCE induction in Chinese hamster V79 fibroblasts in culture (Keane et al., 1991). The synthesis and secretion of surfactants are the most important function of human pneumocytes type II. Surfactant is released into the alveoli by a calcium-dependent exocytosis process (Behrendt et al., 1987). Extracts of the airborne particles stimulated the synthesis of surfactants but only a weak release of them (Behrendt et al., 1987).

In airborne particulate samples from a small industrial town corresponding to "non-pollution level", Pyysalo et al. (1987) observed that the highest SCE frequency in Chinese hamster ovary cells (CHO) was approximately twice the control value.

PM10 extracts of airborne particles of Mexico City were more genotoxic than TSP extracts for *Drosophila melanogaster* in the somatic mutation and recombination test in the wings (Delgado-Rodríguez et al., 1999).

PAHs of PM2.5 formed DNA adducts through covalent binding in human lymphocytes. These adducts correlated with exposures, but at higher doses no-linearity was found (Lewtas et al., 1997). Mermelstein et al. (1981) considered that the direct mutagenic activity of nitropyrenes resulting from the atmospheric nitration of PAHs, was due to adduct formation.

Some human populations of the Cezch Republic and Poland were evaluated for DNA adducts in lymphocytes of non-smoker males exposed to high levels of pollutants (Chorazy et al., 1994; Lewtas, 1994). In the population of Silesia, Poland, the DNA adducts in winter exceeded summer values (Chorazy et al., 1994; Möller et al., 1996).

DNA adducts in human placenta were found in a population of a polluted industrial area of Bohemia, Cezch Republic, and differences with another "non-polluted area" were only significant in winter (Topinka et al., 1997). Binková et al. (1998) treated the calf thymus DNA with extracts of organic matter from the same sites in order to detect the adducts production and found the highest values in the neutral aromatic fraction when they added S9 mix (oxidative activation) and the highest values in the acidic fraction when they added xanthine oxidase (reductive conditions).

5. MUTAGENICITY

Organic extracts from particles caused strong frameshift-type of mutagenic activity on TA98 strain in the Ames assay (Pitts et al., 1977) and it has been shown that respirable particles contained potentially cytotoxic, mutagenic and carcinogenic chemicals (Seemayer et al., 1987). Möller and Alfheim (1983), Alsberg et al. (1985), Sicherer-Roetman et al. (1988) and DeMarini et al. (1994, 1996) found in the polyaromatic fraction the major contribution to mutagenic effects. Chuang et al. (1992)

associated the highest risk mainly to the PAHs with more than 4 rings in the complex mixtures. For other authors (Reali et al., 1984; Lewtas, 1988) the PAHs only account for part of the carcinogenicity and mutagenicity of the whole organic extract.

The mutagenic activity of PAHs appears to be restricted to the particulate phase of ambient air (de Raat et al., 1987), although Pyysalo et al. (1987) found high mutagenicity in some fractions in the organic extracts of the vapor phase. Unsubstituted PAHs have been observed as possible markers of air quality in terms of mutagenicity (Garner et al., 1986; de Raat et al., 1987; de Raat, 1988).

DeMarini et al. (1994) calculated that an average sized person inhaled the equivalent per day to 92.5 μg of the organic matter extracted from particles with dichloromethane and this amount was mutagenic in the *Salmonella* assay.

5.1. Mutagenic Potency and Mutagenic Ratio

Möller and Alfheim (1980) considered that revertants of *Salmonella* per cubic meter of air might be a better expression of mutagenicity than revertants per mg of collected particles. Most of the work published on organic extracts from airborne particles on mutagenicity involving the Ames assay used mutagenic potency based on the slope of the regression line of the concentration-response which permitted the evaluation of revertants per cubic meter of air.

To obtain the mutagenic ratio, the frequency of revertants induced by the complex mixture was divided by the frequency of revertants induced by the negative control and the mutagenic effect was considered positive when the ratio was >2, that means at least a doubling of the spontaneous reversion (De Flora et al., 1989a; Villalobos-Pietrini et al., 1995). In the study made by Villalobos-Pietrini et al. (1995) in Mexico City, most points sampled exceeded the value 2. Therefore, it could be deduced that mutagens were contained in the majority of the complex organic mixtures associated to airborne particles. The places with highest mutagenicity were different for TSP (Northeast) and for PM10 (Downtown). In the first site (high TSP mutagenicity) the major contribution was afforded by vehicle exhausts, industries and soil-erosion particles and in the second (high PM10 mutagenicity) the main source of pollution was the vehicle exhaust with dense traffic conditions. Chow et al. (1996b) and Harrison et al. (1997), correlated PM10 with traffic emmisions.

For Alfheim (1982) and Alfheim et al. (1983), the motor vehicle exhaust was the most important source of mutagenic particles in urban areas. Scarpato et al. (1993) found a correlation between mutagenicity in a rural area and the number of circulating vehicles, indicating traffic emissions as the most probable source of mutagens.

In an urban area with dense traffic, the level of revertants per cubic meter in TA98 strain was higher than in industrial sites (Athanasiou et al., 1986). Sato et al. (1995) showed much higher mutagenicity where vehicle emissions were the major pollution source than in other industrial areas. From several sites, Barale et al. (1989) and De Flora et al. (1989a) considered the vehicle exhausts as the major source of air pollution.

Mutagenic effects were higher for PM10 than for TSP, even through TSP samplers collected small and large particles simultaneously (Villalobos-Pietrini et al., 1995). This work was in accordance to the results obtained by Lewtas (1980) who considered the large particles diluted the overall mutagenicity of the sample. Also for Chrisp and Fisher (1980), large particles might prevent the detection of the mutagenicity of the small ones. The higher mutagenic activity in a low concentration of particles was explained by Möller and Alfheim (1980) and van Vaeck and van Cauwenberghe (1984) considering that particles might adsorb compounds in the gas phase or larger particles might have been removed or the large specific surface area of the particles <0.5 μm.

Teranishi et al. (1977) and Pitts et al. (1978b) used high-volume cascade impactors to separate the airborne particles by size using series of plates with either holes or slots and found that the organic extracts from particles <1.1 μm were significantly more mutagenic than those from larger particles. Alfheim and Möller (1980) noted major mutagenicity with the organic extracts of particles <2.7 μm, obtained also from cascade impactors in TA98, in presence and absence of S9. Monarca et al. (1997) found that inhalable particles collected in a high volume sampler with a cascade impactor for particles fractionation were mainly made of fine particles (<0.5 μm) which afforded for most of PAHs and mutagenicity. Talcott and Harger (1980), Möller et al. (1982), Sorenson et al. (1982) and Kado et al. (1986) considered that mutagens could be mainly found in particles <3 μm.

All these studies have indicated that the mutagenic effects were found mainly in the smaller particles that had a diverse composition depend from their emission sources. According to the morphology and size distribution of atmospheric particles and their deposition on foliage and inert surfaces, Coe and Lindberg (1987) classified them in two types: spherical fly ash, released during fossil fuel burning, and non-fly ash, which were non-spherical particles (soil material) consisting of both organic (as pollen, bacteria, fungal spores and hyphae, etc) and mineral matter. The first, composed mainly by Si, Al, Fe, K, Ca, and S, occurred in two aggregate forms, one that agglomerated particles of similar size or larger particles that carried several smaller ones attached to them and the second was composed mainly by Ca, Si, Fe, Ti, Al, K, Cl, S, P and Mg, and occured in several irregular forms.

There are several techniques to determine the composition and the morphology of the airborne particles as electron microprobe X-ray emission spectrometry (EMP), electron spectroscopy for chemical analysis (ESCA), Auger electron spectrometry (AES) and secondary ion-mass spectrometry (SIMS). The elements found by Keyser et al. (1978), with ESCA, in surface airborne particulate matter were predominantly C, N, Pb and S.

5.2. Direct- and Indirect-acting Mutagens

When the metabolic fraction of rat liver, the S9 mix obtained from 9,000 × g supernatant, was added to *Salmonella* strains cultures, the mutagenic activity of some metabolites of the extracts of airborne particles appeared. So they were considered indirect-acting mutagens, also known as promutagens. When the original chemical compound obtained from the extract produced the mutagenic effects, it was called direct-acting mutagen (Ames et al., 1973, 1975; McCann et al., 1975; Maron and Ames, 1983). Al-Khodairy and Hannan (1997) noted the conversion of promutagens into direct-acting mutagens on TA98 from extracts exposed to the sunlight.

Courtois et al. (1988, 1992) noted that maximum mutagenicity in air particles extracts was reached with 2.5% of S9 in the S9 mix and a decrease in mutagenicity with more than 10% of S9 in the mix. Tokiwa et al. (1980) described higher mutagenicity in *Salmonella* in the presence of rat lung S9 fraction than with liver S9 fraction. Krökje et al. (1991) found higher mutagenicity in the same biological test system of the extracts of airborne particles with rat liver S9 than with lung and kidney activation. Van Houdt et al. (1988) observed that the capacity of activation of the homogenates of normal liver and lung had minor metabolic differences with the extracts of the airborne particles. Therefore, they considered the respiratory tract as well as the liver as important sites for *in vivo* activation of respirable particles.

When no modification of the mutagenic activity in *Salmonella* extracts was noted after the addition of S9 (de Wiest et al., 1982; Ohsawa et al., 1983; Kado et al., 1986), the action of the main agents have been direct (Yamanaka and Maruoka, 1984; Viras et al., 1990; Sato et al., 1995).

In general, air particles contain both direct and indirect-acting types of mutagens and they have been described in many countries: USA (Pitts et al., 1977, 1978a,b, 1982; Talcott and Wei, 1977; Chrisp and Fisher, 1980; Daisey et al., 1980; Lewtas 1980; Talcott and Harger, 1980, 1981; Jungers et al., 1981; Kolber et al., 1981; Kado et al., 1986; Ong et al., 1986, 1989; Keane et al., 1991; Arey et al., 1992; DeMarini et al., 1992, 1994, 1996; Finlayson-Pitts and Pitts, 1997), Japan (Tokiwa et al., 1977, 1980; Teranishi et al., 1978; Fukino et al., 1982; Ohsawa et al., 1983; Iwado et al., 1991; Goto et al., 1992; Matsushita et al., 1992), Canada (Salamone et al., 1979; Legzdins et al., 1995), Norway (Alfheim and Möller, 1980; Möller and Alfheim, 1983), Belgium (de Wiest et al., 1982), Greece (Athanasiou et al., 1986; Viras et al., 1990), Finland (Pyysalo et al. 1987), Netherlands (van Houdt et al., 1987, 1989), France (Courtois et al., 1988), Italy (Crebelli et al., 1988, 1991; De Flora et al., 1989a; Barale et al., 1991; Nardini and Clonfero, 1992, Scarpato et al., 1993), Denmark (Nielsen, 1992), Chile (Sera et al., 1991; Adonis and Gil, 1993), Brazil (Sato et al., 1995), and Mexico (Villalobos-Pietrini et al., 1995, 1999), among others.

Indirect mutagen activity was higher than the direct one with the extractable organic matter of air particles from an urban place, but was the same or higher than the first in a rural site (Alfheim and Möller, 1980). Talcott and Wei (1977) observed the presence of direct- and indirect-acting mutagens from airborne particulate in Buffalo and New York but only detected direct-acting mutagens in Berkeley samples.

On the other hand, PAHs could react with the atmospheric oxidants present in urban air like O_3, PAN, NO_x, SO_x, among others, and water, temperature and solar radiation might have intervened in the reactions to form mainly nitro, hydroxy, oxy and sulphonate derivatives, compounds that frequently had higher genotoxicity than those of the original compounds (Pitts, 1987; Alebic-Juretic, 1990; Finlayson-Pitts and Pitts, 1997). As a result of these reactions, direct-acting mutagens, as the nitroarenes, were produced (Pitts et al., 1978a; Pitts, 1987; Arey et al., 1992; Houk et al., 1992).

OH and NO_3 radicals, and O_3 play a main role in the formation and destinity of PAHs and fine particles (Finlayson-Pitts and Pitts, 1997). Fukino et al. (1982) found that monthly variations of atmospheric NO, NO_2, NO_2^- and NO_3^- concentrations were similar to the direct-acting mutagenic activity and they observed that atmospheric concentrations of heavy metals were also related to this kind of mutagens, suggesting that automobiles emissions, home heaters and power plants were primary sources of direct-acting mutagens. Secondary direct-acting mutagens might be partially formed by the nitration of PAHs with NO_2. Particulate matter extracts stimulated the production of reactive oxygen species that affected lipid peroxidation and provoked oxidative DNA damage (Hitzfeld et al., 1997). These reactive oxygen species could be inhibited by superoxide dismutase, catalase and sodium azide (Hitzfeld et al., 1997).

Atmospheric reactions caused transformation of non polar compounds in the primary emission to more oxygenated mutagenic compounds (Alfheim et al., 1983). Athanasiou et al. (1986) suggested that the formation of secondary direct-acting mutagens due to the oxidation by ozone or nitration by NO_2 of PAHs was favored by an urban atmosphere and was higher in the late spring months.

The PAHs reactions with NO_x were of particular interest because the combustion sources emitted both classes of compounds simultaneously (Pitts et al., 1978a; Hughes et al., 1979; Siak et al., 1985). Augmented concentrations of NO_x in the polluted air, increased the probability of conversion of PAHs to nitroderivatives (Butler and Crossley, 1981) catalysing the reactions with O_3, SO_2 and HNO_2 (Pyysalo et al., 1987). The reaction rates depended of the PAHs structure and of the nature of the substrate on which they were adsorbed (Korfmacher et al., 1979; Nielsen et al., 1983), i.e. B(a)P and benz(a)anthracene were more reactive than benzo(e)pyrene and chrysene respectively and benzo-fluoranthenes were more resistant toward O_3 attack (van Vaeck and van Cauwenberghe, 1984).

Although reactions of PAHs with SO_x produced direct-acting mutagen derivatives in the atmosphere (Pitts et al., 1978a; Pitts, 1987; Houk et al., 1992), these reactions were uncertain because sulphonic acids were water soluble and, therefore, were difficult to characterize in the organic extracts of the particulate matter (Olufsen and Björseth, 1983).

Higher activity of indirect-acting mutagens compared to direct-acting ones was observed by Butler et al. (1985) in extracts obtained from air particles collected in a site of Mexico City. The same behavior was noted in five sites of Mexico City sampled in 1989–1990 by Villalobos-Pietrini et al. (1995). However, the opposite was shown in air particles collected in the same city in 1992 (Villalobos-Pietrini et al., 1999). This might have been related to changes made in the gasoline formulation and a wider employment of emission control devices in the automobiles' exhaust systems.

5.3. Day and Night Mutagenic Variation

Pitts et al. (1982) found higher mutagenicity in air samples from Los Angeles and Riverside taken during the morning "rush hours" (6 to 9 h), which was enhanced by S9 mix. Kado et al. (1986) found that fine particles ($\leq 2.5\,\mu m$) had a three times higher mutagenic response with S9 than without S9. Although the diurnal patterns of both were similar, the morning peak occurred between 10 and 12 noon, but appeared later than the traffic flow (7 to 8 h). Viras et al. (1990) noted the maximum value of mutagenic activity during morning hours (09 to 11 h) in Athens, but later than the "rush hours". Villalobos-Pietrini et al. (1999) observed the highest mutagenicity (–S9) between 10 to 13 h, 16 and 19 h (evening "rush hours") and 22 and 01 h and when S9 was added, the highest response was found between 19 and 01 h. The volatile organic compounds and NO_2 oxidation at night depended on the NO_3 radical. During daytime, it was related to the OH radical as O_3 appeared during day and night. When NO_2 and O_3 were present in enough concentrations in the atmosphere, the formation of NO_3 radical and N_2O_5 took place (Finlayson-Pitts and Pitts, 1997). When N_2O_5 reactioned with PAHs, nitroarenes were produced. These products were also formed by reactions between PAHs and OH radical in the presence of NO_x (Zielinska et al., 1989).

As direct-acting mutagens, nitropyrenes are the class that include most mutagenic chemicals (Mermelstein et al., 1981). The main source of 1-nitropyrene and 6-nitrobenzo(a)pyrene are combustion processes (Lee and Schuetzle, 1983; Jacob et al., 1991), while 2-nitropyrene, hydroxynitropyrenes, 2-nitrofluoranthene and 10-nitrobenzo(a)anthracene are products of the hydroxyl radical-mediated atmospheric nitration reactions on pyrene, fluoranthene and benz(a)anthracene, respectively (Nielsen et al., 1984; Arey et al., 1986, 1987; Gibson et al., 1986; Nielsen and Ramdahl, 1986; Ramdahl et al., 1986; Atkinson et al., 1990a,b).

5.4. Seasons

Seasonal variation in the particle size was also observed (Katz and Chan, 1980). In winter, shifts from large to small particles occur (Peltonen and Kuljukka, 1995). Thus, during cold weather, major concentrations of PAHs were found in smaller particles by condensation mechanisms (De Wiest and Della Fiorentina, 1975; van Vaeck and van Cauwenberghe, 1985). Baek et al. (1991a) reported that PAHs of molecular weights of 252 or less were 20% in gaseous phase, even in summer seasons, but those lower than 234 were 50% in gaseous phase in summer and decreased to the half in winter.

The PAHs relationships with particulate and gaseous phases in urban and rural zones varied considerably with the season of the year (Ramdahl et al., 1982; McVeety and Hites, 1988; Tuominen et al., 1988; Freeman and Cattell, 1990) and were attributed to changes in emission profiles and differences in ambient temperature, vapor pressure, adsorption surface and particulate concentrations (Masclet et al., 1986; Chuang et al., 1987; Coutant et al., 1988; McVeety and Hites, 1988).

The higher mutagenicity of airborne particles in winter observed in several studies (Daisey et al., 1979; Löfroth, 1980, 1981; Möller and Alfheim, 1980; Alfheim and Möller, 1980; Alfheim et al., 1983; Alink et al., 1983; Reali et al., 1984; van Houdt et al., 1987; Torok et al., 1988, 1989; Tuominen et al., 1988; Barale et al., 1989; Scarpato et al., 1993; Chorazy, 1994) could be due to domestic heating, decreased volatilization, less photochemical degradation or increased deposition of mutagens onto airborne particulate matter that are typical of the cold season (Atherholt et al., 1985).

A fall-winter maximum of *Salmonella* TA98 and TA100 mutagenesis and a spring-summer minimum per cubic meter of air observed in TSP of New York City were explained by the fuel oil combustion for space heating (Daisey et al., 1980). Sato et al. (1995) obtained the same for TA100 in the extracts of Sao Paulo, Brazil, but TA98 had the lowest response with the winter extracts.

Atmospheric concentrations of nitro-substituted compounds were higher in the colder months (Talcott and Harger, 1981) but some authors found that mutagenicity due to nitroderivatives was slightly greater in summer than in winter (Crebelli et al., 1988; De Flora et al., 1989a; Nardini and Clonfero, 1992), probably due to photoactivation of aromatic polynuclear compounds (De Flora et al., 1989b).

During spring and summer, higher levels of mutations in *Salmonella* TA98 (–S9) were found in Brazil and the addition of S9 did not modify the mutagenicity, but in winter the highest values were for TA100 (+S9) at all sites (Sato et al., 1995).

5.5. Meteorology and Air Pollution Parameters

Mutagenicity of airborne extracts of two sites of the Netherlands correlated with temperature, wind direction, SO_2 and NO_2 but not with wind speed, rainfall, relative humidity, global radiation and atmospheric pressure (van Houdt et al., 1987).

Alink et al. (1983) in Wageningen, Barale et al. (1989) in Pisa and De Flora et al. (1989a) in Genoa did not find correlations between mutagenicity and meteorological conditions such as intensity and wind direction and rainfall. Barale et al. (1989) however, obtained an inverse relationship between mutagenicity of unfractionated organic extracts of airborne particles and temperature.

On the other hand, mutagenicity of TA98 +S9 from extracts of airborne particles in Mexico city correlated with NO_2 and the same strain, but –S9 correlated with SO_2, NO_2 and CO. This effect did not correlate with relative humidity, temperature, wind speed or direction (Villalobos-Pietrini et al., 1995). Kado et al. (1986) obtained high correlation between mutagenicity and lead content and less correlation with NO_2, O_3 and SO_2. Wind direction and speed were associated with mutagenicity of PM10 extracts (Morris et al., 1995).

Absence of contrast in mutagenicity between urban and non-urban air, could be related to the extention of the polluted zone (Ohsawa et al., 1983) or to how far particles are transported (Alink et al., 1983; Masclet et al., 1988). Strong correlation among H^+, SO_4^{2-}, PM2.5 and PM10 were found by Suh et al. (1997) in USA and Brook et al. (1997) in Canada. The correlation found by these last authors, could be due to the influence of meteorology at synoptic scale and to the length of the atmospheric lifetime of fine particles.

6. ANTIMUTAGENICITY

Calf serum extracts reduced both direct- and indirect-acting mutagenicity of *Salmonella* TA98 and TA100 strains obtained in benzene extracts of airborne particles collected in Tokyo, maybe because the serum was antimutagenic or less effective to extract the mutagens from the particles than benzene or extracted weaker mutagens (Ohsawa et al., 1983).

Chlorophyllin inhibited mutagenicity in 50 to 90% in extracts of complex mixtures of airborne particles (Ong et al., 1986; Warner et al., 1991). Vitamin C enhanced the mutagenic effects of these extracts (Ong et al., 1989). Chili added to organic matter extracts of airborne particles of Mexico City produced antimutagenic effects (Espinosa-Aguirre et al., 1993).

A methanol extract of airborne particles collected in Okayama City, Japan, showed mutagenicity but also antimutagenicity. The antimutagenic factors were long chain fatty acids as palmitic, stearic, oleic and linoleic acids (Iwado et al., 1991).

On the other hand, Haughen and Peak (1983) showed that complex mixtures of aromatic compounds suppressed the indirect-acting mutagenicity of PAHs inhibiting metabolic activation by the microsomal monooxigenase system and mentioned that the failure to detect the mutagenic activity of these complex mixtures did not necessarily show the absence of mutagenic components.

7. CONCLUDING REMARKS

Humans are continually more aware of their daily exposure to one of the most hazardous pollutants, air particles. Standards have been generated in order to establish the permissible levels of exposure. Studies of their association to morbility and mortality have permitted the establishment of apparently safe levels but standards are changing rapidly, referring lower concentration and also lower size of particles. First, total suspended particles (TSP) were the standard, later the particles equal or lower than 10 μm (PM10) and now, a new standard has been proposed and involves particles equal or lower than 2.5 μm (PM2.5). Fine particles were more significantly related to human morbility and mortality than soil-derived coarse particles. The actual existence of a safety threshold level is unsure and therefore, the problem becomes a first order concern.

The filters most commonly used to collect particles have been glass fiber filters. The use of soxhlet apparatus to extract organic matter proved to be as suitable as ultrasound agitation and solvents applied were non-polar to polar, although the most frequently used was a moderately polar solvent, dichloromethane. Several authors used mixtures of solvents to concentrate the samples. The most commonly used systems were rotatory evaporator and Kuderna-Danish apparatus. To fraction the samples, the most suitable technique was the high performance liquid chromatography and for organic compounds' identification and quantification, the most often used was gas chromatography-mass spectrometry.

The presence of PAHs in some fractions afforded the mutagenic activity. Several strains of *Salmonella typhimurium* have been tested for mutagenicity. The most frequently used were TA98 and TA100, that have been handled to increase their sensitiveness or resistance to airborne mutagens. The addition of rat liver S9 fraction demonstrated the presence of indirect-acting mutagens that needed to be activated, as the PAHs, and then its metabolites produced the effects. When S9 was not added, the mutagens involved were direct-acting, namely nitro-, hydroxi-, and oxy-PAH derivatives. The high metabolic activation that produces indirect-acting mutagenicity is similarly effective with the S9 fraction from rat liver and rat lung.

Mutagenic potency based on the slope of the regression line of revertant frequencies and the concentration of the sample, are the parameters used to express mutagenic activity per cubic meter of air or per mg of extractable organic matter. The polyaromatic fraction of the organic matter extracts afforded the major contribution to mutagenic effects.

Other biological test systems as Chinese hamster cells in culture, as well as human lymphocytes in culture, have been used, specially for sister chromatid exchanges. DNA adducts were induced by the extracts in human lymphocytes and placenta and were found in populations exposed to high levels of pollutants.

In some places of the world higher mutagenicity has been shown at "rush hours", but in others it did not happen. There were examples of high mutagenicity at nighttime. During winter, airborne particles were more mutagenic than in other seasons. In relation to meteorological or air pollution parameters, mutagenicity has a lot of variation.

Some substances as chlorophyllin, chili, long chained fatty acids and calf serum have been considered as antimutagenic, because they reduced mutagenicity.

8. ACKNOWLEDGMENTS

We want to thank to Dr. Bettina Sommer of the Instituto Nacional de Enfermedades Respiratorias for her kind review of the English version, to Dr. Ernesto Jáuregui of the Centro de Ciencias de la Atmósfera, UNAM and Dr. Javier Espinosa of the Instituto de Investigaciones Biomédicas, UNAM for the valuable discussion of some parts of the paper, to Lic. Arturo F. Méndez of the Biblioteca Conjunta del Instituto de Geofísica y del Centro de Ciencias de la Atmósfera, UNAM for his hemerographical review of the field of the article and to Q.I. Zenaida Munive for her technical assistance. This work was performed under a contract with CONACyT (code 3727PN).

9. REFERENCES

Adonis, M., and L. Gil. 1993. Mutagenicity of organic extracts from Santiago (Chile) airborne particulate matter. Mutat. Res. 292: 51–61.

Alfheim, I., and M. Möller. 1980. Mutagenicity of airborne particulate matter in relation to traffic and meteorological conditions. In: *Short-Term Bioassays in the Analysis of Complex Environmental Mixtures II*, Eds.: Waters, M.D., S.S. Sandhu, J.L. Huisingh, L. Claxton, and S. Nesnow, Plenum Press, New York, pp. 85–99.

Alfheim, I. 1982. Contribution from motor vehicle exhaust to the mutagenic activity of airborne particles. Prog. Clin. Biol. Res. 109: 235–248.

Alfheim, I., G. Löfroth, and M. Möller. 1983. Bioassay of extracts of ambient particulate matter. Environ. Health Perspect. 47: 227–238.

Alink, G.M., H.A. Smitt, J.J. van Houdt, J.R. Kolkman, and J.S.M. Boleij. 1983. Mutagenic activity of airborne particles at non-industrial locations. Mutat. Res. 116: 21–34.

Alebic-Juretic, A., T. Cvitas, and L. Klasinc. 1990. Heterogeneus polycyclic aromatic hydrocarbon degradation with ozone on silica gel carrier. Environ. Sci. Technol. 24: 62–66.

Al-Khodairy, F., and M.A. Hannan. 1997. Exposure of organic extracts of air particulates to sunlight leads to metabolic activation independence for mutagenicity. Mutat. Res. 391: 71–77.

Allen, J.O., N.M. Dookeran, A.K. Smith, and A.F. Sarofim. 1996. Measurement of polycyclic aromatic hydrocarbons associated with size-segregated atmospheric aerosols in Massachusetts. Environ. Sci. Technol. 30: 1023–1031.

Alsberg, T., U. Stenberg, R. Westerholm, M. Strandell, U. Rannug, A. Sundvall, L. Romert, V. Bernson, B. Pettersson, R. Toftgard, B. Franzén, M. Jansson, J.A. Gustafsson, K.E. Egebäck, and G. Tejle. 1985. Chemical and biological characterization of organic material from gasoline exhaust particles. Environ. Sci. Technol. 19: 43–50.

Ames, B.N., W.E. Durston, E. Yamasaki, and F.D. Lee. 1973. Carcinogens are mutagens: a simple test system combining liver homogenates for activation and bacteria for detection. Proc. Natl. Acad. Sci. (USA) 70: 2281–2285.

Ames, B.N., J. McCann, and E. Yamasaki. 1975. Methods for detecting carcinogens and mutagens with the *Salmonella*-mammalian-microsome mutagenicity test. Mutat. Res. 31: 347–364.

Anderson, H.R., A. Ponce de Leon, J.M. Bland, J.S. Bower, and D.P. Strachan. 1996. Air pollution and daily mortality in London: 1987–1992. Medicina, 312: 665–669.

Arey, J., B. Zielinska, R. Atkinson, A.M. Winer, T. Ramdahl, and J.N. Pitts Jr. 1986. The formation of nitro-PAH from the gas-phase reactions of fluoranthene and pyrene with the OH radical in the presence of NO_x. Atmos. Environ. 20: 2339–2345.

Arey, J., B. Zielinska, R. Atkinson, and A.M. Winer. 1987. Polycyclic aromatic hydrocarbons and nitroarene concentrations in ambient air during a Winter time high-NO_x episode in the Los Angeles basin. Atmos. Environ. 21: 1437–1444.

Arey, J., W.P. Harger, D. Helmig, and R. Atkinson. 1992. Bioassay-directed fractionation of mutagenic PAH atmospheric photooxidation products and ambient particulate extracts. Mutat. Res. 281: 67–76.

Athanasiou, K., L.G. Viras, and P.A. Siskos. 1986. Mutagenicity and polycyclic aromatic hydrocarbons analysis of ambient airborne particles collected in Athens, Greece. Sci. Total Environ. 52: 201–209.

Atherholt, T.B., G.J. MacGerrity, J.B. Louis, L.J. McGeorge, P.J. Lioy, J.M. Daisey, A. Greenberg, and F. Darack. 1985. Mutagenicity studies of New Jersey ambient air particulate extract. In: *Short-Term Bioassays in the Analysis of Complex Environmental Mixtures IV.* Eds: Waters, M.D., S.S. Sandhu, J.L. Huisingh, L. Claxton, G. Strauss, and S. Nesnow, Plenum Press, New York, pp. 211–231.

Atkinson, R., J. Arey, B. Zielinska, and S.M. Aschmann. 1990a. Kinetics and nitro-products of the gas-phase OH and NO_3 radical-initiated reactions of naphtalene-d_8, fluoranthene-d_{10} and pyrene. Int. J. Chem. Kinet. 22: 999–1014.

Atkinson, R., E.C. Tuazon, and J. Arey. 1990b. Reactions of naphtalene in N_2O_5 -NO_3 -NO_2 -air mixtures. Int. J. Chem. Kinet. 22: 1071–1082.

Baek, S.O., M.E. Goldstone, P.W.W. Kirk, J.N. Lester, and R. Perry. 1991a. Phase distribution and particle size dependency of polycyclic aromatic hydrocarbons in the urban atmosphere. Chemosphere 22: 503–520.

Baek, S.O., M.E. Goldstone, P.W.W. Kirk, J.N. Lester, and R. Perry. 1991b. Methodological aspects of measuring PAH in the urban atmosphere. Environ. Technol. 12: 107–129.

Baker, J.E., and S.J. Eisenreich. 1990. Concentrations and fluxes of polycyclic aromatic hydrocarbons and polychlorinated biphenyls across the air-water interface of lake Superior. Environ. Sci. Technol. 24: 342–352.

Barale, R., D. Zucconi, F. Giorgelli, A.L. Carduci, M. Tonelli, and N. Loprieno. 1989. Mutagenicity of airborne particles from a non industrial town in Italy. Environ. Mol. Mutagen. 13: 227–233.

Barale, R., L. Giromini, G. Ghelardini, C. Scapoli, N. Loprieno, M. Pala, F. Valerio, and I. Barrai. 1991. Correlations between 15 polycyclic aromatic hydrocarbons (PAH) and the mutagenicity of the total PAHS fraction in ambient air particles in La Spezia (Italy). Mutat. Res. 249: 227–241.

Behrendt, H., N.H. Seemayer, A. Holle, and W. Dehnen. 1987. Effect of extract of airborne particulates and of Ca^{++}- ionophore A23187 on formation and release of surfactant from human type II pneumocytes. J. Aerosol Sci. 18: 705–708.

Binková, B., J. Lenícek, I. Benes, P. Vidová, O. Gajdos, M. Fried, and R.J. Srám. 1998. Genotoxicity of coke-oven and urban air particulate matter in in vitro acellular assays coupled with ^{32}P-postlabeling and HPLC analysis of DNA adducts. Mutat. Res. 414: 77–94.

Biran, R., Y. Tang, J.R. Brook, R. Vincent, and G.J. Keeler. 1996. Aqueous extraction of airborne particulate matter collected on hivol teflon filters. Intern. J. Environ. Anal. Chem. 63: 315–322.

Brook, J.R., A.H. Wiebe, S.A. Woodhouse, C.V. Audette, T.F. Dann, S. Callaghan, M. Piechowski, E. Dabel-Zlotorzynska, and J.F. Dloughy. 1997. Temporal and spatial relationships in fine particle strong acidity, sulphate, PM_{10}, and $PM_{2.5}$ across multiple Canadian locations. Atmos. Environ. 24: 4223–4236.

Burton, R.M., H.H. Suh, and P. Koutrakis. 1996. Spatial variation in particulate concentrations within metropolitan Philadelphia. Environ. Sci. Technol. 30: 400–407.

Butler, J.D., and P. Crossley. 1981. Reactivity of polycyclic aromatic hydrocarbons adsorbed on soot particles. Atmos. Environ. 15: 91–94.

Butler, J.P., T.J. Kneip, F. Mukai, and J.M. Daisey. 1985. Interurban variations in the mutagenic activity of the ambient aerosol and their relations to fuel use patterns. In: *Short-Term Bioassays in the Analysis of Complex Mixtures IV.* Eds.: Waters, M.D., S.S. Sandhu, J. Lewtas, L. Claxton, G. Strauss, and S. Nesnow, Plenum Press, New York, pp. 233–246.

Cautreels, W., and K. van Cauwenberghe. 1976. Extraction of organic compounds from airborne particulate matter. Water Air Soil Pollut. 6: 103–110.

Cautreels, W., and K. van Cauwenberghe. 1978. Experiments on the distribution of organic pollutants between airborne particulate matter and the corresponding gas phase. Atmos. Eviron. 12: 1133–1141.

Cederlöf, R., R. Doll, B. Fowler, and L. Friberg. 1978. Air pollution and cancer: risk assessment methodology and epidemiological evidence. Report of a task group. Environ. Health Perspect. 22: 1–12.

Chan, Y.C., R.W. Simpson, G.H. Mctainsh, P.D. Vowles, D.D. Cohen, and G.M. Bailey. 1997. Characterization of chemical species in $PM_{2.5}$ and PM_{10} aerosols in Brisbane, Australia. Atmos. Environ. 22: 3773–3785.

Chorazy, M., J. Szeliga, M. Strozky, and B. Cimander. 1994. Ambient air pollutants in upper Silesia: partial chemical composition and biological activity. Environ. Health Perspect. 102: 61–66.

Chou, M.C., and H. Lee. 1990. Mutagenicity of airborne particles from four cities in Taiwan. Proc. Natl. Sci. Counc. Repub. China. 14: 142–150.

Choudhury, A.H., M.E. Gordian, and S.S. Morris. 1997. Associations between respiratory illness and PM10 air pollution. Arch. Environ. Health. 52: 113–117.

Chow, J.C., J.G. Watson, Z. Lu, D.H. Lowenthal, C.A. Frazier, P.A. Solomon, R.H. Thuillier, and K. Magliano. 1996a. Descriptive analysis of $PM_{2.5}$ and PM_{10} at regionally representative locations during SJVAQS/AUSPEX. Atmos. Environ. 30: 2079–2112.

Chow, J.C., J.G. Watson, D.H. Lowenthal, and R.J. Countess. 1996b. Sources and chemistry of PM_{10} aerosol in Santa Barbara County, CA. Atmos. Environ. 30: 1489–1499.

Chrisp, C.E., and G.L. Fisher. 1980. Mutagenicity of airborne particles. Mutat. Res. 76: 143–164.

Chuang, J.C., S.W. Hannan, and N.K. Wilson. 1987. Field comparison of polyurethane foam and XAD-2 resin for air sampling for polynuclear aromatic hydrocarbons. Environ. Sci. Technol. 21: 798–804.

Chuang, J.C., S.R. Cao, Y.L. Xian, D.B. Harris, and J.L. Mumford. 1992. Chemical characterization of indoor air of homes from communes in Xuan Wei, China, with high lung cancer mortality rate. Atmos. Environ. 26A: 2193–2201.

Churg, A., and M. Brauer. 1997. Human lung parenchyma retains $PM_{2.5}$. Am. J. Respir. Crit. Care Med. 155: 2109–2111.

Coe, J.M., and Lindberg, S.E. 1987. The morphology and size distribution of atmospheric particles deposited on foliage and inert surfaces. JAPCA. 37: 237–243.

Commoner, B., P. Madyasta, A. Brondson, and A.J. Vithayathil. 1978. Environmental mutagens in urban particles. J. Toxicol. Environ. Health. 4: 59–77.

Courtois, Y.A., S. Min, C. Lachenal, J.M. Jacquot-Deschamps, F. Callais, and B. Festy. 1988. Genotoxicity of organic extracts from atmospheric particles. In: *Living in a Chemical World. Occupational and Environmental Significance of Industrial Carcinogens*. Eds.: Maltoni, C., and I. Selikoff, Ann. N.Y. Acad. Sci. Vol. 534, pp. 724–740.

Courtois, Y.A., M.L. Pesle, and B. Festy. 1992. Activation of promutagens in complex mixtures by rat liver S9 systems. Mutat. Res. 276: 133–137.

Coutant, R.W., L. Brown, J.C. Chuang, R.M. Riggin, and R.G. Lewis. 1988. Phase distribution and artifact formation in ambient air sampling for polynuclear aromatic hydrocarbons. Atmos. Environ. 22: 403–409.

Crebelli, R., S. Fuselli, A. Meneguz, G. Aquilina, L. Conti, P. Leopardi, A. Zijno, F. Baris, and A. Carere. 1988. In *vitro* and in *vivo* mutagenicity studies with airborne particulate extracts. Mutat. Res. 204: 565–575.

Crebelli, R., S. Fuselli, G. Conti, L. Conti, and A. Carere. 1991. Mutagenicity spectra in bacterial strains of airborne and engine exhaust particulate extracts. Mutat. Res. 261: 237–248.

Daisey, T., I. Hawryluk, T. Kneip, and F. Mukai. 1979. Mutagenic activity in organic fractions of airborne particulate matter. In: *Conference on Carbonaceous Particles in the Atmosphere*. U.S. Dept. of Commerce, Washington, D.C., pp. 187–192.

Daisey, T., T.J. Kneip, I. Hawryluk, and F. Mukai. 1980. Seasonal variations in the bacteria mutagenicity of airborne particulate organic matter in New York City. Environ. Sci. Technol. 14: 1487–1490.

Dams, R., J. Billiet, C. Block, M. Demuynck, and M. Janssens. 1975. Complete chemical analysis of airborne particulates. Atmos. Environ. 9: 1099–1106.

De Flora, S., M. Bagnasco, A. Izzotti, F. D'Agostini, M. Pala, and F. Valerio. 1989a. Mutagenicity of polycyclic aromatic hydrocarbon fraction extracted from urban air particulates. Mutat. Res. 224: 305–318.

De Flora, S., A. Camoirano, A. Izzotti, F. D'Agostini, and C. Bennicelli. 1989b. Photoactivation of mutagens. Carcinogenesis 10: 1089–1097.

Delgado-Rodríguez, A., R. Ortiz-Marttelo, R. Villalobos-Pietrini, S. Gómez-Arroyo, and U. Graf. 1999. Genotoxicity of organic extracts of airborne particles in somatic cells of *Drosophila melanogaster*. *Chemosphere* 39: 33–44.

De Raat, W.K., S.A. Kooijman, and J.W. Gielen. 1987. Concentrations of polycyclic hydrocarbons in airborne particles in The Netherlands and their correlation with mutagenicity. Sci. Total Environ. 66: 95–114.

De Raat, W.K. 1988. Polycyclic aromatic hydrocarbons and mutagens in ambient air particles. Toxicol. Environ. Chem. 16: 259–279.

De Wiest, F., and H. Della Fiorentina. 1975. Suggestions for a realistic definition of an air quality index relative to hydrocarbonaceous matter associated with airborne particles. Atmos. Environ. 9: 951–954.

De Wiest, F., D. Rondia, R. Gol-Winkler, and J. Gielen. 1982. Mutagenic activity of non-volatile organic matter associated with suspended matter in urban air. Mutat. Res. 104: 201–207.

DeMarini, D.M., A. Abu-Shakra, R. Gupta, L.J. Hendee, and J.G. Levine. 1992. Molecular analysis of mutations induced by the intercalating agent ellipticine at the *hisD3052* allele of *Salmonella typhimurium* TA98. Environ. Molec. Mutagen. 20: 12–18.

DeMarini, D.M., M.L. Shelton, D.A. Bell, J.G. Levine, and A. Abu-Shakra. 1993. Molecular analysis of mutations induced at the hisD3052 allele of *Salmonella* by single chemicals and complex mixtures. Environ. Health Perspect. 101: 207–212.

DeMarini, D.M., M.L. Shelton, and D.A. Bell. 1994. Mutation spectra in *Salmonella* of complex mixtures: comparison of urban air to benzo(a)pyrene. Environ. Molec. Mutagen. 24: 262–275.

DeMarini, D.M., M.L. Shelton, and D.A. Bell. 1996. Mutation spectra of chemical fractions of a complex mixtures: role of nitroarenes in the mutagenic specificity of municipal waste incinerator emissions. Mutat. Res. 349: 1–20.

Dideren, H.S., R. Guicherit, and J.C.T. Hollander. 1985. Visibility reduction by air pollution in the Netherlands. Atmos. Environ. 19: 377–383.

Dockery, D.W., J. Schwartz, and J.D. Spengler. 1992. Air pollution and daily mortality associations with particulates and acid aerosols. Environ. Res. 59: 362–373.

Dockery, D.W., C.A. Pope III., X. Xu, J.D. Spengler, J.W. Ware, M.E. Fay, B.G. Ferris Jr., and F.A. Speizer. 1993. An association between air pollution and mortality in six U.S. cities. N. Engl. J. Med. 329: 1753–1759.

Dockery, D.W., and C.A. Pope. 1994. Accute respiratory effects of particulate air pollution. Annu. Rev. Pub. Health. 15: 107–132.

EPA,1990. Method TO13, determination of benzo(a)pyrene (BaP) and other polynuclear aromatic hydrocarbons (PAHs) in ambient air using gas chromatographic (GC) and high performance liquid chromatographic (HPLC) analysis. In: *Methods for Determination of Toxic Organic Compounds in Air, EPA methods*. Eds.: Winberry, Jr. W.T., N.T. Murphy, and R.M. Riggan, Noyes Data Corporation, Park Ridge, New Jersey, pp. 370–466.

EPAQS. 1995. Particles. Department of the Environment Expert Panel on Air Quality Standards. HMSO., London.

Espinosa-Aguirre, J.J., R.E. Reyes, J. Rubio, P. Ostrosky-Wegman, and G. Martinez. 1993. Mutagenic activity of urban air samples and its modulation by chili extracts. Mutat. Res. 303: 55–61.

Fairley, D. 1990. The relationship of daily mortality to suspended particulates in Santa Clara County, 1980–1986. Environ. Health Perspect. 89: 115–168.

Falk, H.L., I. Markul, and P. Kotin. 1956. Aromatic hydrocarbons IV. Their fate following emission into the atmosphere and experimental exposure to washed air and synthetic smog. AMA Arch. Ind. Health 13: 13–17.

Finlayson-Pitts, B.J., and J.M. Pitts, Jr. 1997. Tropospheric air pollution: ozone, airborne toxics, polycyclic aromatic hydrocarbons and particles. Science 276: 1045–1052.

Fitch, W.L., and D.H. Smith. 1979. Analysis of adsorption properties and adsorbed species on commercial polymeric carbons. Environ. Sci. Technol. 13: 341–346.

Freeman, D.J., and F.C.R. Cattell. 1990. Woodburning as a source of atmospheric polycyclic aromatic hydrocarbons. Environ. Sci. Technol. 24: 1581–1585.

Fukino, H., S. Mimura, K. Inoue, and Y. Yamane. 1982. Mutagenicity of airborne particles. Mutat. Res. 102: 237–247.

Funcke, W., J. Köning, E. Balfanz, and T. Romanowski. 1981. The PAH profiles in airborne particulate matter of five german cities. Atmos. Environ. 15: 887–890.

Garner, R.C., C.A. Stanton, C.N. Martin, F.L. Chow, W. Thomas, D. Hubner, and R. Herrmann. 1986. Bacterial mutagenicity and chemical analysis of polycyclic aromatic hydrocarbons and some nitro derivatives in environmental samples collected in West Germany. Environ. Mutagen. 8: 109–117.

Gibson, T.L., P.E. Korsog, and G. Wolff. 1986. Evidence for the transformation of polycyclic organic matter in the atmosphere. Atmos. Environ. 20: 1575–1578.

Golden, C., and E. Sawicki. 1975. Ultrasonic extraction of total particulate aromatic hydrocarbons (TPAHs) from airborne particles at room temperature. Int. J. Environ. Anal. Chem. 4: 9–15.

Gordon, R.J., and R.J. Bryant. 1973. Patterns in airborne polynuclear hydrocarbon concentration in four Los Angeles sites. Environ. Sci. Technol. 7: 1050–1053.

Gordon, R.J. 1976. Distribution of airborne polycyclic aromatic hydrocarbons throughout Los Angeles. Environ. Sci. Technol. 10: 370–373.

Goto, S., O. Endo, and H. Matsushita. 1992. Results of a comparative study on the *Salmonella* pre-incubation and plate incorporation assays using test samples from the IPCS collaborative study. Mutat. Res. 276: 93–100.

Griest, W.H., L.B. Yeatts Jr., and J.E. Caton. 1980. Recovery of polycyclic aromatic hydrocarbons on fly ash for quantitative determination. Anal. Chem. 52: 199–201.

Grimmer, G. 1983. Profile analysis of polycyclic aromatic hydrocarbons in air. In: *Handbook of Polycyclic Aromatic Hydrocarbons*. Ed.: Björseth, A., Marcel Dekker, New York, Vol. 1, pp. 149–182.

Grosjean, D. 1983. Polycyclic aromatic hydrocarbons in Los Angeles air from samples collected on teflon, glass and quartz filters. Atmos. Environ. 17: 2565–2573.

Hadnagy, W., N.H. Seemayer, and R. Tomingas. 1986. Cytogenetic effects of airborne particulate matter in human lymphocytes *in vitro*. Mutat. Res. 175: 97–101.

Hadnagy, W., and N.H. Seemayer. 1987. Comparative investigation on the genotoxicity of city smog and automobile exhaust particulates. J. Aerosol. Sci. 18: 697–699.

Harrison, R.M., A.R. Deacon, M.R. Jones, and R.A. Appleby. 1997. Sources and processes affecting concentrations of PM_{10} and $PM_{2.5}$ particulate matter in Birmingham (U.K). Atmos. Environ. 24: 4103–4117.

Haughen, D.A., and M.J. Peak. 1983. Mixtures of polycyclic aromatic compounds inhibit mutagenesis in the *Salmonella*/microsome assay by inhibition of metabolic activation. Mutat. Res. 116: 257–269.

Helmes, C.T., D.L. Atkinson, J. Jaffer, C.C. Sigman, K.L. Thompson, M.I. Kelsey, H.F. Kraybill, and J.I. Munn. 1982. Evaluation and classification of the potential carcinogenicity of organic air pollutants. J. Environ. Sci. Health. A17: 321–389.

Hileman, B. 1981. Particulate matter: the inhalable variety. Environ. Sci. Technol. 15: 983–986.

Hitzfeld, B., K.H. Friedrichs, J. Ring, and H. Behrendt. 1997. Airborne particulate matter modulates the production of reactive oxygen species in human polymorphonuclear granulocytes. Toxicology 120: 185–195.

Houk, V.S., S. Goto, O. Endo, L.D. Claxton, J. Lewtas, and H. Matsushita. 1992. Detection of direct-acting mutagens in ambient air: a comparison of two highly sensitive mutagenicity assays. Environ. Molec. Mutagen. 20: 19–28.

Hughes, M.M., D.F.S. Natusch, D.R. Taylor, and M. Zeller. 1979. Chemical transformations of particulate polycyclic organic matter. In: *Polynuclear Aromatic Hydrocarbons: Chemistry and Biological Effects*. Eds.: Björseth, A., and A.J. Dennis., Battelle Press, Columbus, OH., pp. 1–7.

Iwado, H., M. Naito, and H. Hayatsu. 1991. Mutagenicity and antimutagenicity of airborne particulates. Mutat. Res. 246: 93–102.

Jacob, J., W. Karcher, J.J. Belliardo, R. Dumler, and A. Boenke. 1991. Polycyclic aromatic compounds of environmental and occupational importants—their occurence toxicity and the development of high-purity certified reference materials. Fres. J. Anal. Chem. 340: 755–767.

Jauregui, E. 1986. The urban climate of Mexico City. Proceedings of the WMO Technical Conference on Urban Climatology, Ed.: Oke, T.R., WMO No. 652, Geneva, pp. 63–86.

Jauregui, E. 1997. Heat island development in Mexico City. Atmos. Environ. 31: 3821–3831.

Jeltes, R. 1974. Fingerprinters techniques as aids in the analysis of composite chemical pollutants in the environment. J. Chromatogr. Sci. 12: 599–605.

Jungers, R., R. Burton, L. Claxton, and J. Lewtas Huisingh. 1981. Evaluation of collection and extraction methods for mutagenesis studies on ambient air particulate. In: *Short-Term Bioassays in the Analysis of Environmental Complex Mixtures II*. Eds.: Waters, M.D., S.S. Sandhu, J.L. Huisingh, L. Claxton, and S. Nesnow, Plenum Press, New York, pp. 45–65.

Kado, N.Y., N.G. Guirguis, C.P. Flesse, R.C. Chan, K.I. Chang, and J.J. Wesolowski. 1986. Mutagenicity of fine (<2.5 μm) airborne particles: diurnal variation in community air determined by a *Salmonella* (micro preincubation /microsuspension) procedure. Environ. Mutagen. 8: 53–66.

Katz, M., T. Sakuma, and A. Ho. 1978. Chromatographic and spectral analysis of polynuclear aromatic hydrocarbons. Quantitative distribution in air of Ontario cities. Environ. Sci. Technol. 12: 909–915.

Katz, M., and C. Chan. 1980. Comparative distribution of eight polycyclic aromatic hydrocarbons in airborne particulates collected by conventional high-volume sampling and by size fractionation. Environ. Sci. Technol. 14: 838–843.

Keane, M.J., S.-G. Xing, J.C. Harrison, T. Ong, and W.E. Wallace. 1991. Genotoxicity of diesel-exhaust particles dispersed in simulated pulmonary surfactant. Mutat. Res. 260: 233–268.

Kelly, T.J., R. Mukund, C.W. Spicer, and A.J. Pollack. 1994. Concentrations and transformations of hazardous air pollutants. Environ. Sci. Technol. 28: 378A–387A.

Keyser, T.R., D.F.S. Natusch, C.A. Evans Jr., and R.W. Linton. 1978. Characterizing the surfaces of environmental particles. Environ. Sci. Technol. 12: 768–773.

Kolber, A., T. Wolff, T. Hughes, E. Pellizari, C. Sparacino, M. Waters, J. Lewtas Huisingh, and L. Claxton. 1981. Collection, chemical fractionation, and mutagenicity bioassay of ambient air particulate. In: *Short-Term Bioassays in the Analysis of Complex Environmental Mixtures II*. Eds.: Waters, M.D., S.S. Sandhu, J.L. Huisingh, L. Claxton, and S. Nesnow, Plenum Press, New York, pp. 21–43.

König, J, W. Funcke, E. Balfanz, B. Grosch, and F. Pott. 1980. Testing a high volume air sampler for quantitative collection of polycyclic aromatic hydrocarbons. Atmos. Environ. 14: 609–613.

Korfmacher, W.A., D.F.S., Natusch, D.R. Taylor, E.L. Wehry, and G. Mamantov. 1979. Thermal and photochemical decomposition of particulate PAHS. In: *Polynuclear aromatic hydrocarbons*. Eds.: Jones, P.W., and P. Leber, Ann. Arbor Science Pub., Michigan, pp. 165–170.

Korfmacher, W.A., E.L. Wehry, G. Mamantov, and D.F.S. Natusch. 1980. Resistance to photochemical decomposition of polycyclic aromatic hydrocarbons vapor-adsorbed on coal fly ash. Environ. Sci. Technol. 14: 1094–1099.

Korfmacher, W.A., G. Mamantov, E.L. Wehry, D.F.S. Natusch, and T. Mauney. 1981. Nonphotochemical decomposition of flourene vapor-adsorbed on coal fly ash. Environ. Sci. Technol. 15: 1370–1375.

Krishna, G., T. Ong, W.Z. Whong, and J. Nath. 1983a. Mutagenicity studies of ambient airborne particles. I. Comparison of solvent systems. Mutat. Res. 124: 113–120.

Krishna, G., J. Nath, W.Z. Whong, and T. Ong. 1983b. Mutagenicity studies of ambient airborne particles. II. Comparison of extraction methods. Mutat. Res. 124: 121–128.

Krishna, G., J. Nath, and T. Ong. 1984. Correlative genotoxicity studies of airborne particles in *Salmonella typhimurium* and cultured human lymphocytes. Environ. Mutagen. 6: 585–592.

Krishna, G., J. Nath, L. Soler, and T. Ong. 1986. Comparative in vivo and in vitro genotoxicity studies of airborne particle extract in mice. Mutat. Res. 171: 157–163.

Krökje, A., R. Schmid, and K. Zahlsen. 1991. Liver, lung and kidney homogenates used as an activation system in mutagenicity studies of airborne particles and of expectorate and urine samples from exposed workers in a coke plant. Mutat. Res. 259: 49–65.

Lee, M., M. Novotny, and K. Bartle. 1976. Gas chromatography/mass spectrometry and nuclear magnetic resonance determination of polynuclear aromatic hydrocarbons in airborne particulates. Anal. Chem. 48: 1566–1572.

Lee, S.F., and D. Schuetzle. 1983. Sampling, extraction, and analysis of polycyclic aromatic hydrocarbons from internal combustion engines. In: *Handbook of Polycyclic Aromatic Hydrocarbons*. Ed.: Björseth, A., Marcel Dekker, New York, Vol. 1, pp. 27–94.

Lee, H., S.M. Law, and S.T. Lin. 1991. The effect of extraction solvent on the mutagenicity of airborne particles. Toxicol. Lett. 58: 59–67.

Legzdins, A.E., B.E McCarry, C.H. Marvin, and D.W. Bryant. 1995. Methodology for biossay-directed fractionation studies of air particulate material and other complex environmental matrices. Intern. J. Environ. Anal. Chem. 60: 79–94.

Lewtas, H.J. 1980. Bioassay of particulate organic matter from ambient air. In: *Short-Term Bioassays in the Analysis of Environmental Complex Mixtures II*. Eds.: Waters, M.D., S.S. Sandhu, J.L. Huisingh, L. Claxton, and S. Nesnow, Plenum Press, New York, pp. 9–19.

Lewtas, J. 1988. Genotoxicity of complex mixtures: strategies for the identification and comparative assessment of airborne mutagens and carcinogens from combustion sources. Fundam. Appl. Toxicol. 10: 571–589.

Lewtas, J. 1990. Experimental evidence for the carcinogenicity of air pollutants. In: *Air Pollution and Human Cancer*. Ed.: Tomatis, L., Springer-Verlag, Berlín, pp. 49–61.

Lewtas, J. 1994. Human exposure to complex mixtures of air pollutants. Toxicol. Lett. 72: 163–169.

Lewtas, J., D. Walsh, R. Williams, and L. Dobias. 1997. Air pollution exposure-DNA adduct dosimetry in humans and rodents: evidence for non-linearity at high doses. Mutat. Res. 378: 51–63.

Li, Y., and H.D. Roth. 1995. Daily mortality analysis by using different regression models in Philadelphia County, 1973–1990. Inhalation Toxicol. 7: 45–58.

Lioy, P.J., J.M. Waldman, T. Buckley, J. Butler, and C. Pietarinen. 1990. The personal, indoor and outdoor concentrations of PM-10 measured in an industrial community during the Winter. Atmos. Environ. 24B: 57–66.

Löfroth, G. 1980. Comparison of the mutagenic activity in diesel and gasoline engine exhaust and in carbon particulate matter. USEPA Second Symposium on the Application of Short-Term Biossays in the Fractionation and Analysis of Complex Environmental Mixtures, Williamsburg, V.A.

Löfroth, G. 1981. Comparison of the mutagenic activity in carbon particulate matter and in diesel and gasoline engine exhaust. In: *Short-Term Bioassays in the Analysis of Complex Environmental Mixtures II*. Eds.: Waters, M.D., S.S. Sandhu, J.L. Huisingh, L. Claxton, and S. Nesnow, Plenum Press, New York, pp. 319–336.

Maron, D.M., and B.N. Ames. 1983. Revised methods for the *Salmonella* mutagenicity test. Mutat. Res. 113: 173–215.
Masclet, P., G. Mouvier, and K. Nikolaou. 1986. Relative decay index and sources of polycyclic aromatic hydrocarbons. Atmos. Environ. 20: 439–446.
Masclet, P., P. Pistikopoulos, S. Beyne, S., and G. Mouvier. 1988. Longe range transport and gas/particle distribution of polycyclic aromatic hydrocarbons at a remote site in the Mediterranean Sea. Atmos. Environ. 22: 639–650.
Matsushita, H., O. Endo, S. Goto, H. Simizu, H. Matsumoto, K. Tamakawa, T. Endo, Y. Sakabe, H. Tokiwa, and M. Ando. 1992. Collaborative study using the preincubation *Salmonella thyphimurium* mutation assay for airborne particulate matter in Japan. A trial to minimize interlaboratory variation. Mutat. Res. 271: 1–12.
McCann, J., N.E. Spingarn, J. Kobori, and B.N. Ames. 1975. Detection of carcinogens as mutagens: bacterial tester strains with R factor plasmids. Proc. Natl. Acad. Sci. (USA) 72: 979–983.
McCann, J., and B.N. Ames. 1976. Detection of carcinogens as mutagens in the salmonella/microsome test: assay of 300 chemicals: discussion. Proc. Natl. Acad. Sci. (USA) 73: 950–954.
McCann, J., and B.N. Ames. 1977. The *Salmonella*/microsome mutagenicity test: predictive value for animal carcinogenicity. In: *Origins of Human Cancer*. Eds.: Hiatt, H.H., J.D. Watson, and J.A. Winston, Cold Spring Harbor Laboratory, Book C. pp. 1431–1450.
McLachlan, M., D. Mackay, and P.H. Jones. 1990. A conceptual model of organic chemical volatilization at waterfalls. Environ. Sci. Technol. 24: 252–257.
McVeety, B.D., and R.A. Hites. 1988. Atmospheric deposition of polycyclic aromatic hydrocarbons to water surfaces: a mass balance approach. Atmos. Environ. 22: 511–536.
Mermelstein, R., D.K. Kiriazides, M. Buttler, E.C. McCoy, and H.S. Rosenkranz. 1981. The extraordinary mutagenicity of nitropyrenes in bacteria. Mutat. Res. 89: 187–196.
Möller, M., and I. Alfheim. 1980. Mutagenicity, and PAH-analysis of airborne particulate matter. Atmos. Environ. 14: 83–88.
Möller, M., I. Alfheim, S. Larssen, and A. Mikalsen. 1982. Mutagenicity of airborne particles in relation to traffic and air pollution parameters. Environ. Sci. Technol. 16: 221–225.
Möller, M., and G. Löfroth. 1982. Genotoxic components in polluted air. In: *Mutagens in Our Environment*. Eds.: Sorsa, M., and H. Vainio, Alan Liss, New York, pp. 221–234.
Möller, M., and I. Alfheim. 1983. Mutagenicity of air samples from various combustion sources. Mutat. Res. 116: 35–46.
Möller, L., E. Grzybowska, M. Zeizig, B. Cimander, K. Hemminki, and M. Chorazy. 1996. Seasonal variation of DNA adduct pattern in human lymphocytes analyzed by 32P-HPLC. Carcinogenesis 17: 61–66.
Monarca, S., R. Crebelli, D. Feretti, A. Zanardini, A. Fuselli, L. Filini, S. Resola, P.G. Bonardelli, and G. Nardi. 1997. Mutagens and carcinogens in size-classified air particulates of a northern Italian town. Sci. Total Environ. 205: 137–144.
Monn, C., A. Fuchs, D. Högger, M. Junker, D. Kogelschatz, N. Roth, and H.-U. Wanner. 1997. Particulate matter less than 10 μm (PM_{10}) and fine particles less than 2.5 μm ($PM_{2.5}$): relationships between indoor, outdoor and personal concentrations. Sci. Total Environ. 208: 15–21.
Moolgolvkar, S.H., E.G. Luebeck, T.A. Hall, and E.L. Anderson. 1995. Air pollutiom and daily mortality in Philadelphia. Epidemiology 6: 476–484.
Moolgolvkar, S.H., and E.G. Luebeck. 1996. Particulate air pollution and mortality: A critical review of the evidence. Epidemiology 7: 420–428.
Moriske, H.J., I. Block, H. Schleibinger, and H. Ruden. 1985. Polar neutral organic compounds in urban aerosols. I. Chemical characterization and mutagenic effect in relation to various sources. Zentralbl Bakteriol Mikrobiol Hyg. 181: 240–271.
Morris, W.A., J.K. Versteg, A.W. Bryant, A.E. Legzdins, B.E. McCarry, and C.H. Marvin. 1995. Preliminary comparisons between mutagenicity and magnetic suceptibility of respirable airborne particulate. Atmos. Environ. 29: 3441–3450.
Mosiño, P.A.A., and E. García. 1973. The climate of Mexico. In: *Climates of North America*. Eds.: Bryson, R.A., and F.K. Hare, Elsevier, Amsterdan, pp. 345–404.

Motykiewics, G., J. Michalska, J. Szeliga, and B. Cimander. 1988. Mutagenic and clastogenic activity of direct-acting components from air pollutants of the Silesian industrial region. Mutat. Res. 204: 289–296.

Motykiewics, G., W. Hadnagy, N.H. Seemayer, J. Szeliga, A. Tkocz, and M. Chorazy. 1991. Influence of airborne suspended matter on mitotic cell division. Mutat. Res. 260: 195–202.

Nardini, B., and E. Clonfero. 1992. Mutagens in urban air particulate. Mutagenesis 7: 421–425.

Natusch, D.F.S. 1978. Potentially carcinogenic species emitted to the atmosphere by fossil-fuelled power plants. Environ. Health Perspect. 22: 79–80.

Natusch, D.F.S., and B.A. Tomkins. 1978. Theoretical consideration of the absorption of the polynuclear aromatic hydrocarbons vapor onto fly ash in a coal-fired power plant. In: *Polynuclear aromatic hydrocarbons.* Eds.: Jones, P.W., and R.I. Fruedenthal, Raven Press, New York.

Nielsen, T., T. Ramdahl, and A. Björseth. 1983. The fate of airborne polycyclic organic matter. Environ. Health Perspect. 47: 103–114.

Nielsen, T., B. Seitz, and T. Ramdahl. 1984. Occurrence of nitro-PAH in the atmosphere in a rural area. Atmos. Environ. 18: 2159–2165.

Nielsen, T., and T. Ramdahl. 1986. Discussion on: determination of 2-nitrofluoranthene and 2-nitropyrene in ambient particulate matter: evidence for atmospheric reactions. Atmos. Environ. 20: 1507.

Nielsen, P.A. 1992. Mutagenicity studies on complex mixtures: selection of solvent system for extraction. Mutat Res. 276: 117–123.

Nishioka, M.G., C.C. Howard, D.A. Contos, L.M. Ball, and J. Lewtas. 1988. Detection of hydroxylated nitro aromatic and hydroxylated nitro polycyclic aromatic compounds in an ambient air particulate extract using bioassay-directed fractionation. Environ. Sci. Technol. 22: 908–915.

Ohsawa, M., T. Ochi, and H. Hayashi. 1983. Mutagenicity in *Salmonella typhimurium* mutants of serum extracts from airborne particulates. Mutat. Res. 116: 83–90.

Ohta, S., M. Hori, S. Yamagata, and N. Murao. 1998. Chemical characterization of atmospheric fine particles in Sapporo with determination of water content. Atmos. Environ. 32: 1021–1025.

Olufsen, B.S., and A. Björseth. 1983. Analysis of polycyclic aromatic hydrocarbons by gas chromatography. In: *Handbook of Polycyclic Aromatic Hydrocarbons.* Ed.: Björseth, A., Marcel Dekker, New York, Vol. 1, pp. 257–300.

Ong, T.M., W.Z. Whong, J.D. Stewart, and H.E. Brockman. 1986. Chlorophyllin: a potent antimutagen against environmental and dietary complex mixtures. Mutat. Res. 173: 111–115.

Ong, T., W.-Z. Whong, J.D. Stewart, and H.E. Brockman. 1989. Comparative antimutagenicity of 5 compounds against 5 mutagenic complex mixtures in *Salmonella typhimurium* strain TA98. Mutat. Res. 222: 19–25.

Ostro, B.D. 1993. The association of air pollution and mortality. Examining the care for inference. Arch. Environ. Health, 48: 336–342.

Ostro, B., and L. Chestnut. 1998. Assessing the health benefits of reducing particulate matter air pollution in the United States. Environ. Res. 76: 94–106.

Ozkaynak, H., and G.D. Thurston. 1987. Associations between 1980 U.S. mortality rates and alternative measures of airborne particle concentration. Risk Anal. 7: 449–461.

Peltonen, K., and T. Kuljukka. 1995. Air sampling and analysis of polycyclic aromatic hydrocarbons. J. Chromatogr. 710: 93–108.

Pistikopoulos, P., H.M. Worthman, L. Gomes, S. Masclet-Beyne, E. Bon Nguyen, P.A. Masclet, and G. Mouvier. 1990. Mechanisms of formation of particulate polycyclic aromatic hydrocarbons in relation to the particle size distribution: effects on meso-scale transport. Atmos. Environ. 24A: 2573–2584.

Pitts, Jr. J.N., D. Grosjean, T.M. Mischke, V.F. Simmon, and D. Poole. 1977. Mutagenic activity of airborne particulate organic pollutants. Toxicol. Lett. 1: 65–70.

Pitts, Jr. J.N., K.A. van Cauwenberghe, D. Grosjean, J.P. Schmid, D.R. Fitz, W.L. Belser, G.B. Knudson, and P.M. Hynds. 1978a. Atmospheric reactions of polycyclic hydrocarbons: facile formation of mutagenic nitroderivatives. Science 202: 515–519.

Pitts, J.N., K.A. van Cauwenberghe, D. Grosjean, J.P. Schmid, D.R. Fitz, W.L. Belser, G.B. Knudson, and P.M. Hynds. 1978b. Chemical and microbiological studies of mutagenic pollutants in real and simulated atmospheres. In: *Application of Short-Term Bioassays in the Fractionation and Analysis of*

Complex Environmental Mixtures. Eds.: Waters, M.D., S. Nesnow, J.L. Huising, S.S. Sandhu, and L. Claxton, Plenum Press, New York, pp. 353–379.

Pitts, Jr. J.N., W. Harger, D.M. Lokensgard, D.R. Fitz, G.M. Scorziell, and V. Mejia. 1982. Diurnal variations in the mutagenicity of airborne particulate organic matter in California's south coast air basin. Mutat. Res. 104: 35–41.

Pitts, Jr. J.N. 1987. Nitration of gaseous polycyclic aromatic hydrocarbons in simulated and ambient urban atmospheres: a source of mutagenic nitroarenes. Atmos. Environ. 21: 2531–2547.

Ponciano, R.G. 1996. Cáncer pulmonar y contaminación atmosférica. Existe una asociación?. In: *Riesgos Ambientales para la Salud en la Ciudad de México*. Eds.: Serrano, O.R., and R.G. Ponciano, Programa Universitario del Medio Ambiente, UNAM, México, D.F., pp. 127–171.

Pope, C.A.III., J. Schwartz, and M.R. Ransom. 1992. Daily mortality in PM10 pollution in Utah Valley. Arch. Environ. Health. 47: 211–217.

Pope, C.A.III., M.J. Thun, M.M. Namboodiri, D.W. Dockery, J.S. Evans, F.E. Speizer, and C.W. Heath. 1995. Particulate air pollution as a predictor of mortality in a prospective study of US adults. Amer. J. Resp. Crit. Care Med. 151: 669–674.

Pyysalo, H., J. Tuominen, K. Wickström, E. Skyttä, L. Tikkanen, S. Salomaa, M. Sorsa, T. Nurmela, T. Mattila, and V. Pohjola. 1987. Polycyclic organic material (POM) in urban air. Factionation, chemical analysis and genotoxicity of particulate and vapour phases in an industrial town in Finland. Atmos. Environ. 21: 1167–1180.

Ramdahl, T., I. Alfheim, S. Rustad, and T. Olsen. 1982. Chemical and biological characterization of emissions from small stoves burning wood and coal. Chemosphere 11: 601–611.

Ramdahl, T., B. Zielinska, J. Arey, R. Atkinson, A.M. Winer, and J.N. Pitts Jr. 1986. Ubiquitous occurrence of 2-nitrofluoranthene and 2-nitropyrene in air. Nature 321: 425–427.

Reali, D., H. Schilitt, C. Lohse, R. Barale, and N. Loprieno. 1984. Mutagenicity and chemical analysis of airborne particulates from a rural area in Italy. Environ. Mutagen. 6: 813–823.

Redmon, S.K., B.R. Srobino, and R.H. Cypress. 1976. Cancer experience among coke-by-product workers. Ann N.Y. Acad. Sci. 217: 102–115.

Rodes, C.E. y E.G. Evans. 1985. Preliminary assessment of 10 µm particulate sampling at eight locations in United States. Atmos. Environ. 19: 293–303.

Rosenkranz, H.S., and R. Mermelstein. 1980. The *Salmonella* mutagenicity and the *E. coli* Pol A^+/Pol A_1^- repair assays: evaluation of relevance to carcinogenesis. In: *The Predictive Value of In Vitro Short-Term Screening Tests in the Evaluation of Carcinogenicity*. Eds.: Williams, G.M., R. Kroes, H.W. Waaijers, and K.W. van der Poll, Elsevier/North-Holland, Amsterdam, pp. 5–26.

Salamone, M.F., J.A. Heddle, and M. Katz. 1979. The mutagenicity activity of thirty polycyclic aromatic hydrocarbons (PAH) and oxides in urban airborne particulates. Environ. Int. 2: 37–43.

Samet, M.J. 1992. Indoor radon and lung cancer. Estimating the risk. Wester J. Med. 156: 25–29.

Sasaki, Y., T. Kawai, K. Ohyama, A. Nakama, and R. Endo. 1987. Carcinogenicity of extract of airborne particles using newborn mice and comparative study of carcinogenic and mutagenic effect of the extract. Arch. Environ. Health 42: 14–18.

Sato, M.I., G.U. Valent, C.A. Coimbrao, M.C. Coelho, P. Sanchez-Sanchez, C.D. Alonso, and M.T. Martins. 1995. Mutagenicity of airborne particulate organic material from urban and industrial areas of Sao Paulo, Brazil. Mutat. Res. 335: 317–330.

Scarpato, R., F. Di Marino, A. Strano, A. Curti, R. Campagna, N. Loprieno, I. Barrai, and R. Barale. 1993. Two years, air mutagenesis monitoring in a northwestern rural area of Italy with an industrial plant. Mutat. Res. 319: 293–301.

Schuetzle, D., and J. Lewtas. 1986. Bioassay-directed chemical analysis in environmental research. Anal. Chem. 58: 1060a–1072a.

Schwartz, J., and A. Marcus. 1990. Mortality and air pollution in London: a time series analysis. Am. J. Epidemiol. 131: 185–194.

Schwartz, J. 1991a. Particulate air pollution and daily mortality: a synthesis. Public Health Rev. 19: 39–60.

Schwartz, J. 1991b. Particulate air pollution and daily mortality in Detroit. Environ. Res. 56: 204–213.

Schwartz, J., and D.W. Dockery. 1992a. Particulate air pollution and daily mortality in Steubenville, Ohio. Am. J. Epidemiol. 135: 12–19.

Schwartz, J., and D.W. Dockery. 1992b. Increased mortality in Philadelphia associated with daily air pollution concentration. Am. Rev. Respir. Dis. 145: 600–604.

Schwartz, J., D.W. Dockery, and L.M. Neas. 1996. Is daily mortality associated specifically with fine particles?. J. Air Waste Manag. Assoc. 46: 927–939.

Seinfeld, T.H. 1989. Urban air pollution. State of the science. Science. 243: 745–762.

Sera, N., M. Kai, K. Horikawa, K. Fukuhara, N. Miyata, and H. Tokiwa. 1991. Detection of 3,6-dinitrobenzo[a]pyrene in airborne particulates. Mutat. Res. 263: 27–32.

SEDUE (Secretaría de Desarrollo Urbano y Ecologia. 1987. Indice metropolitano de calidad del aire (IMECA). México, D.F.

Seemayer, N.H., W. Hadnagy, and R. Tomingas. 1987. Mutagenic and carcinogenic effects of airborne particulate matter form polluted areas on human and rodent tissue cultures. In: *Advances in Aerobiology*. Eds.: Boehm, G., and R.M. Leuschner, Birkäuser Verlag, Basel, pp. 231–234.

Shaw, R.W., and R.K. Stevens. 1980. Trace element abundances and chemistry of atmospheric aerosols: current techniques and future possibilities. Ann. N. Y. Acad. Sci. 338: 13–25.

Sheu, H.L., W.-J. Lee, S.J. Lin, G.-C. Fang, H.-C. Chang, and W.-C. You. 1997. Particle-bound PAH content in ambient air. Environ. Pollut. 96: 369–382.

Siak, J., T.L. Chan, T.L. Gibson, and G.T. Wolff. 1985. Contribution to bacterial mutagenicity from nitro-PAHS compounds in ambient aerosols. Atmos. Environ. 19: 369–376.

Sicherer-Roetman, A., M. Ramlal, C.E. Voogd, and H.J.T. Bloemen. 1988. The fractionation of extracts of ambient particulate matter for mutagenicity testing. Atmos. Environ. 22: 2803–2808.

Sideropoulos, A.S., and S. Specht. 1994. Mutagenicity of airborne particulates in Allegheny County: unmasking of potential carcinogenicity by size class fractionation. Microbios 77: 167–179.

Sorenson, W.G., W.Z. Whong, J.P. Simpson, F.J. Heart, and T.M. Ong. 1982. Studies of the mutagenic response of *Salmonella typhimurium* TA98 to size-fractionated particles: comparison of the fluctuation and plate incorporation tests. Environ. Mutagen. 4: 531–541.

Suh, H.H., Y. Nishioka, G.A. Allen, P. Koutrakis, and R.M. Burton. 1997. The metropolitan acid aerosol characterization study: results from the summer. Environ. Health Perspect. 105: 826–834.

Sweet, C.W., and D.F. Gatz. 1998. Short communication summary and analysis of available $PM_{2.5}$ measurements in Illinois. Atmos. Environ. 32: 1129–1133.

Takada, H., T. Onda, and E. Wei. 1990. Determination of polycyclic aromatic hydrocarbons in urban street dusts and their source materials by capillary gas chromatography. Environ. Sci. Technol. 24: 1179–1186.

Talcott, R.E., and E. Wei. 1977. Airborne mutagens bioassayed in *Salmonella typhimurium*. J. Natl. Cancer Inst. 58: 449–451.

Talcott, R.E., and W. Harger. 1980. Airborne mutagens extracted from particles of respirable size. Mutat. Res. 79: 177–180.

Talcott, R.E., and W. Harger. 1981. Chemical characterization of direct-acting airborne mutagens: the functional group. Mutat. Res. 91: 433–436.

Teranishi, K., K. Hamada, N. Tekeda, and H. Watanabe. 1977. Mutagenicity of the tar in air pollutants. Proc. 4th Int. Clean Air Congress, Tokio., pp. 33–36.

Teranishi, K., K. Hamada, and H. Watanabe. 1978. Mutagenicity in *Salmonella typhimurium* mutants of the benzene-soluble organic matter derived from air-borne particulate matter and its five fractions. Mutat. Res. 56: 273–280.

Thrane, K.E., and A. Mikalsen. 1981. High-volume sampling of airborne polycyclic aromatic hydrocarbons using glass fiber filters and polyurethane foam. Atmos. Environ. 15: 909–918.

Tokiwa, H., K. Morita, H. Takeyoshi, K. Takahashi, and Y. Ohnishi. 1977. Detection of mutagenic activity in particulate air pollutants. Mutat. Res. 48: 237–248.

Tokiwa, H., S. Kitamori, K. Takahashi, and Y. Ohnishi. 1980. Mutagenic and chemical assay of extracts of airborne particulates. Mutat. Res. 77: 99–108.

Topinka, J., B. Binková, G. Maracková, Z. Stávková, I. Benés, J. Dejmek, J. Lenícek, and R.J. Srám. 1997. DNA adducts in human placenta as related to air pollution and to GSTM1 genotype. Mutat. Res. 390: 59–68.

Torok, G., K. Bejczi, M. Csik, E. Fay, M. Kertesz, A. Pinter, A. Surjan, and M. Borzsonyi. 1988. Mutagenicity and PAH analysis of particulate matter from urban atmosphere and various industrial emission. Proc. XVIII Annual meeting of EEMS, Varna, Bulgaria, October 3–8, p. 266.

Torok, G., M. Csik, M. Kertesz, E. Fay, T. Somorjay, M. Borzsonyl, A. Surjan, Z. Kelecsenyi, and A. Pinter. 1989. Mutagenicity and PAH content of airborne particulates and of fallen dusts from two

Hungarian towns and emission samples from aluminum reduction and power plants. Ann. Ist. Super Sanita 25: 595–599.

Tuch, T.H., P. Brand, H.E. Wichmann, and J. Heyder. 1997. Variation of particle number and mass concentration in various size ranges of ambient aerosols in eastern Germany. Atmos. Environ. 24: 4193–4197.

Tuominen, J., S. Salomaa, H. Pyysalo, E. Skyttä, L. Tikkanen, T. Nurmela, M. Sorsa, V. Pohjola, M. Sauri, and K. Himberg. 1988. Polynuclear aromatic compounds and genotoxicity in particulate and vapor phases of ambient air: effect of traffic, season, and meteorological conditions. Environ. Sci. Technol. 22: 1228–1234.

United States National Academy of Science. 1972. Particulate polycyclic organic matter. Committee on Biological Effects of Atmospheric Pollution, National Research Council, National Academy Press, Washington D.C., 375 p.

USEPA. 1982. Reference method for the determination of suspended particulate matter in the atmosphere (high-volume method). Federal Register 47: 723–736.

USEPA. 1987. Reference method for the determination of particulate matter as PM10 in the atmosphere. Federal Register 52: 24664.

USEPA. 1996. National ambient air quality. Standards for particulate matter: proposed decision. Federal Register 61: 65638–65711.

Utell, M.J., and M. Samet. 1993. Particulate air pollution and health. New evidence on an old problem. Am. Rev. Respir. Dis. 147: 1334–1335.

van der Meulen, A., B.G. van Elzakker, and G.N. van den Hooff. 1987. PM10: results of a one-year survey in the Netherlands. JAPCA 37: 812–818.

van Houdt, J.J., G.M. Alink, and J.S.M. Boleij. 1987. Mutagenicity of airborne particles related to meteorological and air pollution parameters. Sci. Total Environ. 61: 23–36.

van Houdt, J.J., P.W. Coenen, G.M. Alink, J.S. Boleij, and J.H. Koeman. 1988. Organ specific metabolic activation of five extracts of indoor and outdoor particulate matter. Arch. Toxicol. 61: 213–217.

van Houdt, J.J., L.H. de Haan, and G.M. Alink. 1989. The release of mutagens from airborne particles in the presence of physiological fluids. Mutat. Res. 222: 155–160.

van Vaeck, L. y K. van Cauwenberghe. 1978. Cascade impactors measurements of the sizes distribution of the major classes of organic pollutants in atmospheric particulate matter. Atmos. Environ. 12: 2229–2239.

van Vaeck, L., G. Broddin, and K. van Cauwenberghe. 1979. Differences in particle size distribution of major organic pollutants in ambient aerosols in urban, rural and seashore areas. Environ. Sci. Technol. 13: 1494–1502.

van Vaeck, L., K. van Cauwenberghe, and J. Janssens. 1984. The gas-particle distribution of organic aerosol constituents: measurement of the volatilization artefact in hi-vol cascade impactor sampling. Atmos. Environ. 18: 417–430.

van Vaeck, L., and K. van Cauwenberghe. 1985. Characteristic parameters of particle size dsitribution of primary organic constituents of ambient aerosols. Environ. Sci. Technol. 19: 707–716.

Villalobos-Pietrini, R., S. Blanco, and S. Gómez-Arroyo. 1995. Mutagenicity assesment of airborne particles in Mexico City. Atmos. Environ. 29: 517–524.

Villalobos-Pietrini, R., S. Blanco-Jiménez, and S. Gómez-Arroyo. 1999. Direct- and indirect-acting mutagens of airborne particles from south-western Mexico City. Toxicol. Environ. Chem. 70: 103–115.

Viras, L.G., K. Athanasiou, and P. Siskos. 1990. Determination of mutagenic activity of airborne particulates and of the benzo(a)pyrene concentrations in Athens atmosphere. Atmos. Environ. 24: 267–274.

Wallace, L.A. 1991. Comparison of risk from outdoor and indoor exposure to toxic chemicals. Environ. Health Perspect. 95: 7–13.

Warner, J.R., J. Nath, and T.M. Ong. 1991. Antimutagenicity studies of chlorophyllin using the *Salmonella* arabinose-resistant assay system. Mutat. Res. 262: 25–30.

Westerholm, R., J. Almen, H. Li, U. Rannug, and A. Rosen. 1992. Exhaust emission from gasoline-fuelled light duty vehicle operated in different driving conditions: a chemical and biological characterization. Atmos. Environ. 26B: 79–90.

Whong, W.-Z., J.D. Stewart, D.C. Adamo, and T. Ong. 1983. Mutagenic detection of complex environmental mixtures using the *Salmonella*/arabinose-resistant assay system. Mutat. Res. 120: 13–19.

Yamanaka, S., and S. Maruoka. 1984. Mutagenicity of the extract recovered from airborne particles outside and inside a home with an unvented kerosene heater. Atmos. Environ. 18: 1485–1487.

Yamasaki, H., K. Kuwuata, and H. Miyamoto. 1982. Effects of ambient temperature on aspects of airborne polycyclic aromatic hydrocarbons. Environ. Sci. Technol. 16: 189–194.

Zielinska, B., J. Arey, R. Atkinson, and A.M. Winer. 1989. The nitroarenes of molecular weight 247 in ambient particulate samples collected in southern California. Atmos. Environ. 23: 223–229.

25

ABSTRACTS

MONITORING SYSTEMS PRESENTED AT THE SYMPOSIUM ARE INCLUDED BY ABSTRACT ONLY

ASSESMENT OF THE ENDANGERING POTENTIAL OF ENVIRONMENTAL CHEMICALS OF FISH BY ANALYSIS OF LONG-TERM BEHAVIORAL CHANGES

D. Baganz,[1] G. Staaks,[1] O.H. Spieser,[2] and W. Scholz[3]
[1]Institute of Freshwater Ecology and Inland Fisheries, Berlin. [2]GSF, Institute of Toxicology, Munich. [3]Metacom GmbH, Munich, Germany

Behavioral changes of zebrafish (*Danio rerio*) were quantified to assess and predict long-term sublethal effects of environmental chemicals. The BehavioQuant®, a pattern analysis system by computerized image processing was applied to record and quantify the main behavioral parameters. One of the test substances investigated was the cyanobacteria toxin microcystin-LR (MCYST-LR) at nominal concentrations of 0.5, 5, 15 and $50\,mg\,L^{-1}$. Whereas the two lower exposure concentrations (0.5 and $5\,mg\,L^{-1}$) caused an increase in day-time motility, elevated exposures led to a significantly decreased motility. The highest exposure ($50\,mg\,L^{-1}$) also reduced spawning activity and spawning success of fish. In contrast to day-time activity, night-time swimming activity was significantly increased at the higher MCYST-LR exposures. The chronobiological analysis indicated a phase shift of maximum swimming activities and a lowered reaction on trigger points like feeding, dusk and dawn. Furthermore, the results indicate some adverse consequences for the reproductive success and for the spatial and temporal fit of fish into its habitat.

CONTROLLED STUDIES ASSESSING ESTROGENIC ACTIVITY OF ENVIRONMENTAL COMPOUNDS USING A TURTLE SEX-DETERMINATION ASSAY

J.M. Bergeron[1] and D. Crews[2]
[1]University of Texas at Austin Marine Science Institute, Port Arkansas, TX, 78373.
[2]Institute of Reproductive Biology, Austin, TX, 78712

Numerous reports describe environmental compounds that induce hormone-regulated responses. Because of the plasticity of their reproductive development, reptiles with temperature-dependent sex determination (TSD) present a population at risk for the adverse effects of this hormone-mimicking activity. TSD reptiles may also provide a method of monitoring hormonal effects of environmental compounds in controlled laboratory settings. TSD is a system of sexual development in which the incubation temperature of the egg determines the individual's sex. When applied to the eggshell during a critical period of development, estrogens are capable of overriding temperature and producing female hatchlings at temperatures that normally produce all males. Because we know the effects of estrogens on reproductive development in this model, we can use it to conduct comparative experiments to examine estrogenic effects of many compounds. Past studies have examined sex ratio effect of a variety of xenobiotic compounds including PCBs, dioxin, phytoestrogens, and organochlorines. These studies allow assessment of estrogenic activity of the substances tested using a whole organism, which is valuable as a monitoring system for environmental compounds. Information gained from testing chemicals using a sex-determination assay may also contribute to elucidation of a mechanism of action of hormone mimics.

5-BROMO-2′-DEOXYURIDINE (BrdU) AS A BIOMARKER OF CHEMICALLY-INDUCED HEPATOCELLULAR PROLIFERATION IN THE JAPANESE MEDAKA (*Oryzias Latipes*)

L.M. Brennan,[1] M.W. Toussaint,[1] E.M. Boncavage-Hennessey,[2] S.J. Gunselman,[2] M.J. Wolfe,[3] and H.S. Gardner[4]

[1]Geo-Centers, Inc., 5678 Doughten Dr., Fort Detrick, MD, USA 21702-5010. [2]Geo-Centers, Inc., Colorado State University, Fort Collins, CO, USA 80523. [3]Experimental Pathology Laboratories, P.O. Box 474, Herndon, VA, USA 20172. [4]U.S. Army Center for Environmental Health Research, 568 Doughten Dr., Fort Detrick, MD, USA 21702-5010

Cellular proliferation has been shown to be a necessary step in the formation of tumors in mammalian models. A method has been developed for detection of enhanced hepatocellular proliferation in the Japanese medaka fish. When introduced directly into aquarium water, the thymidine analog 5-bromo-2′-deoxyuridine (BrdU) is incorporated into the DNA of hepatocytes in S-phase. Proliferation or labeling indices are then calculated from the immunohistochemically-stained livers. Fourteen-day old medaka fry were exposed to 10 or 100 mg/L diethylnitrosamine (DEN) for 48 hr, followed by continuous exposure to 0.1 or 1.0 mg/L trichloroethylene (TCE) for 6 months. At 7, 17, 27 and 94 days post-DEN exposure, subgroups of fish were exposed to 50–75 mg/L BrdU for 72 hr and then sacrificed. The remaining fish received continuous TCE until sacrifice at 6 months. Significant increases in hepatocellular proliferation corresponded with hepatocellular neoplasms in high dose DEN fish. TCE concentration had no effect on cellular proliferation or neoplasia incidence. BrdU-labeling of S-phase hepatocytes was an effective biomarker of cellular proliferation and predictor of liver neoplasia in Japanese medaka as early as 7 days after exposure to a carcinogenic dose of DEN.

BRYOPHYTES AS MONITORS OF ENVIRONMENTAL CONTAMINATION: A REVIEW AND ASSESSMENT OF AVAILABLE METHODS

M.A.S. Burton

Department of Environmental Sciences, University of Hertfordshire, Hatfield, Hertfordshire AL109AB, United Kingdom

Bryophytes, particularly mosses, have been widely used in monitoring the distribution of contaminants from point and diffuse sources, on land and in freshwater. Their very diverse habitats, ranging from very dry to wetland and from arctic to tropical regions allows comparison to be made of their interception and retention of contaminants over a wide geographical area. Most such studies have examined concentrations of the contaminants, for which bryophytes have a high capacity for accumulation, by chemical analysis. For some species it is possible to measure accumulation in shoots of known age, thus enabling temporal trends to be detected. The use of transplanted mosses, to sites with different contamination, allow estimates of rates of accumulation to be made together with concurrent measurements of other relevant parameters. There have been various attempts to measure the responses of bryophytes to contaminants: differences in species sensitivity, resulting in changes in community composition have been observed in impacted terrestrial and aquatic environments; other methods have been more detailed, including measuring changes in, for example, enzyme activity, metabolism and pigment composition. A review of recent methods to detect bryophyte responses to a variety of stresses will be discussed in relation to their applications in environmental assessment.

A NOVEL FIBRE OPTIC SENSOR FOR THE DETERMINATION OF ENVIRONMENTAL TURBIDITY

R.P. Chandy,[1] P.J. Scully,[1] C. Whitworth,[2] and D. Fearnside[2]

[1]Optical Fibre Sensor Research Group, School of Engineering (Electronics), Liverpool John Moores University, Byrom Street, Liverpool L3 3AF, UK. [2]Research and Development, Yorkshire Water Plc, Bradford, UK

Turbidity is used as a measure of the clarity of water. Suspended particles obstruct the transmittance of light through water. Documented in this paper is the development and test results of a novel optical fibre sensor to measure turbidity in the range 0–1 FTU that water treatment regulations require potable water to have. Higher turbidity measurements should also be possible with proper calibration. The sensor consists of a commercially available extended diffused light source and a sensitized optical fibre placed inside a length of glass tubing through which the sample under investigation is caused to flow. As the turbidity of the sample changes the light output that is scattered by the particles in the suspension changes which changes the intensity of the light that is detected by the optical fibre. This change in intensity is measured by a power meter. Preliminary measurement indicate that the sensor has a sensitivity of 3.3 nW/unit FTU and an accuracy of 0.1 to 0.15 FTU.

INTERPRETATION OF BIOLOGICAL COMMUNITY, PHYSICAL HABITAT AND WATER QUALITY DATA USING SPATIAL COMPARISONS, BIOLOGICAL RESPONSE SIGNATURES, AND MULTIVARIATE ANALYSIS

D.J. Chappie, P.C. Fuchsman, T.R. Barber, D.J. Duda, and J.C. Sferra
McLaren/Hart Inc., 5900 Landerbrook Drive, Suite 100, Cleveland, Ohio 44124

Fish and macroinvertebrates were collected from three segments of the Tuscarawas River, Ohio: an upstream segment relatively unimpacted by urban and industrial sources, a middle segment impacted by these sources, and a downstream segment adjacent to a RCRA facility. Values were calculated for the Index of Biotic Integrity and Modified Index of Well-Being for fish and the Invertebrate Community Index for macroinvertebrates. Physical habitat was evaluated using the Qualitative Habitat Evaluation Index. Data were interpreted through area of degradation value calculation, biological response signature evaluation, and multivariate statistical analysis. The calculation of area of degradation values relative to upstream conditions as well as relative to biocriteria provided an indication of the extent of potentially site-related impacts. Biological response signature analysis showed that while macroinvertebrates appeared to be impacted by chemical stressors both upstream and adjacent to the RCRA site, the fish community response was not characteristic of typical "complex toxic" effects. Multiple linear regression models explained approximately 70% of the variation in fish community quality within the study areas as a function of total dissolved solids and instream cover quality. These analyses provided a consistent characterization of the extent and apparent causes of impacts to biological communities in the river.

DETERMINATION OF FUGACITY GRADIENTS IN VARIOUS GUT SECTIONS OF THE RING DOVE FED A DIET CONTAINING 2,2=,4,4=,5,5=-HEXACHLOROBIPHENYL

K.G. Drouillard[1] and R.J. Norstrom[2]
[1]Watershed Ecosystems Program, Trent University, Peterborough, Ontario, K9J 7B8.
[2]Canadian Wildlife Service, 100 Gamelin Blvd., Hull, Quebec, J8Y 1V9

Recent bioaccumulation models, calibrated to fish, have employed a gastrointestinal fugacity gradient mechanism to account for the phenomenon of biomagnification. The application of such a model in the avian system was explored using the Ring Dove (*Streptopelia risoria*) fed a diet of pigeon pellets containing 2,2=,4,4=,5,5=-hexachlorobiphenyl (CB#153), a highly persistent and negligibly metabolized compound. Ring doves were fed diet#1 containing CB#153 at a concentration of 50 ug/kg for 25 d. On day 26, the birds were fed a new diet containing $^{13}C_{12}$-CB#153 at similar concentration for a further 60 d. Replicate birds were sacrificed at 4 h, 8 h and 18 h after feeding the new diet and over geometric time intervals for the remainder of the experiment. The ratios of $^{13}C/^{12}C$-CB#153 along the gastrointestinal tract (subdivided into five sections), the maximum fugacity gradient achieved in the gut tract and relative contribution of endogenous contaminant vs. recently assimilated ^{13}C-

CB#153 in blood plasma was determined for each time interval. Results indicate variations in the fugacity of CB#153 along the gut tract on the order of 1 to 9 depending on the location of feces and duration of fasting. Maximum predicted lipid normalized BMFs, predicted from the fugacity achieved in gut relative to food, were on the order of 7 to 12, below typical BMFs of 30 observed for this congener in avians.

HISTORIC ENVIRONMENTAL CHANGE IN THE WORLD'S GREAT LAKES-A COMPARISON OF LAKE BAIKAL AND THE LAURENTIAN GREAT LAKES

M.B. Edlund and E.F. Stoermer
Center for Great Lakes and Aquatic Sciences, University of Michigan, Ann Arbor, Michigan 48109-1090

The sedimentary record in large lakes preserves signals of environmental change that have occurred both within and outside the lake basin. In particular, changes in the primary producer community are often well-preserved in lake sediments and may reflect large-scale impacts (e.g. climate change) and/or small-scale changes (e.g. annual variability) in a lake system. Lake Baikal contains a sedimentary record spanning nearly 25 million years. Analyses of cores covering the last 250,000 years have shown that the Baikal phytoplankton community has undergone resolvable changes that reflect major and minor climatic oscillations, speciation events, development of endemism, and recent introduction of exotic species. Compared to Lake Baikal, the sediment record from the Laurentian Great Lakes preserves a contrasting history of impacts on a much younger lake system. Recent past changes (<200 years) in the Laurentian Great Lakes primary producer have been dramatic with instances of endemics driven to extinction, loss of entire phytoplankton functional groups, wide scale response to nutrient loadings, and introduction of eurytopic non-natives. The paleolimnological method provides a high resolution analytical tool for interpreting historic change and monitoring recent impacts in large lakes.

THE USE OF *SALMONELLA* STRAINS LACKING CLASSICAL NITROREDUCTASE OR ACETYLTRANSFERASE ACTIVITY, FOR THE PRIMARY CHARACTERIZATION OF ENVIRONMENTAL COMPLEX MIXTURES

J.J. Espinosa-Aguirre,[1] M. Yamada,[2] and T. Nohmi[2]
[1]Instituo de Investigaciones Biomédicas, Universidad Nacional Autónoma de México, Apartado Postal 70228, 04510 México, D.F. México. [2]Division of Genetics and Mutagenesis, Biological Safety Research Center, National Institute of Heath Sciences, 1-18-1, Kamiyoga, Setagaya-Ku, Tokyo 158, Japan

N-hydroxyarylamine *O*-acetyltransferase (OAT) and "classical nitroreductase" (CNR), are enzymes involved in the metabolic activation of environmental aromatic amines and nitroarenes. The *oat* and *cnr* genes of *Salmonella typhimurium*-Ames strains TA1538 and TA1535 were specifically disrupted on their chromosome DNA and plasmid pKM101 was introduced into these mutants producing the *oat* (YG7130, YG7126) and *cnr* (YG7132, YG7128) null mutant derivatives. The sensitivities of the new strains to several nitroarenes and aromatic amines were compared with those of the wild-type TA98 an TA100 Ames strains. Both CNR- and OAT-deficient strains were resistant to the mutagenic actions of 2-nitrofluorene (2-NF) and 1-nitropyrene (1-NP), whereas 1,8-dinitropyrene, 3-amino-1-methyl-5*H*-pyrido[4,3-b]indole (Trp-P-2), 2-amino-3-methyl-3*H*-imidazo[4,5-*f*]quinoline (IQ), and 2-amino-6-methyldipyrido[1,2-a:3′,2′-*d*]imidazole (Glu-P-1) produced a reduced mutagenic response in the *oat* deleted strains. 2-nitronaphthalene (2-NN) and furylfuramide were ineffective against the *nr* deleted strains. These results shown that the new NR- and OAT-deficient strains could be useful for the characterization of environmental complex mixtures and to elucidate the role of classical nitroreductase and *O*-acetyltransferase in the metabolic activation of environmental nitroaromatics and aminoaromatics.

PHOTOTACTIC BEHAVIOR, DEFORMITIES AND SPECIES COMPOSITION OF TANYPODINAE LARVAE (Chironomidae) BELOW AN INDUSTRIAL EFFLUENT IN SWEDEN

A.E. Gerhardt and L.G. Janssens de Bisthoven

Department of Ecology, Ecotoxicology, University of Lund, Solvegatan 37, 22362 Lund, Sweden; present address, LimCo International, Oststrasse 24, D-49477 Ibbenburen, Germany

The Ybbarspan (Scania, Sweden), a mesosaprobic brownwater forest stream on silicate bedrock, receives an effluent from a big industrial area in Perstorp, which contains metals and persistent organic xenobiotics. The macrozoobenthos below the effluent was degraded and the Chironomidae were dominated by Tanypodinae larvae (*Procladius* sp.: 64%, *Psectrotanypus* sp.: 26.7), whereas upstream of the factory complex larvae of the Orthocladiinae, Tanytarsini and *Polypedilum* sp. were found. The *Procladius* sp. Larvae exhibited buccal and antennal deformities (>10%). Last instar *Procladius choreus* and *Psectrotanypus* sp. From 100 m below the effluent were kept in the laboratory for 14 days and mortality, emergence, cephalic deformities and behavior (measured with impedance converter) such as reaction time to a light stimulus and time spent on locomotion were monitored. The behavior showed a typical dichotomy, with some animals being very active and reacting directly to the stimulus, and others being passive. Some larvae (17.6%) showed weak deformities. The mortality was 51 ± 16.5 (SD)% and the emergence 35.7 ± 8.4%.

METALLOTHIONEIN, AS A BIOMARKER FOR TRACE METAL EXPOSURE AND EFFECTS IN THE FRESH WATER OLIGOCHAETE *Tubifex tubifex*

P.L. Gillis,[1] T.B. Reynoldson,[2] D.G. Dixon,[1] and M.D. Dutton[1]

[1]Department of Biology, University of Waterloo, Waterloo, ON. [2]National Water Research Institute, Burlington, ON

Oligochaetes were chosen as a representative benthic organism for the development of a biomarker for trace metal exposure. They are continually exposed to any toxicants in sediment both through bodily contact and ingestion of the sediment. The mercury saturation assay of Dutton et al., (1993) was used to measure the concentrations of the metal binding protein, metallothionein, in oligochaetes following exposure to trace metals. This study relates the levels of metallothionein in oligochaetes to ecologically relevant effects at higher levels of organization. Whole sediment bioassays were used to determine if *T. tubifex* produces metallothionein in a dose responsive manner and to illustrate the relationship between metallothionein and reproductive output of the exposed worms. Organisms were exposed to a range of cadmium concentrations in artificially contaminated field sediment. Following exposure, adult survival and reproductive output were determined as well as cadmium body burden and the concentration of metallothionein produced. The critical concentration of Cd which induced metallothionein synthesis and the concentration which resulted in decreased reproductive output were determined. Initial results indicate that *T. tubifex* produces metallothionein in a dose-responsive manner when exposed to cadmium and that the amount produced is inversely related to the reproductive output of the worms.

USE OF INFRARED SPECTRAL MODELS IN CANCER RESEARCH AND THEIR POTENTIAL CLINICAL APPLICATIONS

S.J. Gunselman,[1] N.L. Polissar,[2] and D.C. Malins[3]

[1]Geo-Centers, Inc., CETT/CSU Foothills Campus, Fort Collins, CO 80523, [2]The Mountain-Whisper-Light Statistical Consulting and Dept. of Biostatistics, University of Washington, Seattle, WA 98195. [3]Pacific Northwest Research Foundation, 720 Broadway, Seattle, WA 98122

Statistical analysis to Fourier transform infrared (FT-IR) spectra has been shown to be a sensitive tool to assess DNA modifications thought to be involved in the development of carcinogenesis. DNA was extracted from normal and cancerous prostate tissues and analyzed by FT-IR. Subtle modifications in DNA structure were readily identified by principal components analysis (PCA) of FT-IR spectra that were not discernible using traditional spectroscopic techniques. After baselining and normalizing each spectrum, the samples underwent statistical analysis including unequal variance *t*-tests, PCA and discriminant analysis. The FT-IR/PCA methodology provided a perspective of the changes in the entire DNA structure—nucleotide bases, phosphodiester groups and the deoxyribose moiety. Results indicated distinct structural differences in DNA of normal and cancerous tissues. Comparisons of by P-values showed statistically significant ($P \leq 0.05$) differences in the areas assigned to C-O stretching, NH_2 bending and deoxyribose vibrations. PCA of the samples revealed clustering between tissue types and discriminant analysis classified the samples with high sensitivity and specificity. The FT-IR/PCA methodology was also applied to human breast tissues and wild fish exposed to environmental carcinogens with similar results, evidence that the discriminatory ability of the technology is not limited to a single tissue type.

CONTINUOUS WATER MONITORING WITH MUSSELS: VALVE MOVEMENT BEHAVIOR AS AN INDICATOR OF POLLUTANTS

M. Hoffmann, E. Blübaum-Gronau, and F. Krebs
Federal Institute of Hydrology, Kaiserin-Augusta-Anlagen 15-17, D56068 Koblenz, Federal Republic of Germany

In the presence of water contaminants mussels change the normal movement pattern of their valves that is characterized mainly by respiration and feeding. The movements of the valves can be recorded and evaluated by means of a biological early warning system named MOSSELMONITOR®. A sensor system with small coils which are fixed to the valves of the mussels measures at short intervals the current width of the valve opening. Sublethal effects are identified by means of three assessment criteria: "valves closed", "increasing valve movement activity" and "decreasing width of valve opening". The biomonitor using eight zebra mussels as test organisms was tested in spiking experiments with water from River Rhine and sodium pentachlorophenolate.

BIOMONITORING: FROM WATER TO SEDIMENT

N.H.B.M. Kaag and M.C.T. Scholten
TNO Institute of Environmental Sciences, Energy Research and Process Innovation. Dept. for Ecological Risk Studies, P.O. Box 57, 1780 AB, Den Helder, The Netherlands

Since the early 60s, mussels are used for the assessment of ambient water quality. In order to obtain a standardized measure of ambient water quality, a method was developed for Active Biological Monitoring (ABM), in which mussels of a homogeneous size class were placed at selected locations for a fixed period of time, suspended in a basket. This method is now routinely used by the Dutch government. Mussels were also used in an early warning device, called the Mussel monitor. When the mussels close their valves in reaction to changes in water quality, an alarm is given. During the 80s, when the general water quality started to improve, more and more attention was given to contaminated sediments. As will be shown, the mode of feeding of organisms is the predominant (biological) factor determining the bioavailability of sediment-bound contaminants. For use in biomonitoring of sediments, organisms should, therefore, be selected that actually feed on the sediment, such as some annelids or amphipods. This aspect needs also be considered when the effects of contaminated sediments are tested in laboratory bioassays, because the choice of test-species, or test-conditions, may have a profound influence on the compartment that is actually tested.

THE USE OF FETAX FOR ASSESSING THE QUALITY OF RIVER SEDIMENTS

L.G. Marquez,[1] J.A. Bantle,[2] A.A. Lerdo de Tejada,[1] A.M. Sandoval,[1] A. Ordoñez,[1] R. López,[1] E. Bandala,[1] and S. Pérez[1]

[1]Mexican Institute of Water Technology, Blvd, Cuauhnahuac 8532, Jiutepec, Morelos. C.P. 62550. México. [2]Oklahoma State University, Department of Zoology, 430 Life Sciences West, Stillwater, Oklahoma 74078, U.S.A.

The importance of sediment quality assessment has been recognized as part of the water quality monitoring programs. Because rivers are experiencing increasing agricultural and industrial discharges, these programs must be capable of detecting changes in water and sediments quality that can impose a hazard to the ecosystem and humans. FETAX (Frog Embryo Teratogenesis Assay *Xenopus*) test will be used to assess the developmental toxicity of the solvent extractable compounds of river sediments. Aroclor-1254-induced-microsomes facilitates the identification of groups of compounds making the test more relevant to mammals. Validation studies shown that the test can be used to predict development toxicity in mammals with a precision of 85%. In this report, results in percentages of mortality, malformations and growth inhibition of frog embryos are presented and related with concentration of organochlorine pesticides and polyaromatic hydrocarbons. A seasonal pattern is observed with higher effects during rainy season. Due to the sensitivity of the test, it is being proposed to be included as part of the National Network Monitoring Program to assess the effectiveness of the performance of basin management policies that are being implemented in México, mainly in those rivers where the exact nature of compounds being disposed are unknown.

CONCEPTUAL AND EMPIRICAL APPROACHES TO THE ROLE OF FREE RADICALS, OXIDATIVE STRESS, AND RELATED MECHANISMS IN DISEASE SYNDROMES

T. Muir

Environment Canada—Ontario Region, Canada Centre for Inland Waters, P.O. Box 5050, Burlington, Ontario L7R 4A6

The role of the free radicals, and oxidative stress (OS) in the toxicity of chemicals is long recognized in biochemical studies of diseases syndromes and aging. These reactive species interactions, particularly involving a continuous and progressive degeneration of DNA, or other key macromolecules, including antioxidants, can lead to a variety of lesions and disease states. Until recently, the research program of the "dioxins" has obscured the role of free radicals. Currently, "endocrine disruption" has moved to the forefront. Concurrently, oxidative stress and related mechanisms are being rediscovered TEQs augment the induction of the cytochromes P450, and accelerate the production of radical species and OS from the metabolism of the complex mixtures of real world pollutant exposures. A new paradigm, called "radical-induced DNA disorder", results from the redox cycling of estrogens and xenoestrogens, and the generation of hydroxyl radical damage to DNA bases, and to mutagenesis and carcinogenesis, with specific links to DNA damage in wild fish, and to breast and prostate cancer. To what extent do the synthetic chemical, identified and unidentified as xenoestrogens, also participate in the generation of free radicals and/or oxidative stress? What is the primary risk factor—the hormone activity, or the metabolic byproducts of the hormone cycling? While some argue that synthetic chemicals are not potent enough to have a hormonal activity effect, many synthetic aromatic chemicals and inorganic compounds are metabolized to a reactive-free radical by at least one enzyme. The ability of compounds to modulate free-radical defense systems, and the toxicity, supports the hypothesis.

REMOTE, AUTOMATED, CONTINUOUS WATER QUALITY MONITORING FOR VOLATILE ORGANIC COMPOUNDS DOWNSTREAM OF A LARGE PETROCHEMICAL COMPLEX

T.S. Munro,[1] R. Brooks,[2] and E. Kuley[2]
[1]Lambton Industrial Society, Suite 111, 265 Front St, N, Sarnia, Ontario N7T 7X1.
[2]ORTCH Corporation, 1133 Vanier Rd., Sarnia, Ontario N7S 3Y6

Since 1986, The Lambton Industrial Society, an environmental co-operative, has operated an unattended, remote monitoring system for ambient water quality in the St. Clair River. Located immediately downstream of a complex of petroleum refineries, petrochemical production facilities and support industries, the monitor provides hourly analysis of 20 volatile organic compounds potentially released from those operations. Data are available through telemetry and an intranet in real time. Samples are processed through a purge-and-trap system, a standard gas chromatograph, and split between electron capture and flame ionization detectors. Method detection limits for the aliphatic, aromatic and chlorinated hydrocarbons range between 0.04 ppb for cyclohexane and 0.27 ppb for tetraethyl lead. System reliability is excellent, providing 98.5% valid data. The system routinely reports data below detection limits except during spill events. The instrument is equipped with an automated alerting system, and automated grab sampling for confirmatory analysis by GC/MS can be initiated remotely. During incidents, the system provides peak concentrations and time duration of the spill plume, allowing informed decisions on the need to close water intake facilities. A performance history of the instrument, along with typical spill profiles will be provided.

RECOMBINOGENIC EFFECT OF CIGARETTE SMOKE CONDENSATE IN SOMATIC CELLS OF *Drosophila Melanogaster*

P. Ramírez-Victoria,[1] J. Guzmán-Rincón,[1] U. Graf,[2] and L. Benitez-Bribiesca[4]
[1]Instituto Nacional de Investigaciones Nucleares, Km. 36.5 Carretera México-Toluca, Salazar, Edo. Méx. México. [1]ETH Zurich, Switzerland. [3]Centro Médico Nacional, México

The Somatic Mutation and Recombination Test (SMART) is able to detect a wide spectrum of genetic endpoints such as point mutations, deletions, and certain types of chromosomal aberrations as well as mitotic recombination. The assay was conducted with larvae derived either from the standard (ST) or the high bioactivation (HB) cross. Furthermore, the genotoxic effects were recorded in two different genotypes, inversion-free flies and inversion-heterozygous flies, in order to determine quantitatively the contribution of recombination. Two-day-old larvae were pre-treated for 24 h either with water (negative control), chlorophyllin, ascorbic acid, or retinoic acid, respectively, and then treated for 48 h with cigarette smoke condensate (CSC). The wings of the resulting flies were scored for the occurrence of mutant spots. The CSC was weakly positive in both crosses, the frequencies of total spots per wing were 0.53 for the ST and 0.68 for the HB cross, respectively. The pre-treatments with chlorophyllin, ascorbic acid, or retinoic acid did not modify the effect of the CSC treatment giving total spot frequencies of 0.64, 0.62, and 0.61, respectively, in the HB cross. In both crosses, about 50% of all the spots were due to induced recombinational events.

NON-LINEAR INDICATORS OF ENVIRONMENTAL CHANGE

M. Vahl
Nijenrode University, Centre for Corporate and Community Renewal (CORE), Straatweg 25, 3621 Breukelen. The Netherlands

The relation between environmental change and human activity involves both autonomous and interactive sources of variation. The use of stories can help to transform part of this variation into useful feedback mechanisms, and thereby make the relation transparent. The issue is how to achieve closure over observations. An example is presented.

EFFECT OF CADMIUM ON THE MORPHOLOGY OF *Vallisneria Americana* AND THE DEVELOPMENT OF A NEW MEASUREMENT TECHNIQUE

J.J. Vanderwall and L. Lovett Doust
Department of Biological Sciences, University of Windsor, 401 Sunset, Windsor, Ontario N9B 3P4

This study assesses the value of a submerged aquatic macrophyte, *Vallisneria americana*, as a biomonitor of a heavy metal, cadmium. It also presents a new method of measuring the leaf-to-root surface area ratios which have previously been used as indicators of overall site quality and of plant response to the organochlorine, trichloroethylene. Plants collected from Mitchells Bay, Ontario, were transported to a greenhouse where they were transplanted into aquaria containing one of several concentrations of cadmium in solution. The plants were destructively harvested at the end of six weeks, biomass distribution to different tissues and surface area measurements were made, in addition to analyses of cadmium concentrations. The new method for determining plant surface area measurements uses a scanner and an image analysis program, UTHSCSA Image Tool 2.00. Measurements are made in a quarter of the time needed for manual measurements of the surface areas. This new method for determining plant surface area measurements could be used as a quick and inexpensive way to field managers to prioritize sites for remediation, particularly with reference to organochlorine contamination and overall site quality.

MULTI-LOCUS DNA FINGERPRINTING IN HERRING GULLS (*Larus argentatus*) AS A BIOMARKER FOR HERITABLE GENETIC MUTATION IN INDUSTRIALIZED AREAS ON THE GREAT LAKES

C.L. Yauk and J.S. Quinn
Department of Biology, McMaster University, Hamilton, Ontario L8S 4K1

Tandemly-repeated sequence elements known as minisatellite DNA may provide sensitive markers for induced heritable mutations in animals exposed to ambient levels of contaminants. By using DNA fingerprinting, several minisatellite loci can be screened simultaneously decreasing the time and cost for mutation analyses. Additionally, the high germ-line mutation rate at minisatellite loci allows the detection of induced mutations in relatively small sample sizes. We used multi-locus DNA fingerprinting in full families of herring gulls (*Larus argentatus*) to show that germ-line mutation rates in birds inhabiting heavily industrialized areas are elevated compared to gulls from rural locations. We found mutation rates 0.0092 and 0.0089 per minisatellite band scored in herring gulls from Hamilton Harbour, Lake Ontario and East Chicago, Lake Michigan. These sites are areas that are heavily industrialized with steel operations and also near other types of industries and highways. The rates found at steel sites were significantly greater than rates fund at rural colonies in the Great Lakes (0.0038 and 0.0035 per band scored for Presqu'ile Provincial Park, Lake Ontario and Chantry Island, Lake Huron, respectively; $x^2 = 9.335$, $p = 0.002$). We suggest that these minisatellite mutations are important biomarkers for heritable genetic changes resulting from exposure to environmental genotoxins.

INDEX

MIX
Papier aus verantwortungsvollen Quellen
Paper from responsible sources
FSC® C105338

If you have any concerns about our products,
you can contact us on
ProductSafety@springernature.com

In case Publisher is established outside the EU,
the EU authorized representative is:
Springer Nature Customer Service Center GmbH
Europaplatz 3, 69115 Heidelberg, Germany

Printed by Libri Plureos GmbH
in Hamburg, Germany